OECOLOGY OF PLANTS

OECOLOGY OF PLANTS

Eug[enius] Warming

ARNO PRESS

A New York Times Company

New York / 1977

Editorial Supervision: LUCILLE MAIORCA

————◆————

Reprint Edition 1977 by Arno Press Inc.

This reprint has been authorized by
  the Oxford University Press

Reprinted from a copy in
  The University of Illinois Library

HISTORY OF ECOLOGY
ISBN for complete set: 0-405-10369-7
See last pages of this volume for titles.

Manufactured in the United States of America

————◆————

Library of Congress Cataloging in Publication Data

Warming, Eugenius, 1841-1924.
  Oecology of plants.

  (History of ecology)
  Rev. English version of the author's Plantesamfund.
  Reprint of the 1909 ed. published by Clarendon Press,
Oxford.
  Bibliography: p.
  Includes index.
  1. Botany—Ecology. 2. Plant communities.
I. Warming, Eugenius, 1841-1924. Plantesamfund.
II. Title. III. Series.
QK901.W3 1977   581.5   77-74254
  ISBN 0-405-10423-5

OECOLOGY OF PLANTS

AN INTRODUCTION TO THE STUDY OF PLANT-COMMUNITIES

By EUG. WARMING, Ph.D.

PROFESSOR OF BOTANY IN THE UNIVERSITY OF COPENHAGEN

ASSISTED BY

MARTIN VAHL, Ph.D.

PRIVATDOCENT IN THE UNIVERSITY OF COPENHAGEN

PREPARED FOR PUBLICATION IN ENGLISH BY

PERCY GROOM, M.A., D.Sc., F.L.S.

ASSISTANT PROFESSOR OF BOTANY IN THE IMPERIAL COLLEGE OF SCIENCE AND
TECHNOLOGY, LONDON

AND

ISAAC BAYLEY BALFOUR, M.A., M.D., F.R.S.

KING'S BOTANIST IN SCOTLAND, REGIUS KEEPER OF THE ROYAL BOTANIC GARDEN
AND PROFESSOR OF BOTANY IN THE UNIVERSITY, EDINBURGH

OXFORD
AT THE CLARENDON PRESS
1909

HENRY FROWDE, M.A.
PUBLISHER TO THE UNIVERSITY OF OXFORD
LONDON, EDINBURGH, NEW YORK
TORONTO AND MELBOURNE

NOTE

IT is now some years since expectation became prevalent of an English edition of Professor Warming's book—*Plantesamfund*. Nothing need now be said about the difficulties opposing its production, because Professor Warming has solved them happily by writing for the Delegates of the Oxford Press this present book, founded upon his original Danish work. To the manuscript, as it has been prepared by and received from Professor Warming at intervals, Professor Groom has applied with untiring patience his skill in interpretation and in apt expression, and the book as it now appears is therefore not an English edition of a foreign book—as are others of the botanical series issued by the Delegates of the Oxford Press, but is ' practically a new work ', as the author himself designates it.

The book is a valuable addition to botanical literature, and will appeal to a wide audience. Its subject, Oecology, is the field in which the botanical morphologist, physiologist, and systematist happily meet, and to them this statement of the views of a pioneer and leader in oecological work will be welcome. Those whose interests lie in the practical application of a knowledge of plant life in the several domains of rural economy—Agriculture, Horticulture, Forestry—will find in the matters treated in the book the clue to many of the problems which they meet with. Teachers within whose sphere it lies to encourage a study of Nature will find its pages full of information and suggestion to guide them. Students of Botany will glean from it sound instruction in a subject which now occupies a prominent place in botanical teaching. Every one, indeed, for whom the varying aspects of vegetation have interest will obtain by perusal of the book new lights by which that interest may be increased. Perhaps in no way will the value of the book be greater than as a stimulus to accurate observation and inquiry, through which Oecology will be advanced from the stage of 'infancy', in which, as Professor Warming says, it now is.

The character of the book and the method of its production have necessarily placed a restriction upon editorial functions, and have not allowed of modifications that might have brought more directly home some of its teachings to readers in Britain— for instance, by the introduction of a greater number of illustrative references to vegetation in the British Isles. If, as we believe will be the case, a new edition of the book is called for, attention may be given in it to matters of this kind.

I. B. B.

AUTHOR'S PREFACE

IN 1895 I published a Danish work entitled *Plantesamfund*, which was based upon lectures that I had delivered in the University of Copenhagen. I never imagined that the book would appeal to more than a few readers outside my audience, and was therefore greatly surprised shortly after its publication to receive from Dr. E. Knoblauch a request for permission to prepare a German version of my book—an act of courtesy, since Denmark had not subscribed to the Bern Convention, and my book was thus public property. Thanks to Dr. Knoblauch's energy the German edition was published in 1896. In the short time available I found it impossible to introduce more than trifling changes into his edition, and was forced to postpone the more important modifications that I contemplated.

In 1902 the publishers, Gebrüder Bornträger of Berlin, issued a second edition of this German translation. It was edited by Dr. P. Graebner. With this edition I had nothing whatever to do. The book was unchanged as regards plan and arrangement of the subject-matter.

I had always entertained grave doubts as to the arrangement of the contents of *Plantesamfund*. When I wrote it I had no models to study ; mine was the first attempt to write a work on Oecological Plant-geography, of which the very name was then all but new. The present book is practically a new one ; for, not only have I myself introduced a number of new features, but I have also invoked the aid of the young Phyto-geographer, Dr. M. Vahl, in order that he might deal critically with purely geographical and climatic considerations.

The following changes are amongst the most important that appear in the book as it is now presented in English :—

Chapter II contains fresh subject-matter dealing with growth-forms, as well as an entirely new classification of these. The parts of the book referring to adaptations of water-plants and land-plants have been combined to form Section III ; and in the same section I have given my views on oecological classification in a more comprehensive and detailed manner.

For these alterations I am mainly responsible, but the new classification of the formations is largely due to Dr. Vahl, who has thus materially remodelled parts of the book. In place of the four Sections discussing hydrophytic, xerophytic, halophytic, and mesophytic communities respectively, thirteen Sections (IV–XVI) have been devoted to thirteen oecological classes established on the basis of edaphic and climatic distinctions. The arrangement of the subject-matter dealing with the several Formations is new in many respects, the changes involved being due partly to myself (for instance, in connexion with halophytes and lithophytes) and partly to Dr. Vahl (notably in Sections XI–XV).

So far as my other varied work, including administrative duties, would permit, I have endeavoured, with the assistance of Dr. Vahl, to take into consideration the vast amount of pertinent literature issued since 1895. Since that year there have been published, not only the large general works by *Schimper* (1898, English Edition 1903), *Solms-Laubach* (1905), and *Clements* (1904, 1905, 1907), which contain much that is original and suggestive, but also an immense number of original papers in various periodicals and countries. So far as possible, recognition has been made of all important contributions issued up to the present moment, and their titles will be found in the appended Bibliography. But there is considerable difficulty in selecting the most important from such a vast accumulation of literature.

In many places I have felt the lack of definite, detailed, and truly oecological information concerning various questions, and, as in 1895, I must confess that my ideal is far from being realized. The oecology of plants is a subject still in its infancy ; numerous investigations must be made before the foundations can be truly and rightly laid, and before a consistent, clear, and natural classification of plant-communities is achieved.

In conclusion, I must express my thanks to Dr. Martin Vahl for the great interest he has shown in efforts to improve the book, and to my colleagues in Britain for the exceeding care which they have bestowed upon the production of the book in English.

EUG. WARMING.

COPENHAGEN,
March, 1909.

CONTENTS

INTRODUCTION

SECTION I

OECOLOGICAL FACTORS AND THEIR ACTION

SECTION II

COMMUNAL LIFE OF ORGANISMS

SECTION III

ADAPTATIONS OF AQUATIC AND TERRESTRIAL PLANTS. OECOLOGICAL CLASSIFICATION

SECTION IV

CLASS I. HYDROPHYTES. FORMATIONS OF AQUATIC PLANTS

SECTION V

CLASS II. HELOPHYTES. MARSH-PLANTS

SECTION VI

CLASS III. OXYLOPHYTES. FORMATIONS ON SOUR (ACID) SOIL

SECTION VII

CLASS V. HALOPHYTES. FORMATIONS ON SALINE SOIL

SECTION VIII

CLASS VI. LITHOPHYTES. FORMATIONS ON ROCKS

SECTION IX

CLASS IV. PSYCHROPHYTES. FORMATIONS ON COLD SOIL

SECTION X

CLASS VII. PSAMMOPHYTES. FORMATIONS ON SAND AND GRAVEL

SECTION XI

CLASS IX. EREMOPHYTES. FORMATIONS ON DESERT AND STEPPE

SECTION XII

CLASS VIII. CHERSOPHYTES. FORMATIONS ON WASTE LAND

SECTION XIII

CLASS X. PSILOPHYTES. SAVANNAH-FORMATIONS

INTRODUCTION

CHAPTER I. FLORISTIC AND OECOLOGICAL PLANT-GEOGRAPHY

PLANT-GEOGRAPHY deals with the distribution of plants upon the earth, and with the principles determining this. We may regard this distribution from two different standpoints, and accordingly may divide the subject into two branches, *floristic plant-geography* and *oecological*[1] *plant-geography ;* but these are merely different aspects of the same science, touching at many points and occasionally merging into one another.

Floristic plant-geography is concerned with—

1. The compilation of a ' Flora ', that is, a list of species growing within a larger or smaller area. Such lists form the essential basis of the subject.

2. The division of the earth's surface into natural floristic tracts (floristic kingdoms and so on[2]) according to their affinities, that is, according to the numbers of species, genera, and families common to them.

3. The sub-division of the larger natural floristic tracts—floristic kingdoms—into smaller natural tracts, regions, and districts, and the precise definition of these.

4. The discussion of the limits of distribution of species, genera, and families (their ' area ') ; of their distribution and frequency in different countries ; of endemism ; of the inter-relations between the floras of islands and of continents, and between those of mountains and of lowlands ; and so forth.

The thoughtful investigator will not remain content with the mere recognition of facts ; he will seek after their *causes*. These are, in part, *modern* (geognostic, topographical, and climatic), and, in part, *historical*. The limits of distribution of a species may depend upon prevailing conditions, upon barriers now existing in the form of mountain, sea, soil, and climate, which oppose its spread ; but they may also depend upon geohistoric or geological and climatic conditions of ages long past, and upon the whole evolutionary history of the species, the site of this, and the facilities for and means of migration. In addition, problems must be dealt with concerning centres of development, the rise and age of species and genera ; and behind these lies the question of the origin of species.

To deal with the yet undescribed floristic plant-geography of Denmark,

---

[1] Haeckel (1886) defined Oecology (οἶκος, a house, λόγος, theory) as the science treating of the reciprocal relations of organisms and the external world. Reiter (1885) employed the term in the same sense; see MacMillan, p. 950, 1897.
[2] Drude, 1884, 1886–7, 1890.

for instance, it will be necessary to investigate the following items : the distribution of the species present, their arrangement in the country, the sub-division of Denmark into natural floristic sections, Denmark as a floristic portion of a larger natural district or its floristic affinity with Sweden and Norway, Germany, and other lands, the problems as to when and whence the species immigrated after the glacial epoch, the routes of their migrations and their means of migration, the problem of species left behind (vestigial plants), and the like.[1]

With these interesting and far-reaching results of floristic plant-geography we shall not deal in this work. This subject has been treated by Wahlenberg, Schouw, A. de Candolle, Grisebach, Engler, Drude, and others.

Oecological plant-geography has entirely different objects in view :—

It teaches us how plants or plant-communities adjust their forms and modes of behaviour to actually operating factors, such as the amounts of available water, heat, light, nutriment, and so forth.

A casual glance shows that species by no means dispose their individuals uniformly over the whole area in which they occur, but group them into communities of very varied physiognomy. Oecology seeks—

1. To find out which species are commonly associated together upon similar habitats (stations). This easy task merely involves the determination or description of a series of facts.

2. To sketch the physiognomy of the vegetation and the landscape. This is not a difficult operation.

3. To answer the questions—

Why each species has its own special habit and habitat,

Why the species congregate to form definite communities,

Why these have a characteristic physiognomy.

This is a far more difficult matter and leads us—

4. To investigate the problems concerning the economy of plants, the demands that they make on their environment, and the means that they employ to utilize the surrounding conditions and to adapt their external and internal structure and general form for that purpose. We thus come to the consideration of the *growth-forms* of plants.

CHAPTER II. GROWTH-FORMS

EVERY species must be in harmony, as regards both its external and internal construction, with the natural conditions under which it lives ; and when these undergo a change to which it cannot adapt itself, it will be expelled by other species or exterminated. Consequently one of the most weighty matters of oecological plant-geography is to gain an understanding of the *epharmony* of species.[2] This may be termed its *growth-form* in contradistinction to its *systematic form*. It reveals

---

[1] Warming, 1904.

[2] Vesque (1882 a) defines 'L'épharmonie' as 'l'état de la plante adaptée'. *Epharmosis*, a term also invented by Vesque, on the other hand, denotes the act of adaptation (or the behaviour) of organisms exposed to new conditions.

itself especially in the habit, and in the form and duration, of the nutritive organs (in the structure of the foliage-leaf and of the whole vegetative shoot, in the duration of life of the individual, and so forth), but shows to a less extent in the reproductive organs. This subject leads us into deep morphological, anatomical,[1] and physiological investigations ; it is very difficult, yet very alluring ; but only in few cases can its problems be satisfactorily solved at the present time. Thus we impinge upon the problem of the origin of different species.

But difficulty is imparted to the question under discussion by the circumstance that, not only is a species changed in form by external factors and capable of adapting itself to these, but each species is also endowed with certain hereditary tendencies, which, for inherent but unknown causes, evoke morphological characters that cannot be correlated with the present environment and are consequently inexplicable. These inherent tendencies, differing as they do according to systematic affinity, render it possible for different species, in their evolution under the influence of identical factors, *to achieve the same object by the most diverse methods.* While one species may adapt itself to a dry habitat by means of a dense coating of hairs, another may in the same circumstances produce not a single hair, but may elect to clothe itself with a sheet of wax, or to reduce its foliage and assume a succulent stem, or it may become ephemeral in its life-history.

On the one hand, in very few families of flowering plants (e.g. Nymphaeaceae) have the different species assumed approximately the same growth-form, or in other words acquired in harmony with the same environment the same external form, and similar adaptations and habits of life. As a rule, the members of a family differ widely from one another, both in form and in their demands upon the environment. On the other hand, species belonging to families systematically wide apart may be extraordinarily like one another in regard to the structural features of the vegetation-shoot. A striking example of this is afforded by Cactaceae, cactus-like species of Euphorbia and of Stapelia. These furnish an admirable example of a single, marked growth-form which is clearly adapted to definite external conditions, appearing in three families that are distant in affinity (*epharmonic convergence*). Another illustration is seen in the case of Hydrocharis, Limnanthemum, and others, which display so puzzling a likeness in form of their leaves to the Nymphaeaceae.

The term *growth-form* used in this work nearly corresponds with the term *vegetative form* employed by some botanists, but involves more rigid scientific definition. The term vegetative form was introduced by Grisebach and has been employed in literature in various senses, so that it requires explanation. Within the same vegetative form were included all those species that are more or less closely similar in design and appearance, whether they be of close or very distant affinity. The design

---

[1] Anatomy, particularly stimulated by Haberlandt, has recently been greatly enriched by numerous researches dealing with the question of the harmony between structure and function or environment. Duval-Jouve (1875) had already defined work of this kind in the following words : ' L'objet de la présente étude est de constater les principales dispositions des tissus dans les feuilles des Graminées, et de déterminer, autant que possible, le rapport de certaines dispositions avec les fonctions imposées par le milieu.'

expresses itself not only in external features (form of the vegetative shoot, and of the leaves, position of the renewal buds, and so forth), but also in the anatomical structure and the behaviour of the plant in life (defoliation, duration of life, and the like). In this regard it is the vegetative organs, especially the vegetative shoots, that are of significance, whereas in systematic botany it is the floral structure that is of import. The vegetative shoot adapts itself to the conditions prevailing in regard to its nutrition ; but the flower follows other laws, other aims, and particularly adopts very diverse methods of pollination. In the morphology and anatomy of the vegetative shoot are reflected the climatic and assimilating conditions ; whereas floral structure is scarcely or not at all influenced by climate, but preserves the impress of phyletic origin under very different conditions of life.

An examination of the catalogues or ' systems ' of vegetative forms compiled from time to time will further elucidate the matter.

Humboldt (1805) was the first to lay stress upon the significance of plant-physiognomy in relation to the landscape : ' Above sixteen different forms of vegetation are principally concerned in determining the aspect or physiognomy of Nature.' [1] He treated the following nineteen forms in greater detail : those corresponding to the palm, banana, malvaceous and bombaceous plants, Mimosa, heath, Cactus, orchid, Casuarina, conifer, Pothos (aroid), liane, aloe, grass, fern, lily, willow, myrtle, melastomaceous plant, laurel. This is, of course, merely a superficial distinction among physiognomic and systematic types ; each of these ' forms ' in reality includes plants with very diverse modes of life. A purely physiognomic system is devoid of scientific significance, which is introduced only when physiognomy is founded upon physiological and oecological facts.

Grisebach [2] made the next important attempt in this direction. He established fifty-four, and subsequently sixty, vegetative forms, arranged in a physiognomic ' system ', and he endeavoured to prove that this demonstrated a connexion between the external form and the environment, in particular the climatic conditions ; according to him a physiognomic type is for the main part also an oecological one. Whilst Grisebach clung in the main to physiognomy and entered into such minutiae as to distinguish, for instance, the laurel-form with stiff, evergreen, undivided, *broad* leaves, from the olive-form with stiff, evergreen, undivided, *narrow* leaves, and the liane-form with *reticulate-veined* leaves, from the rattan-form with *parallel-veined* leaves ; yet with his sixty forms he distinguished by no means all growth-forms, but rather, as he himself pointed out, only such as could serve to indicate country or climate by reason of their growing in numbers together. Furthermore, he did not in the least take anatomical structure into consideration, nor did he adequately appreciate epharmony. [3]

In 1884, Warming, having in view the North-European Spermophyta, gave a general survey of growth-forms which he arranged in fourteen chief groups with many sub-groups, based upon morphological and biological characters ; among other characters the vegetative methods of migration occupied a prominent position. Drude, in 1895, justly

---

[1] Humboldt, 1805, vol. ii, p. 18.　　　[2] Grisebach, 1872.
[3] See Reiter, 1885 ; Warming, 1908.

remarked that these did not take sufficient note of geographical considerations.

Reiter, in 1885, was the next to devote detailed consideration to the subject. His standpoint was sound and he laid stress upon internal structure, particular observation of truly adaptive features, and a due regard for *all* the types of a characteristic mode of life and of special design—as opposed to a regard for merely such of these types as occur in great numbers. Nevertheless Reiter's ' system ' is capable of improvements.

Subsequently Drude[1] dealt with the question. He adopted the ' biological-geographical ' standpoint as resting on the answers to the two questions :

What functional rôle does any particular species of plant play in the vegetation of a definite country ?

How does it complete the whole of its periodic life-cycle under the conditions prevailing in its habitat ?

As features of the greater importance he denoted, ' the duration of organs and the protective measures against injuries during the resting period,' also ' the position of the renewal-shoot on the main axis in relation to hibernation '.[2] In his later work he divided plants into thirty-five classes of vegetative forms.

Krause,[3] and later Pound and Clements,[4] gave the main outlines of systems. That of Pound and Clements approaches, as a whole, Drude's. It ranges plants in the following main groups : Woody plants, half-shrubs, pleiocyclic herbs, hapaxanthic herbs, water-plants, hysterophytes, and thallophytes, and it contains thirty-four sub-groups.

Raunkiär[5] devised a system, in which, like Drude, he laid greatest stress upon the adaptation of plants to enable them to live through the unfavourable seasons, as particularly evinced by the degree and kind of protection afforded to the dormant buds and shoot-apices. His five chief groups were phanerophytes, chamaephytes, hemicryptophytes, cryptophytes, and therophytes.

The most recent treatment of this subject is due to Warming, who, since 1890, has published various papers dealing with the structure of growth-forms and the parts they play in formations. In 1908[6] he attempted to map out the main lines of a system of which the following is an outline :—

Just as species are the units in systematic botany, so are growth-forms the units in oecological botany. It is therefore of some practical importance to test the possibility of founding and naming a limited number of growth-forms upon true oecological principles. It cannot be sufficiently insisted that the greatest advance, not only in biology in its wider sense, but also in oecological phyto-geography, will be the oecological interpretation of the various growth-forms : from this ultimate goal we are yet far distant.

It is an intricate task to arrange the growth-forms of plants in a genetic system, because they exhibit an overwhelming diversity of forms

[1] Drude, 1887, 1889, 1890, 1896.   [2] Drude, 1890, p. 69 ; 1896, p. 46.
[3] Krause, 1891.   [4] Pound and Clements, 1898.
[5] Raunkiär, 1903, 1905, 1907.   [6] Warming, 1908.

and are connected by the most gradual intermediate stages, also because it is difficult to discover guiding principles that are really natural. Nor is it an easy task to find short and appropriate names for the different types. Genetic relationships, and purely morphological or anatomical characters, such as the venation and shape of leaves, the order of succession of shoots, monopodial and sympodial branching, are of very slight oecological or of no physiognomic significance. Oecological and physiological features, particularly the adaptation of the nutritive organs in form, structure, and biology, to climate and substratum or medium, are of paramount importance. Cases, however, are not wanting in which oecological grouping runs parallel with systematic classification.

Growth-forms may be arranged in the following six main classes, namely :—

1. Heterotrophic.   4. Lichenoid.
2. Aquatic.   5. Lianoid.
3. Muscoid.   6. All other autonomous land-plants.

Heterotrophic growth-forms are shown by all holosaprophytes and holoparasites, which are undoubtedly derived from autophytes and are degenerate in form and structure. Hemi-saprophytes and hemi-parasites, on the contrary, are under the dominance of chlorophyll and exhibit the same rich diversity of form as other green plants. (See Chapter XXV.)

Aquatic growth-forms differ from those shown by land-plants so widely as regards their morphology, anatomy, and physiology, that they must be regarded as constituting a separate class. (See Section IV.)

The muscoid and lichenoid growth-forms are seen, almost only, in mosses and lichens. Their powers of enduring extreme loss of water and of rapidly replacing this by means of absorption over the whole free surface, are oecologically very important. Associated with these characters are a number of others. The distinction between the muscoid and lichenoid types lies in the method of nutrition, as autotrophic and symbiotic respectively. (See Chapter XXV.)

The lianoid growth-form is mainly determined by social conditions, and shows peculiar oecological and physiological characters. (See Chapter XXV.) Epiphytes, on the contrary, form an edaphic community of autotrophic land-plants including many different types.

The sixth class includes the growth-forms adopted by all the remaining autotrophic land-plants that contain chlorophyll and, as regards nutrition, are independent of other plants, and are thus *autonomous*. The growth-forms of Pteridophyta are included here, although these differ so widely from those of Spermophyta as regards their reproductive organs.

These main classes may, in turn, be divided into sub-classes. In particular, the growth-forms of the autotrophic plants of the sixth class admit of grouping in categories the foundation of which is the duration of the individual plant and of its parts. Upon this basis these plants are divided into *monocarpic* and *polycarpic :* [1] the former produce flower and fruit (or spores) once, and then die ; the latter may produce fruit repeatedly before death claims them.

[1] In recent times these Candollean terms have been suppressed often in favour of A. Braun's ' hapaxanthic' (ἅπαξ, once ; ἄνθος, flower) and Kjellman's ' pollakanthic' (πολλάκις, several times ; ἄνθος, flower).

We adopt the following sub-division :

I. MONOCARPIC HERBS,

which include the following groups :—

(a) **Aestival annual plants.** The whole cycle of life is completed within one vegetative period, varying from a few weeks, as in ephemeral desert-plants, to several months. Shoots foliaged with elongated internodes (monocyclic). No vegetative organs of storage. The unfavourable season passed in the form of seed.[1] Adaptation to dry climates and localities, and to disturbed soil (littoral sand, cultivated soil, and the like).

(b) **Hibernal annual plants.** These germinate in autumn, and conclude their existence with the production of fruit in spring. Rosette-shoots are usually prevalent. Otherwise like those belonging to the preceding group (a).

(c) **Biennial-perennial (dicyclic, pleiocyclic[2]) herbs** produce in their first vegetative period or in several successive ones rosettes of leaves, and in the following period the flowering shoot, which usually bears foliage. The foliage-leaves often live through winter. Buds open. Form of the shoot as in (b). Reserve-food often stored in tuberous axial organs (Beta vulgaris, Daucus Carota). Occurring in cold-temperate climates on open soil, also as cultivated plants.

II. POLYCARPIC PLANTS

In the case of the polycarpic plants it is necessary to consider, first, their adaptation to climate, and in particular the season unfavourable to plant-life; secondly, the vegetative season; and, finally, the conditions prevailing in regard to the soil, which Schimper terms *edaphic* conditions. Of greatest importance is—

1. *Duration of the vegetative shoot:* lignified axes of trees, shrubs, and undershrubs; perennial herbaceous shoots; herbaceous shoots deciduous after a short period.

And closely associated with this is—

2. *Length and direction of the internodes :* whether the shoots have short internodes (rosette-shoots) or long internodes, and whether the latter are erect (orthotropous) or prostrate and creeping (plagiotropous).

3. *Position of the renewal-buds* during the unfavourable season [high up in the air, near the soil, under the surface of the soil, or buried in the soil (geophilous)].[3]

Of less importance is—

4. *Structure of the renewal-buds or of buds in general.* All stem-apices and very young leaves are protected by leaves ; this protection is accomplished in some cases merely by older foliage-leaves (open buds), in other cases by specially differentiated protective organs, which are either parts of foliage-leaves or definite bud-scales. This depends less upon climate than upon the form of the assimilatory shoots ; short-jointed shoots with leaves in rosettes usually have open buds ; long-jointed shoots are more varied. These differences in the shoot are physiognomically important, not only in themselves, but because the former shoots are branched little or not at all, while the latter are usually richly so.

[1] See Ascherson, 1866.   [2] Warming, 1884.
[3] See the nomenclature in the paper by Raunkiär, 1905, 1907.

5. *Size of the plant* is of some moment, not only because in the struggle for existence the taller plants are enabled to establish a supremacy more easily, but also because they are more exposed to inclemency of climate ; shrubs reach greater altitudes and latitudes than trees, while dwarf-shrubs and herbs extend even farther than shrubs.

6. *Duration of the leaves* varies ; some live for only a few months, others for years. In all climates deciduous (summer-green, rain-green) and evergreen plants are met with side by side. This distinction is associated with edaphic conditions, and can be utilized in the classification of sub-divisions.

7. *The adaptation of the assimilatory shoot to the conditions of transpiration,* is determined by the substratum, and by the climate. Some plants assimilate mainly by their leaves, which exhibit very great variety in shape and structure ; but others depute their assimilatory function to the stem, and reduce their leaves. The shapes of leaves (their venation, division, and so forth) depend partly upon systematic affinity, and partly upon the surrounding medium and climate ; they are probably of but slight value as a basis of oecological classification.

8. *The capacity for social life* is of great importance in the struggle between species, and consequently in the composition and physiognomy of the plant-community. This capacity is due in some cases to the prolific production of seed, but usually to more vigorous vegetative multiplication by means of travelling shoots, or shoots given off from the root. And this latter is to some extent determined by the soil (moist or wet soil, loose sandy soil, and so forth).

In accordance with these considerations polycarpic plants may be grouped under four sub-classes :—

    (*a*) **Renascent (Redivivus) herbs.**

    (*b*) **Rosette-plants.**

    (*c*) **Creeping plants.**

    (*d*) **Plants with erect long-lived shoots.**[1]

We will now proceed to discuss these sub-classes, and, as a final step in the process of sub-division, we shall be able to define types that can be named after definite species or genera, e. g. Primula-type, Bromelia-type, Cycas-type, and so forth.

(*a*) **Renascent herbs.** Polycarpic herbs whose assimilating and flowering shoots develop at a definite climatic time. The plant therefore passes through a resting period, during which its hypogeous or epigeous renewal-shoots are protected by scale-leaves. When the favourable season arrives, the plant once more reveals itself (and is thus renascent). The photophilous shoots are aestival-annual and usually have long internodes and mesophilous leaf-structure. The perennating hypogeous parts are necessarily provided with reserve food.

A great variety of types is included amongst these herbs. Some are ' spot-bound ' [stationary, sedentary]; others are travelling plants. Among them are the following :—

---

[1] This classification approximates to that proposed by Krause (1891). For the further sub-division of these sub-classes, and for the distinction of the various types included among these fundamental forms, we must refer to the characters mentioned in the paragraphs numbered 4–8.

Multicipital rhizomes. Herbs with a multicipital rhizome [1] ordinarily have axes with short internodes which are hypogeous, but often lie well above the soil, and, at the bases of the connected stems, usually bear irregularly placed buds, which are protected with bud-scales, and give rise to erect flowering long-shoots and inflorescences. Growth is caespitose :—

On grassland and savannah : Vincetoxicum officinale, Silene inflata, Rheum, Dahlia variabilis (with food-storing roots).

In savannah-plants the rhizome often becomes a thick, irregularly lignified ' xylopodium '.[2]

Mat-geophytes.[3] Perennial spot-bound herbs, mostly monocotylous. The renewal-buds often deeply embedded in the soil on a short-jointed, feebly branched shoot, which usually contains a large amount of food-material. Occurring especially in hot dry lands. Rest during summer. Epigeous vegetative shoot with long internodes or, in some species, with a rosette of leaves—

with stem-tubers : Crocus, Arum maculatum, Amorphophallus, Eranthis hyemalis, species of Corydalis ;

with root-tubers : Ophrydeae ;

with bulbs : many Liliaceae and Amaryllidaceae ;

with perennial tuberous stem : Cyclamen.

Travelling geophytes (Rhizome-geophytes). Geophytic perennial herbs with horizontal hypogeous branched scaly shoots, from which are emitted either foliage-leaves or erect epigeous shoots that bear foliage and flowers. The renewal-buds are inserted on the hypogeous shoots, and have bud-scales. Great variety is exhibited in the length of the internodes and other details :—

On loose soil of dunes : Ammophila arundinacea, Carex arenaria.

On loose humus soil in the forest : Polygonatum multiflorum, Anemone nemorosa, Asperula odorata, Zingiber officinale.

On mud in water or swamp : Phragmites communis, Equisetum limosum, Hippuris vulgaris.

In some cases the subterranean shoots become more or less tuberous food-reservoirs, and their elongated thin parts are then very short-lived : Solanum tuberosum, Stachys tuberifera.

Special forms are where—

Epigeous parts are exclusively foliage-leaves : Pteris aquilina.

Hibernation is by roots : Cirsium arvense.

(b) **Rosette-plants.** The erect foliaged shoot has short internodes and consequently closely set leaves ; it is usually epigeous and evergreen with naked buds, although often only the youngest leaves remain fresh and, in winter, are protected by the old faded foliage. The type occurs on—

Open land (grassland, moors, arctic heaths, and so forth) :

Arctic and alpine fell-fields : Papaver nudicaule and Draba.

In the cold-temperate climate there are many rosette-herbs, including those :—

With leaves sessile elongated : Plantago major, Armeria vulgaris, Taraxacum vulgare ;

[1] ' Crown-formers ' (Hitchcock) (1898); also see Pound and Clements, 1898, p. 106 ; Drude, 1896, p. 48.
[2] Lindman, 1900 ; also see Warming, 1892.   [3] See Raunkiär, 1905, 1907.

With leaves long-stalked, broad, more or less cordate : Soldanella
    alpina, Anemone Hepatica ;
With leaves succulent, but thin in most cases and destroyed in winter,
    except the youngest : Crassulaceae.
With runners, by which they can multiply freely : Ranunculus repens,
Fragaria, Potentilla anserina, and Hieracium Pilosella.
With flowers upon—
Foliaged shoots with long internodes : Alchemilla, Geum urbanum ;
Leafless scapes : Drosera vulgaris, Primula officinalis, Taraxacum,
    and others.
Among rosette-plants must be reckoned a number of Graminaceae,
Cyperaceae, Eriocaulonaceae, and other Monocotyledones, with grass-
like leaves that are crowded together, close to the ground, in dense com-
pound rosettes. These plants are particularly found in open country :
grassland, steppe, savannah, and moor.
On high mountains and in arctic lands there are numbers of rosette-
plants, whose perennating leaves are more or less coriaceous and fleshy,
as in species of Saxifraga. On hot, sunny, rocky sites the leaves become
thicker and fleshy, so that there develop such forms as Sempervivum,
Echeveria, and other Crassulaceae. These lead to such types as Aloë,
Agave, Mesembryanthemum, Bromeliaceae, and others, with undivided,
fleshy or leathery, often thorny, long-lived leaves.
Musa-form. It is well to include among rosette-plants the types
belonging to the *Musa-form :* Gigantic tropical herbs with a perennial,
epigeous, evergreen, false stem, composed of the involute leaf-sheaths,
and arising from a subterranean rhizome, and large leaf-blades of
characteristic venation.
Most of the species are stemless, but others have tall stems, and thus
lead to tuft-trees.
Tuft-trees.[1] Shoots with short internodes ; leaves densely set on the
end of the shoot, large, and few ; buds usually naked. Stem unbranched
or with only a few thick branches, none of which are thrown off :—
    1. Trunk unbranched and usually exhibiting no secondary growth
in thickness ; leaves large and divided : tree-ferns, palms, cycads.
    2. Trunk sometimes sparsely branched, undergoing secondary thicken-
ing ; leaves undivided, linear : arborescent Liliaceae (Yucca, Dracaena,
Cordyline, Xanthorrhoea, Vellozia).
    3. Strelitzia-form.
    (c) **Creeping plants.** The assimilating shoots are prostrate, plagio-
tropous, rooting, often long-jointed, and sometimes bear short erect
branches. Buds naked, or encased in scales :—
Some are herbs : Lycopodium clavatum, Lysimachia Nummularia,
Hydrocotyle vulgaris, Menyanthes trifoliata, and Ipomoea Pes-caprae.
Others are woody : Arctostaphylos Uva-ursi, A. alpina (deciduous),
Empetrum, Vaccinium Oxycoccos, and Linnaea borealis.
It is difficult to distinguish this group from the rosette-plants, such
as Ranunculus repens, Potentilla anserina, and Fragaria, which possess
epigeous means of travelling.
Jungermannia-form. In the tropics there are many epiphytic creeping

---

[1] Drude's (1896) *Schopfbäume.*

plants : for instance, species of Philodendron, ferns, and others. The leaves are often short-stalked, more or less orbicular, and distichous (so that the species often bears such names as ' nummulariaefolia ' and ' serpyllifolia ') ; this constitutes the *Jungermannia-form*.

(*d*) **Land-plants with long erect long-lived shoots.** To this sub-class belong very many species that possess more or less woody stems, or, less frequently, herbaceous stems. Among them are the following :—

Cushion-plants. Shoot-system richly ramified, often with the branches densely packed to form hemispherical cushions. Foliage-leaves usually small, more or less evergreen, remaining attached for a long time in a faded condition, and decaying slowly. Buds open. Adaptation to a cold, physiologically dry climate, or to cold dry air and a hot soil : dicotylous plants such as Azorella, Raoulia, Silene acaulis, species of Saxifraga, Draba, Dionysia, Aretia. Mosses assume similar cushion-shapes.

Undershrubs (Suffrutices). There are a great many kinds of these which are transition-forms between herbs and shrubs, and have incompletely lignified stems, or lignified stems that soon perish. The yearling shoot often is branched. The buds are often naked. They include various types :—

Labiate type. Considerable parts of the flowering shoots die after blossoming. Many are Mediterranean plants, adapted to a mild winter during which rain falls, and are particularly found in Continental steppes and *maquis*: species of Salvia, Lavandula, Thymus, Helianthemum, Artemisia, Ruta.

Acanthaceous type. Erect, weakly lignified, tropical forest-plants, with thin leaves : species of Acanthaceae, Rubiaceae, Verbenaceae, Piperaceae.

Rhizomatous undershrubs having subterranean runners : Vaccinium Myrtillus and V. Vitis-Idaea.

Cane-undershrubs,[1] with lignified but commonly monocarpic shoots : Rubus Idaeus, other species of Rubus.

Then there are also—

Soft-stemmed plants. Stems thick, green, soft, scarcely lignified ; leaves usually very large. Plants essentially belonging to tropical forest and marsh, and epiphytes : Araceae specially.

Succulent-stemmed plants : Cactus-form. Stem lignified, green, juicy, unbranched or feebly branched, often thorny. Foliage-leaves suppressed. Buds sunken, often protected by hairs. Adaptation to a hot climate, with prolonged drought. On deserts and rocks, usually forming very open associations. Varying in size from trees to prostrate forms : Cactaceae, species of Euphorbia, Stapelia.

Woody plants with long-lived, lignified stems. Buds naked or scaly. Evergreen or deciduous. To these belong—

Canopy-trees.[2] Dicotylous and gymnospermous trees, with well-branched crown and many small leaves. The crown increases in size from year to year, and the stem necessarily exhibits corresponding secondary increase in thickness. Buds scaly, or at any rate not typically open. Leaves, deciduous or evergreen ; extremely varied in form, venation, and structure ; large, broad, and thin, simple or compound ; or small, broad and coriaceous (sclerophyllous trees) ; pinoid (Coniferae) ;

[1] Drude's (1896) *Schösslingssträucher.*   [2] Drude's (1896) *Wipfelbäume.*

ericoid (Ericaceae) ; cupressoid (Cupressus) ; scale-like in aphyllous trees (Casuarina, Halimodendron). Dwarf-forms occur, some of them having succulent leaves. In certain tropical forms the leaves are aggregated at the ends of long, feebly branched twigs : for instance, in Cecropia and Carica.

Shrubs (*frutices*) and *dwarf-shrubs* (*fruticuli*). Low trunkless woody plants, with the variety in the construction of the leaf and shoot seen in canopy trees :—

Switch-shrubs : erect, long assimilating stems and small caducous leaves.

Succulent-leaved shrubs also belong here : species of Crassulaceae, Mesembryanthemum, Chenopodiaceae, and others.

Gramineous shrubs represent a peculiar type. In the bamboo-form there arise from the subterranean stem tufts of many richly and characteristically branched evergreen stems, which undergo no secondary thickening. Leaves grass-like. Particularly a tropical growth-form forming forest and bush.

Aphyllous shrubs : which sometimes have equisetoid or salicornioid shoots.

The sub-classes of the other classes of growth-forms will be referred to later, when the environmental conditions relating to them are discussed.

CHAPTER III. PLANT-COMMUNITIES

OECOLOGICAL Botany has further to investigate the natural plant-communities, which usually include many species of extremely varied growth-form.

Certain species group themselves into natural associations, that is to say, into communities which we meet with more or less frequently, and which exhibit the same combination of growth-forms and the same facies. As examples in Northern Europe may be cited a meadow with its grasses and perennial herbs, or a beech-forest with its beech-trees and all the species usually accompanying these. Species that form a community must either practise the same economy, making approximately the same demands on its environment (as regards nourishment, light, moisture, and so forth), or one species present must be dependent for its existence upon another species, sometimes to such an extent that the latter provides it with what is necessary or even best suited to it (Oxalis Acetosella and saprophytes which profit from the shade of the beech and from its humus soil) ; a kind of symbiosis seems to prevail between such species.[1] In fact, one often finds, as in beech-forests, that the plants growing under the shade and protection of other species, and belonging to the most diverse families, assume growth-forms that are very similar to one another, but essentially different from those of the forest-trees, which, in their turn, often agree with one another.[2]

Oecological plant-geography has also to inquire into the kinds of natural communities in existence, their special methods of utilizing their resources, and the frequent intimate association together of species

---

[1] See Chap. XXVI regarding Unlike Commensals.   [2] Warming, 1901.

differing in growth-form and economy. The physical and other characters of the habitat play a fundamental part in these matters, and,
for this reason, form the introductory subject-matter of Section I in this
work.

The oecological analysis of a plant-community leads to the recognition
of the growth-forms composing it as its ultimate units. From what has
just been said in regard to growth-forms it follows that species of very
diverse physiognomy can very easily occur together in the same natural
community. But beyond this, as already indicated, species differing
widely, not only in physiognomy but also in their whole economy, may
be associated. We may therefore expect to find both great variety of
form and complexity of inter-relations among the species composing a
natural community ; as an example we may cite the richest of all types
of communities—the tropical rain-forest. It may also be noted that
the physiognomy of a community is not necessarily the same at all
times of the year, the distinction sometimes being caused by a rotation of
species.

In countries far apart there are to be found communities identical in
type, but entirely different in *floristic* composition. Meadows in North
America and in Europe, or the tropical forest in Africa and in the East
Indies, may show the same general physiognomy, the same kinds of
constituent growth-forms, and the same type of natural community,
though of course their species are entirely different and thus introduce
slight physiognomic differences.

The different communities, it need hardly be stated, are scarcely
ever sharply marked off from one another. Just as soil, moisture, and
other external conditions are connected by the most gradual transitions,
so likewise are the plant-communities, especially in cultivated lands.
In addition, the same species often occur in several widely different communities ; for example, Linnaea borealis grows not only in coniferous
forests, but also in birch-woods, and even high above the tree-limit on
the mountains of Norway and on the fell-fields of Greenland. It appears
that different combinations of external factors can replace one another
and bring into existence approximately the same community, or at
least can satisfy equally well one and the same species, and that, for
instance, a moist climate often completely replaces the forest-shade of dry
climates.

It is evident that all these circumstances render very difficult the
correct scientific interpretation, delimitation, diagnosis, and systematic
classification of plant communities, especially when we consider the
condition of our present knowledge—for we have only just commenced
to investigate growth-forms and communities, and what we do not know
seems infinite. Another difficulty, to which allusion has already been
made, is to assign suitable names to the more or less comprehensive,
principal or subordinate, plant-communities occurring on the Earth and
imparting to the landscapes entirely different physiognomies. Nor is it
easy to estimate the true significance of floristic distinctions.

CHAPTER IV. PLAN OF THIS BOOK

WHEREAS *geography* proper has to give information in regard to the species and the distribution of associations of plants in various parts of the earth, *oecological plant-geography* treats of the following :—

1. The external factors affecting the plant's economy ; the effects of these factors upon the external and internal structure of the plant, upon the duration of life and other biological relations, and upon the topographical distribution of species. These factors and their effects are discussed in Section I.

2. The grouping and diagnosis of the plant-communities occurring on the Earth. In connexion with each class the endeavour must be made to discover the determinant factors, the modes in which they are combined, and in which they possibly may replace one another. In Sections II and III the communities are treated generally, and their special treatment follows in Sections IV to XVI.

3. The struggle between plant-communities. This is dealt with in Section XVII.

The various factors require to be dealt with separately, although this is a disadvantage, partly because they never work singly but work in complex combination, and partly because it is by no means clear in all cases what must be ascribed to one factor and what to another. Following Schouw, one may divide the factors into *directly* and *indirectly* operating factors.

Among the *direct* factors are—

(*a*) *Geographical factors.* It is thus that Drude describes the factors that work over great areas, because they are dependent upon the Earth's course round the sun, and upon the latitude : composition of the air, light, temperature, atmospheric precipitations and humidity, movements of the air.

(*b*) *Topographical factors* Those that operate within smaller, more localized, limits : the chemical and physical nature of the soil—the ' *edaphic factors* ' as Schimper[1] termed them.

In the first section of this work the following arrangement is adopted :

Atmospheric factors are treated in Chapters V–VIII, as follows :—

Chapter   V. Light.
      ,,   VI. Temperature.
      ,,   VII. Atmospheric humidity and precipitations.
      ,, VIII. Movements of the air.

To the atmospheric factors likewise also belongs the *composition of the atmosphere.* Excepting as regards the variable humidity of the air, this is very constant over the whole world. The two gases playing the greatest part in plant-life, oxygen and carbon dioxide, are present nearly everywhere in the same relative proportion. The more recent investigations have proved that the relative amount of carbon dioxide at great altitudes and on lowlands is the same. And between the amount of carbon

---

[1] Schimper, 1898.

dioxide in the air of forests and of open land there is no difference.[1] The composition of the atmosphere is therefore devoid of geographical significance.

Edaphic factors are treated in Chapters IX–XVII as follows :—

Chapter	IX.	The nutrient substratum—its constitution.
„	X.	Its structure.
„	XI.	Its air.
„	XII.	Its water.
„	XIII.	Its temperature.
„	XIV.	Its dimensions.
„	XV.	Its nutriment.
„	XVI.	The kinds of soil.
„	XVII.	The problem as to the chemical or physical action of soil.

Indirect factors are—
The contour of the Earth's surface, configuration of land and sea, altitude, latitude, and other active and modifying factors. These are treated as follows in Chapters XVIII–XXI :—

Chapter XVIII.		The influence of a non-living covering over vegetation.
„	XIX.	The influence of a living vegetable covering on soil.
„	XX.	The activity of animals and plants in the soil.
„	XXI.	Exposure. Orographic, and other factors.

The atmospheric factors coincide approximately with Drude's 'geographically operating' factors, because they are usually almost constant over large parts of the Earth's surface. Consequently these are the ones that more than all others make their impress upon the vegetation of a country, for all plant communities are subject to their influence. Edaphic and indirect factors, on the contrary, determine the topographical differences in the vegetation.[2]

[1] Hann, 1897, vol. i, p. 76.
[2] Further particulars are to be found in the works by Sachsse, Dehérain, Vallot, Ramann, Drude, Gräbner, Schimper, and Clements.

SECTION I

OECOLOGICAL FACTORS AND THEIR ACTION

CHAPTER V. LIGHT

RADIANT energy is a marked geographical factor the intensity of which varies according to the season, latitude, altitude, atmospheric humidity, and cloudiness. For the sake of brevity it will be designated by the term ' light '.

Both the intensity and the duration of light vary, and therefore are of import. The intensity may be measured by the aid of the eye or, better, by the action on silver salts.[1] The most powerful action is exerted on the eye by yellow rays, on silver salts by the violet and ultra-violet rays, and upon plants by the red and blue rays.

Conclusions as to the action of light upon plants cannot be drawn directly from observations upon its luminosity and chemical intensity.

Light plays a part—

1. *By its chemical action on chlorophyll.* Without light there would be no production of chlorophyll, no assimilation of carbon dioxide, and no life upon the globe. Commencing at a certain minimum intensity of light (which varies according to the species) assimilation increases as the intensity of light rises, until an optimum is attained. Light that is too strong is injurious in action.[2]

2. *By its heating action.* A plant exposed to insolation attains a considerably higher temperature than does the surrounding air ; while shaded organs, by reason of radiation, become colder than the air.

3. *By promoting transpiration through rise of temperature.* In this case also we must assume the existence of an optimum, which likewise varies with the species and generally does not coincide with the optimum for nutrition.[3] Against excessive transpiration the plant makes various provisions.

4. *By influencing growth movements,* the lie of foliage-leaves, and nearly all vital phenomena. And in these cases, too, the composition of the light as regards the admixture of rays of short and long wavelength (especially a clear or a clouded sky) is of great import.[4]

5. *By influencing the distribution of plants.* The earth, viewed as a whole, has scarcely a spot from which plant-life is excluded by insufficiency of light ; for although the light may be too weak at certain seasons (e. g. during the polar night), yet it becomes at other times strong enough to call forth life. But when we descend to the depths either of the solid earth or of the water, life, dependent as it is upon

---

[1] See Wiesner, 1876 *a*, 1876 *b*, 1893, 1895 *b*.
[2] Wiesner, 1898, 1900, 1904, 1905, 1907 ; K. J. V. Steenstrup, 1901.
[3] Sachs, 1865.
[4] Kissling, 1895 ; Sachs, 1865 ; and, in regard to the physiological action of light especially, Wiesner's many papers.

light, soon ceases, and only some of the most lowly organized plants reach any considerable depth.

The intensity of light plays a leading part in determining the distribution of species and abundance of individuals of a community, because the optimum is very different for different species. With inadequate illumination plants do not thrive ; they become etiolated and undergo degeneration or die. The distinction between *light-plants* and *shade-plants* (for instance, in the forest) is well known. Stebler and Volkart,[1] in Switzerland, have made comparative measurements of the intensity of light beneath trees, and investigations concerning the light required by meadow-plants ; these they classify into light-demanding, light-loving, indifferent, light-avoiding, and light-dreading species. In accordance therewith the distribution of species is different in different localities. In arctic countries the nature of the sky (the number of sunny days, the frequency of clouds and mists) certainly causes the contrast mentioned by many travellers between the rich flora and vegetation in the seclusion of the fjords and those of exposed coasts as well as of the islands of the region.[2]

6. *The development of plants* depends not only upon the intensity but also upon the duration of the light to which they are exposed. For instance, in Finland or the north of Norway barley ripens its grain in eighty-nine days from the day of sowing, but in Schonen (in Sweden, 55–7° N.) it requires one hundred days, despite the higher temperature and the more intense light ; and the explanation of this must in part be that in the former places prolonged illumination promotes anabolism. In the north the periodic vital phenomena of plants set in much more rapidly in summer than in spring, because of the longer duration of the daylight. Arnell states that, going northward from Schonen, for each degree of latitude anthesis is later by 4·3 days in April, 2·3 days in May, 1·5 days in June, and 0·5 of a day in July.

7. *Direct light promotes the production of leaves and flowers.* The side of a tree facing the source of light often acquires foliage before the reverse side : Brazilian Ficus-trees may be seen to be in leaf on their north side, whilst still leafless on their south side ;[3] tufts of Silene acaulis in arctic countries on their south side may be decked with flowers, which also point towards the south, while they are devoid of flowers on their north side.[4]

8. *The vegetative shapes of plants* are greatly influenced by the intensity and direction of the light. The effects of *insufficiently intense light* are revealed not only in the phenomena of etiolation, which are essentially pathological in nature, but also in connexion with healthy normal individuals. Of this, forest-trees furnish admirable examples. Light, in the first place, determines the shape of the individual tree. The duration of life of the branches depends partially upon the intensity of light. The shade cast by younger branches retards the assimilatory activity of the leaves on older branches, and thus renders impossible the normal development of buds and the ripening of wood. The branches

---

[1] Stebler und Volkart, 1904.
[2] In regard to Spitzbergen and East Greenland respectively, consult Nathorst, 1883, and Hartz, 1895. [3] Warming, 1892.
[4] Rosenvinge, 1889–90 ; Stefánsson, 1894.

die off, become brittle, and break by reason of their weight or of storms ; it is because of their suppression that the central parts of trees and shrubs are leafless and have so few twigs. A spruce standing out in the open is conical and bears branches from its summit to its base ; whereas one standing in a dense forest has only a small green crown, and outside this no branches, or only leafless dead ones, because its illumination is different. Dicotylous trees, such as the oak or beech, standing in the open, have a full ovoid or conical crown, but when growing in dense woods have a small crown with upwardly directed branches.[1]

Light plays an important part in the struggles between trees that are growing in company. Forest trees may be divided into—

    (a) *Light-demanding trees*, which demand much light and endure but little shade ;

    (b) *Shade-enduring trees*, which are content with less light and can endure deeper shade.

The reasons for these distinctions must be sought for in the specific distinctions in the chlorophyll, rather than in any difference in the architecture (structure of the shoot, phyllotaxy, and form of the leaf) of the species. Arranging our commonest forest-trees in accordance with their demands for light when individuals of the same age are competing with one another, we arrive at the following series, the order of which approximately denotes decreasing requirements as regards light :—

    1. Larch, birch, aspen, alder.

    2. Scots pine, Weymouth pine, ash, oak, elm, sycamore.

    3. Pinus Montana, Norway spruce, lime, hornbeam, beech, silver fir.

It is worthy of note and biologically important that nearly all trees can endure deeper shade in early youth than they can later in life. It may be added that the power to endure shade also depends upon the fertility of the soil.

Distinction between Sun-plants and Shade-plants.

Between *heliophilous* or *photophilous* plants, which prefer sunlight, and *heliophobous* or *sciophilous* plants, which prefer the shade, there are great differences in external form and internal structure.

1. *Intense light retards the growth of the shoot ;* consequently, heliophytes are compact and have short internodes, but sciophytes have elongated internodes ; species clothing the forest soil are mainly tall and long-stemmed. The leaves of heliophytes are often small, narrow, of linear or some similar form ; but those of sciophytes under the same conditions are large and broad, longer in proportion to the width,[2] and thinner. The leaves of Maianthemum Bifolium in sunlight attain scarcely one-third of the size that they reach in the shade.[3]

The leaves of many species, especially of cultivated plants, are larger in northern lands than in lower latitudes ; this is possibly due to the poverty of the light of high latitudes in rays of short wave-length. In gardens on the west coast of Norway, for instance, the flowers of Tropaeolum majus lurk almost hidden beneath the mass of large leaves.[4]

[1] See Vaupell, 1863.   [2] Warming, 1901.   [3] Kissling, 1895.
[4] Bonnier et Flahault, 1878 ; Schübeler, 1886–8, and others.

2. *Intense light decomposes chlorophyll.* A whole series of structural peculiarities have been interpreted as affording protection against too intense light,[1] and among these are the following :—

3. *The leaves of heliophytes are often folded* (grasses, palms, screw-pines), or wrinkled and bent (Myrtus bullata), while those of sciophytes are flat and smooth ; this feature is well seen in many plants growing in hot, dry places in the West Indies.[2]

4. *The lie of the leaves is different* in heliophytes and sciophytes. Leaves of heliophytes are often directed sharply or even almost vertically upwards (Lactuca Scariola in sunny spots, and other ' compass-plants '),[3] or hang vertically downwards, particularly when young (mango and other tropical plants) ; whereas the leaves of sciophytes are extended horizontally, as we may see in the case of dicotyledons in our beech-woods. The sun's rays strike the leaves of heliophytes at acute angles, and therefore lose in efficiency, but the weaker light in the forest strikes the leaves of sciophytes at right angles. Young leaves are directed vertically or obliquely. In dicotylous sciophytes, a *leaf-mosaic*[4] is often formed by the juxtaposition of large and small leaves in such a manner that the interstices are reduced to a minimum (Fagus, Trapa, Trientalis, Mercurialis, and many other forest-herbs). In plants with acicular and linear leaves, such as Juniperus and Calluna, a great difference exists between heliophytes and sciophytes ; the former have erect ad-pressed leaves, and the latter have spreading leaves ; the former assume a permanent profile-lie, and the latter display their full surface ; these orientations are necessarily assumed by the plants during the young and growing stage.

5. *The photometric movements* exhibited by the leaves of many plants as a consequence of change in illumination may be mentioned here ; to light that is intense (or rich in rays of short wave-length) leaves oppose their edges, to light that is weaker (or composed mainly of rays of long wave-length) they oppose their faces.[5]

6. *The histology of leaves* produced in the sunlight and shade respec-tively is not less different. Heliophylls are often isolateral, namely, when they are erect and their two surfaces are consequently equally illuminated ; sciophylls are universally dorsiventral.[6] Heliophylls have a thick pali-sade tissue, which owes its thickness either to the length of the palisade cells, or to the presence of additional layers of them, or to both of these characters (stems with little or no foliage likewise have a thick palisade tissue extending completely round them) ; sciophylls have a thinner palisade tissue or none at all. Palisade cells are often directed obliquely in reference to the surface ; this appears to be associated with the direc-tion of the rays of incident light.[7] Spongy parenchyma is relatively more developed in sciophylls than in heliophylls. Heliophylls are thicker than sciophylls. Heliophylls have small intercellular spaces, sciophylls have large ones. Heliophylls respire and assimilate more rapidly than do sciophylls of the same species.

---

[1] Wiesner, 1876 *b*.   [2] Johow, 1884.   [3] Stahl, 1881, 1883.
[4] Kerner, 1887 ; Warming, 1901.
[5] See Section III, Chapter XXX.   [6] Heinricher, 1884.
[7] Pick, 1881; Johow, 1884; Heinricher, 1884; Haberlandt, 1886; Warming, 1897.

The *epidermis* of the *heliophyll* is thick ; usually contains no chloro-phyll, at least on its upper face ;[1] sometimes it is converted by periclinal divisions into an aqueous tissue several layers in thickness (e. g. in Ficus elastica and some other tropical plants); its cuticle or cuticular layers are thick. The epidermis of the sciophyll is thin, one cell in thickness, sometimes contains chlorophyll, and its cuticle is thin. The heliophyll is often very glossy and a good reflector of light, as is demonstrated by many tropical examples ;[2] the sciophyll is dull in surface and, when subjected to dry air, fades much more easily than the heliophyll. The epidermal cells of heliophylls have less sinuous lateral walls than those of sciophylls. Stomata of the dorsiventral helio-phyll are confined to the lower face, or are more numerous there than on the upper face (except in some alpine plants), and are often sunk below the level of the surface ; those of the sciophyll are on both faces, but perhaps on the whole more numerous on the lower face, and are inserted at or above the level of the surface. Many tropical sciophytes have velvety leaves that are beset with refractive papillae, which serve to collect the obliquely incident rays of light.[3]

7. *Lignified parts* are more general in heliophytes than in sciophytes, for example, the production of thorns is more frequent. Heliophylls are often stiff and coriaceous (sclerophyllous plants), partly from lignification, partly because of their thickness, and partly because of the nature of the epidermis ; sciophylls are thin and, if large, flaccid (many herbs in European forests, such as species of Corydalis and Circaea, Lappa nemo-rosa, Lactuca muralis, Oxalis Acetosella, many ferns ; and in the tropics, Hymenophyllaceae, mosses, and others).

8. *In the production of hairs* variety is exhibited. Heliophylls often have a dense covering of hairs, a grey tomentum, a silvery coating, or are hairy in divers ways, especially on the lower face (e. g. many plants on rocks, heaths, and steppes) ; sciophylls are universally much less hairy, sometimes quite glabrous.

9. *In the sensitiveness of chlorophyll* to light, great differences probably exist, for presumably the chlorophyll of sciophylls is more sensitive than is that of heliophylls, and is consequently better able to utilize weaker light. This suggestion harmonizes well with the fact that an alcoholic solution of the chlorophyll of sciophylls is very easily decolorized in the presence of light.

10. *Light influences the coloration of plants* by its action in regard to the production not only of chlorophyll but also of red cell-sap (antho-cyan or erythrophyll). This pigment occurs especially in young parts of plants (in young shoots and seedlings), in autumn leaves, in alpine[4] and arctic[5] plants, in tropical sciophytes.[6] Engelmann has demonstrated that it absorbs the rays of light complementary to those absorbed by chlorophyll ; red leaves exposed to radiant heat acquire a higher tem-perature than green leaves do. The red pigment provides the means of storing up heat, which is available in connexion with metabolism when the temperature of the air is relatively low.[7] It may also be men-tioned that the colours of leaves, flowers, and fruits become deeper in

[1] Stöhr, 1879.   [2] Volkens, 1890.   [3] Stahl, 1896 ; Haberlandt, 1905.
[4] Kerner, 1887.   [5] Th. Wulff, 1902.   [6] Stahl, 1896.
[7] Stahl, 1896; Buscalioni et Pollacci, 1903; Jönsson, 1903; Overton, 1899.

high latitudes;[1] while a whole series of heliophytes (Myrtus bullata, Perilla nankinensis, Prunus Pissardi and others) growing naturally possess the dark-red to blackish-red tints characteristic of the familiar copper-varieties of the beech, hazel, and other trees.

The features described above will be treated in greater detail in subsequent chapters that deal with xerophytes.

That light is of great significance in influencing the external and internal construction of plants is beyond doubt. This follows, not only from what has been already said, but also from the fact that many, perhaps most, plants can *adjust* their *anatomical structure*, especially of their leaves ('plastic leaves'), according to the intensity of the light. A beech-leaf exposed to sunlight is structurally different from a beech-leaf in the shade.[2] The arrangement and movements of chloroplasts in cells, and therefore the tint of foliage, depend upon the light ;[3] stronger light causes the leaf to be paler in tint, weaker light causes it to become darker green. As to the exact method in which light acts physiologically our notions are very hazy. Some (Stahl, Pick, Mer, and Dufour) opine that it is *light itself* which determines, according to its intensity, the above-mentioned structural differences in the chlorenchyma ; but these investigators fail to explain how light acts. Others (Areschoug, Vesque and Viet, Kohl, and Lesage) suggest that the cause may be increased *transpiration* due to increased light. Still others (Wagner and Mer) are inclined to lay chief stress upon the strong *assimilation* following upon more intense light.

The action of light of different composition upon the activity of protoplasm and the arrangement of chloroplasts is treated in papers by Sachs and Kissling.[4]

It is scarcely open to doubt that the structural differences between heliophytes and sciophytes must be regarded as affording an example of self-regulation (direct adaptation [5]) on the part of the plant. We see this taking place before our eyes in plastic plants which adjust their structure to light ; opposed to this are other cases in which the structure probably has been modified during the course of phyletic development and become fixed by heredity in successive generations. Among the *uses* of the various structural features are the following : Protection of the chlorophyll from decomposition by intense light,[6] protection of the protoplasm itself (intense light can injure protoplasm, as is demonstrated by its destructive action on bacteria, its use as a means of disinfection, and so forth), protection against excessive transpiration, and regulation of assimilation. When we consider that the volume of the palisade tissue is increased not only by more intense illumination, but also, as research has proved, by stronger transpiration, as well as by various factors (salts in the substratum, injury to the roots, and so forth) that influence the absorption of water from the soil and consequently affect transpiration ; and when we further consider that the palisade tissue increases in all stations where great atmospheric aridity prevails, then we shall be inclined to regard regulation of transpiration as the most

[1] Bonnier et Flahault, 1894 ; Schübeler, 1886.
[2] Stahl, 1880, 1883 ; Hesselman, 1904 ; Woodhead, 1906.
[3] Stahl, 1880, and others.   [4] Kissling, 1895.
[5] See Section XVII, Chapter C.   [6] Wiesner, 1876.

essential reason for the structural differences in question. Transpiration increases with an increase of insolation. Thus light is one of the most important factors influencing transpiration, and the plant regulates the latter according to the intensity of the former. But for further information in this matter we must look to the future.[1]

CHAPTER VI. HEAT

HEAT is to a far higher degree than is light an oecological and geographical factor, not only in general, but also in detail.

Each of the various vital phenomena of plant-life takes place only within definite (minimum and maximum) limits of temperature, and most actively at a certain (optimum) temperature ; these temperatures may even differ in respect to the different functions of one species. Heat is of import in the manufacture of chlorophyll, in the processes of assimilation, of respiration, and of transpiration, the functional activity of the root, germination, the production of foliage and blossom, growth, movement, and so forth. It is therefore clear that conditions as regards heat determine the boundaries of the distribution of species on the Earth.

As the *lower and upper critical temperatures* vary greatly in different species, we can only say generally that the lower critical point ('specific zero' of the species) descends to 0° C., or slightly lower in certain rare cases, which include many arctic and alpine species, mostly of low organization. Algae in the Arctic Ocean, off the coast of Spitzbergen, at about 80° N., grow and fructify vigorously during winter, in darkness at a temperature of −1.8° C. to 0° C. ; of twenty-seven species Kjellman observed twenty-two with reproductive organs. Usually the functions do not commence activity until a temperature several degrees above 0° C. is reached, in some cases (especially in tropical plants) not before 10° C. or 15° C. The upper critical temperature does not attain 50° C., and generally not even 45° C.

The different organs of a plant usually have different capabilities of enduring extreme temperatures. A species may therefore thrive in a country and produce blossom, but its seeds may not ripen, or if they do so may not be able to resist frost, or the seedlings may suffer from the cold. Such a species would be dependent upon vegetative propagation for its permanent existence in such a country.

Heat has also an indirect significance, in that the relative humidity of the air and transpiration depend upon temperature.

Temperatures outside those that are critical to species are not necessarily equally lethal ; in this respect there is a certain amount of latitude, which is greatest below the specific zero—that is to say, plants can, without

---

[1] For further information reference should be made to the works of Areschoug, Stahl, Pick, Dufour, Haberlandt, Heinricher, Vesque, Viet, Mer, Lothelier, Johow, A. Nilsson, Eberdt, Schimper, Gräbner, Wiesner, Hesselman, Woodhead, Stebler, and Volkart. In regard to photometry the works by Wiesner, K. J. V. Steenstrup (1901), and Hesselman (1904) should be consulted.

suffering death, be exposed to low temperatures that are more degrees below the minimum than high, actually fatal, temperatures are above the maximum. (Possibly the sole exception is provided by many bacteria.) Moreover, temperatures below the minimum and above the maximum are not always devoid of significance to plant-life, even if they be not of direct utility.

On the Earth there is scarcely a spot from which plant-life is absolutely excluded by reason of the thermal conditions ; for even in places where the temperature remains for months below the minima of species, or above the maxima (e. g. in parts of Africa), plants thrive at certain seasons of the year. Yet it may be necessary for plants to guard against extreme temperatures and against that which these involve, *change of temperature*. To the latter many plants (e. g. palms) are much more sensitive than to low temperatures. Sudden thawing is injurious to many plants because the tissue is ruptured ; forests often suffer from night-frosts on the east side, on eastern slopes, and on similar spots where the sun's rays strike them early in the day.

The following means are adopted as affording *protection against extreme temperatures*, and particularly against such as are *too low*[1] :—

1. The *cell-contents* of some plants have certain (hitherto unexplained) characters in virtue of which they can withstand extreme temperatures for a long time : in phyto-geography, extremes of cold almost alone have to be considered. These resistant characters may be due to the proto-plasm itself, or to the admixture of sugar, oils, or resinous bodies, with the protoplasm or cell-sap. Protection of this kind is apparently exempli-fied by the snow-alga, Sphaerella nivalis, whose thin-walled isolated cells can endure the cold of arctic snow-fields and ice-fields.[2] Likewise Coch-learia fenestrata is evidently protected ; for on the north coast of Siberia, in the winter of 1878–9, this plant endured unsheltered a temperature remaining lower than $-46°$ C., and in the following spring it continued its flowering which had been interrupted by winter.[3] In a number of trees at autumn time the starch changes to fat ;[4] this is probably of use, in that fatty oil in the form of emulsion prevents sub-cooling and increases the power of resistance to frost. Fat-storing trees (birch, conifers) are precisely the ones that grow in the coldest lands. The change that takes place during winter of solid reserve substances into dissolved substances, namely sugar, also prevents the under-cooling of plant-tissue and death of the plant.[5]

2. *Amount of water.* The water contained in plant-parts plays the leading rôle in regard to their power of enduring extreme temperatures ; the power of endurance is inversely proportional to the amount of con-tained water. Consequently the young shoots of North-European trees often suffer from late frosts, while the older shoots are not damaged ; also seeds, which are always poor in water—for instance those of the wheat—can hibernate uninjured for many years in arctic countries. The smallness of the amount of water possibly also explains the perennation

---

[1] For more recent work on the endurance of plants in winter, and on the effect of freezing, see Mez (1904–5), and Lidforss (1907).
[2] Wittrock, 1883 ; Lagerheim, 1892.   [3] Kjellman, 1884.
[4] A. Fischer, 1891 ; O. G. Petersen, 1896.
[5] Mez, loc. cit. ; Lidforss, loc. cit

of many mosses, lichens, and other lowly organized plants. Lignified parts endure cold better than do herbaceous parts,[1] consequently many arctic and alpine species are woody (dwarf-shrubs). Southern shrubs cultivated in North-European gardens, and trees and shrubs on the Färöes,[2] often do not receive sufficient heat to enable them to ripen their wood ; the ends of their branches are killed by the cold in winter ; and they dwindle from shrubs to sub-shrubs. In places with a longer vegetative season the same species (Broussonetia, Tamarix, and others in Hungary) endure, despite equally severe winter-cold. The forests in arctic Siberia withstand temperatures descending to $-70°$ C. (in Verchoyansk during January the mean temperature is $-51·2°$ C., the mean minimum $-63·9°$ C., and the lowest recorded temperature $-69·8°$ C.) When a plant is frozen to death this is usually associated with a formation of ice, and thus with a drying up of the cell-sap.

3. *Bad conductors of heat*, for example bud-scales or hairs, often envelop the parts requiring protection ; their cells are mostly filled with air or have between them air-containing intercellular spaces, and contain as little water as possible. Very many protective devices are displayed by young shoots when in foliage.[3] Many arctic and alpine plants have a grey cottony or white woolly coat of hairs (Leontopodium alpinum, which is ' edelweiss ') ; ' Frailejon,' which are Compositae occurring on the *paramos* of South America and belong to the genera Culcitium and Espeletia,[4] the shoots of these plants are encased by old faded leaves which hang on and envelop them,[5] just as in Central Europe tender garden plants in autumn are artificially clothed with straw, hay, leaves, and the like. It must be pointed out that though these devices hardly exclude extreme cold (for this may reach the interior of the plant), yet they ward off three contingencies—rapid change of temperature, rapid thawing, and precarious transpiration. Experience and research have shown that, though cold itself is sometimes responsible for the death of plant-members (potatoes, petals, tropical plants in high stations of Brazil, and so forth) which have been fatally frozen, yet the act of thawing is the critical matter in the case of many plants which can be frozen solid without injury. Hence thawing ought to take place *slowly*, and in this direction assistance is rendered by the above-mentioned structural features, which are therefore particularly met with in sub-glacial communities. Repeated sudden freezing and thawing cannot be endured by the majority of the plants of Central Europe (e.g. beech, oak, and others). In contrast with this, Kihlman [6] asserts that ' the extraordinary power of enduring great and rapid oscillations of temperature, and even of withstanding the recurrence of freezing-point several times within twenty-four hours, is an outstanding peculiarity ' of the tundra-vegetation of Lapland.

Submerged aquatic plants are well protected by the *surrounding water* ; many of them sink to the bottom or possess buds that become detached and sink in like manner.

The devices mentioned also serve as a protection against rapid *transpiration*—against the desiccating action of dry, cold winds, which are

[1] Mohl. 1848.   [2] Börgesen, 1905.
[3] See Grüss, 1892.   [4] Göbel, 1892.
[5] See p. 75 and Chap. LXVI.   [6] Kihlman, 1890.

dangerous to plant-life when the soil is cold and physiologically dry and the activity of the roots is arrested.

To species both in respect of their conditions of life and their distribution it is by no means without import which of the efficient *temperatures* (those *between the maximum and minimum*) prevail. The life of the individual is influenced not only by the height of the temperatures to which it is exposed, but also by the amount of efficient heat received or the duration of efficient temperatures. Annual mean temperatures are devoid of significance to plant-life. Only the season during which useful temperatures prevail is of import.[1] Thus in Northern Siberia, where the mean annual temperature is below $-15°$ C., forests occur, yet on Kerguelen Island, where the mean temperature even of the coldest month is above freezing-point, the vegetation is arctic.

In most regions of the Earth change of season causes plant-life to undergo a *period of rest*. The cause in north-temperate climates is change of temperature, and particularly lowness of temperature ; in the tropics it is lack of water. The time during which efficient temperatures are available may be so short, sometimes only a few *weeks* in length, that many species are excluded because they cannot obtain sufficient heat. This certainly explains why *annual species* are so *rare* in arctic latitudes and at alpine altitudes ; they require for the completion of their life-cycle more time than is available.

Perennial herbs in arctic countries and on high mountains display much variety in their adaptation to climate. For example, they may have perennating foliage-leaves, which sometimes contain reserve food, and with the aid of these they can utilize each passing moment that is favourable to assimilation, and lose no fraction of the vegetative season in producing new assimilating organs.[2] They display another adaptation in that they initiate their flowers in the year before these open, so that they can burst into blossom at the immediate commencement of the succeeding spring, and thus have as long as possible a period of blossoming and fertilization, and can utilize the warmest season for maturing the seed.[3]

The temperature and length of the vegetative season affect the *physiognomy* of the individual plant and of the whole vegetation. At one extreme are equatorial countries, where resting seasons are all but imperceptible, and where high temperature is linked with humidity ; here is developed the evergreen tropical plant-life whose luxuriantly growing species clothe the soil with the densest of vegetation. At the other extreme are arctic countries and regions on high mountains, where Nature doles out her gifts with niggardly hand perhaps for only three months in the year ; here the plants developed are, in places, not sufficient to cover the soil ; here, too, dwarf-forms present themselves, because, among other reasons, the vegetative season is too short and the efficient heat too feeble. With increasing heat the rate of growth accelerates until an optimum is attained. But in the two last-named sites, low vegetation, condensed shoots, rosette-plants, small leaves, and caespitose habit, inevitably result. In the tropics, dwarfed growth may result when a high temperature is combined with drought.

[1] Köppen, 1884.   [2] Kerner, 1896.   [3] See Warming, 1908 *a*.

Phenologists have frequently endeavoured to estimate the *accumulated temperature* that species presumably require for their various functions. This reveals its existence most distinctly in spring, at which time the opening of flowers and leaves is clearly dependent upon the conditions prevailing in regard to heat and takes place in one year at one time, in another year at another time, and in one place earlier than in another. The number of days of vegetative activity commencing from a certain date having been estimated, and the temperatures prevailing at various places having been ascertained, endeavours have been made to explain upon this basis the differences in development and the facts of distribution ; but in details there has been great diversity of treatment. Some investigators have sought to estimate the accumulated temperature by the addition of the daily mean temperatures ; others have multiplied the mean temperature of a certain period (particularly the period of growth) by the number of days ; others, again, have relied upon the square of the mean temperature or of the number of days ; still others opined that the daily maxima above 0° C., registered by a thermometer exposed to the sun (insolation-maxima), should be added together. These investigations absolutely demand the support of strictly scientific experimental determinations of the temperatures important in relation to the vital phenomena of the different species. But the results of these estimations will not suffice to explain the extremely difficult and complex question of the relations between heat and the distribution of species, or between phenological phenomena, because other conditions, such as light, the temperature of the soil, the after-effects of the preceding vegetative season, are perhaps capable of replacing higher temperature. A general source of error is that shade-temperatures, and not temperatures resulting from insolation, are used in estimations; but even the sum of the insolation temperatures would hardly give a correct account of the temperatures prevailing during a definite period.

In the following **morphological features** heat indubitably plays a part:

1. *Many sub-glacial plants*, particularly woody plants (Salix, Betula, Juniperus, and others) assume the *espalier-shape*: that is to say, their stems lie on the ground, are pressed against it, and concealed more or less between other plants (mosses and lichens), stones, and such like ; only their tips are directed upwards, sometimes almost at right angles, and rise above the ground only a few centimetres. By this mode of growth the plants doubtless receive a greater amount of heat than they would were they erect ; but it is a question if it be not rather evaporation resulting from the dry cold winds that induces this change of shape.

The same form of growth is exhibited by many littoral plants (Atriplex, Suaeda, Salicornia, Matricaria inodora, in Northern Europe ; Frankenia pulverulenta, on the Mediterranean shores) ; it is not the lateral shoots alone that lie prostrate and radiate in all directions ; but the main shoot itself bends down, sometimes almost at right angles, prone on the soil.[1] Again this habit reveals itself on the desert and on sandy soil that is strongly heated by the sun (e.g. Aizoon canariense, Cotula cinerea, and Fagonia cretica, in Africa ;[2] Artemisia campestris and Herniaria glabra in Northern Europe). In the hot dry climate of the lowland of Madeira

—————————
[1] Warming, 1906.   [2] Volkens, 1887.

prostrate forms are rare, and such forms are protected by succulence, strong coatings of hair, and the like.[1]

These familiar forms of growth have beyond doubt a common cause. With the popular explanation that the plant wishes to ' bend before the wind ', science cannot content itself. Probably the cause must be sought in the difference of temperature of the air and soil at the time when the shoots are developing. There are often to be met, growing side by side, erect and prostrate individuals, e.g. of Atriplex, Salicornia, Suaeda, and others on the northern coasts of Europe ; this fact denotes that the decisive cause is no general factor prevailing at all seasons in a definite place. Neither the wind nor its direction can be responsible, since individuals of one species growing on the same shore may have their main shoots pointing in different directions, as may easily be observed on the North-European coasts. The explanation must, apart from individual peculiarities, probably be sought in the different degree of heating of the plants during their development above ground, and the consequent execution of thermotropic movements. Krasan [2] suggests that plants on a warm soil, particularly in a climate with a warm atmosphere, acquire vigorous erect shoots, but on a cold soil, and especially with an alpine climate, prostrate ones. That psychrocliny in reality is responsible for the espalier-like prostrate growth of plants in various cases, is confirmed by the admirable investigations of Vöchting and Lidforss.[3] Henslow [4] also expresses the view that thermotropism plays a part.

On roads and soil frequently trodden down, prostrate forms, such as Polygonum aviculare, are frequently found. Here the cause is perhaps mostly strong vegetative heliotropism.

2. *Rosette-plants.* Many herbs have their basal leaves more or less horizontally expanded in the form of a rosette ; even when they have elongated rhizomes or subterranean stolons their shoots, upon reaching the surface, become condensed. What factors are responsible for this is scarcely known ; but presumably heat plays some part. Bonnier [5] has experimentally demonstrated that great changes in temperature are among the most efficient factors in determining the character of alpine plants ; those plants that were exposed to intense cooling action at night acquired condensed stems, smaller, thicker, and harder leaves, and they blossomed earlier. Conditions of illumination exerted less influence. Rosette-herbs occur in great numbers in temperate countries and are particularly characteristic of sunny meadows which are covered by low vegetation ; they are found in great numbers in arctic countries and on high alpine situations on open grassy or rocky expanses, yet they occur in still greater quantity on the meadows of lowlands, but rarely in forests.[6] In hot dry climates rosette-plants are less common, some occurring only on moist soil,[7] some others being confined to hot dry spots and, in such cases, characterized by succulence (e.g. with succulent leaves, Echeveria, Aizoon, Agave, and Bromeliaceae) or by other potent means of protection against drought.

3. *Tufted* (caespitose) *growth and scrub-plants* are common in climates

[1] Vahl, 1904 b.   [2] Krasan, 1882, 1884.
[3] Vöchting, 1898 ; Lidforss, 1902, 1906 ; compare C. Schröter, 1904–8.
[4] Henslow, 1894.   [5] Bonnier, 1898.   [6] Warming, 1901.
[7] Meigen, 1893, 1894 ; Vahl, loc. cit.

with extreme temperatures, and in arctic and alpine situations are due to cold, but in deserts are called forth by great dryness, and evaporation arising from intense heat. The shoots become short and curved, in the former places because the warmth requisite for growth is wanting, and in the latter site because heat robs them of moisture.[1]

From what has been said it follows that various structural features apparently must be interpreted as due to the action of heat upon plants. Attention will be directed later in this work to the great importance of the temperature of the air in relation to atmospheric humidity and to transpiration ; these exercise great influence upon plant-form.

The distribution and habitats of species in their main features (vegetative zones of the earth, vegetative regions of mountains) are determined by heat. In terrestrial species of wide geographical distribution the difference between the maximum and minimum temperatures is as a rule especially wide (this is not true of aquatic species). But, above all, heat exerts an influence upon the *habitat, economy, and struggles of plant-communities.* It partly determines the distinction between the climates and vegetation of the coasts and interiors of countries ; this is most clearly shown in arctic countries, where the feeble vegetation of the cold coasts contrasts with that of the interior, which is relatively rich in species and individuals and displays more vigorous individuals.[2]

Arctic countries also exhibit great contrasts between the feebler vegetation of lowlands and the richer, more luxuriant vegetation of sunny mountain-slopes ; for the angle of incidence of the sun's rays is far more acute in the plains than in the declivities. If steep mountains should occur near the Poles they certainly must have a relatively rich vegetation. The angle of inclination and the exposition of mountain-slopes obviously influence the result, since the soil and consequently the air will be heated to different degrees when these differ. But as these and other conditions depend intimately on the temperature of the soil, their treatment is relegated to Chapter XIII. That the contour of the earth's surface may influence details in the geographical distribution of plants may often be seen in situations where, on calm frosty nights, cold air remains suspended over depressions and valleys and causes the plants to be frost-bitten.

CHAPTER VII.   ATMOSPHERIC HUMIDITY AND PRECIPITATIONS

THE oecological importance of *water* to the plant is fundamental, and almost surpasses that of light or heat. Without water, vital activity is possible neither to plant nor animal. Its significance to the plant in a condition of full vital activity is as follows :—

1. As *imbibition-water* it is necessarily present in all protoplasm and all cell-walls.

2. As *cell-sap* it occurs in vacuoles and plays a part in turgidity and normal growth.

3. As a *nutritive substance* it is directly assimilated.

---

[1] Further details are given in Chapter VIII, dealing with the effects of wind.
[2] In regard to the influence of light see Chap. V.

4. *All absorption of nutriment* from the soil, *all osmosis,* all *transference of substance,* take place solely through the aid of water. The germination of seeds and spores and the sprouting of sclerotia demand a supply of water for their initiation. The mineral nutriment of a plant must be present in a dissolved condition.

5. The *assimilation of carbon dioxide* depends upon water; it is retarded in a plant that is not fully turgescent, because for one reason the stomata are closed, and it ceases in a fading plant.

6. *Respiration* ceases when the amount of water in a plant sinks below a certain limit.

7. The open or closed condition of the stomata, and consequently *transpiration,* or the evaporation of water from the plant, depends upon moisture. A moist condition of the leaves increases transpiration.

8. All *movements,* whether due to swelling or to irritability, take place only through the agency of water.

9. The amount of water in a plant is the factor determining *life or death* when the temperature lies outside the critical ones. Dry parts of plants are the most resistant.[1]

It is therein not remarkable that death may ensue from lack of water or from desiccation; yet many plants or their parts can withstand long and severe drought. The limits of desiccation vary greatly; only very few, mostly lowly-organized plants such as lichens, mosses, Selaginella lepidophylla and allies, appear capable of withstanding almost complete desiccation.

It is likewise not surprising that *no other influence impresses its mark to such a degree upon the internal and external structures* of the plant as does *the amount of water present in the air and soil* (or medium), and that no other influence calls forth such great and striking differences in the vegetation as do differences in the supply of water. It has been demonstrated by Hellriegel and others that a larger supply of water yields a richer crop (of leaves, straw, fruit, roots); if the plant be supplied with but little water, dwarf growth (*nanism*) ensues.[2] But it may be noted that the vigour of an ordinary terrestrial plant is not proportional to the water supplied up to an indefinite amount, for there is an optimum which varies according to the nature, aeration, and other characters of the soil. For the purpose of ridding itself of any excess of water absorbed the plant exhibits certain devices (water-pores, guttation, internal bleeding); but there is a limit to the amount of water that can be endured by a plant, for instance, plants that prefer drought mostly perish soon when supplied with large quantities of water (e.g. heath-plants).

Water is conveyed to the plants by two channels—the *air* and the *soil* (or the water in case of aquatic plants). The power of absorbing and retaining water derived from the soil will be discussed in Chapter XII; here moisture in the air and atmospheric precipitations alone will be considered.

Atmospheric humidity.

In the atmosphere there is always some water present in an invisible gaseous state, but the amount of this varies greatly : it rises and

[1] See p. 23.   [2] See Kraus, 1906 *a*.

falls with the temperature of the air—the amount of the water that air is capable of retaining in a gaseous condition varying with the temperature. Cold air does not take up so much water as does hot air before becoming saturated ; consequently great fluctuations occur at different times of the day and year. It is not the absolute humidity of the atmosphere that is of greatest moment to plant-life, but the *saturation-deficit*. The saturation-deficit is the difference between the maximum and the observed vapour-pressure at a given temperature, and therefore indicates the additional amount of water which the atmosphere is capable of taking up, or the amount that it requires to become saturated. Evaporation from the surface of water at the same temperature as the air is nearly proportional to the saturation-deficit. Consequently, the saturation-deficit is one of the indices of the evaporating action of a climate [1] (if temperature be also taken into consideration). The saturation-deficit is, as a rule, smallest during the night and greatest during the day time. But among mountains this condition is often reversed owing to the daily alternation of winds blowing up and down the valleys. Transpiration is largely dependent upon the saturation-deficit, yet even in a very moist atmosphere the transpiration may be very considerable, because the stomata remain open and the plant is heated by rays of light.

The amount of evaporation also depends upon several other conditions, including the temperature, the extent and other qualities of the evaporating surface, so that plants very naturally have produced many morphological and anatomical adaptations, enabling them to flourish under various conditions of humidity.[2] The plant strives in some cases to depress transpiration to a certain low limit, in other cases to promote it ; certain plants, for example many sciophytes (mosses, ferns, particularly Hymenophyllaceae, and others) clothing forest-soil, can assimilate only in very moist air ; others are adapted to very dry air. The structural features guarding against dry air and depressing transpiration are, in part, identical with those guarding against intense light.[3] It must be noted that it is very difficult to decide which features are to be correlated with atmospheric humidity, and which with other factors co-operating at the same time. The peculiarities of sciophytes mentioned on pages 19–20 can scarcely be attributed solely to the greater atmospheric humidity that usually prevails in the shade when compared with the open, but must be partially caused by weaker light, just as the peculiarities of heliophytes are caused by not only intense heat and intense transpiration but also by intense light. Sorauer, Mer, Vesque and Viet, Lothélier, and others have found that the effects of moist air are like those of weak illumination. Plants become more elongated, long-jointed, thinner, paler ; their leaves are smaller and thinner, more transparent, and have their dorsiventral anatomy obliterated because of the absence or feeble development of palisade tissue ; the vascular bundles become weaker, the intercellular spaces larger, and the mechanical tissue weaker or even suppressed. It is the *difference in transpiration* in the one case as in the other that is responsible for the difference in structure.

Mosses and lichens presumably can absorb aqueous vapour from the atmosphere ; but it is uncertain to what extent spermophytous plants

[1] See Hann, 1901.   [2] See Section III.
[3] See Chap. V.

can accomplish this, for instance, by employing hairs or the velamen of aerial roots to condense the vapour. Possibly the cases in which such absorption has been assumed may be due to the deposit of liquid water taking place in or on plant-parts as a consequence of change in temperature. The fact that plants drooping on a warm day become turgescent at night, does not imply that water-vapour has been condensed from the damper night air, but is certainly to be ascribed to the circumstances that transpiration is decreased owing to the smaller saturation-deficit, and that the roots, which may have been continuously conveying water to the plant, supply at night more water than is transpired.

Certain desert-plants excrete hygroscopic salts which at night-time abstract water from the moister air ; but it is not certain that this water, which moistens the surface of the plant, is absorbed and utilized by the cells.[1]

Atmospheric precipitations.

If from any cause the air be cooled to dew-point, so that it cannot hold in a vaporous state the water which it contains, the water is deposited in one of the three known forms of atmospheric precipitation, mist (clouds), rain (snow), or dew (hoar-frost). *Atmospheric precipitations* in part are absorbed by the soil and thus become a source of profit to the plant,[2] and in part are retained by epigeous portions of the plant,[3] with which they come into direct contact and which in certain cases seem to be adapted to absorb them. Many plants (epiphytes and lithophytes) have no source of water other than direct atmospheric precipitations.

Adaptation for the absorption of atmospheric precipitations.[4]—In order that a plant may absorb atmospheric precipitations it is necessary that the cell-wall shall be permeable, the superficial cells shall contain osmotically active substances, and that the water shall not flow off the surface too rapidly. There are plants such as lichens, mosses, and certain algae, which can absorb liquid precipitations easily and rapidly over the whole surface, and thus become turgescent ; these plants also endure extreme desiccation. Other plants have on the surface definite parts which can be wetted and absorb water, but have other parts which do not permit this, or, at all events, can be wetted with difficulty (owing to thick cuticle, coating of wax, and the like). For the purpose of absorbing water from atmospheric precipitations a number of plants have special organs (aerial roots with peculiar absorbing tissue ; spongy old plant-remains that greedily suck in water ; hairs, like those in Bromeliaceae, capable of taking up water ; characteristic leaf-cells with perforated walls, and so forth.[5]

But it must be assumed that water is absorbed by epigeous organs, as a rule, only when these are in a half-faded condition, and when the root can provide no water and the plant contains no reserve-supply ; absorption by the root is a matter of necessity to the normal land-plant.[6]

---

[1] Volkens, 1887 ; Marloth, 1887 ; J. Schmidt, 1904 ; Massart, 1898 a.
[2] See Chapter XII.   [3] See Burgerstein, 1904.
[4] See subsequent Sections, especially those dealing with xerophytes (Chap. XXXI.)
[5] See subsequent Sections, especially those dealing with xerophytes.
[6] Böhm, 1863 ; Detmer, 1877 ; Tschaplowitz, 1892 ; Kny, 1895 ; Wille, 1887 ; see Burgerstein, 1904.

The deposition of *dew* is of very great importance in tracts where there is but little rain ; many, especially subtropical tracts, would be almost devoid of vegetation were it not for the strong deposition of dew in the dry season. The deposition of dew is much greater in lower than in higher latitudes. It plays a remarkable part in plant-life, for instance, in the Egypto-Arabian desert ;[1] it must be the dew that in many places evokes in spring-time the phenomena of plant-life, which take place despite the fact that no rain has fallen for months.[2] According to Mez,[3] some epiphytic Bromeliaceae, Tillandsia usneoides for example, are adapted to take in dew especially by the aid of their loose, chaff-like, scaly hairs ; when the dew-absorbing leaves have an aqueous tissue of considerable extent, this is situated on the lower side of the leaf, but is on the upper side in species adapted for the absorption of rain. Moreover, in European heath-moors dew is of the greatest importance to plant-life, and especially to bog-mosses, indeed it is the solitary source of water during the season of scanty rain. In temperate countries the deposition of dew may be very considerable, but it is of significance not so much as a source of water-supply as an influence depressing transpiration.

It must be assumed that everywhere plants are adapted to the given mean supply of water. But as regards this, great specific differences exist. Wiesner's[4] researches have shown that many terrestrial plants are adapted to a definite average amount of rain, which in general varies with the species. He discovered the existence of two extreme kinds of plants, which he termed respectively *ombrophilous* (rain-loving) and *ombrophobous* (rain-hating) plants, according to their power of enduring without injury the action of rain for a long period (often several months) or only for a short one. Xerophytes are mostly ombrophobous ; mesophytes are ombrophilous or ombrophobous. Ombrophoby is usually associated with unwettability of the leaf-surface, ombrophily with wettability.

Many features have been regarded as adaptations for the *removal of rain*. Jungner and Stahl[5] have demonstrated in plants from rainy climates several characteristic structural features which serve to conduct rain rapidly from the leaves, so that transpiration may not be hindered by the blocking of the stomata, the plant may not be overloaded, fungal spores may be washed off, and so forth. Subserving this purpose are *drip-tips*, which are abnormally long, sudden, apical attenuations especially possessed by entire leaves of tropical plants, such as Ficus religiosa, Theobroma Cacao, and species of Dioscorea ; such tips facilitate the rapid removal of rain-water from the leaf.

Whether certain other features to which Lundström[6] has directed attention subserve the same purpose is perhaps dubious ; lines of hairs, for example in Stellaria media and Veronica Chamaedrys, have been interpreted as affording means of carrying away water, so likewise have *furrowed nerves* and *petioles* of Lamium album, Humulus Lupulus, and Aruncus silvester[7] ; and likewise *velvety leaves* in the tropical forest.[8]

It has been believed that *falling rain*, and particularly violent torrents

[1] Volkens, 1887.   [2] Warming, 1892.   [3] Mez, 1904.
[4] Wiesner, 1894, 1897.   [5] Jungner, 1891 ; Stahl, 1893, 1896.
[6] Lundström, 1884.   [7] Stahl, 1893.
[8] See Section XVI, p. 346.

of rain descending during storms, can mechanically damage parts of plants, and especially young delicate parts. The danger of injury arising from falling rain has certainly been greatly over-estimated. According to Wiesner [1] the weight of a drop of artificially produced rain is ·62 gramme, but that of the largest rain-drops observed was only ·16 gramme. The velocity of descent of rain is small and approximately constant. The maximum kinetic energy of a falling rain-drop is estimated by Wiesner as ·0004 kilogramme-metre.

As a means of protection in this regard, the following devices have been supposed to serve :—

1. The leaves of many, especially tropical, plants are directed upwards or downwards, so that the rain strikes them at acute angles and thus acts less violently [2]; in particular, young parts, either individual leaves or whole twigs, are pendulous and do not erect themselves until they have acquired a firmer texture (many tropical plants, Picea, and others).

2. Foldings, or corrugations of the leaf-blades, may operate in like manner.

3. Other plants having compound leaves execute paratonic movements when the sky darkens, before the rain itself descends ; consequently the rain impinges upon the leaflets at more acute angles.

4. Finely compound leaves of many tropical trees expose, as a whole, a less easily assailable blade than do broad and undivided leaves.

5. The possession by the leaves of most plants of a certain amount of free mobility, due to their stalks or other causes, is probably the very best defence against the impact of falling rain-drops. Nothing beyond a shaking of the foliage or branches as a direct mechanical effect can be assumed to take place. [3]

Hail can be very injurious to plants ; but there can scarcely be said to be adaptations protective against the damages threatened by hail-storms, though the contrary opinion has been expressed.

Mist (clouds) absorbs light and thus can obstruct the assimilation of carbon dioxide. [4] It also retards the heating of the soil. Against it there can scarcely be said to exist any protection. Consequently the coasts of Spitzbergen, Greenland, and other northern lands are barren and poor in vegetation when compared with land distant from the coast.

The vital and morphological significance of water to the plant in other respects can best be dealt with later on, partly in connexion with the different communities, but a few matters may be noted here :—

A moist climate lengthens *the life of individuals and of leaves.* Aridity, on the contrary, shortens the vegetative period, hastens blooming, the inception of fruits, the maturation of seed ; also brings into existence a marked resting-season, and in steppes and deserts, very numerous annual species.

The *geographical significance* of water is still greater than that of heat, because its distribution is still more uneven ; this is true not only in the main, but also and especially in details. Water is of all factors the most pregnant in relation to kind and distribution of plant-communities.

The relation between rainfall and the amount of water needed by the plant is of great import in regard to differences in the vegetation. Upon this depends the development of equatorial forest-zones, where the rainfall is very great, of desert-zones near the two tropics when the rainfall

---

[1] Wiesner, 1895.
[2] In this and other devices subsequently mentioned, illumination also plays a part, see p. 19.
[3] Wiesner, loc. cit.   [4] See p. 16.

is very scanty, and finally of the great temperate forest-zones. The rain-fall and its distribution during the seasons determine *the great regional distribution* of types of vegetation, while differences in the water-capacity of the various kinds of soil and the various conditions controlling the course of water above ground determine *the finer topographical* shades of distinction. On high mountains the regions are correlated with the distribution of the rainfall. There are often three regions : a lower one with scanty rainfall; a middle one, the cloudy region, with much mist and rain, and consequently clothed with forest ; an upper dry one above the clouds (e. g. Tian Shan, Madeira, Teneriffe). Mountains often show a dry lee-side and luxuriant rainy weather-side. The coast mountains of a country may arrest the rain so that in the interior, steppe, savannah, and the like, develop on the drier soil, whilst the coast-land yields a rich forest-vegetation (e. g. the coast of Brazil and the *campos* of the interior).

The distribution of atmospheric precipitations. With the same rainfall there is a very great difference according as the rain falls *uniformly* throughout a long period (Central Europe), or falls for only a very short time in the form of heavy storms ; the number of rainy days is, in so far, therefore, of greater import than is the amount of rain. In the former case the rain is capable of being much more beneficial to the vegetation ; in the latter case the parched soil is not in a condition to absorb all the water, most of which, flooding and denuding the soil, flows away over its surface or percolates to its depths. Under the former conditions we find growth-forms and plant-communities quite different (mesophilous) from those under the latter, which are more extreme.[1]

It is remarkable that even in smaller districts relatively slight differences in the amount of atmospheric precipitations are capable of evoking great distinctions in the vegetation. Thus, in the rainier parts of North Germany, especially in the north-west, heath dominates, and in its company grow a whole series of typical Atlantic plants that are wanting in the less rainy east. In this latter there is, consequently, a flora much richer in species which prefer dryness and which (also in cultivation) show themselves very sensitive to great humidity, especially in spring and autumn.[2]

Small quantities of rain are of little or no use to vegetation, because evaporation is rapid, and the water evaporates before it has time to sink into the soil.

The time of the atmospheric precipitations (according to the season of the year) is of very great importance. Where in the tropics a heavy rainfall is distributed throughout the whole year, evergreen rain-forest prevails ; where the rainfall is likewise very heavy but is confined to a few months in the year, whilst the rest of the year is dry, high-forest may be present but it will consist of deciduous trees. Rain that is essentially winter-rain, as in Mediterranean countries or in South-West Australia[3], obviously favours a type of vegetation entirely different from that favoured by rain falling in summer. The Mediterranean district and South-West Australia are consequently poor in forest, rich in steppe and bushland ; whilst the vegetation of districts with summer-rain, for example East Australia, is characterized by rain-forest, and savannah rich in trees. In

[1] Wöikof, 1887 ; Köppen, 1900.   [2] Gräbner, 1895, 1901.
[3] Diels, 1906.

the former countries hot and dry seasons coincide, and the vegetation consequently has a xerophilous impress ; in districts with summer-rain, where the same quantity of rain falls, the vegetation has a more mesophilous impress.

Against dry seasons plants may protect themselves by *shedding the strongly transpiring surfaces* (*defoliation*). Other plants that do not shed their foliage necessarily retain, also during the moist season, the features required to protect them from dryness during the dry season. In the tropics, where there is a prolonged rainless season, deciduous foliage is the rule. In sub-tropical (warm-temperate) zones, evergreen trees and shrubs predominate, while many herbs dry up during the dry season. In subtropical districts with summer-rain, we may perhaps regard the reduced evaporation during the cooler winter as being responsible for the evergreen nature of the trees. In districts with winter-rain, trees and shrubs that are leafless during winter subsist but badly, because the summer is too dry. Nor are shrubs that shed their leaves in summer so common as evergreen ones. In dry districts with a very short vegetative season (steppes and deserts), the herbs dry up during the dry season, and the shrubs, for the most part, shed their leaves. It should be noted that in many steppes (e. g. South Russia and Hungary) the summer months during which the vegetation is dried up, are the rainiest ; the rainfall in summer is not sufficient to supply the amount of water required by plants, as it is not great and is not sufficient to atone for the intense evaporation during the hot summer months when the air is dry ; the rain falling in spring, though still less in quantity is more efficient. In cold-temperate zones winter is to be regarded as a ' physiologically dry season ',[1] because, while low temperatures prevail the plants cannot absorb water from the soil. The trees and shrubs are either deciduous or have perennial protection against drought. According to Grisebach,[2] deciduous trees have effective protection against evaporation during winter, but the means employed are not economical, because a not inconsiderable portion of the vegetative season is consumed in the issue of foliage. Consequently, evergreen Coniferae preponderate immediately that the length of the vegetative season sinks below a certain minimum. According to Köppen [3] the southern boundary of the predominant coniferous forests is parallel with the lines denoting equal duration of the warm season. Herbs in the cold temperate zone are mostly evergreen. During the frosty period they find protection under cover of the snow. This is also true of the herbs and dwarf-shrubs in the Arctic zone.

It is clear that matters influencing the amount, distribution, and other distinctive features of atmospheric precipitations, are of indirect significance to oecological plant-geography. Such matters are especially topographical, and include : relief of the earth's surface, altitude above the sea-level, proximity to the sea, prevailing winds and their humidity.

Heat and moisture may be the two weightiest factors determining the development of vegetation. According to the different quantitative proportions in which plants receive and are adapted to them, A. de Candolle[4] has ranged plants into the following six groups :—

I. **Hydromegathermic :** Plants making the greatest demands as regards

---

[1] Schimper, 1898.　　　　[2] Grisebach, 1872.
[3] Köppen, 1900.　　　　[4] A. de Candolle, 1874.

heat (a mean temperature of at least 20° C.) and water ; their present home lies particularly in moist tropical tracts, but at an earlier date they were certainly widely distributed.

2. **Xerophilous** : Plants calling for much heat, but making the most modest demands for water. Here belong plants of the desert, steppe, and savannah.

3. **Mesothermic** : Plants calling for an annual mean temperature of 15°–20° C., and, at least during certain periods, abundant moisture. In the Tertiary epoch these extended up to the North Polar lands.

4. **Microthermic** : Plants requiring an annual mean temperature of 0–15° C., little of the sun's heat, uniformly distributed atmospheric precipitations, and a period of rest caused by the cold.

5. **Hekistothermic** : Plants living beyond the limits of tree-growth, where the annual mean temperature sinks below 0° C. ; they endure prolonged lack of light.

6. **Megistothermic** plants existed in earlier ages of the world's history, and demanded high uniform temperatures (above 30° C.). They were especially Crypto-gamia.

A. de Candolle's groups suffer from the defect that no plants are dependent upon the mean annual temperature, but that plants depend upon the duration and temperature of the vegetative season, and upon certain minima of temperature and humidity which must not be transgressed. As an emendation more consonant with nature, the following arrangement is proposed.[1]

1. **Hydromegathermic** plants : mean temperature of the coldest month being more than 16° C.

2. **Xerophilous** plants : the rainiest month having less than twelve days with rain.

3. **Mesothermic** plants : mean temperature of the coldest month being below 16° C. yet not below 0° C. for long together.

4. **Microthermic** plants : winter having periods of prolonged frost (with snow remaining on the ground).

5. **Hekistothermic** plants : mean temperature of the warmest month being less than 10° C.

CHAPTER VIII.   MOVEMENTS OF THE AIR

Wind exerts an influence upon both the configuration and the distri-bution of plants. This is to be seen most clearly when it blows over large stretches where its force is not broken by mountains, forest, or town, as is the case on sea-coasts and on extensive plains, such as the Asiatic steppes, the Sahara, and the like ; it is also seen where a definite wind, the trade-wind, prevails.

The effects are revealed on tracts with a loose sandy soil and a scanty covering of plants, for instance, on many coasts and the Sahara, in the *formation of dunes*, with which is associated a highly characteristic vegetation.

They are furthermore revealed in mountain-districts by the higher *atmospheric humidity* and greater rainfall, which are caused by the daily valley-winds.[2] Likewise on lofty mountain-chains they are reflected in the *distribution of the atmospheric precipitations*, because the windward side catches the moisture brought by the wind (e. g. the east and south-east coasts of Australia, the east side of the Andes), while the lee-side remains dry. Dependent on these circumstances is the distribution of the different plant-communities according to the amount of moisture that they require, many species and whole formations being restricted as to their altitude above sea-level, and as to their environmental bounds.

[1] See Köppen, 1900.   [2] Hann, 1897, 1901 ; Vahl, 1904 b ; Scott-Elliot, 1900.

Eminently worthy of note is the significance of the föhn-wind[1] to vegetation. The valleys where the föhn prevails are well known for the vegetation, which is that of a warmer climate.

In places sheltered from the wind the vegetation shows a development different from that in unsheltered situations. Wind, when strong and much inclined to prevail in one direction, exercises a remarkable influence on the form of tree-growth and on the whole character of the landscape. Distorted growth and nanism are the consequence. Trees display the following peculiarities in shape :—

(i) They are low in stature.

(ii) The trunk is often bent in the direction towards which the prevailing wind blows, and the boughs are curved and bent in the same sense.

(iii) The shoots are short, often irregularly branched and interlaced.

(iv) Many shoots are killed on the windward side, and sometimes one finds new shoots and fresh leaves only on the lee-side.

(v) The crown assumes a peculiar shape by unilateral branching (Picea excelsa), or, because it inclines from the windward side, appears as if clipped and rounded off, and exposes a very close-set surface.[2]

(vi) The whole forest or bush-wood inclines in like manner away from the windward side.

(vii) Sometimes on the most exposed side the shoots springing from the root or from the base of the stems are the only ones to maintain a fair existence : so that on the windward side a forest may dwindle to scrub, and this in turn may be resolved into scattered or isolated cushion-like individuals (e. g. on the heaths of Jutland).

(viii) The leaves become smaller than usual, and often are more or less brown in patches, or reddish (as if burnt), particularly at the margins.

Like effects of the föhn-wind in East Greenland, on dwarf shrubs and perennial herbs, have been described and figured by Hartz[3] ; in this case the masses of sand and stone carried by gales have an erosive and destructive influence on the windward side.[4]

Various explanations of the effects of wind have been given :—

Borggreve assumed that all effects are essentially due to the mechanical action of wind, in that the shoots and leaves are beaten against one another, shaken, rubbed, and lashed. It is certain that wind can exert a direct mechanical action upon plants, and, for example, cause trunks of trees to slope and their branches to be eroded and barked.

Other authorities ascribed to wind an indirect physiological action of some kind or other.

Focke expresses the opinion that the injuries done to plants may be wrought by particles of salt conveyed by sea-breezes, but the same changes in form are to be observed far inland : for instance, in oak-scrub in central Jutland, or in the centre of Switzerland.

Still others regarded cold as the cause, but on tropical coasts, for instance in the West Indies and at the mouth of the Amazon, we note, under the influence of the trade-wind, the re-appearance of shapes identical with those of our own latitudes, and the cessation of the effects of the wind when any sheltering object intervenes.

---

[1] In regard to the theory of the föhn-wind consult Hann, 1897.
[2] Früh, 1901 ; Klein, 1899, 1905.   [3] Hartz, 1895.
[4] See also Bernátzky, 1901.

The truth is probably that the effects of the wind are *largely* due to the consequent increase in *transpiration* leading to *desiccation*, as was suggested by Wiesner in 1887, by Kihlman in 1890, and by Warming in his lectures in 1889.[1] Wind has a desiccating action which increases with its force. It dries the *soil :* places very exposed to the wind acquire a relatively xerophilous vegetation. It dries.*plants*, so that these must adapt themselves to their conditions in order to avoid desiccation. In a calm atmosphere the air adjacent to plants becomes humid, so that transpiration is obstructed. By even weak movements of the atmosphere air is constantly carried away, and fresh, less-humid portions of it come into contact with the plant. Even when the atmosphere is very humid its uninterrupted renewal will lead to strong transpiration. The drier the air and the stronger the wind, the greater will be the drying action. Evaporation is proportional to the square root of the wind-velocity. The shaking of the plant-organs also operates in the same direction. By this transpiration the growth in length of axes and leaves is decreased (nanism), many leaves and whole shoots are killed, so that irregular branching results : thus all the observed phenomena receive an explanation that is not forced. The deviation of the shoots in the direction of the prevalent wind is possibly almost without exception caused by a kind of sympodial growth. That the crown on the lee-side acquires a gradually ascending shape, is caused by the circumstances that the shoots, both living and dead, on the windward side screen the parts on the lee-side from too rapid a renewal of the air. In this matter once more we see the fundamental significance of water to plant-life.

The force of the wind is far less on the ground than at some distance above it ; consequently short plants are much better protected from the wind than are taller ones.[2]

The danger arising from the wind is increased when at the same time the activity of the root is decreased by *coldness of the soil*, so that the loss of water is covered with difficulty or not at all ; hence in Central Europe when there is too little snow in winter cereals and other plants perish. This circumstance is of particular importance in arctic and alpine situations. The *espalier-shape*, mentioned in p. 26, as assumed by shrubs growing in these places, may be caused by wind, and we often see the shoots directed straight away from the windward side.

The *cushion-like* growth of the herbs (Draba,[3] Androsace helvetica,[4] and others) living under similar unfavourable conditions in windy cold places, may obviously arise in the same way. Even arctic mosses show similar construction.[5] Herbs of this kind acquire, for want of water, short shoots and small leaves, become as a whole very stunted pygmy-forms ; they are richly branched, consequently often of extraordinarily dense growth, and are very like miniature shrubs found in scrub. Often cushion-plants, for instance Silene acaulis, are dried up on the windward side. That dryness can really bring into existence such cushion-like forms is confirmed by plants growing in arid, hot, but tolerably calm, desert places.

---

[1] See Warming, 1884, p. 99.
[3] See illustration in Kjellman, 1884, p. 474.
[4] Kihlman, 1890 ; G. Andersson, 1902.

[2] Wiesner, 1887.

[5] Öttli, 1903.

The transverse section of tree-trunks is also influenced by the wind, since it becomes excentric—the diameter in the direction of the wind being longer than that at right angles to it.

Plants vary according to their species in power of resistance to the wind.[1] Of the trees common in Denmark the hardiest in this respect are the following : Pinus montana, P. austriaca, Picea alba, as well as some species of willow and poplar, and these are consequently also the species of greatest value in afforesting dunes and heaths.

The importance of *protection against wind* has thus been made clear. Such protection is provided by elevations of the land, as well as by other natural or artificial protective barriers : careful study will often show that vegetation differing widely as regards density, stature, structure, development, and admixture of species, can arise respectively on the windward and lee sides of such a barrier, even when this is only an insignificant rock, stone, or shrub. The hills of Central Jutland appear when viewed from the east to be clothed with forest, but when viewed from the west to be clad with heath. In beech forests the vegetation clothing the soil of places where light and wind can penetrate is quite different from that where these are excluded. In this case the wind has *inter alia* an indirectly injurious action, in that it removes the carpet of dead leaves which protects and variously affects [2] the nature of the soil, and in that it leads to the conversion of mild humus into acid humus, or prevents the production of humus. Arctic and alpine vegetation, as was shown by Kihlman [3], receive very material protection from snow, and where this remains lying, particularly in calm and sheltered depressions, the vegetation is consequently of a stamp different from that on more elevated, windy spots.[4]

The defences against wind that have hitherto been mentioned are topographical, but many plants have by adaptation acquired *special structural features*, both morphological and anatomical, by which they are protected. To this category belong bud-scales, covering hairs, remnants of leaves and stems that are long persistent, and other features which will be dealt with later in this work.[5]

Distribution of vegetation. It may be added that though the absence of trees from many places on the Earth is mainly due to wind, yet it is also due to cold and other conditions unfavourable to growth. Wind is partially responsible for the delimitation of the boundaries of forest in Polar lands, and of forest and bushland up high mountains. On mountains, forest ceases where the mountain commences to divide into separate peaks. Above this limit forest can still occur where there is local shelter from the wind, for instance, within the crater-valleys of Java.[6] Also, it is in valleys sheltered from the wind that forest extends farthest north in arctic lands ; for instance, along the Lena and Mackenzie rivers. Middendorff [7] was the first to recognize the significance of the wind in assigning limits to the extension of forest.[8]

[1] Illustrations are given in L. Klein, 1905.   [2] See Chapter XVIII.
[3] Kihlman, 1890.
[4] For further particulars see C. Schröter, 1904–8, and Chapter XVIII.
[5] Also see Chapter VI, p. 24; and Chapter XXX.
[6] Schimper, 1893.   [7] Middendorff, 1867.
[8] The importance of wind has been treated in an attractive and detailed manner by Kihlman, 1890 ; and more recently by Bernátzky, 1901 ; Buchenau, 1903;

The *utility of the wind* to vegetation is especially shown in the conveyance of fresh supplies of carbon dioxide. The transport of pollen to anemophilous plants, such as coniferous and dicotylous trees of North and Central Europe, and in the dispersal of seeds ; many of our common trees have their seeds scattered by the wind.[1]

CHAPTER IX. NATURE OF THE NUTRIENT SUBSTRATUM

THE nature of the nutrient substratum, or edaphic[2] conditions, largely determines the habitats of plants and their topographical distribution; and among all characteristics of the substratum the most important is the *amount of water* contained.

There are two different forms of nutrient substratum available to autophytes : *water and soil.* Both these have to provide the plant with space and nutriment, as well as with external conditions suitable for the absorption and preparation of nutritive material : they make these provisions by entirely different methods and must, therefore, be treated separately.[3]

The *air* is not in itself a nutritive medium in which plants habitually live and feed, but is merely a temporary resort for organisms that are nearly all of them microscopic but are present in countless numbers, which vary according to time and space, being greatest in the vicinity of human dwellings, particularly in large towns, and being least over oceans, high mountains, and in forests. The weightiest geographical rôle of the air is that by its currents it affords ways and means for the transport of countless organisms from one place to another.

The discussion on water and its characters that are of greatest import to oecological plant-geography will be deferred until hydrophilous communities are treated in Section IV. But the characters of soil will be dealt with at once : they depend upon the physical and chemical attributes of the soil-constituents.

CHAPTER X. STRUCTURE OF SOIL

THE term soil is used here in a wide sense and includes—
1. Solid rock ;
2. Loose soil produced *in situ* by weathering ;
3. Secondary, loose, transported soil that is the product of weathering at some other spot.

Solid rock. The characters of *solid rock* depend upon its mineralogical nature. It varies greatly in hardness, porosity, specific heat, and power of radiation, as is shown by contrasts such as granite, shale, and limestone.

Loose soil. By the mechanical disintegration and chemical decom-

Früh, 1901 ; Norton, 1897 ; Ganong, 1899 ; L. Klein, 1899, 1905 ; Kraus, 1905 ; Klinge, 1890 ; Schimper, 1898 ; see Schenck, 1905.
[1] Cp. Warming, 1887 ; Sernander, 1901 ; P. Vogler, 1901.
[2] το ἔδαφος, the soil.   This term was introduced by Schimper (1898).
[3] See Chapters XXVII–XXXIII.

position of rock there arises *loose soil :* the active agencies are, particularly, changes of temperature, congelation of water as well as the chemical action of water and of the oxygen and carbon dioxide of the atmosphere. In certain cases lowly organized plants, such as lichens and bacteria, also play a part. Chemical decomposition and mechanical disintegration always go hand in hand.

Secondary soil. This owes its origin to the transport, and partially to the separation of the different constituents, of soil produced by weathering ; the transporting agents are, particularly, currents of water, movements of glaciers, and wind. Rivers (Po, Nile, Ganges, and others) heap up at their mouths masses of loose substance that has been conveyed from mountains ; glaciers during the Glacial Epoch transported vast masses of soil to distant spots (for instance, from Norway and Sweden to Denmark and North Germany) and continue to do so in the present ; seas in their currents carry with them other masses of substance. Wind deposits sand from the seashore and from inland sandy soil in the form of dunes ; it also carries away fine particles from the surface of soil and deposits them in sheltered places (as loess).

The characters of loose soil depend upon many and divers features, and particularly upon the fineness, chemical nature, arrangement, and cohesion of its constituents, as will be detailed in the sequel.

From loose soil there often arise new kinds of rock, for instance sandstone, shale, and conglomerate, which differ in character from the original rock and play a different rôle in plant-economy.

Loose soil has the following structure. It is a mixture of—
1. Solid constituents ;
2. Air (Chap. XI) ;
3. Water (Chap. XII).

SOLID CONSTITUENTS OF THE SOIL

The solid constituents of soil are :—

(*a*) *Larger mineralogical constituents,* stones varying in quantity and size down to extremely small grains of sand ; if the soil be shaken up in water and allowed to stand, these constituents rapidly sink to the bottom. In chemical composition they vary greatly, but quartz is most common.

(*b*) *Very minute, powder-like particles,* which remain suspended in water for a long time when the soil is shaken up in water and allowed to stand. By this process they can be easily separated from sand. They likewise vary greatly in chemical composition, but are mainly composed of aluminium silicate, and compounds of iron and calcium ; they have an essential influence on the amount of nutriment in the soil, on its power of absorption, and on its physical characters.

(*c*) *Humus-substances* arise from the corpses and by-products of plants or animals. They are destroyed by combustion. Many humus-substances clearly show their organic origin and mostly impart to the soil a black or dark brown colour.

These three kinds of constituents occur in nearly all soils.

All constituents that are too large to pass through a sieve with meshes 0·3 millimetre in width are termed by W. Knop the *soil-skeleton* (coarse sand, grit, and stones, which can be further sub-divided into groups by the aid of the sieve) ; the remaining constituents are termed *fine earth.* The fine earth plays a special part on plant-life, directly as food material, and indirectly because of its power of absorbing important nutritive substances and because of its purely physical attributes. An admixture of stones and grit nevertheless considerably modifies physical relations in the soil.

Pore-volume. The commixture, the relative amounts and the arrangement of the solid constituents enumerated is very different in different soils. Between the solid constituents there are small cavities termed *pores.* The sum of these spaces not occupied by solid constituents in a given volume of soil is termed its *pore-volume.* Soil is very rich in continuous spaces which become the more capillary the narrower they are.

These pores are filled with *air* and *water,* the relative proportions of which depend upon the size of the pores and other circumstances. In the region of the ground-water, the pores are nearly completely filled with water ; at the surface of a sand-dune that has been exposed to prolonged drought we find the converse, a maximum of air and a minimum of water.

Some kinds of soil are more or less crumbly, or capable of becoming so, that is, their individual particles do not remain separate, but combine to form larger particles, which may be termed *compound particles* or *compound grains.* Compound particles are especially found in humus ; according to Darwin,[1] P. E. Müller,[2] and others, they are often the excrementa and casts of subterranean animals, especially earthworms and insect-larvae.[3] Soil having these compound particles acquires characters other than those of a soil consisting of simple particles : it is looser, more easily aerated, takes up water more readily, and allows roots to penetrate more freely. In the practise of horticulture and agriculture an endeavour is made to promote the formation of compound particles in the soil by turning over and ploughing the soil so that its bulk is easily changed by physical factors (especially frost), and by adding other kinds of soil or substance, such as sand, humus, marl, so that its tenacity is changed.

Tenacity of the soil. The force with which particles of soil are held together varies greatly. As contrasts may be mentioned the dune, whose grains of sand are quite loose in a dry condition, and clay ; humus likewise, has little tenacity. Soils may be distinguished into such as are *rigid, stiff (heavy), mellow (mild), lax, loose, shifting.* Rigid soil on drying becomes hard, fissured, and crustaceous, so that the subterranean parts of plants may be ruptured ; the particles of shifting soil on drying become separated from one another, and are so light that they may be carried away by the wind. Tenacity depends, *inter alia,* upon the size and chemical constitution of the particles ; the smaller the particles are, the greater in general is the tenacity.

Plant-form and *vegetation as a whole* are clearly influenced by the tenacity of the soil. In loose soil (sand, mud, humus in the forest, bog-moss, and so forth) the production of long, richly-branched roots and long, horizontal, subterranean stems (runners and rhizomes) with long internodes is favoured, doubtless because the resistance to be overcome

---

[1] Darwin, 1881.   [2] P. E. Müller, 1887*a.*   [3] See Chapter XX.

during growth is small [1] ; in this way social growth is promoted, and the landscape may even acquire a special uniform physiognomy, for example, through Ammophila and Elymus on dunes, or Phragmites and Scirpus in swamps. Firm, very tenacious clay, on the other hand, on drying becomes hard and cracked, and therefore is not well suited to such plants ; in it are found plants with a vertical, short, thick root-stock (tuber or bulb), or with a multicipital rhizome and caespitose habit, for instance, on the *campos* of Brazil.[2] Rigid plastic clay is no favourable soil for plants, indeed (when it occurs beneath other layers) it may form an impenetrable obstacle to plants. Solid rock (without any deposit of loose soil) does not in the least suit plants of the former habit, but may permit the entertainment of plants of the second kind in its splits and clefts (chasmophytes), and beyond this only such plants can settle on its surface (lithophytes) as have special organs of attachment.[3]

It remains to be said that the root-structure of the various species is very little known, and distinctions in this may perhaps often afford an explanation of the distribution of species.

The capillary action of soil plays a very important part in the physical constitution. It depends especially upon the size and arrangement of the particles. The smaller are the particles and the more closely are they packed, the greater is the capillary action ; soil with compound particles has less capillary action than if consisting of simple particles ; stones and coarse grit in the earth likewise depress capillary action.

CHAPTER XI. AIR IN THE SOIL

AIR in the soil is of most fundamental significance to plant life ; all living subterranean parts, like all other living parts, require air (oxygen) for respiration. In very wet soil, normal plants, adapted to soil rich in air, are suffocated ; alcoholic fermentation, the evolution of carbon dioxide, and consequently death and putrefaction [4] ensue ; in soil poor in oxygen decomposition takes place in a manner different from that in aerated soil; humous acids are formed in great quantities, so that the soil becomes 'sour'. The aeration of soil depends essentially upon the structure ; the more porous and loose the soil is, the more free is the aeration. Farmers and gardeners break up the soil with plough and spade, and drain and harrow it, so that, among other reasons, air may be freely admitted. In order to aerate the soil, the Dutch farmer causes the water-table of his meadows to sink to a depth of one metre during autumn and winter, but during the remaining months only to a depth of half a metre ; this is also the practice in the meadows of Söborg in Denmark.[5] A production of acid humus in the forest leads to an exclusion of the air, and consequently to an extinction of the forest.

Air in the soil is somewhat different *in composition* from that in the atmosphere ; it contains more carbon dioxide and less oxygen, parti-

[1] Henslow, 1895.
[2] Warming, 1892 ; Lindman (1900) terms certain woody subterranean tuberous structures ' xylopodia '.
[3] Öttli, 1903. See Section VIII.   [4] See Sorauer, 1886.
[5] See Feilberg, 1890.

cularly in the deeper layers, because of the respiration of subterranean organs, plants (bacteria) and animals, and because of the decomposition of organic bodies. The amount of carbon dioxide varies with the quantity of organic matter in the soil, the vegetation, the contour and humidity of the land, the size of the soil-particles, the depth of stratum (the upper-most layers of soil have less carbonic acid than have the lower ones), and the temperature (season).

The *internal structure of the plant* is correlated with the amount of air contained in the soil ; in very wet soil, as a rule (with the special excep-tion of bacteria), only such plants can thrive as have large internal air-spaces, which are in communication with one another throughout the whole plant, and can convey air from the atmosphere itself to the most distant root-tips and parts of the rhizome (aquatic and paludal plants ; horse-tails in firm clay contrasting with plants in heath-moors, which contain much more air).[1]

CHAPTER XII.   WATER IN THE SOIL

WATER is the third component of soil.   It is attracted by the solid particles of soil, and surrounds them as a thinner or thicker film.[2]

The amount of water varies greatly in different places and at different times in the same place.   After Norlin we may distinguish the following grades which, as a rule, are only approximately estimated : 1 = very dry, 2 = moderately dry, 3 = moderately fresh, 4 = fresh, 5 = some-what moist, 6 = moist, 7 = very moist, 8 = moderately wet, 9 = wet, 10 = very wet.[3]   In more detailed scientific research, the amount of water must be expressed in percentages of the weight or volume of soil. The quantity of water in soil is practically indicated best of all by the plants growing on it ; for no factor has such an influence upon the disposi-tion of species as the amount of water in the soil.

The amount of water in the soil is one of the *most important direct factors* operating on plant life : this follows from the statements in Chap. VII respecting the fundamental significance of water in plant-economy. Water must be present in certain proportions, which are definite for each species (in cultivated plants usually not more than sixty per cent. for any prolonged period) ; too much or too little is injurious in this as in other cases.   The significance to plant life of the quantity of water in soil is demonstrated, for example, by Fittbogen's investigations on oats : on soil, the humidity of which varied between forty and eighty per cent., there was no great difference to the resulting crops; but with a humidity of twenty per cent. the crop was halved, and with one of ten per cent. the crop was reduced to an eighth.   Lack of water in the soil causes the plants to be ill-nourished, because roots can obtain nutriment from such a soil only with difficulty.

Water is also of *indirect* significance, as it affects animals and bacteria living in the soil ; a certain degree of humidity is essential to the production of humus.

[1] See Sections III, IV, V.   [2] See Sachs, J. von, 1865, p. 171 ; Hedgcock, 1902.
[3] See Hult, 1881.

Water in the soil is (1) chemically combined water, which for the most part plays no considerable part in plant-economy ; (2) water absorbed from aqueous vapour of the atmosphere ; (3) water received from atmospheric precipitations and retained by capillary action ; (4) ground-water, or water sucked up from it.

Ground-water is that collected above the impermeable stratum of soil, and moving according to the laws of gravity or remaining in the soil in sheets, just as does water exposed above the surface of soil. A layer of clay mostly serves as the substratum of ground-water ; sand and gravel permit the passage of water. Ground-water may contain many soluble substances, especially calcic salts ; but when it lies deep it is, as a rule, poor in substances nutritive to plants (it is pure), because these have been retained by the over-lying layers ; it is also devoid of bacteria, because the upper layers of soil have acted as a filter.

The *level* of the *ground-water* and fluctuations in this according to the seasons of the year depend upon the amounts of atmospheric precipitation and evaporation, and are of very considerable oecological significance, and play a most important part especially in the desert. In many cases ground-water lies too high for certain plants ; in other cases it is so far below that the roots cannot utilize it directly or indirectly ; in still other cases it is at such a depth as to be reached by the roots at certain seasons, but not at others. In these cases the height to which the water can be raised by capillary action is an important item.

The level of ground-water obviously influences the temperature of soil.[1]

The ability of plants to utilize water is very diverse, because the roots penetrate to different depths. Dry summers acquire great significance in relation to different species, some of which suffer or die sooner than others.[2] Trees with deep roots can thrive even in a dry climate when they are able to reach ground-water. It may be noted that, according to Ototzky, the level of ground-water invariably sinks in the vicinity of forest, and always lies higher in an adjoining steppe than in a forest ; forest consumes water.

The significance of the level of ground-water is very clearly demonstrated in Denmark. Here chemical differences in the soil, which has been pulverized and deposited by glaciers, are scarcely so great as in mountainous countries where the rock lies near the surface and possibly reacts on the vegetation by reason of its chemical nature. A case in point, according to Feilberg,[3] is provided by the sandy plains near Skagen in Jutland. When the ground-water in summer is at a depth of three inches, Juncus-vegetation and meadow - moor prevail ; at six inches mosses (Hypnaceae) and Cyperaceae still play a part, but grasses begin to occur ; at nine inches these latter become dominant ; at twelve inches, normal grass-growth occurs in ordinary summers ; at fifteen inches, cereals thrive in somewhat warm summers ; at from eighteen to twenty-four inches, cereals thrive in cold or moist summers ; at from thirty to forty inches, the soil is unsuitable to cereals, and xerophytes reign. Other examples are given by Feilberg, who lays greater stress, and rightly so, than perhaps the majority of other investigators do, on the importance of

---

[1] See Chapter XIII.   [2] See Dehérain, 1892, and others.
[3] Feilberg, 1890.

the level of ground-water ; he gives, for instance,[1] one as showing how the vegetation of a district gradually changes with the descent of the water-table. Many trees assume a peculiar shape, or cannot grow at all on soil with ground-water near the surface. Warming [2] gives additional examples ; but in these cases further investigation is required to demonstrate what part is played by the level of ground-water, and what by other properties of the soil, including its power of raising water.

In addition, periodic fluctuations in the level of ground-water embracing a number of years (Brückner's thirty-five year periods) has been recognized. These observed fluctuations of climate cannot be associated with Blytt's theory of alternating moist and dry epochs, and corresponding changes in the vegetation.

In the layer of *soil lying above the water-table* the amount of water is influenced by the following important characters : facility of percolation in soil, hygroscopic character of soil, its power of raising water, its water-capacity, as well as the amount of atmospheric precipitations [3] and the influx of surface-water.

Facility of percolation in the soil. Atmospheric precipitations do not penetrate all kinds of soil with equal facility, as may readily be seen if water be poured on sand, clay, and humus. The following factors play a part in this matter : the water-capacity of soil, also the kind and dryness of the particles of soil.

The greater the water-capacity the more slowly does water sink in the soil. Very fine-grained soil, especially clay and certain humus-soils, are almost impermeable to atmospheric precipitations when the particles are densely packed ; whereas the more coarse-grained and loose is the soil the more freely is it penetrated by atmospheric precipitations. If the soil be rich in good-sized stones or in crevices and cavities, such as the burrows of earthworms, then the velocity of penetration is modified by these : it is decreased by stones, but increased by crevices and cavities.

Water penetrates most readily into quartz-sand, less readily into humus, and least fully into clay. Clay soil permits the percolation of water with difficulty, not only because of the small size of its particles, but also because of their other characters.

If the uppermost layers of soil be very dry, some time elapses before they are so wetted as to allow the infiltration of water to commence.

The *hygroscopic character of soil.* All porous and dry soil can absorb aqueous vapour, though to a very varied extent. The absorbed water vapour is invariably available to the plant, because it is taken in only when the earth is dry ; it can never provide too much water. On the other hand it is not capable of alone supplying dry soil with water sufficient for the needs of plants ; these wither before the amount of water in the soil has decreased to such a degree that absorption of aqueous vapour takes place.

Power of soil to raise water. The power that soil possesses of raising water from the deeper layers is obviously of importance to plant life. But we must distinguish between the heights to which, and the velocity with which, water is raised. These depend, *inter alia*, upon capillarity

---

[1] Feilberg, 1891, p. 270.   [2] Warming, 1887, 1890, 1891.
                    [3] See Chap. VII.

and upon the nature of the particles. Quartz-sand raises water rapidly ; clay and other very fine grained soils raise it slowly ; calcareous sand and humus fairly rapidly. But the height to which it is raised is least in sand (only about forty centimetres above the water-table in fine sand, according to Ramann[1]), is greater in clay, and greatest in peat. (The widely accepted view that bog-mosses in heath-moors raise water out of the ground is nevertheless incorrect.)[2] If the grains of soil be more than two to three millimetres in size, the pores are too large to act as capillary tubes.

The power of the soil to raise water is particularly of importance when evaporation from the surface of the soil is great. It may be added that it is more advantageous for a soil poor in water to have a small than to have a great power of raising water, because in the former case the soil is not so easily dried up.

By the *water-capacity of soil* we mean its power to take up and retain liquid, so that none of it sinks into deeper layers of soil. This is measured by the quantity of water that a given weight, or better a given volume, of soil can retain. It depends upon the adhesion of water to the particles of soil, and varies with the capillary power of the soil and with the nature of the particles.

The water-capacity is greater the more numerous and narrow are the capillary spaces in soil and the more uniform their size, because the adhesion-surface is thereby increased. Quartz-sand with particles one to two millimetres in size can retain only one-tenth of the amount held when the size of the particles is ·01--·07 millimetre.[3]

Research[4] has shown that the water-capacity is smallest in quartz-sand, greater in calcareous sand, still greater in clay or in fine, pure calcareous soil, and greatest in humus soil. In the last, the amount of water is reinforced by the presence of *imbibition-water*, which occurs in organic bodies ; of all soils peat has the greatest water-capacity.

Some kinds of soil display so strong an adhesion to water that when this is added the interstices between their solid components are widened and thus their volume is increased, that is to say, these soils swell ; on the contrary, when deprived of some water they shrink ; in this way a modification in the characters of these soils takes place ; when wet they are soft and partially plastic, when dry they are hard and brittle. These statements hold good in reference to clay and peat.

In general, soil is not saturated with water (with the obvious exception of swamps and similar spots in the vicinity of ground-water) ; in soil clad with vegetation the maximum capacity is never attained because the plants are continuously expending water in transpiration.

The drying of soil depends on various factors : the above-mentioned characters of the soil, the consumption of water by plants and animals, and evaporation.

Evaporation obviously has a profound influence on the amount of water in soil and consequently on the economy and constitution of vegetation. Soil retains a certain quantity of water when exposed to the most intense natural evaporation. The force with which water is held fast

[1] Ramann, 1893, 1905.   [2] See P. Gräbner, 1901 ; C. A. Weber, 1902.
[3] See Livingston, 1901, 1903, 1905.
[4] Schübeler, 1886-8.

is of high significance to vegetation. In this respect the various kinds of humus soils are very instructive. Heath-peat (from heath-moors, often composed of the remains of bog-mosses) dries uniformly, and for a long time remains moderately moist internally. The soil of meadow-moors may be as dry as powder at its surface yet greasily wet at a slight depth, for it does not readily permit the equable distribution of water within itself. This property renders it unfit for horticultural usage.[1] The factors operating on evaporation are partly *internal* and partly *external*.

Internal factors are those which depend upon soil itself, such as : the structure of soil, the form of the soil-surface (uneven or even), and so forth. From loose soil less water evaporates than from compact soil, because its power of raising water is less ; the formation of compound particles depresses evaporation. Soil with medium-sized particles permits the greatest evaporation ; large-grained soil permits less.

The *colour* and *kind* of soil are of influence. From a darker soil more water is evaporated than from a paler one, because dark soil absorbs more radiant heat ; the order of gradation is : black, grey, brown, yellow, red, white. From quartz-sand and humus soil evaporation is most rapid, from calcareous sand and clay it is slowest ; Masure was able to render sand and humus completely dry in three days, clay and calcareous soil in seven days. But the amount of water evaporated in a given time is greater, the greater is the water-capacity of the soil; in this respect humus stands at the top and quartz-sand at the bottom. In one experiment by Masure, humus retained 41 per cent., but sand only 2·1 per cent. Evaporation from a soil saturated with water is greater than from an equal water-surface.

Among the *external* factors operating on evaporation from the soil must be reckoned : the saturation-deficit of the atmosphere,[2] the slope and exposure of the surface, the strength of the wind,[3] as well as the vegetation clothing the soil.

Plants clothing the soil increase the surface exposed, and uninter-ruptedly extract from the soil water which is dissipated by evaporation from their leaves and other organs above ground. A field under cultiva-tion becomes more rapidly parched than does a fallow field (of course, the remaining conditions being the same). Plants clothing the soil rob it of moisture during the vegetative season, but to degrees that vary with the temperature and the kinds of plants present. The temperature of soil determines how much water is taken up by roots.[4] Herbs parch the soil more than do trees, and grass is particularly active in this respect. Colding's observations showed that at Copenhagen, from April to Sep-tember short grass consumed an amount of water greater than the rainfall. Feilberg[5] estimated the daily amounts per 0·55 hectare of land during May, June, July, and August, at about 400, 500, 350, and 300 cubic feet respectively ; these estimations are, of course, only approximate, and vary with the conditions. The amount of water in soil therefore dimi-nishes from spring to autumn ; at this time of the year it is at its lowest, and may be from five to seven per cent. less than in spring ; subsequently it increases during winter until plant-life awakens. The

---

[1] Gräbner, 1901.   [2] See Chapter VII.   [3] See Chapter VIII.
[4] See Chapter XIII.   [5] Feilberg, 1890.

differences between species of plants depend upon the sum of the leaf-surfaces and on the leaf-structure, the nature of the root-system, and whether this last is shallow or deep ; thus, in the forest various species act as weeds because they consume water before this can reach the tree roots. We may thus explain how it is that some species are less protected than others in the same habitat. In forests the surface of the soil is protected by the tree-trunks, and consequently remains moist ; but the sub-soil, on the contrary, is robbed of its moisture to a greater extent than when under herbaceous vegetation, because of the activity of the roots of the trees.

Roots can utilize water present in the soil only to a certain degree. The more the water in the soil decreases in amount the more firmly is the remaining water held fast, until a point is reached at which the plant can obtain no more, although a large quantity may still be left behind. Sachs was the first to demonstrate this by investigations on the tobacco plant.[1] A young plant began to wither when the soil (a dark humus) still contained water equivalent to 12·3 per cent. of its dry weight ; the water-capacity of the soil was determined by drying it at 100° C., as being 46 per cent. of its weight ; the plant was able to take up only 33·7 per cent., and the rest of the water was unavailable to it. Under the same conditions the plants withered on loam and on sand when the percentages of water remaining were 8 and 1·5 per cent. respectively. According to Heinrich's experiments, plants first began to wither in coarse sand when the amount of water had sunk to 1·5 per cent., but in peat when the amount was still 47·7 per cent.

A soil from which a species is incapable of extracting water may be described as *dry* to that species, even though a large quantity of water may be present in it (*physiologically dry*).[2] Physiological dryness alone plays a part in distribution of plants.[3]

A dead *vegetable* covering also influences evaporation.[4]

A soil of considerable humidity may partially replace a moist climate. In tropical savannahs the banks of streams are clothed with forest. Furthermore, in steppes and deserts, trees occur where there is running water or where ground-water approaches the surface. Many perennial herbs which in Europe favour a dry sandy soil, occur in the hot dry lowlands of Madeira exclusively on wet soil in the vicinity of springs and water-courses.[5] But it is worthy of note that a moist soil cannot always replace atmospheric humidity. Most species of Erica flourish on very dry soil, but are excluded from places with dry air. On the other hand, Tamarix gallica clings to the banks of rivers not only in the Sahara, but also in Central Europe.

The significance of water in the soil to plant-form. In addition to what has been said in Chapter VII, concerning the significance of water, it may be mentioned that the production of *adventitious roots* on prostrate shoots is evidently promoted by moisture ; nowhere else is the production of adventitious roots so abundant and common as in moist places.[6]

[1] Sachs, J. von, 1865, p. 173.   [2] Schimper, 1898 ; see p. 134.
[3] Schimper, 1898 ; also see Kihlman, 1890 ; Hedgcock, 1902 ; Clements, 1904 ; Burgerstein, 1904.
[4] See Chapter XVIII.   [5] M. Vahl, 1904 b.
[6] Warming, 1884, 1892.

This feature reacts on the duration of life of individuals ; in such spots annual species are rare.[1]

Moreover, roots branch more freely in moist than in dry soil.   Upon the production of root-hairs water also exerts an influence.[2]

As regards the forms of roots, many ' water-roots ' are known to assume peculiar forms, but we are ignorant of the actually operating causes.[3]

CHAPTER XIII. TEMPERATURE OF SOIL

THE temperature of soil is a geographical factor of paramount significance. In addition to what has already been recorded in Chapter VI regarding the general significance of heat, it may be mentioned that the *functional activity of the root* depends upon the temperature of the soil, and that it increases as the temperature rises up to a certain optimum. A plant may wilt in a soil saturated with water if the temperature of the soil sinks below a certain degree, because in such circumstances the roots can absorb no water (the soil is *physiologically dry*) ; and a plant may be frozen to death by a soil-temperature that is too low, although it be capable of withstanding a far lower air-temperature ; beech, oak, and ash can withstand an air-temperature of −25° C., but their finer roots succumb to cold at from −13° C. to −16° C.[4] Many places on high mountains and in Polar countries would be certainly devoid of vegetation were it not for the temperature of the soil, for this may considerably exceed that of the air.   The temperature of soil may rise exceedingly high in deserts.   Bonnet observed a temperature of 59° C. in desert-sand between low plants, when that of the air was 33° C.   Pechuel-Loesche observed a temperature of 75°–82° C. in the soil in Loango.   The temperature of soil and its fluctuations form the subject-matter of a considerable number of recent papers.

The effect of the temperature of soil upon *plant-form* is but little understood. Vesque[5] has experimentally shown that a high temperature of the soil gives rise to an abundance of sap (short, thick roots, stems, and leaves), possibly because the activity of the roots suffers from the heat. These features may also be regarded as affording protection against increased transpiration. Prillieux also concluded that a high soil-temperature directly induces the production of tubers. In this way it becomes easier to understand why succulent plants often grow on rocks between stones, or on soil that is easily heated.

Nanism may result from a low soil-temperature, if it causes a diminution in the amount of water absorbed and consequently of mineral nutriment taken in ; this factor probably co-operates in inducing the dwarfed growth generally prevailing in subglacial vegetation. It has already been mentioned on p. 26 that cold soil calls into existence prostrate shoots and rosette-like growths, whereas warm soil brings forth slender, tall plants, as Krasan[6] has proved in Pinus, Juniperus,

---

[1] Hildebrand, 1882.   [2] F. Schwarz, 1888 ; also see Section III.
[3] For further information on the general influence of moisture in the soil, Gain, 1893, 1895, should be consulted.
[4] Mohl, 1848.   [5] Vesque, 1878.   [6] Krasan, 1882-7.

Asperula longiflora, and others. Cold soil would appear to give rise to glaucous shoots, an abbreviation of the vegetative season, and other characters.

The main *sources of the soil's heat* are two in number : (1) heat from the sun ; (2) chemical processes (especially decomposition) in the soil. These processes acquire particular importance in cold countries.

The heating or cooling of soil, and consequently plant-life, is obviously greatly influenced not only by those factors (radiation, evaporation, heat-conduction, and so forth) that promote or retard cooling, but also by other factors which we may now consider briefly. Of these the first three to be considered concern the sun's heat, and the remainder relate to the soil itself.

1. *Availability of the sun's heat.* Particularly in Polar countries direct sunlight plays a leading part, as is clearly shown by the arrangement of the plant-communities over the landscape. In determining this a greater part is played by the heat of the soil than by the heat of the atmosphere.[1]

2. *Angle of incidence of the sun's rays.* The nearer this is to a right angle the greater is the heating power of the rays (their power being proportional to the cosine of the angle of incidence). Latitude, slope, and exposure of the land, all affect the result. In northern latitudes, south-west, south, and south-east slopes are warmest, while north-east, north, and north-west slopes are coldest.

The relationships indicated in the two preceding paragraphs evoke great differences in the distribution of plant-communities in all latitudes. We see not only in Greenland, for example, that the southern slopes of a mountain-chain may have an open xerophytic vegetation appearing as if burnt up, while the northern slopes are at the same time covered with a dense fresh green, mossy carpet, in which flowering plants are scattered and which in summer are moistened by the slowly-melting snow ;[2] but, also in north-temperate latitudes we note, for instance, that the different faces of thatched roofs support different vegetation ; and again, in Mediterranean countries we observe on the southern mountain slopes, and ascending high up them, xerophilous Mediterranean vegetation with its characteristic forms and its early flowering season, whereas the northern and cooler slopes are stamped with the impress of Central European vegetation, with its more tardy development.[3] Even close to the equator, for example in Venezuela (less than 10° N.), we observe most marked distinctions between northern and southern slopes ; near Caracas, stretching from east to west, there are shallow erosion-valleys or folds in the land which, on their southern slopes, are so poor in vegetation that the red clay almost entirely determines the colour of the land, but on their northern. slopes are clad with denser and taller vegetation. In lower latitudes (South Europe and the tropics) it must be remembered that prevailing north winds convey more moisture to northern than to southern slopes. This circumstance is perhaps of greater import than is the exposure to the sun's rays, because when the sun is overhead a steep southern declivity is exposed to less insolation than is a more gently ascending northern slope : the lower the sun stands the more dependent is the intensity of insolation upon exposure.

[1] See Chapter VI.   [2] See Warming, 1887.
[3] See Flahault, 1893.

It may be added that the snow-line may be at very different levels on the north and south sides of a mountain, and that the altitudes reached by many plants depend upon exposure. Selecting as an example the beech in the Alps, according to Sendtner,[1] the maximum altitude attained by it in South Bavaria is greatest on the south-eastern, and least on the north-eastern side. In the northern hemisphere, species ascend far higher up the southern side of mountains than up the northern (in the Pyrenees, for example, according to Bonnier). The statements above will suffice to show to what an extent heat—in this case the soil's heat (though atmospheric heat and radiation cannot be dissociated from this)—depends upon the relationships enumerated.

3. *Duration of radiation.* In this duration the tropics and Polar lands are very different, at least as regards the distribution of light according to seasons of the year.

4. *The specific heat of soil* varies with the mineralogical composition. The most easy to heat is quartz-sand, and the most difficult peat; between these extremes stand calcareous sand, clay, and others. The specific heat of quartz-sand is 0·2, that of peat about 0·5 (water = 1). The amount of humus in soil is of special importance in this relation.

5. *Colour of soil.* Darker soil is more readily and strongly heated than is that of lighter colour, other conditions remaining constant. Humboldt found that black basalt-sand on the island of Graciosa attained a temperature of 51·2° C., whilst white quartz-sand in the same circumstances attained only 40° C. In radiation the conditions are reversed; darker soil cools more rapidly at night-time than does lighter-coloured, but does not become colder than the latter.

6. *Porosity of soil.* A very porous, gravelly soil absorbs the sun's heat rapidly and becomes intensely heated at its surface, but the heat absorbed is equally readily lost by radiation. Soil rich in air conducts heat slowly, the more slowly the greater is the amount of air, because air is a bad conductor of heat. In a rock substratum the conductivity of heat is greater, and varies in velocity with the nature of the rock. Karst limestone, for example, is an excellent conductor of heat, because of its uniform density and its dryness. Granite, basalt, and other crystalline rocks are likewise good conductors.

7. *The amount of water in soil* is, of all the factors, the one that perhaps has the greatest influence on the temperature of soil, because heat is consumed in the heating and evaporation of the water. Water has a specific heat far greater than that of any kind of soil. The more abundant the water the colder is the soil; dry soil is more easily heated than wet soil, but soil containing much water, on the other hand, retains heat longer than does dry soil, and for this reason is warmer than the latter in autumn. Sandy soils are 'warm' because they rapidly lose water and become heated; clay soils are 'cold'. Soil containing abundant water conducts heat to the subsoil better than dry soil. All these relationships are of profound significance to the development of vegetation in spring for instance. A rock soil is the warmest of all kinds of soil, because no heat is expended in the evaporation of water. Heat penetrates rapidly and deeply into a rock soil, because this is a good conductor of

heat. In the deeper layers the temperature extremes are great, whereas in loose soils only the superficial layers are heated.[1]

Frozen soil, which extends more or less deeply below the surface in Polar lands and on high mountains, naturally plays an important part in relation to vegetation, partly because roots bend away from it as from rock soil (also perhaps by reason of the thermotropism of the roots), and partly because the cold depresses the functional activity of the roots.

8. The *texture of vegetation*, in particular its density, influences the temperature of soil, because it more or less screens this from direct insolation and evaporation, and intervenes in radiation from the soil.[2]

9. *Internal heat of the Earth.* According to Tabert's estimation the mean temperature of the earth's soil is raised by conduction from the internal heat of the earth by 0·1° C.—an indifferent quantity. Krasan's view of the great importance to vegetation of conduction of the earth's internal heat, is based upon an inadequate appreciation of the climatology of soil.

In this connexion it may be mentioned that at Zwickau, owing to the heat liberated from the anthracite which undergoes slow subterranean combustion, it has been found possible to cultivate sub-tropical plants in the open.

10. The *cooling of soil by wind* is in many cases capable of playing an important part on vegetation. For instance, the vegetation on the coasts of the North Sea may suffer the greatest injury from the north-west wind, and this must be partly due to depression in the activity of the roots, caused by this cold wind cooling the soil.

Concerning the *relations between the heat of the soil and of the atmosphere*, it may be stated that in winter the surface of the soil is warmer than the air only during a few hours at the middle of the day, but at other hours it is a little colder than the air. Nevertheless the daily mean temperature of the soil is higher than that of the air. Only where there is a covering of snow is the temperature of this surface, also its daily mean temperature, lower than that of the air. In summer, the temperature of the soil during the day is considerably higher than that of the air, but during the night a little lower or rarely higher. Consequently the annual mean temperature of the soil greatly exceeds that of the air. On mountains the maximal temperatures of the soil are nearly as high as in the lowland at their base, whilst the minima are not correspondingly lower, so that the excess of the temperature of the soil over that of the air increases with the altitude.

The daily fluctuations of the temperature affect the soil to a depth of one metre, descending most deeply in compact kinds of soil that are good conductors of heat. Annual fluctuations penetrate much deeper : in Denmark, for example, to a depth of twenty-five metres, where approximately the mean temperature of the soil remains constant.

Thus it follows that the temperature of the soil undergoes greater fluctuations than does that of the air. The fluctuations are greater in warmer kinds of soil. But plants rapidly adjust their vital processes to variable temperatures. A variable temperature that often approximates to the optimum is more beneficial to plants than is a constant temperature that remains far below the optimum.

[1] See Homén, 1897.   [2] See Chapters XII and XIX.

CHAPTER XIV. DEPTH OF THE SOIL. THE UPPER LAYERS OF THE SOIL AND THE SUBSOIL

DEPTH of soil, that is, the thickness of the layers of incoherent soil above the solid rock is obviously of great import to plants. Great distinctions in the vegetation denote a *shallow soil* where rock lies at a very slight depth, whilst *deep soil* is indicated when this is not the case. Depth of soil affects the temperature, supply of water, amount of nutriment, growth of the roots, and so forth. On shallow soil vegetation is more adapted to dryness and is more dependent upon climatic changes than it is in deep soil; shallow soil produces no such vigorous vegetation as does similar but deep soil, and the vegetation suffers more easily at seasons of drought. A transition from one formation to another may be caused by depth of the soil alone; for instance, Rikli,[1] dealing with Corsica, writes: ' When the soil, poor in humus, becomes still more shallow and consequently drier, the open *maquis* and *garigues* gradually give way to typical fell-heath.'

In soil a distinction is made between the *upper layers of soil* (' the soil,' in the narrower sense), and the *subsoil.* In the former must be included the completely weathered uppermost portion of the soil, which, as a rule more or less intermixed with humus, is subject to the activity of plants and animals, is more influenced by light, heat, and air, and is richer in nutriment, partly because of the absorbent faculty of the soil. By *absorbent faculty* we mean the character of the soil, and particularly of fine soil, in virtue of which it retains, partly by chemical attraction and partly by surface tension (physical attraction) certain nutritive substances, which are soluble in water, and are filtered by it in such a way that they cannot be washed out by rain-water, or only with great difficulty. These nutritive substances are precisely the ones that are least abundant but most important: phosphoric acid, potash, ammonia; on the contrary, nitric acid, and for the main part, lime are easily washed out by rain-water. Soil has a noteworthy power, that of regulating the nature of the aqueous solution in itself. This solution is usually very dilute, and its concentration varies according to circumstances. Different kinds of soil have different absorbent faculties. Certain soils, clay for instance, can even abstract nutriment from the atmosphere, in that they can absorb ammonia.

The relationship between the upper layers of soil and the subsoil are very important. The depth of the upper layers of soil, the amount of water in them, and their other characters, all play a part; broadly speaking, it seems that the relationship to plant-life is the more favourable the more opposed are the characters of subsoil and soil as regards power of raising water and as regards the amount of water contained. Dehérain established the following series :—

Light soil with a permeable subsoil is entirely dependent on climate. If this be dry the soil may be extremely sterile; in a number of places in France there are on such soil coniferous forests, which transpire but little. If the atmospheric precipitations be abundant or the soil be irrigated, it can support tall vegetation.

[1] Rikli, 1903.

Light soil with an impermeable subsoil. In a moderately moist climate such soils are of very variable value, according as they are sloping so that water readily flows away, or are horizontal ; the former soils often sustain a rich vegetation, but the latter are very marshy and useless for cultivation.

Heavy soil with a permeable subsoil is as a rule fertile, as the excess of water percolates into the subsoil.

Heavy soil with an impermeable subsoil supports marsh-vegetation, and requires draining before it can be cultivated.

As the constitution of the subsoil often changes from place to place with extreme suddenness, we see the character of the vegetation undergoing entire change frequently at very short distances. The slope of the ground may essentially modify the significance of the subsoil, and in general it greatly affects the quality of the soil.

CHAPTER XV.   NUTRIMENT IN SOIL

THE plant obtains its nutritive substances partly from air, and partly from the substratum. It is therefore clear that differences in the substratum (edaphic differences) must play a leading part in plant-economy. Water will be discussed in Sections III and IV ; in this chapter we shall deal with soil.

Soil in co-operation with the specific activity of the roots, which must be regarded as differing in different species, *prepares* nutriment, which contains three kinds of constituents :

1. Solid mineral particles ;
2. Salts dissolved in water ;
3. Humus substances.

Soil, as was mentioned in Chapter XIV, *collects* nutriment in its upper layers by means of absorption.

Essential nutritive substances in soil. Some substances are indispensable for the completion of the whole normal development of the plant. In the Higher plants hitherto investigated, the elements required are invariably only ten : oxygen, hydrogen, carbon, nitrogen, phosphorus, sulphur, iron, potassium, calcium, magnesium. If one of these in chemical form available to the plant be lacking, then the plant enters into a pathological condition, or entirely refuses to grow. Beyond this, all plants absorb various other substances that are of unknown utility, and yet cannot be regarded as devoid of significance ; for instance, when present they may so act that certain essential substances are used in smaller quantities than would be the case if they were absent.

Amount of nutritive substances in soil. Not only the nature of nutritive substances, but also their amount, is decisive. If a substance be present in a quantity less than a certain minimum the plant will not thrive ; but species vary greatly in their demands ; different species take in different amounts (one of the reasons for the farmer's adoption of a rotation of crops). The practical man distinguishes between poor and rich soil.

The amount of soluble salt in soil depends upon—

1. The minerals capable of being weathered present in the soil.

2. The absorbent faculty of the soil, which has already been explained.

3. The climate.

Where but little rain falls, the soluble salts produced by weathering, being incompletely washed out, accumulate and may crystallize out, especially on clay surfaces.[1]

An inadequate supply of soluble salts is unfavourable to plant-growth, but too large an amount of them is also fatal to most species, because the osmotic absorption of water is thereby impeded. The same effect is produced by an abundance of humous acids in the soil. Such types of soil belong to those described by Schimper [2] as being *physiologically dry*. Plants on physiologically dry soil are often identical with those on arid (*physically dry*) soil, or are guarded from excessive transpiration by the same protective devices.

The *quantity and quality of nutriment influence plant-form*. *Defective nutriment* (that is an inadequate supply of one or more substances) may be the cause of *dwarf-growth* (nanism) ; this has been demonstrated by many physiological investigations, and is shown in natural vegetation, for instance, on sand-fields and other poor soils. *Dwarf-shrub* is a growth-form characteristic of soil poor in nutriment, and particularly of heath. The amount of a single substance may determine the issue. It is a general rule that the size of a crop, in so far as it is dependent upon nutriment, is determined by that nutritive substance which is available to the species concerned *in relatively the smallest quantity* (Liebig's Law of the Minimum).

When a nutritive substance occurs in so small a quantity that the crop is decreased for this reason, then, according to Atterberg's rule, the substance in question is present in the plant in relatively smaller amount than are those nutritive substances of which there is no deficiency ; it is then easy to surmise that other, morphological distinctions, may also thereby arise.

The *form* of the *root* is adjusted to the characters of the soil. According to the investigations of Sachs [3] the more concentrated the nutrient solution the shorter are the roots. Roots are mostly long and feebly branched in poor soil, for example in plants on sand, especially on dunes, in Central Europe ; nevertheless the majority of heath-plants show the contrary. Roots branch very copiously and form dense clumps in rich soil. If roots encounter strata of soil with different quantities of nutriment, the contrasts in the ramification within the different strata are striking. " Roots search for food as if they possessed eyes."[4]

The *chemical constitution* of the nutritive substratum in certain cases evokes formal differences. This is particularly true of one substance, *common salt*. It is recognized that all halophytes are distinguished by a special configuration ; they have fleshy leaves, transparent tissue, and so forth.[5] The effects of calcium carbonate and other substances are less obvious.

Distinctions in soil have probably led to the separation of *new species :*

The calamine violet (Viola calaminaria) is presumably a form that has arisen from Viola lutea by the action of zinc in the soil.[6]

[1] Hilgard, 1892.   [2] Schimper, 1898.
[3] Sachs, J. von, 1865, p. 177.   [4] Liebig.
[5] See Section VII.   [6] Schimper, 1898 (1903, p. 93).

On serpertine, a silicate of magnesium, there grow two species of Asplenium, A. Serpentini and A. adulterinum, which are closely allied with A. Adiantum-nigrum and A. viride.[1] These new forms of Asplenium have not yet become fixed, but in other cases fixation has probably taken place, so that only very prolonged action could reform them, if indeed it could do so at all.

According to Kerner's [2] researches in the Alps, there exists a wide difference between parallel species occupying limeless slate and limestone mountains; such parallel species, or, better perhaps, races, are the following :—

Calcicolous	Not Calcicolous
Hutchinsia alpina	Hutchinsia brevicaulis
Thlaspi rotundifolium	Thlaspi cepeaefolium
Anemone alpina	Anemone sulphurea
Juncus monanthos	Juncus trifidus
Primula Auricula	Primula villosa
Ranunculus alpestris	Ranunculus crenatus

Dolomite	Not Dolomite [3]
Androsace Hausmanni	Androsace glacialis
Asplenium Seelosii	Asplenium septentrionale
Woodsia glabella	Woodsia hyperborea.[4]

Since such species as replace each other on different soils are certainly derived from one parent species, it becomes of interest to ascertain wherein they differ from each other, because the effects of the soil will presumably be revealed. The first experimental investigations on the action of calcium were, according to Schimper, made by Bonnier.[5] Observations in the open air were conducted by Fliche and Grandeau, and others.[6] Kerner observed in parallel forms the following distinctions :

Calcicolous	Not Calcicolous
Plants more strongly and densely clothed with hairs; often clothed with a white or grey felt.	Hairs glandular.
Leaves often bluish green.	Leaves grass-green.
Leaves more divided and more deeply so.	
If the leaves be entire	Then the leaves not uncommonly glandular-serrate.
Corolla larger	
Flowers mostly with duller surface but lighter hue.	

---

[1] See Schimper, 1898 (1903, p. 93); also Pfeffer, 1897–1904.
[2] Kerner, 1869. [3] Kerner, 1863 b.
[4] Blytt doubts if the Norwegian Woodsia glabella be the dolomite form of W. hyperborea ; it occurs not only on dolomite, but also on slate.
[5] Bonnier, 1894. [6] See Schimper, 1898 (1903, p. 95).

Although the characteristics of the lime-flora are clear and distinct, yet in the past the influence of lime upon vegetation has been over-estimated. Indeed, a distinction has been made between calciphilous and calciphobus plants.[1] Recently it has been definitely established that the amount of lime in itself, in so far as it does not operate physically, cannot be the cause of differences in the flora, for not only can calcicolous plants be cultivated in soil that is poor in lime, but silicicolous plants, and even bog-mosses, which are regarded as pre-eminently calciphobous, can grow vigorously in pure lime-water[2] if the aqueous solution be otherwise poor in dissolved salts. It has been overlooked that nearly all lime soils are rich in soluble mineral substances, and this wealth excludes plants belonging to poorer soils; beyond this the important physical characters of calcareous soil, compared with granite soil, come into play.

Geographical significance of nutriment in soil. The nutritive substances indispensable to the higher plants occur in nearly all soils—certain ones, such as quartz-sand, being excepted—in quantities so considerable that in this respect there is no obstacle to prevent any plant growing almost anywhere on earth. It must be remembered that even when a nutritive substance is present in the substratum in very small quantity, the plants requiring it can yet absorb it in large quantities; for instance, species of Fucus accumulate a great deal of iodine, though extremely little of this is contained in sea-water. The plant has a certain power of quantitative selection, in that it absorbs various substances in proportions other than those in which they occur within the substratum. There are, however, substances which exert a poisonous action on certain plants and exclude these from soils that contain a large amount of them. This is perfectly comprehensible when it is remembered that the plant can select its nutriment only to a certain extent. The larger the amount of a substance in the soil, the more of it, as a rule, is absorbed by the plant; and in all cases substances that are useful or even essential in small quantities, may be absorbed to excess or act as poisons. Substances of this nature are common salt and ferrous salts. But a certain amount of latitude prevails in this matter, for one and the same species absorbs the various nutritive substances from various soils in different proportions. Individuals of the same species on granite-soil contain much silica, and on calcareous soil much lime. Finally, it may be noted that certain substances, for instance lime and magnesia, can replace each other to some extent.

It is of profound significance to communities of plants that each species has its own peculiar economy, the nature of which is almost unknown to us; for in virtue of its metabolic activity and the attributes of its root-system each absorbs substances in proportions different from those prevailing in other species. For the communal life of species it is also of importance that substances are not absorbed at the same rate and time, or at the same ontogenetic stage. This renders it possible for many species to live side by side on the same soil without entering upon a struggle for food. Partially dependent upon this is also the system of ' rotation of crops '.

[1] Sendtner, 1860; Contejean, 1881.
[2] See C. A. Weber, 1900; Gräbner, 1901; and the critique of this by F. E. Clements, 1904.

CHAPTER XVI. KINDS OF SOIL

IN accordance with the different constitution of soil the following main kinds may be distinguished :—

Rock soil	Clay soil
Sand soil	Humus soil
Lime soil	Saline soil

These are connected with one another by gradual transitions and countless admixtures, so that in reality there exist innumerable varieties of soils of multifarious character. As, however, the kinds of soil above named have extremely different properties, and therefore necessarily support plant-communities very different oecologically, their characteristics are briefly recorded here :

1. **Rock soil.** In this case it is the nature of the rock that determines what vegetation can develop upon it. And in this there come into play distinctions in the hardness, porosity, specific heat, and thermal conductivity. Among the most important kinds of rock are : granite, gneiss, limestone, dolomite, sandstone, slate, basalt.[1]

2. **Sand soil.** Sand consists of loose particles of various minerals, mainly of quartz, but also of felspar, hornblende, mica, even lime (for example, in coral-sand), volcanic products. The *nutritive value* of sand varies with the chemical nature of its particles ; pure quartz-sand is sterile, because particles of quartz are incapable of acting as nutriment ; sands containing lime, mica, felspar, and other minerals, have a greater nutritive value. Humus is formed only with difficulty in dry, loose, sandy soil, because in this organic bodies are easily decomposed and oxidized by the admission of air. In addition, sand, particularly quartz-sand, which is the most frequent kind, has only a *slight absorbent faculty* in regard to substances nutritious to the plant.

Sand is *loose soil*, because its particles have but little cohesion, and this diminishes as the particles increase in size. Atmospheric precipitations easily percolate through sand, and with a facility that is greater the larger are the particles. Sand generally contains only a small *amount of water* ; the coarser are its grains the less water does it retain—approximately from three to thirty per cent. ; dune-sand from Bordrup in Jutland, according to Tuxen, takes up twenty-seven per cent. The power of sand to raise water from the subsoil is as a rule very slight ; water is usually raised at most one-third of a metre.

Sand *dries* as a rule very quickly, and consequently becomes rapidly and intensely *heated* in sunlight, but it also *cools* very rapidly and intensely at night. The surface of a shifting sand-dune becomes covered by a dry layer of sand, which becomes strongly heated in sunlight, yet though this layer is of but slight depth it hinders evaporation from the subjacent sand, which consequently always remains moist and cool ; this consideration is of very great importance for the proper understanding of dune-vegetation. In the deserts of Arizona, according to Livingston,[2] a powdery superficial layer appears to act in like manner. The difference

[1] See Section VIII.   [2] Livingston, 1906.

between the temperature by day and by night may be very wide (from forty to forty-five centigrade degrees). As a result, at night-time *dew* is readily and richly deposited upon sand, whose water-content and vegetation are thus profoundly affected. On the other hand, plants suffer from frost more easily on sandy soil. The sand-flora develops early in spring—a feature that recalls the steppe. Plants habitually growing on sand are usually termed *psammophytes* or *psammophilous* plants.[1]

3. **Lime soil.** Calcareous sand, consisting of calcium carbonate, is less poor in nutriment than is quartz-sand, has a greater water-capacity, and dries less rapidly, but is nevertheless dry and warm. *Marl* is an intimate admixture of calcium carbonate (8 to 45 per cent., in calcareous marl 75 per cent.), clay (8 to 60 per cent), and quartz-sand : the lower diluvial marl from the Mark Brandenburg contains calcium carbonate 12 to 18 per cent., clay 25 to 47 per cent., quartz-sand 38 to 62 per cent. The characters of marl depend upon the relative proportions of the constituents, and generally stand between those of sand and clay.

4. **Clay soil.** This offers an almost complete contrast to sand soil. The particles that are invisible to the naked eye predominate over large grains. Clay consists mainly of kaolin (hydrated silicate of aluminium), and may contain more or less quartz, calcium carbonate, ferric oxide, and so forth. Kaolin is of no nutritive value to the plant, yet the presence of many additional substances may render clay very rich in nutritive bodies ; these are, however, available only with difficulty. With a favourable admixture of sand, lime, and humus, a clay soil is a fertile soil.

Clay soil has a large *absorbent faculty*, and is at the same time *very hygroscopic* ; it can absorb five or six per cent. of aqueous vapour from the atmosphere.

Clay soil is *tenacious* or *heavy*, as its particles have great cohesive power ; aeration is mostly defective, a circumstance unfavourable to vegetation, and leading to the production of acids and swampiness.

Clay soil is *wet* and *cold*, because its water-capacity is great (up to ninety per cent.), and because its capillary power is great ; it raises much water from the subsoil, and is *almost impermeable* to water. If we overload it with water it swells, its volume increases, and its individual particles are forced asunder, so that a paste is formed. Clay soil containing an abundance of water is plastic. After prolonged drought, clay soil acquires a stony hardness, contracts, cracks ; and these occurrences react on vegetation.[2]

The unfavourable characters of clay may be ameliorated by commixture with substances of opposite character, such as sand and lime.

Loam, which may be dealt with in connexion with clay, is weathered marl, the calcium carbonate of which has been more or less dissolved in water, and the ferrous salts of which have been converted into ferric oxide and hydroxide ; the soil consequently becomes brown and essentially contains clay and quartz-sand.

5. **Humus soil.**[3] Humus is produced from the remains and products of plants and animals, often from animal excrement in all stages of decomposition, and is mixed in soil with various proportions of mineral

[1] In Section X.   [2] See page 42, Chapter X.
[3] Consult the important work by Früh and Schröter, 1904.

constituents. Humus is black or brown, and rich in carbon, also sometimes in nitrogen—the Russian ' Black Earth ' contains, according to
Kostytscheff, as much as from four to six per cent. In the production
of humus a prominent part is played by micro-organisms (bacteria,
monerae, and the like), and by larger animals, particularly earthworms.

Humus substances combine with nutritive bodies that are soluble
with difficulty to form easily soluble compounds, and thus they increase
the nutritive value of soil. They also change the physical characters
of soil ; when mixed with mineral soil they increase its absorbent faculty,
its specific heat, its water-capacity, and so forth.

There exist wide distinctions among humus soils, according to the
degree of decomposition, and according to the species of plants and
animals engaged in producing the humus. The first of these soils that
we shall deal with is that which is richest in humus :

(a) *Peat soil.* If water containing oxygen come into contact with
organic bodies it is robbed of its oxygen by these. If then the admission
of oxygen be prevented and the activity of micro-organisms excluded,
or at least restricted, in many cases an incomplete decomposition and
change of the organic substances ensues ; as a result, carbon will accumulate in larger quantities the more the supply of air is restricted, and *free
humous acids* make their appearance : there is a production of *peat.
Peat is humus rich in carbon, brown* (from a light to a black-brown) *in
colour,* and contains *many free humous acids,* and other acids which are
contained in the remains of organisms buried in the peat. The organic
portions (50 to 90 per cent.) of peat are chiefly plant-remains, which are
readily recognizable ; animal-remains, on the other hand, are quite
subordinate. By the removal of water and admission of air, peat can be
changed into a humus that is well suited to plants. Peat contains 1-2
(-3) per cent. of *nitrogen* and 0-4 per cent. of lime (certain peats, for
instance that of moss in Gothland, are stated to have up to 3.21 per cent.
of nitrogen and much lime), but peat contains *very little potash* and *still
less phosphoric acid.* The amount of these important nutritive substances
is so small because the acids in peat combine with alkalis to form soluble
salts, which are washed away.

Peat soil has the following characters :—

Of all soils it has the greatest *water-capacity,* so that it can take up
much more water than its solid parts weigh ; air-dried peat contains
only 15 to 20 per cent. of water ; peat *swells* on the addition of water
to a far greater size, but *contracts* on drying and becomes cracked without
however crumbling to pieces. When it is completely dried it becomes
extremely *loose,* almost *powdery* (peat-dust ; driving peat-dust may be
compared with driving sand). If one reckons the tenacity of clay as
one hundred, that of peat is only nine. It is almost impermeable to
water, and its power of raising water exceeds that of all other soils. It
is powerfully hygroscopic (absorbing up to ten per cent. of aqueous
vapour). In regard to their power of conducting water, different peats
(for example those of heath-moors and meadow-moors) behave very
differently. Sphagnum-peat conducts water rapidly and is therefore
uniformly moist in all cases, but meadow-peat may be dry above and
wet beneath. On account of its dark colour peat-soil is strongly heated
by the sun, but is intensely cooled at night. Despite its dark colour peat

forms a *cold* soil, because it is usually rich in water. Neither bacteria that produce nitrates, nor other bacteria, nor earthworms, can thrive in peat, because of its acid contents. Further details in regard to peat will be given in Section VI.

(*b*) *Raw humus* is a ' production of peat in the dry '[1] a black or black-brown peat-like mass, which is built up of densely interwoven, incompletely decomposed plant-remains, consisting of roots, rhizomes, leaves, mosses, fungal hyphae, and the like. Certain plants in particular give rise to raw humus, because they bear very thin, numerous, richly-branched roots (or rhizoids), which lie at the surface of the soil and weave the plant-remains into a dense felt work ; such species are, for example, Fagus, Calluna, Vaccinium Myrtillus, Picea excelsa. According to the constituent forming the main mass, we speak of heath (Calluna) raw humus, moss raw humus, beech raw humus, spruce raw humus, silver-fir raw humus, oak raw humus, pine raw humus—and so forth.[2] In plants growing on raw humus mycorhiza is frequent. Raw humus may be so rich in plant-remains as to be employed as fuel (heath-peat); it may contain from fifty to sixty per cent. of organic matter. As it forms so dense and tough a felt above the mineral soil, on the one hand, it excludes air (oxygen) from the subjacent layers and, on the other, sucks up water as greedily as a sponge and holds it with great force ; in rainy European climates it is frequently wet for a large part of the year. Consequently in it, as in peat, *free humous acids* are produced in abundance. Like peat, *it has an acid reaction.* There occur in it only few animals, mostly Rhizopoda and Anguillulidae, but no earthworms. The part played by bacteria has not yet been ascertained. Raw humus appears in forest, especially in places exposed to wind, whilst ordinary humus, with its earthworms and other animals, reigns in fresh places sheltered from desiccation ; when ordinary humus in beech-forest has given way to raw humus, because of timber-falls and such like, then the beech, being no longer capable of regenerating, disappears, and is often replaced by Calluna-heath.[3]

The production of raw humus is linked with lowness of temperature, and is promoted by moisture.[4]

The formation of a layer of raw humus also induces in the constitution of the subjacent layers of soil great changes, which have become best known through P. E. Müller's [5] pioneer researches in Denmark, the main results of which are given in the succeeding paragraphs.[6]

From the raw humus, humous acids and their compounds descend with rain-water into the subjacent sand, which has been more thoroughly washed out and is poor in soluble salts ; here they are oxidized by contact with inorganic (particularly ferric) compounds rich in oxygen, and there arise, for example, freely soluble ferrous compounds, which are carried by water containing carbonic acid down from the upper layers of soil. These layers are consequently decolorized, lose their absorbent faculty

[1] P. E. Müller, 1878, 1884, German edition (1887, p. 45).
[2] P. E. Müller, in the German edition of his work, employs the terms heath-peat, moss-peat, beech-peat, etc., in this connexion.
[3] P. E. Müller, loc. cit.   [4] Ramann, 1895, p. 125.
[5] P. E. Müller, loc. cit.
[6] Also see Ramann, 1886, 1905 ; Warming, 1896 ; Früh and Schröter, 1904.

almost completely, become poor in nutriment, and thus there is formed under the raw humus a light-grey or black ' bleached sand '.[1] As raw humus dries, some of the humous substances that were originally freely soluble become scarcely soluble, and are precipitated as carbonized humus.

Movements of the water also carry down particles of clay, ferric oxide, and humus, which are soluble only in water containing but little salt, and convey these through the layer of sand that is poor in salt until, at the lower limit of this washed out sand, they meet with the particles of soil which are still undergoing disintegration, and which therefore still contain soluble salts. The water takes these salts into solution, and the humous acids are precipitated as a gelatinous mass, which at a definite degree of dryness becomes solid, possibly by chemical changes, and insoluble in water. The grains of sand are cemented together, and there is formed a reddish-brown or brown layer of earth known as *hard pan* or *moor-pan*, which may be half a metre in thickness, and when completed is impervious to plant-roots.

The change from ordinary humus soil to raw humus is brought about by the following means :—

(1) Plants with densely interlaced roots occur ;

(2) Animals, particularly earthworms, vanish, so that the soil is not worked ;

(3) Particles of soil, particularly grains of sand, sink down and leave the soil more compact and poorer in air.

(c) *Ordinary humus* (leaf-mould, garden-mould, vegetable mould [2]) is an intimate mixture of sand and clay with completely disintegrated organic ingredients—rarely more than ten per cent. The mixture comes about mainly through the agency of animals and water.[3] It is devoid of free soluble humous acids, and is *neutral or alkaline* in reaction. It contains many fungal mycelia, earthworms, insects, and so forth. That ordinary humus soil is so excellent a nutritive substratum for plants is due to—

Its physical attributes (loose, having compound particles, aerated).

Its chemical characters, as it contains many compounds of carbon and nitrogen.

The humous substances, forming freely soluble compounds with nutritive substances that are otherwise scarcely soluble.

The production of humus in forest partially replaces the manuring and amelioration of the soil in agricultural practice.

Factors that strongly promote decomposition of organic matter hinder the production of humus ; according to Wollny, heat and moisture are the factors of greatest import. In connexion with both these, as in all other physiological processes, there exist a minimum, an optimum, and a maximum. Temperatures above the maximum are almost devoid of significance ; moisture beyond a moderate amount may exclude air from clay and humus soils, and thereby exert a restrictive action upon decomposition even before the soil is saturated with water.

In lower latitudes decomposition proceeds very slowly during the dry season, but at the wet season it is greatly accelerated, and in most spots is so complete that a soil very poor in humus results.[4] In the

[1] See Albert, 1907.   [2] Darwin, 1887.   [3] See Chapter XX.
[4] Hilgard, 1892.

tropics and sub-tropics true humus soil occurs only in shady forest ;[1] peat soil is very rare, but still it occurs where the climate is sufficiently moist ;[2] typical moors are wanting.[3]

In steppe and desert likewise the soil is mostly poor in humus, because plant-remains are scanty, though the soil is sometimes sufficiently moist. Only in most richly clad grass-steppes is humus ('black earth' occurring in South Russia) formed, especially upon closely deposited loess soil.[4]

In cold-temperate lands vegetable mould is the most frequent kind of soil. Only in sunny open localities exposed to wind, such as dunes, do we find the soil with scanty humus. Raw humus is frequent wherever decomposition is restricted from any cause, which according to Ramann[5] may be lack of nutriment, exclusion of air, excess or lack of water, or lowness of temperature. The formation of raw humus is of specially wide distribution in the heaths of maritime Western Europe where the summers are cool, as well as in alpine and arctic situations.[6]

Different species of plants demand very different amounts of humus in the soil. Accordingly Kerner has ranged plants into three groups :

(1) Plants which can settle upon bare rock, the most barren, sandy, or gravelly surfaces, and other spots where there is not a trace of humus ; their seeds or spores are mainly transported by wind : sub-glacial plants, many tundra-plants, desert-plants, and the like.

(2) Plants requiring a moderate amount of humus : for instance, Gramineae and Cyperaceae.

(3) Plants thriving only in rich humus, in the remains of a previous vegetation : many Orchidaceae, species of Pyrola and Lycopodium, Azalea procumbens, Vaccinium uliginosum, a number of other moorland plants, hemisaprophytes, and finally the highly modified holosaprophytes, Monotropa, Neottia, and others.

We may regard it as certain that there is a correlation between the unusual forms of the last-named plants and their method of nutrition, and thus between their forms and the kind of soil upon which they live ; but beyond this we know nothing of the matter.[7]

6. Saline soil is soil of varied constitution (sandy, clayey, and so forth) that is heavily charged with sodium chloride. It will be treated in detail in Section VII.

Soils at the bottom of water. Deposits and varieties of soil are formed here. In the sea, fine particles of mud are accumulated by the action of animals and Blue-green Algae, in places where the water is calmest, and they form the foundations of the fertile marshes on the coasts of the North Sea.[8] Mud of another kind is raised up in mangrove-swamps. On many coasts and at the mouths of many rivers there arise masses of mud that are rendered deep-black by sulphide of iron ; according to Beijerinck and Van Delden,[9] anaerobic bacteria play a part in the production of iron sulphide.

---

[1] See Warming, 1892 a ; Vahl, 1904 b.   [2] Ule, 1901.
[3] See Früh and Schröter, 1904, p. 143.   [4] Albert, 1907.
[5] Ramann, 1893, 1895.   [6] Kerner, 1863 ; Warming, 1887.
[7] For further information on moors and peat, readers should consult the great work by Früh and Schröter (1904), which has already been cited.
[8] Wesenberg-Lund and Warming, 1904, Chapter 58.
[9] Beijerinck, 1895 ; A. van Delden, 1903 ; Wesenberg-Lund and Warming, 1904.

In fresh-water, owing to the advent of material brought by rivers, there arise deposits whose nature is determined by the geological character of the soils traversed by the river. ' Pollen-mud ' arises by the accumulation of pollen derived from anemophilous trees, such as conifers, beech, and the like.

Other deposits are produced by chemical processes, more or less due to the activity of organisms.

Calcic carbonate : In many spots in lakes a portion of the carbonic acid in the water is decomposed mainly, it appears, through consumption by green plants. The result is a precipitation of calcic carbonate, first, on plants (Potamogeton, Characeae, and certain algae), partly in their cell-walls ; and subsequently the precipitated chalk accumulates on the lake-bed. Characeae may contain as much as 80 to 90 per cent. of chalk.

Compounds of iron are very often deposited, with or without the co-operation of bacteria, Blue-green Algae, and other plants.

Deposits mainly constituted of the remains of plants and animals occur particularly in calm lakes, pools, and the like. They appear to be most frequent in cold-temperate regions. Some (termed ' Gytja ' in Scandinavia) are structureless, grey or brown masses, largely composed of the more or less disintegrated excrement of animals, also of the remains of small animals and plants. A large amount of fatty oil occurs in these deposits, according to Potonié. A very important constituent of this kind of soil comes from plankton. In lakes where diatoms abound these are deposited in great quantities. In other cases it is mainly Blue-green Algae, or the chitin of small freshwater crustaceans that are accumulated. By chemical processes these common deposits undergo change as the organic constituents are reduced. This ' Gytja ' is a kind of humus-production under water.

Another kind of deposit that is amorphous, jelly-like and brown when moist, but more black when dry, arises especially in shallow waters where the water is brown with humus and the vegetation is rich ; with this type of deposit nymphaeaceous vegetation is particularly associated.[1]

CHAPTER XVII. ARE THE CHEMICAL OR THE PHYSICAL CHARACTERS OF SOIL THE MORE IMPORTANT ?

WE have already learnt that there are numerous differences in the chemical and physical characters of soils, that is to say, in the amount and kind of the components, and in the water-capacity, tenacity, and so forth. Varied combinations of these bring into existence extremely diversified types of soil.

[1] Our knowledge of these various deposits, connected as they are with one another by very gradual intermediate types, is very defective. They were first studied by H. von Post, 1862 ; and more recently investigated by Ramann, 1895 ; Weber, 1903 ; Potonié, 1905 ; Früh and Schröter, 1904 ; Wesenberg-Lund, 1901 ; Ellis, 1907.

Many species of plants are very *indifferent* as regards soil, inasmuch as they can grow on widely dissimilar kinds, for example: Phragmites communis grows both in very saline and in fresh water; and Typha latifolia is said by Sickenberger to be capable of growing vigorously in the soda-lakes of Egypt, where a nearly saturated solution of salts is present. Many ubiquitous or cosmopolitan species display but little preference, yet most species are confined to soil that has quite definite physical and chemical relationships.

Long ago it was noticed, particularly in mountainous countries with a substratum of varied geognostic nature, that the distribution of species and the whole appearance of the vegetation show a certain correlation with the soil. As an example we may select the case treated by Petry[1] of the Kyffhauser Hill, where there is a sharp contrast between the vegetation on the *rothliegende* and that on the *zechstein*, not only in the forest and in its undergrowth, but also in the weed-flora, the vegetation of the sunny dry heights, and of the copses. The *rothliegende*, as a consequence of its poverty in nutriment, supports a meagre, uniform vegetation, partially agreeing with heath; the *zechstein* tract, on the other hand, has beech-forests and a herbaceous flora of many species. The contrast between the two sub-formations is so sharp that we can detect from the vegetation, either in forest or field, whether we are standing upon the one or upon the other; and the relations are such that this contrast must be attributed to conditions prevailing in the soil.

Similarly in Montpellier,[2] in Switzerland,[3] and in many other mountainous lands we can observe the most emphatic contrast in the vegetation of two contiguous tracts, and even in a moraine-country like Denmark the same can be observed. In Jutland we can see sharply delimited patches with the Weingaertneria-association (Corynephorus-association), containing Weingaertneria canescens, Trifolium arvense, Scleranthus, Hieracium Pilosella, and others, dotted about a tract which likewise has a poor arable soil, but which supports an entirely different vegetation composed of Leontodon autumnale, Jasione, Lotus corniculatus, Erigeron acris, Euphrasia officinalis, Trifolium pratense, T. repens, Achillea Millefolium, Chrysanthemum Leucanthemum, Equisetum arvense, and others: in the former areas there are no mole-hills, whereas in the latter there are many.

The reasons for these generally observed distinctions have been sought chiefly in two directions. Some authorities regarded the *chemical* constitution of soil as the decisive factor, while others laid greatest stress upon the *physical* characters, and particularly upon the relations prevailing in regard to heat and moisture. The main points in the discussion are the following :—

The dominating influence of the chemical constitution of soil.

One of the earliest advocates of the chemical theory was the Austrian, Unger. He directed special attention to the contrast between calcareous and siliceous or slate soils, and he ranged plants in three groups :

Indifferent to soil, are those plants unaffected by the chemical nature of the substratum.

---

[1] Petry, 1889.   [2] Flahault, 1893.   [3] Magnin, 1893.

Partial to a certain *soil*, are those that show a preference for it without however being strictly confined to it.

Restricted to a certain *soil*, are those limited to it.

In accordance with this we can distinguish : calcicolous plants, silicicolous plants, slate-plants, halophilous plants, and so forth.[1]

Of other botanists who likewise assume that the chemical constitution of soil has a controlling influence, we may mention, among Germans, Sendtner, Schnitzlein, Nägeli; and among Frenchmen, Vallot, Fliche, Grandeau, Saint-Lager, Contejean (in later years), Magnin. Upon the whole, French investigators appear, in recent times, mainly to support this view.

Various facts favour this interpretation. On p. 58 it was pointed out that certain substances in excess act on certain plants as poisons. This is seen most clearly in the case of *common salt*.

Halophilous plants. Halophytes are not only of highly characteristic morphological and anatomical architecture, but have an absolutely defined topographical distinction on coasts, and in salt deserts and salt steppes. Common salt in excess has a highly exclusive action ; it acts as sterilizer, and only relatively few species, mainly belonging to definite families (Chenopodiaceae, and others), can endure much of it. Section VII should be consulted for further particulars in reference to halophytes.

Calciphobous plants. In cases of other substances, lime for instance, the matter is more doubtful. Lime is essential to the plant. Certain plants are stated to avoid soil containing much calcium carbonate.[2] Such reputedly calciphobous species are : Castanea sativa, Pinus maritima, Calluna vulgaris, species of Erica, Sarothamnus scoparius, Genista anglica, Ulex europaeus, Pteris aquilina, Rumex Acetosella, and other plants that we often find on heaths and on raw humus ; also Gramineae, Cyperaceae, many lichens and mosses, especially Sphagnum,[3] and among algae the Desmidiaceae. The flowering plants named are reputed to be incapable of carrying on an existence in soil containing more than from 0·02 or 0·03 per cent. of calcium carbonate. But cultures made by C. A. Weber[4] and Gräbner have clearly demonstrated that none of these plants suffer from lime when this is unaccompanied by a large amount of soluble salts.

Calciphilous plants. Other plants that do not desert a soil rich in calcium carbonate are put forward as calciphilous plants, for example : Papilionaceae (Trifolium, Anthyllis, Vulneraria, Ononis Natrix, and others), Rosaceae, Labiatae, many Orchidaceae, Tussilago Farfara, and others. Unger gives a whole array of examples belonging to the lime-flora. According to Blytt[5], Ophrys muscifera and Libanotis montana are the sole vascular plants in Norway that occur exclusively on calcareous soil. Among algae the Mesocarpaceae are calciphilous.

Silicicolous plants. These are brought forward in contrast to calcicolous plants. The calciphobous plants mentioned above are regarded as silicicolous. The truth may perhaps be that they are expelled from

[1] See Chapter XV, pp. 56–8.   [2] See p. 58.
[3] Fliche et Grandeau, 1888 ; see Contejean, 1893.
[4] C. A. Weber, 1900.   [5] Blytt, 1893.

calcareous soil by competition, and are compelled to select a soil containing less calcium, without having any real preference for silica, which is a very neutral substance ; thus Contejean interprets the matter. To the silicicolous plants belong the majority of those growing on sand and moor in north-temperate Europe.

Nitrophilous plants (nitrophytes, ruderal plants). These thrive best in soil where compounds of ammonium and nitric acid are abundant, and therefore especially in the vicinity of human dwellings (dung-heaps, highly-manured soil). They belong to certain special families (Chenopodiaceae, Cruciferae, Solanaceae, and others), and nitrates occur in their cell-sap. Other species develop feebly on such soil, because they take into their tissues more nitrate than they can endure.[1] Certain fungi and mosses (Splachnaceae) flourish only on dung.

The *solfataras* of Java, according to Holtermann's assertions, have a peculiar flora that differs from those of others.[2]

Other substances also can act as poisons if they be supplied in large quantities.

If *gypsum* be scattered over a meadow certain ferns and grasses die off, while clover becomes more luxuriant.

Similarly *iron* (iron sulphate, ferrous oxide) may act injuriously if present in quantity, though it is one of the absolutely indispensable nutritive elements.

Investigations conducted at Rothamsted in England have demonstrated in a particularly clear manner the significance of the chemical constitution of nutriment ; they showed that with nitrogenous manure, especially with nitrates, grasses preponderated and expelled Leguminosae ; whereas, on the contrary, potassic salts favoured Leguminosae. Experimental manuring of high moors has, according to Weber, led to entirely similar results ; certain species of grasses were expelled by others. But one can hardly say that research has yielded any considerable support to the chemical theory we are discussing ; calcicolous and silicicolous plants, the calamine-violet, and even halophytes, are perhaps always capable of flourishing in a soil not containing the respective substances they affect, or practically in any soil, in botanic gardens for instance. On the other hand, the *amount* of nutriment in soil plays a more prominent part.

In course of seven years' wanderings, A. P. de Candolle found nearly all species upon soils of varied chemical nature ; and Blytt came to the conclusion that the very few species which in 1870 he had regarded as restricted to definite soils in Norway must be further reduced in number as a result of his later investigations.

' Every distributional relationship may be due to either a physical or a chemical cause, but the simultaneous presence of both prevents us from clearly distinguishing the part played by either singly '.[3] This is perfectly correct, and the history of botanical science shows that some botanists, in opposition to those previously mentioned, ascribe greater importance to *physical* than to chemical relations.

[1] Schimper, 1890-1.   [2] Holtermann, 1907.
                           [3] Vallot, 1831.

The dominating influence of physical characters of soil.

The protagonist for the dominating importance of physical relationships was Jules Thurmann (1849). His doctrine can be summarized as follows :—

It is the physical structure of soil that regulates the distribution of species ;

Upon this structure depend the amount of water and the thermal conditions in soil.

The same species can grow on very different kinds of soil, if it encounters the same conditions of moisture.

Thurmann discusses the different weathering properties of kinds of rock under the action of air, water, and heat (both frost and warmth), As a result he divides rocks into *eugeogenous* and *dysgeogenous*.

Some kinds of rock are easily weathered and rapidly produce loose masses (grit, sand, and similar detritus) ; these soft types are the *eugeogenous*, and in accordance with the fineness of their products of weathering they are *pelogenous* when the particles are very fine and powdery, especially clay and marl soils ; or *psammogenous*, when the particles are coarser—sand. According as soil is more or less pelogenous or psammogenous, Thurmann employs the prefixes ' per ', ' hemi ', and ' oligo ', to denote sub-divisions, or speaks of pelopsammitic soil.

In opposition to types of rock that are easily weathered, are the hard, resistant types—*dysgeogenous* ; they give rise to scanty or no products of weathering.

Finely comminuted soil absorbs more water than does slightly weathered rock.[1] Eugeogenous rocks therefore bring into existence a moist, cold soil ; dysgeogenous, a dry, warm soil.

To plants that exploit moist soil and eugeogenous land Thurmann applies the term *hygrophilous ;* plants that exploit drier soil and dysgeogenous rock he terms *xerophilous*. His hygrophilous species correspond approximately to the silicicolous plants of Unger and others, and his xerophilous species to their calcicolous ones. Indifferent species of plants occurring on all kinds of soil Thurmann designates *ubiquists*.

According to Thurmann, the obvious distinction between the floras on calcareous and siliceous soils is caused, not by the preference of species for lime or silica, but by the circumstance that calcareous rocks weather with difficulty, and permit water to flow away rapidly through clefts and fissures ; they produce a dry warm soil of slight depth, whereas quartz and felspar produce a loose, deep, moist, and cold soil. When species of rock, with identical chemical composition, in some cases are hard and resistant but in others become easily disintegrated, then calcicolous plants are found on the former soil even when it is siliceous, and silicicolous plants on the latter soil even when it is calcareous. Furthermore, a single species of plant in a definite climate may require a definite soil on account of the physical characters of the latter ; for instance, in a moist climate it may choose a warm dry soil like lime ; but in a different climate it may prefer an entirely different soil; for instance, in a warm dry climate a moist, cold, siliceous soil. A favourable soil may facilitate the existence of a plant in an unfavourable climate. According to Blytt,

[1] See Chapter XII, p. 47.

many species in Norway reach their extreme stations in altitude and latitude on calcareous soil. Eugeogenous and dysgeogenous kinds of rock may bear the same flora. It is thus that the distribution of beech in Southern France must be explained. In Denmark it passes for a calciphilous plant ; yet in Mediterranean countries, according to Flahault,[1] only upon siliceous soil does it form extensive forests, while on dry, warm, calcareous soil it is sporadic, having been expelled by Quercus sessiliflora, excepting in cool valleys with north and east exposure.

As supporting Thurmann's theory we may name Contejean, who however subsequently adopted the rival theory ; and approaching of nearly similar views are A. P. and Alph. de Candolle, Celakovský, Krasan, Hoffmann, Kerner, H. von Post, Blytt, P. E. Müller, Negri, and others.[2] Yet Thurmann's theory certainly cannot explain all cases.

In both theories there is some truth ; both chemical and physical relations operate ; the actual truth seems to be that in some (few) cases, where the soil is specially rich in a chemical substance, it is the chemical characters of the soil, but in other (far more frequent) cases it is the physical characters that are of greatest import. When we consider a country like Denmark or the North German Plain, with a soil produced by corrosion and commixture of multifarious kinds of rock, yet scarcely possessed of any marked chemical characters, then the chemical significance of soil becomes evident in the halophytic vegetation on the coast, but probably there alone, whilst everywhere else conditions as regards moisture play the leading part.[3] Temperature, illumination, air, atmospheric precipitations and humidity, chemical nature of the soil, may all be completely alike, and yet is the vegetation different. One solitary factor is different—*the amount of water in the soil*—and this it is that is decisive.[4] When we further consider that the most important characters of soil (temperature, aeration, amount of water, evaporation) are mainly dependent on its structure, it then appears that *the physical characters of soil are the weightiest*, especially because they react upon the amount of water. Chemical differences are always accompanied by physical ones, and chemical characters seem to be capable of replacement by physical ones, but it would appear that physical attributes are in the last instance most frequently decisive. Yet it must be remarked that the *amount of nutriment* in soil is likewise of grave import—a principle of which Gräbner and A. Nilsson[5] are advocates. But even the amount of nutriment depends upon the physical features of soil, that is, upon its water-capacity and absorbent faculty.

Competition among species as a factor of distribution of plants.

Darwin and Nägeli[6] directed attention to one factor affecting the distribution of species and the production of plant-communities—*the competition among species*—which has not always been taken into consideration, but which must not be overlooked. How trivial a part may be played by chemical distinctions in soil is shown, for instance, by a botanic garden

---

[1] Flahault, 1893.
[2] The most recent literature is cited by Woodhead, 1906.
[3] Consult pp. 45-8, dealing with ground-water. [4] Warming, 1894.
[5] Gräbner, 1898, 1901 ; A. Nilsson, 1902 b. [6] Nägeli, 1865, 1872.

in which there flourish on the same soil plants coming from the most
diverse soils. But if we neglect the garden, only very few (mainly indigenous) plants will emerge as victors from the ensuing struggle. Plants
are evidently, in general, tolerably impartial as regards soil, if we except
certain chemical and physical extremes (abundance of common salt,
of lime, or of water), *so long as they have no competitors ;* only some few
plants may perhaps be regarded as obligatory in the one or the other
respect ; well-nigh all are facultative, and their occurrence depends upon
competitors. If these be present, the one drives back the others, and
the victorious species is the one that can best utilize the given combination of soil, light, climate, and so forth. For instance, according to
Fliche, the Scots pine (Pinus sylvestris) over the whole Champagne is
confined to calcareous soil and is wanting on non-calcareous soil ; the
reason for this is that the Scots pine is an introduced plant, to which
the climate, without being actually hostile, is yet not favourable ; on
non-calcareous soil, upon which it thrives admirably elsewhere, it is
here suppressed by other species, and only on calcareous soil does it
become dominant, even then without developing really well. We should
therefore err were we to describe it as being calciphilous ; like many
other forest-trees it will grow on the most diverse soils, and in Denmark
it is most frequent on sandy soil. When in Denmark we find the oak
growing sometimes on moist compact soil, and sometimes on dry poor
soil, the reason for this is not that it prefers these soils, but that it is
expelled from others by the beech. Similar competitors are ling (Calluna),
and many other species, such as Anthemis Cotula and A. arvensis, Carlina
vulgaris and C. acaulis, Prunella vulgaris and P. grandiflora, Veronica
Teucrium and V. Chamaedrys.[1]   In the Alps, according to Nägeli [2]
Rhododendron ferrugineum and R. hirsutum, as well as Achillea moschata
and A. atrata (silicicolous and calcicolous plants) struggle against one
another. P. E. Müller [3] has brought forward several examples of forest-
trees in mountains driving each other back in the same manner ; lofty
forests of silver-fir, for example, are sharply delimited from lofty forests
of another species without it being a question of inability to thrive at
the boundaries. Moreover, Bonnier [4] and others came to the conclusion
that species restricted to calcareous soil in one district may be calciphobous
in another, and indifferent to soil in a third. In the middle of its distributional area a species often makes no selection as to soil, but outside
this central position it is forced by other species to exercise a choice.[5]

As noteworthy examples of plants being able to flourish luxuriantly
in countries other than their own homes, we may cite : Erigeron canadensis,
Galinsoga parviflora, from tropical Peru ; Oenothera biennis, and other
American weeds that are now common in Central Europe ; Impatiens
parviflora and Elodea canadensis may also be mentioned. On the other
hand, Salsola Kali, a common littoral European plant, has become a
most pestilent weed in the cornfields of North America ; in places it
takes nearly complete possession of the soil.[6]

[1] After Pietsch, according to Ludwig, 1895, p. 121.   [2] Nägeli, 1872.
[3] P. E. Müller, 1871, 1887.   [4] Bonnier, 1879.   [5] See Section XVII.
[6] Among more recent literature on this subject readers should consult the works
of Cowles, 1901 ; Saint-Lager, 1895 ; Schimper, 1898 ; Gillot, 1894 ; Gain, 1895 ;
Ernst, 1907.   The older literature is to be found cited in Engler, 1899, on pp. 164–6.

CHAPTER XVIII. THE EFFECT OF A NON-LIVING COVERING
OVER VEGETATION

A NON-LIVING covering exerts an action that depends upon, *inter alia*, its looseness or compactness ; the looser it is the greater is its action in the following respects :—

1. Water is sucked in, evaporation depressed, and moisture of the soil increased.
2. Radiation is lessened.
3. Fluctuations and extremes of temperature are diminished.

In this relation there come into play two kinds of coverings : (*a*) *snow ;* (*b*) *fallen foliage and withered grass.*

(*a*) Snow

It has long been known that snow can protect vegetation very efficiently, and that it guards winter-crops from being frozen. In high alpine situations falls of snow in summer would appear sometimes to protect plants from exposure to the dry, cold weather, and consequent evaporation that often set in after such falls of snow. In arctic countries every patch of surface from which the snowy covering is blown away by winter storms has vegetation different from that on snow-clad depressions ; on the tundras of Lapland, for instance, Lecanora tartarea dominates in places exposed to wind, whereas in more sheltered places fruticose lichens can exhibit dense and tall growth.[1] The distribution of the covering of snow determines the distribution of entire and definite communities : some are protected at the expense of others ; the spots that in winter are covered with snow usually show in summer the greatest number of species and individuals. *The snow-covering is thus of oecological importance.* Snow seizes upon the countless particles of dust in the air, purifies this, and collects other small organic and inorganic particles that the wind brings. When snow melts these masses of particles are deposited on the ground, and there is formed gradually a fine, fertile soil, which remains stationary in gentle depressions, and entertains special species of plants. This ' snow-patch flora ' is subsequently alluded to in Chapter LXVII.[2]

The covering of snow influences *plant-form.* On the one hand, we may here include the influence exerted by heavy loads of snow at high alpine altitudes upon the shapes of trees and shrubs, whose stems are pressed down into a *prostrate* position on the soil and lie flat on slopes ; for instance, Pinus montana assumes the habit of elfin-tree or contorted shrub, while Juniperus, Alnus viridis, Fagus sylvatica, and other trees dwindle to form scrub, and spruce-birch-scrub develops in South Greenland.[3] On the other hand in Lapland Juniperus and Picea excelsa become scrub,[4] in this case because all the twigs projecting above the snow regularly die off, and the individual plants acquire low, plate-like, or umbrella-like crowns.

[1] Kihlman, 1890.   [2] See Schröter, 1904–8.
[3] Kerner, 1863, p. 512 ; Rosenvinge, 1889 ; C. Schröter, 1904–8, p. 663 ; Szabó, 1907.   [4] See the illustrations in Kihlman, 1890.

The reasons for this significance on the part of a snow-covering are the following :—

(a) The **thermal conditions** in snow play a part, though scarcely the leading one. Snow is white because the spaces between its crystals are occupied by air, which may be very considerable in quantity. And it is this air that mainly renders snow a bad conductor of heat. By reason of its feeble thermal conductivity snow keeps the soil *warmer*, and the deeper one descends into snow the less cold is it, so that soil lying under deep snow is exposed to less cold than is bare soil. But this does not suffice to explain all the observed facts, for even under deep snow plants may be exposed to very extreme cold.[1] Neither can it be of very great importance that *fluctuations in temperature are decreased* so that plants are not exposed to the alternate heat of the day and cold of the night ; snow does serve to prevent too sudden thawing, which may constitute a danger.[2]

Snow acts as a protection against those *changes of volume* in frozen soil, occasioned by hoar-frost, which cause plants to be ruptured and uprooted.

(b) Of far greater importance is the significance of snow in regard to the **amount of water in the plant,** as shown in succeeding paragraphs .

Snow acts as a defence against transpiration. It is to this action that we must attribute the preservation of many species during winter, and, as described by Kihlman, the death of twigs which project above the snow. It is not low temperature that kills those twigs, but the great atmospheric aridity prevailing in arctic countries, and the violent storms which increase transpiration at a time when the roots are incapable of absorbing water. Twigs and whole plants wilt through desiccation[3] ; the shapes of the shrubs serve to show how high the layer of snow stands in winter. The aberrant, sometimes bent and contorted, shapes are occasioned by the death of many twigs and the production of new ones in abnormal positions.

The topographical distribution of species is also affected by relations in regard to water, namely, by the uneven distribution of water in the soil that results from the uneven distribution of snow on its surface. Depressions filled with snow remain moist for a longer time than do more elevated spots bare of snow, in fact, they may be moist throughout the vegetative season.[3]

In some places, for instance, in the steppes of Russia and North America, snow, by reason of its depth, acquires importance as a reservoir of water ; the greater or smaller the supply to the soil, the richer or scantier will be the vegetation in the succeeding vegetative season.

When a covering of snow has an injurious effect, for example, on a dense vigorous winter-crop in depressed spots of fields, the cause may be that it suffocates them by restricting the supply of air.

Snow affects adjoining slopes by wetting them when it melts. As mentioned on p. 39, in Greenland the northern slopes of a mountain chain may be fresh and vivid green (rich in mosses) in summer, while the southern slopes at the same time are dry and scorched ; this is because

---

[1] Kjellman, 1884.   [2] See p. 23.   [3] Kihlman, 1890.

the northern slopes are kept moist by, *inter alia*, the slowly melting snow, which rapidly disappears from the southern slopes.

A covering of snow shortens the vegetative season by preventing the soil-temperature from rising above freezing-point in spring-time, thus hindering plants from awakening into activity as early as on snowless spots. This has a profound effect upon the economy and distribution of plants ; certain species are excluded from places where the snow is wont to lie for a long time, because the vegetative season is *too short* or the soil *too cold* ; other species are actually favoured by these conditions. Blytt records that on Norwegian mountains, around accumulations of snow which melt to some extent during summer but scarcely ever entirely disappear, the flora is high-alpine in nature on account of the short vegetative season, and corresponds to an altitude above sea-level that is greater than those of the places in question. Even in places where the snow melts only in extremely warm summers we can find vegetation. This must have rested under the snow for several years before awakening. Obviously there are many spots where the snow lies so long that vegetation is absolutely excluded.

It is easy to see that orographic and other relations—such as slope and exposure of the soil, nature of the wind, specific heat of the soil, and the like—that influence the melting of the covering of the snow, thereby acquire a phyto-geographical significance.[1]

(b) Fallen Foliage and Withered Grass

The other kind of dead covering to the soil is made by fallen foliage or withered grass ; fallen foliage is met with especially in forest (both deciduous and evergreen), withered grass on meadow and savannah.

These coverings have the same physical action as snow, diminishing the extremes of temperature, keeping the soil moister, and so forth. In the forest many plants can scarcely continue to exist without such protection against desiccation—that the protection is not merely against cold, is even more evident than when the covering is composed of snow.[2]

A covering of leaves over the soil in beech-forest and similar forests, where it is very thick, has a great influence upon vegetation on the ground inasmuch as it suppresses mosses and sundry other plants.

A covering of leaves powerfully affects the production of humus in soil, which it thus improves, and is, further, of deep significance to *animal life in forest soil*, for it concerns moisture, and provides food for animals living on forest soil, among which earthworms seem to be the most important.[3] Both circumstances prevent the humus soil of forest from changing into raw humus, and check all those modifications in the soil-covering that would accompany such a change, and would gravely interfere with the whole economy of forest.[4]

In this connexion may be mentioned the utility of their *old dead*

---

[1] For further information on the significance of snow see Wöikof, 1887, 1889.
[2] Concerning the characters of the various coverings on forest soil see Ramann, 1890, 1893, 1905.   [3] See Chapter XX.
[4] See P. E. Müller, 1878, 1894 ; Ramann, loc. cit.

parts to certain other plants, particularly to arctic, alpine, and desert plants.

A fact that has long been known, and already mentioned on p. 24 of this work, is that the old, dead leaves remain attached in great numbers to the branches of *sub-glacial* plants, and thus envelop them with dense coverings, whose closeness is further increased by the production of condensed short branches. This is evidently a result of the circumstance that the processes of disintegration and decay take place extremely slowly in the cold climate where bacteria and fungi do not thrive ; it is of utility to the plant in obstructing transpiration. Nature ensheaths plants just as a gardener in preparation for winter mulches sensitive forms.

Certain species growing upon dry rock or similar arid spots are, in like manner, enveloped by remains of old twigs and leaves ; in this case it is lack of moisture, not of heat, that arrests the disintegrating action of fungi and bacteria. That the plants concerned derive any benefit therefrom cannot be generally asserted, though it is probable. It is possible to conceive that these old plant-parts serve partly as a protection against transpiration, and partly as organs for the absorption and retention of water. In this connexion attention may be directed to the tunic grasses,[1] also to the envelopes formed by the leaves and roots on the stems of the Velloziaceae, and by the roots in Dicksonia, and some other ferns.[2]

CHAPTER XIX.   EFFECT OF A LIVING VEGETABLE COVERING ON SOIL

EVERY kind of covering formed by vegetation acts upon the physical relations in soil ; and the denser, taller, and longer-lived the vegetation is, the more powerful is its action. Forest therefore acts most powerfully ; and for this reason the vegetation clothing the ground in forest, on the one hand, and the plants forming the high-forest, on the other hand, are subject to entirely different physical relations. The effects partially agree with those wrought by inanimate coverings :

1. **The temperature of the soil is modified.**

A vegetable covering screens the soil, and therefore decreases the action of the sun's heat. But a vegetable covering is a very effective radiator of heat. Fluctuations of temperature, both diurnal and annual, are less considerable ; compared with soil clad with vegetation, bare soil is warmer by day and colder by night, warmer in summer and colder in winter. The maxima of temperature are much higher, but the minima only little lower in bare soil than in shaded soil (clad with vegetation), so that the mean temperature of the latter soil is lower by, at any rate, one or two centigrade degrees in forest than in bare soil. The temperature at the surface of forest soil in Central Europe, according to Ebermayer, is rarely higher than 25° C. Inside forest the dead covering of course contributes to increase these effects.

---

[1] Hackel, 1890 ; Warming, 1892 *a*.   [2] Warming, 1893; see also Section III.

2. **The amount of water in soil is influenced.**
A portion of the atmospheric precipitations is lost to the soil of forest not bare of foliage, because it is deposited on the plants and thence evaporated ; this is specially true of small precipitations. In forest, about fifteen per cent. of the atmospheric precipitations is thus lost, and more in coniferous than in dicotylous forest. The power of soil of the forest to retain the moisture that reaches it is increased, as it is protected against evaporation. Snow melts more slowly, and water derived from melting snow is absorbed in larger quantities by the soil.

On the other hand, a plant-covering tends to dry the layers of soil in which roots occur, and the more completely so the denser the vegetation is, because plants absorb water from the soil and dissipate it by transpiration.[1]

Soil contains less water if it be clothed with vegetation than if it be bare (other conditions remaining the same). A vegetation of weeds may have a very drying effect on soil.

3. **Soil covered with vegetation is less compact than bare soil.**
This is so because descending rain can exercise no excessive mechanical action upon it ; moreover, animals (earthworms) play a more direct part in this connexion.

4. **Light falling upon soil covered by plants is weakened.**

5. **The action of wind is decreased** among dense, and especially among tall, vegetation.

6. **Air underneath a vegetable covering is changed**, especially in forest ; it is cooler and moister.

The air *above soil* occupied by plants, particularly above forest, is also cooler ; and this may perhaps lead to an increase in the deposit of dew, in cloudiness, and in rainfall. It is certain that forest and dense vegetation in general prevents atmospheric precipitations from flowing away rapidly, and thus being lost to plants and causing floods.

7. **A covering of moss** requires special mention, because it differs from any other vegetable covering in its effect on the amount of water in soil.

The effect varies with the species of moss concerned. Some mosses (Hypnum and its allies) produce dense cushions, five to six centimetres in thickness, lying loose on the soil. The stems of other mosses (Polytrichum, Dicranum) are enveloped in a felt of rhizoids ; their protonemata and rhizoids permeate the soil in the form of a dense felt, and promote the formation of raw humus. Mosses therefore must act upon soil in divers ways. But, according to Oltmanns,[2] the general facts of the case are as follows :—

(a) *A carpet of moss acts as a sponge.* The dense, low carpet, with countless capillary spaces between leaves and rhizoids, absorbs capillary and superficial water, but obtains little or none by suction from the soil and internal conduction—the internal structure is an index of this.[3] Consequently, living and dead carpets of moss imbibe and evaporate approximately the same amount of water.

(b) *A carpet of moss does not desiccate soil.* Since mosses, particularly those forming loose-lying cushions, do not take much water from

---

[1] See Chapter XII, p. 48.   [2] Oltmanns, 1885.
[3] Haberlandt, 1904, Absch. vii.

soil, they dry it to a less degree than does other vegetation, and they protect dry easily-heated soil from desiccation. Evaporation takes place more rapidly from a covering of moss than from a dead covering, but the former keeps soil moist and cool upon the whole, and may easily occasion swampiness in wet, shaded soil.

8. **The chemical relations of soil** are influenced by a covering of plants, for different kinds of vegetation differently affect the nutriment in soil and its absorbent faculty, abstracting different inorganic substances and enriching it with organic bodies. A rotation of crops and the application of manure are matters of necessity to the farmer, because with each crop he constantly removes from the soil certain quantities of nutritive substances. The forester does this to a smaller extent, except when, as in Germany, he carries away forest-litter, and the application of manure is as a rule not necessary in forest or, at any rate, has been but little practised. Nevertheless wind blows leaves out of many forests, and consequently brings about great changes in the soil and vegetation. A change in the forest-vegetation is known to have occurred in Denmark during past millenia,[1] yet one would be only partially correct were one to seek to attribute this to a kind of rotation of crops practised by Nature, and due to each species of tree impoverishing the soil in such a manner as to render the soil less suitable for its own maintenance but more so for that of other species. Certain it is that the annual removal of wood from the soil withdraws some of the most essential plant-nutriment— potash for example. Where no forest-litter is taken away, one can trace no deterioration in the soil. The exodus of nutrient substances may be balanced by the advent of suitable salts derived from the weathering of deeper layers of soil. But where the removal of forest-litter is practised, easily satisfied species, e. g. Scots pine, replace species that are more exacting in their demands, e. g. beech and oak.

CHAPTER XX.   THE ACTIVITY OF ANIMALS AND PLANTS IN SOIL

BETWEEN the plant-life and the animal-life of a place there exists an intimate and complex reciprocal relation, which expresses itself in various ways, and promises biological results of the deepest interest. Here we shall consider only two aspects of the matter : the effect on the soil of the tunnelling by animals and the effect of saprophytic plants.

Tunnelling of the soil by animals.

Soil is traversed by many species of animals : terrestrial soil, particularly by earthworms, larvae of insects, millipedes, wood-lice, ants, as well as by animals, such as moles, which search for these ; marine soil, by small Crustacea, Sedentaria or Tubicolae, and others.

Terrestrial soil. The uppermost layer of soil in forest and field usually consists of an intimate mixture of mineral matter, animal remains, and

---

[1] See Sect. XVII, Chapter XCVI.

the remains of previous vegetation in the form of leaves, fragments of twigs, fruits, seeds, and so forth, which occur in various stages of decomposition, and of demolition wrought by animals.

Terrestrial soil, if abounding in animal life, is favourable to vegetation because it is rich in humus bodies ;[1] but if wanting in animal life then the vegetation is usually low and stunted. Animals, especially earthworms, work on the soil and thus on vegetation in four different ways :

1. They comminute vegetable remains by means of their jaws or, in the case of earthworms, by means of their alimentary canal with the aid of ingested stones.

2. In their intestines they mix their food with mineral particles of soil, thus promoting the formation of humus by producing a finely-mixed soil.

3. They bury vegetable fragments in the soil.

4. By the tunnels and passages due to their activity they render soil more porous and better aerated—the soil becomes ' mellow '—thus promoting respiration in the roots and consequently growth in the plants. The excrement deposited likewise serves to render soil friable and porous. In this way animals also facilitate drainage.

Earthworms play a special rôle in ordinary soil. In Denmark two large species, Lumbricus terrester and L. rubellus ; as well as L. purpureus, Allolobophora turgida, and species of Euchytreus, are of significance. They make burrows which descend vertically into soil to a depth of two or more metres, and which reach down to deeply buried roots. The burrows are filled with substances, fragments of leaves and excrement, nutritious to the plant. Five other species live in arable soil. Sometimes they are so numerous that some 400,000 individuals may occur in a hectare of land.

At night, and in moist, dull weather they emerge from their burrows and deposit excrement in the form of friable castings on the surface of soil. They drag leaves into the ground, so that these decompose ; they comminute vegetable remains, acting on these mechanically, swallowing and intimately mixing them with mineral particles which they have likewise swallowed. In addition, their alkaline digestive liquids neutralize humous acids in soil. Shade, shelter from wind, and moist air contribute towards a wealth of animal life in the soil ; shade and shelter from wind are therefore of indirect importance to vegetation. When a forest-soil is exposed to desiccation and the fallen leaves are carried away by wind, the earthworms vanish, the soil becomes dry and hard, and the vegetation suffers. In acid soil (bog, heath,) and dune, earthworms are wanting. Upon their presence or absence depends the occurrence of a humus soil or a raw humus soil in north temperate forest and heath. Conversely they disappear upon the production of raw humus and humous acids. Even upon the growth of rhizomatous plants in the forest do they exert an action ;[2] their presence or absence causes a series of variations in the kinds of soil that corresponds to a series of variations in the plants clothing it.[3]

As an additional example, showing how animals may affect vegetation,

[1] See p. 42.   [2] P. E. Müller, 1894.
[3] The natural history of earthworms has been investigated by C. Darwin (1881), P. E. Müller (1878), and in the tropics by C. Keller (1887).

it may be mentioned that mole-heaps and ant-hills very often support a vegetation somewhat different from that on the surrounding soil.[1]

Marine soil. A rôle obviously of less general significance, though similar to that played by earthworms in relation to plant-life on terrestrial soil, is played by species of Arenicola in relation to Zostera-vegetation on the marine soil of European coasts.[2]

Saprophytic plants in the soil.

A more important part is played by saprophytic plants in the soil, especially by fungi and bacteria, than by animals.

Fungi in soil. In all kinds of soil with abundant humus, fungal mycelia live ; forest-soil in northern Europe in autumn, by its wealth of Basidiomycetes, reveals the extent to which it is permeated by fungal hyphae. But even when few or no fungi exhibit themselves above ground, microscopical examination will certainly demonstrate their presence in all humus-laden soil, even in acid heath-peat ; hyphae of Cladosporium humifaciens occur in this, while the roots of Calluna and other denizens have mycorrhizae, just like the majority of forest-trees and some perennial herbs living on humus.[3] Saccharomycetes hibernate in soil.[4]

Bacteria in soil. These are of still greater importance than fungi. They occur in all soil and in all water, in terrestrial soil, in the various types of mud, in saline and fresh water. In the uppermost layers of soil, especially near human dwellings, they occur in millions upon millions ; in soil occupied by vegetation their number increases with the depth as far down as about half or three-quarters of a metre ; it then rapidly decreases until, at a depth of five to six metres, there are as a rule no bacteria : the soil has filtered them out of the percolating water. The investigations of Adametz, according to Sacchse,[5] gave the following results :—

NUMBER OF BACTERIA IN ONE GRAMME OF SOIL ACCORDING TO ADAMETZ.

Nature of soil	Depth below the surface, in centimetres.	Number of individuals.
Sandy soil	At the surface	380,000
,, ,,	,, 20–25	460,000
Clay soil ,,	,, the surface	500,000
,, ,,	,, 20–25	464,000

Other investigators have found in one gramme of soil up to a million bacteria. The number of course depends upon various conditions.

The number of species concerned is probably exceedingly great, and we know that some of them play a prominent part in the biology of soil. Some are aerobic, others anaerobic. There are present not only ordinary putrefactive bacteria, many of which are of the highest significance in regard to the composition of the air in soil, but also pathogenic species, for example the tetanus germ (Bacillus tetani), as well as others, including nitrifying and denitrifying bacteria, which cause the formation of important chemical compounds in soil. Schlösing and Müntz were

[1] Buchenau, 1876 ; Warming, 1894, 1906 ; P. E. Müller, 1894 ; also see p. 66.
[2] Rosenvinge, 1889-90, see Warming, 1906 ; concerning *Corophium* see Warming, 1906.   [3] See Chapter XXV.
[4] C. E. Hansen, 1881.   [5] Sacchse, 1888.

the first to prove that the formation of nitrates in soil is due to the activity of micro-organisms, since nitrogen-containing soil, in which this process can take place, loses the power of inducing it if heated up to 110° C., but regains that power when non-sterilized soil is added to it, and, furthermore, since chloroform instantly arrests the process in question.

Winogradsky first isolated these organisms. They flourish in a well-ventilated, moderately moist, alkaline soil that contains nitrogen, at temperatures between 10° C. and 45° C. According to Müntz, the nitrate-bacteria play a prominent part in disintegrating rock, by penetrating the finest pores and exercising their chemical activity.[1]

It may be regarded as established that bacteria in the soil enrich this in nitrogen by utilizing free nitrogen from the air. Leguminosae with root-tubercles containing bacteria, Elaeagnaceae, and Alnus also have this power. According to P. E. Müller's [2] investigations the mountain-pine (Pinus montana), which has both mycorhizae and peculiar, coralloid, branched root-tubercles, also belongs to the plants capable of fixing free nitrogen. Experience has shown that where Picea excelsa has been planted on heaths in Jutland it flourishes better in company with Pinus montana than without it. It is probable that in this case the mountain-pine provides the spruce with nitrogen.

Bacteria do not flourish in a soil containing free acids (humous acids); consequently they are scanty, or lacking, in peat and similar soils.

CHAPTER XXI.   EXPOSURE.   OROGRAPHIC AND OTHER FACTORS

THE different factors already considered are in nature so varied and connected by such a number of transitions that the greatest diversity results in the nature of habitats and in the differentiation of vegetation, and it becomes extremely difficult to decide which factors are the most weighty in a given case. But this multiplicity and variety are further increased by modifications occasioned by certain geographical or orographic factors. Among these are included the direction of mountain-chains and valleys, the height of mountain-chains, the steepness and exposure of declivities, and so forth.

The *direction and height of mountain-chains*. These are of paramount climatic significance. They steer wind into definite directions, occasion föhn-winds, capture moisture from the wind on certain sides, and condense aqueous vapour in higher regions in the form of clouds and rain; consequently, on certain sides or at a certain altitude above sea-level luxuriant forests may prevail, whilst on other sides or at a lower level extreme aridity reigns. Thus the coast-mountains of Brazil are rainy and clad with forest; but the interior is dry because the moisture of the trade-wind is condensed and deposited before the interior is reached. In like manner the coast of South Africa is moist, but the Karroos are dry. In the West Indies the more low-lying islands are dry and receive

---

[1] See C. Schröter, 1904–8, p. 558.   [2] P. E. Müller, 1903.

but little rain, whilst the more raised ones receive heavier atmospheric
precipitations and have more luxuriant vegetation. In miniature the
conditions of the surface may produce an effect ; for instance Blytt[1]
mentions that steep walls of rock facing south place the vegetation
concerned under unusual conditions as regards temperature ; beneath
lofty walls of rock at Christiania a vegetation occurs which is rich and
varied, and includes a number of southern species ; intense heat prevails
here on sunny days.

Steepness of slope (angle or inclination to the horizon). This decides
whether the products of weathering and the humus-substances can
remain *in situ* or are carried away, the rapidity with which water
flows away from the surface, the extent to which the surface is soaked
with water, the density and height of the vegetation, and the intensity
with which the sun's rays can heat the soil.[2]

Exposure of slope. This largely determines the kind of vegetation
present. A slope exposed to sun and wind bears vegetation entirely
different from that on one less exposed to either. In addition
to what has been stated on p. 51, it may be noted that in the Russian
east sea-provinces the south-western slopes bear a more mesophilous,
and the north-eastern slopes a more xerophilous vegetation, because
the south-west winds bring humidity, and the north-east winds aridity.[3]
Even in very small concerns, exposure may affect vegetation, for instance
on dunes ; Giltay[4] has made some observations, showing the differences
that can exist in temperature and atmospheric humidity only a few
paces apart on the northern and southern slopes of sand-dunes in Holland.[5]
In like manner the vegetation on the opposite sides of a cutting or the
embankments of a railway may be very different, as Stenström has
pointed out.[6] On the southern side of slopes in the east of North Germany
the flora of the sunny (Pontic) hills is especially characterized by the
development of plants belonging to a Continental climate.

Differences in geognostic structure, for instance in the inclination of
the strata, evoke distinctions in vegetation. Inclination of the strata
acts on the course taken by water, on the emergence of springs, and
therefore on vegetation. Moreover, the nature of the surface itself may
be entirely different, according as to whether it forms an angle with the
dip of the strata or runs approximately parallel with this ; in the former
case the surface may be steep and gravelly as well as dry, so that only
scanty and stunted vegetation can develop, while in the latter case it slopes
gradually, is richer in water, and consequently bears dense and vigorous
vegetation. Examples illustrating this are to be met with in many
districts with slate mountains.[7]

[1] Blytt, 1893. [2] See Chap. XIII. [3] Klinge, 1890.
[4] Giltay, 1886. [5] Warming, 1907 (1909). [6] Stenström. 1905.
[7] The study of oecology will be much promoted by the preparation of maps in
which the type vegetation is denoted by a special colour, and by a comparison
with maps showing the geognostic surface. Excellent detailed studies of the kind
have been made by Woodhead (1906) and W. G. Smith (1903-5). Flahault (1894,
1897, 1901), and Drude (1902, 1908), have published vegetation-maps dealing with
more extensive areas. Clements's (1905) work may also be consulted.

SECTION II

COMMUNAL LIFE OF ORGANISMS

CHAPTER XXII. RECIPROCAL RELATIONS AMONG ORGANISMS

THE non-living (physical, chemical) and other factors dealt with in Section I do not suffice to impart a full comprehension of the production of communities in the vegetable kingdom.

On p. 84 mention is made of another factor—the *competition* among species of plants—the importance of which is so great that many species are excluded from great areas on the Earth, not by direct interference on the part of non-living factors, but by the indirect interference involved in competition for food with other stronger species.

Another factor, *animal-life*, also has a powerful influence upon the kind and the economy of vegetation. We have discussed the parts played by earthworms, insects, and other small animals in causing physical and chemical changes in soil, but animal-life affects the existence of plants in many other ways, and among all living beings man stands in the foreground as inducing the greatest modifications in plant-communities and in their reciprocal struggles.

The manifold, complex, mutual relations subsisting among organisms are matters of such profound import to plant-life and plant-communities that this Section of our book is set apart for their consideration.

CHAPTER XXIII. INTERFERENCE BY MAN

VERY diversified are the reciprocal relations between the plant-world and mankind. Although the plant-world affects the human race, it is itself to a far greater extent influenced by mankind; indeed, vegetation is the result of man's influence to such an extent that soon there will be but few places upon earth where he has not modified or destroyed the vegetation by directly turning it to his own use or by indirectly interfering with it. Here we merely draw attention to the extent to which man alters the condition and economic status of the original plant-communities by ameliorating the soil, also by tending cultivated plants and domestic animals, and further point out how, by introducing new cultivated plants (such as forest-trees) and new weeds, he voluntarily or involuntarily brings in fresh forms to compete with native plants. Old plant-communities are eradicated by man, and new ones inaugurated for instance, when we see in South America an abandoned plantation filled with weeds in the form of bushes, this is a new, secondary, community which did not naturally occur before the soil had been drawn into the service of man ; and the species which now occur in vast numbers, and form a community with its own special stamp and economy, must previously have been scattered singly at the edge of the forest or in other open places.[1]

Further information concerning interference by man will be given in Section XVII.

CHAPTER XXIV. SYMBIOSIS[2] OF PLANTS WITH ANIMALS

MODERN biological investigations,[3] to which Darwin's works gave the impulse, have elucidated the manifold and complex relations subsisting between the plants and animals that form one community, and have demonstrated the adaptations of plants to animals and the converse.

From a floristic standpoint we may note between the distributional area of certain plants and animals a connexion which is due to exact reciprocal adaptation : as examples may be cited : Aconitum and Bombus[4]; Vanilla, which was introduced into Mauritius at the commencement of the eighteenth century, but could be made to bear fruit only by artificial pollination, because the proper pollinating insects were lacking ; Angraecum sesquipedale, which is undoubtedly adapted to a moth with an immensely long proboscis ; Yucca filamentosa, dependent for pollination upon Pronuba yuccasella.[5]

Attention may also be drawn to the utterly different parts played by entomophilous and anemophilous flowers in the physiognomy of the whole plant-community and the landscape. Trees of the northern forests are anemophilous, those of the tropical ones are mainly entomophilous, and there thus arise those differences in floral beauty that give to the forest an entirely distinct appearance.

Many oceanic isles, the Galapagos Isles for instance, are poor in spermophytes with highly coloured blossoms, but abound in ferns and plants with small or inconspicuous flowers : apparently this is to be correlated with the scantiness of the insect fauna.[6]

But other matters have also to be taken into consideration. All structural features serving to protect plants against animals : poisons, bitter bodies, raphides, stinging hairs, sharp bristles, and so on ; [7] the reciprocal adaptations between insects and flowers ; structural features enabling plants to utilize animals as agents dispersing their fruits (endozoic dispersal of seeds in juicy and· coloured fruits, epizoic dispersal of fruits and seeds provided with hooks and glandular hairs, myrmecochorous plants)[8] or even buds and parts of shoots ; symbiosis of ants and plants

---

[1] Warming, 1892.

[2] [A somewhat extended significance is here given to the term Symbiosis.]

[3] In this connexion may be mentioned the names of Axell, Beccari, Briquet, Burkill, Delpino, Scott-Elliot, Hildebrand, Keller, Knuth, Lindman, Löw, Ludwig, MacLeod, H. Müller, A. F. W. Schimper, Schumann, Warming, Willis, and many others.

[4] Kronfeld, 1890.   [5] Riley, 1873, 1891 ; see also Knuth, 1904, iii, p. 130.

[6] Wallace, 1880 ; but see M. G. Thomson, 1880.

[7] Stahl, 1904.   [8] Sernander, 1901, '1906.

to their mutual advantage (Myrmecodia, Cecropia, Acacia, and Triplaris, according to Belt,[1] Delpino,[2] Schimper,[3] Schumann,[4] and Warming[5]; the symbiosis of acari and plants in which the domatia (acaro-domatia) are constructed for occupation by the former[6]; the symbiosis which, according to Cienkovsky, Entz, Brandt, and Geddes, prevails between green or yellow algae (Zoochlorella, Zooxanthella) and animals (Radiolaria, Infusoria, Flagellata, Spongilla, and Hydra viridis), and which must be regarded as mutualistic, since the alga supplies carbonaceous food and oxygen, while the animal provides shelter and constantly fresh supplies of water containing carbon dioxide. Reference may be made here to the adaptations of insectivorous plants in accordance with their peculiar method of feeding ; also to the fact that an important oecological and geographical part is played by certain animals which search for and utilize certain plants as food, e.g. stags, hares, mice, and the like in forest, also large ruminants in savannah and desert. In this way certain species of plants are favoured at the expense of others, so that the whole stamp of the plant-community is changed. The manner in which plant-shape may be changed by animal bites has been illustrated and explained by L. Klein.[7]

CHAPTER XXV. SYMBIOSIS OF PLANTS WITH ONE ANOTHER. MUTUALISM

VARIOUS kinds of bonds of very various strengths can knit plants together ; in some cases the symbiosis is very intimately bound up with the existence of the species concerned, in others the connexion is far looser, even quite casual.

In what follows we shall deal first with those types of symbiosis in which species are most intimately and firmly linked, that is to say, organically united (symbiosis, in the strict sense, oecological guilds in Schimper's sense),[8] and shall gradually pass on to the looser types until we conclude with the great plant-communities which include many associated species, and which will be the special subject of our consideration. The various types of symbiosis are not sharply delimited from one another.

PARASITISM.

Parasitism is a form of symbiosis in which the two symbionts are associated in the most intimate manner. One species provides the other with nutriment ; the parasite lives on or in its host, and at the expense of the living tissue of the host.

[1] Belt, 1874.   [2] Delpino, see Schimper, 1898 (1903, p. 155); Raciborski, 1898.
[3] Schimper, 1898 (1903, pp. 140–53).   [4] Schumann, 1888, 1889, 1891 a, b.
[5] Warming, 1893 ; see also Ule, 1900.
[6] Lundström, 1887 ; Penzig and Chiabrera, 1903.
[7] L. Klein, 1899. The interdependent and reciprocal relations between plants and animals have been dealt with by Ludwig, 1895. Reference should also be made to C. Schröter, 1904–8.   [8] Schimper, 1898.

There are, however, stages in the degree to which parasite is dependent upon host and requires to abstract food from it :

Most dependent of all are many rust-fungi, the species of Cuscuta and Orobanche, which are not only *holoparasites*, that is to say, incapable of utilizing inorganic food-material, but are able to live only upon one definite species of host.

Less dependent are those species of parasites that can thrive equally upon several or many kinds of host belonging either to one or to several families. Cuscuta Epithymum (holoparasitic) is one such species, as it lives on Calluna, Labiatae, Papilionaceae, and even on Monocotyledones and Equisetum. While Viscum album (hemiparasitic) is another such species, of which one race can be parasitic upon about fifty species of dicotylous trees, and other races upon several kinds of coniferous trees ; these do not pass from dicotyledon to conifer or the reverse ; they are *physiological races* (the ' habitation-races ' of Magnus, the ' specialized forms ' of Eriksson, the ' biological races ' of Rostrup).

While certain species are *obligate* parasites and therefore can exist only as parasites, there are others, less exacting, which on occasion flourish as saprophytes, for instance the honey-fungus (Armillaria mellea). Nectria cinnabarina and some other fungi are perhaps always saprophytes at first, but subsequently pass from dead stumps of branches into living tissue.

Between parasite and host the relationship is hostile (a one-sided antagonism) : the parasite attacks the host and robs it of energy. The host may be so weakened as to perish—for instance, orange-trees are killed by Loranthaceae ; in such cases the parasite of course likewise perishes.

The struggle between a species and its parasites has a highly important bearing upon the composition of the plant-community. Many forest-trees succumb to fungal attack—for example, Scots pine in Denmark attacked by Lophodermium pinastri—and the nature of the forest-vegetation of whole countries may be thus influenced.

Pure forests are much more exposed to parasitic attack than are mixed forests, because parasites spread more easily through a homogeneous assemblage of plants than through a heterogeneous one. Attack by parasites, as well as climatic conditions, is often the cause for one species giving way to another.

HELOTISM

The symbiosis between lichen-fungi and algae is obviously most correctly interpreted as *helotism*. A lichen is a dual organism consisting of a fungus and an alga, which is enveloped by the fungal hyphae and is incorporated with the fungus. The relationship is usually described as *mutualistic*, that is to say, the two organisms are said to be of mutual service to each other ; and this is to an extent true, since the alga by means of its chlorophyll provides the carbonaceous food and directs the metabolism to the best advantage of the community, while the fungus secures all else that is required. But the reciprocity is not equal, and the term *consort* seems inadequate, because though the fungus requires to combine with the alga before it can develop into its completest condition, yet the alga has not the least need of the fungus, and indeed prefers to live apart from it. That the alga grows vigorously, multiplies rapidly, and may even acquire larger cells than when free, may be nothing more than an example of hypertrophy—a pathological condition. It has been suggested that the alga finds protection from desiccation within the fungal mass ; but this seems to be scarcely necessary, as the algae in question are certainly capable of enduring desiccation admirably ; moreover, it is not the case that they secure real protection against desiccation, for under given circumstances the lichen dries up so completely as to become brittle. Besides this, the alga is prevented from multiplying in its most efficient manner—for instance, by zoospores. The alga is in a condition of slavery in relation to the fungus, which is a kind of parasite differing from ordinary parasites in incorporating the host and in providing a portion of the food consumed in the host's maintenance. There is therefore a certain likeness to hemiparasitism, but we must assume that green hemiparasites provide their own carbonaceous nutriment, whereas the lichen-fungus needs merely to secure for itself the non-carbonaceous food-material.

In this case, too, the bond of union between the two organisms may be exceedingly close, in that the fungus selects definite species of algae.

MYCORHIZA AND ENDOPHYTES

A mutualism characterized by complete reciprocity, in which the symbiosis is equally advantageous to both partners, may or may not occur. Even in the most familiar forms of symbiosis the relations subsisting between the symbionts are not sufficiently understood to permit of our completely explaining the nature of their connexion. This is true of *mycorhiza*, in which the root of a highly organized plant enters into intimate connexion with fungal hyphae, which are either *ectotrophic*, and mainly form a sheath enveloping the distal surface, or are *endotrophic*, and live inside the cortical cells.

Mycorhiza has been found in the majority of Amentaceae, Coniferae, Ericaceae, and many other plants, especially in perennial herbs growing on humous soils, such as acid humus, peat, and mould. Mycorhiza is present in the humicolous herbs, whether these do or do not contain chlorophyll ; plants belonging to the latter category (saprophytes) seem certainly to be dependent on the fungus for nutrition. The fungus possibly derives some benefit from the phanerogam, and there is scarcely a doubt that it is of use to the latter.

As regards the endophytic mycorhizal fungus, Percy Groom[1] has proved that in Thismia it indulges in an interchange of nutritive material with the root, and promotes the production of protein bodies. The fungus abstracts from the root certain substances and provides others in exchange, and is itself perhaps digested to some extent. In the ectotrophic form the relations are different.

Stahl has expressed the opinion that the mycorhizal fungus in all cases undertakes the absorption of water and ash-constituents from the soil; accordingly, it would be of special importance in soils poor in nutriment. This view harmonizes well with the distribution of mycorhiza.

---

[1] Percy Groom, 1895 *b*.

Perhaps we have here a notable example of one plant being aided by another to settle in a habitat and to secure nutritive material on a soil from which it would otherwise be excluded ; Calluna-heath, spruce-forest, and the like, would then to a certain extent owe their existence to such symbiosis. But a very great deal concerning this form of symbiosis is completely unexplained.[1]

Apparently similar in some respects to endotrophic mycorhiza is the symbiosis of Leguminosae and bacteria which has already been mentioned on p. 80. The small root-tubercles of Leguminosae are entered and occupied by bacteria which manufacture nitrogenous food and finally perish, becoming changed into ' bacteroids ' and utilized as food by the Leguminosae. It is not definitely established that the bacteria profit by this symbiosis (they presumably acquire carbon-compounds for their host) ; but if they did not profit it would be remarkable that they, like endotrophic fungi, should enter roots.

Going one step farther, we come to plants (algae) which inhabit others without, so far as we know, doing any service in return. They do not live at the expense of the host, in fact perhaps absorb nothing whatever from it, but have free quarters. In this category may be placed the cyanophyceous Anabaena living in the under-side of the leaves of Azolla within special cavities ; these seem to exist only on the alga's account, as they occur in all four species of Azolla which are never free from Anabaena. The alga can flourish quite apart from Azolla.

In like manner other algae live as endophytes, that is to say, inside other plants : in Sphagnum, whose leaves are occupied by Nostoc, which enters the colourless cells by way of the pores in their walls ; in certain liverworts, or in algae—for instance, Entoderma viride living in the cell-wall of Derbesia Lamourouxii. But perhaps the last is an example of parasitism.

To some extent the same is presumably true in the case of those Cyanophyceae which enter the erect dichotomous roots of cycads, and stimulate a definite layer of parenchyma to grow in a special manner so as to provide space for themselves ; also, in the case of Nostoc punctiforme, which penetrates the stems of Gunnera but can live equally well apart from roots or stems.[2] The present state of our knowledge does not permit us to define the exact nature of the symbiosis in every case.

EPIPHYTES.

From those endophytes that only seek for accommodation in other plants but do not absorb food from these, it is but a slight step to epiphytes, or plants living on others but abstracting no food from the living parts of the latter, and at most deriving sustenance from the dead tissue of these. Still it is not always permissible to say that epiphytes do not live at the expense of the supporting plants, for they may occur on these in such quantities as to necessitate the assumption that they do injury by this very quantity, by causing excessive humidity, or by diminishing respiration, as, for instance, in the case of lichens on trees.

---

[1] Those who have worked on mycorhiza include Kamiensky, 1881 ; Frank, 1887 ; Sorauer, 1893 ; Percy Groom, 1895 ; W. Magnus, 1900; Mazé, 1899; Stahl, 1900 ; P. E. Müller, 1886, 1902, 1903 ; and many others.   [2] B. Jönsson, 1894.

The bond between epiphyte and the species upon which it rests is usually less close than in the preceding cases ; most epiphytes can grow upon various kinds of plants, some even upon rock or on the ground. Yet some are confined to definite species, because the nature of the cortex of the latter is of importance. There are epiphytes on aquatic as well as on terrestrial plants. Many kinds of algae live on other algae, or even on phanerogams, and some algae only upon quite definite species—for instance, Elachista fucicola on Fucus, E. scutulata on Himanthalia lorea.[1]

Epiphytes upon terrestrial plants thrive best where atmospheric humidity and precipitations are rich. Yet a change of seasons, and movements of the air, seem to promote their welfare. Meyen (1836) devoted attention to this subject, and subsequently A. F. W. Schimper dealt with it in greater detail in his papers upon epiphytes.[2] In cold and temperate climes epiphytes mostly belong to algae, lichens, and mosses ; but in warm countries they also include a number of ferns and phanerogams (Orchidaceae, Araceae, Bromeliaceae, Cactaceae, Piperaceae, and others) ; and in moist tropical forests there are many epiphyllous species of algae and lichens living upon leaves.[3]

Peculiarities in the habitat have resulted in a number of biological adaptations, which Schimper, Göbel, Raciborski, Mez, Treub, Karsten, and Beccari have explained in reference to flowering plants. The particular details in question are described in the succeeding paragraphs.

The *seeds* (and spores) are calculated to gain double object—dispersal and fixation to the substratum. Either they are scattered by wind, in which case they are so small and light, or are so provided with long hairs (and the like), as to be easily conveyed by the wind on to trunks and branches where they encounter a fissure or some other depression in which they can become firmly lodged. Or the seeds are contained in fleshy fruits (e. g. Araceae, Bromeliaceae, and Cactaceae) which are eaten by birds, in whose excrement they are dispersed and fastened on to branches. An entirely exceptional method of multiplication characterizes the rootless Tillandsia usneoides, detached fragments of which easily become twisted round twigs of trees by means of their long slender shoots.

Fixation of the epiphyte to parts of plants is accomplished either by rhizoids (as in the case of mosses and lichens) which penetrate the substratum (dead cortex) to some extent, or by attaching-roots which are irritable and sometimes adhere firmly to the substratum with the aid of fixing-hairs and secretions. There is often a division of labour between attaching-roots and absorbing-roots.

Provision of water is a problem of difficulty to the epiphyte, because rain-water soon flows away. Some epiphytes obtain the necessary water rather from dew and mist than from rain ; others do the converse.[4] Many are able to seize the momentary opportunity, and in their dry condition can instantly absorb moisture over their whole surface, e. g. algae, mosses, lichens, Tillandsia usneoides and other Bromeliaceae

---

[1] Concerning the significance attached by Fritsch to the term ' consortium ', see Fritsch, 1906.
[2] A. F. W. Schimper, 1884, 1888 ; compare also Mez, 1904 a.
[3] Göbel, 1889–92; Raciborski, 1898; Mez, 1904 a; G. Karsten, 1894; Treub, 1888.
C. Jennings. [4] Mez, 1904 a.

with peculiar absorbing-hairs.[1] Others (Orchidaceae, Araceae) have aerial roots with a special velamen adapted to absorb water ; yet others, for example Tillandsia bulbosa, have their leaves so constructed as to facilitate the retention of water among them ; others, again, possess two kinds of leaves, some of which, as ' pocket-leaves ',[2] are pressed closely against the substratum so that water is held by capillarity between them and the supporting-stem, or is actually taken up by them, as G. Karsten suggests in the case of the fern Teratophyllum aculeatum. Epiphytes are much exposed to *desiccation*. Against this, certain species (algae, lichens, and mosses) have no evident protection ; they can endure without injury existence in a dry condition for a long period, and awaken into life again at the first fall of rain or dew. But others have fashioned for themselves water-receptacles of various kinds : aqueous tissue in leaves and stems (in Orchidaceae, Peperomia, and others), water-storing cells in leaves (in Orchidaceae and others), urn-shaped water-bags or con-cavities of other shapes (as in liverworts,[3] Dischidia, and others).

Food-material is obtained by epiphytes as follows : Carbon is taken from the air, as all epiphytes are photophilous and evergreen ; some also accumulate humous and mineral bodies among their roots or with the aid of specialized leaves (' pocket-leaves ', ' mantle-leaves'), as in the cases of ferns such as Asplenium Nidus, Polypodium quercifolium, and Platycerium alcicorne.[4]

The *construction* of the shoot and the whole architecture of the epiphyte varies widely. Some species, like Tillandsia usneoides, are rootless, while the vegetative organs of others, such as the orchid Poly-rrhiza (Aeranthus) funalis, consist almost entirely of green roots. With Schimper [5] we may divide epiphytes into four groups :

1. Those that find their nutriment on the cortex of their support.
2. Those that send aerial roots into the soil.
3. Those that collect moist humus within the large interwoven mass of roots, and, in some cases, behind or between their leaves.
4. Those whose leaves take over the functions of roots and absorb water and nutritive salts.[6]

Epiphytes have many structural features in common with terrestrial *xerophytes*, for, like these, they must be adapted to endure prolonged drought. They form in fact a group of xerophytes, and it is consequently easy to understand how it comes that certain species, such as Rhipsalis Cassytha and other Cactaceae, can live upon either trees or rocks. The characteristic structural features of epiphytes will therefore be considered in detail in Section III, which deals with xerophytic vegetation.[7]

SAPROPHYTES

In the case of many epiphytes it must be assumed that they derive nutriment from dead parts of plants (bark) upon which they grow ; they thus feed upon dead organic substance, and are saprophytes.

Larger numbers and more pronounced forms of saprophytes are,

[1] Schimper, 1884, 1888a ; Mez, 1904a. [2] Göbel, 1891.
[3] Göbel, 1889–93 and 1898–1901. [4] Göbel, 1889–93.
[5] A. F. W. Schimper, 1884, 1888, 1898. [6] See G. Karsten, 1894.
[7] See Wittrock, 1894 ; Willis and Burkill, 1904 ; Ule, 1904 ; Cockayne, 1901.

however, met with on the ground, especially in the forest, where all kinds of fallen fragments (withered leaves, twigs, flowers, and fruits) accumulate year after year and produce humus. Saprophytes are therefore bound up with other plants, but the bond is different from that associated with parasitism ; for it is the cast-off parts and redundant individual plants that they utilize. Some saprophytes select special kinds of vegetable remnants and are therefore inseparably associated with definite species of plants ; others are less narrow in their choice. Clavaria abietina, Lactarius deliciosus, and certain other fungi are only met with in coniferous forest, others select dicotylous forest, while others again grow only upon dung, as is the case with Poronia, Coprinus, Pilobolus, and Sordaria among Fungi, and Splachnum among mosses.

Among cryptogamic and phanerogamic plants alike, saprophytes display a very varied degree of adaptation to the saprophytic mode of existence.[1] Every kind of humus is permeated with fungal mycelia and bacteria, and the soil in forest at autumn reveals hosts of pileate fungi. Phanerogamic plants that are most completely adapted to a saprophytic life, that is to say, *holosaprophytes*, exhibit the following characters :—

1. They have little or no chlorophyll, being yellow, red, or brown in tint.

2. Their leaves are upwardly directed, and more or less reduced to adpressed scales.

3. Stomata are usually absent.

4. The root-system is more or less reduced, some forms like Corallorrhiza being quite rootless ; the roots are short, thick, and but feebly branched, and in the vast majority of cases form mycorhiza.

5. The vascular bundles are reduced.

As examples may be cited : Neottia, Corallorrhiza, Epipogum, Pogonopsis, and other orchids ; some Burmanniaceae ; Triuridaceae ; Monotropa and Sarcodes among Pyrolaceae ; Voyria in the Gentianaceae.[2]

Hemisaprophytes have the external appearance and structure of normal plants that assimilate carbon dioxide. Their needs as regards organic nutriment probably differ extremely, for while some cannot exist away from a soil rich in humus, such as a forest soil, others, like many orchids and species of Pyrola, are tentatively to be regarded as facultative saprophytes.[3]

LIANES

While the want of humus-containing food constitutes the bond uniting saprophytes with other plants, lianes are linked with fellow plants by reason of the want of support for their weak long-jointed stems by which they reach the light. The term *liane* is here employed in the widest sense, and includes twining plants as well as the various kinds of other climbers. Lianes owe their origin to the grouping of plants into communities in the form of forest and bush ; the shade due to dense vegetation has caused them in the past to elongate, to produce long internodes, and in the course of time to adapt themselves in various ways not only

---

[1] See p. 85.   [2] See Johow, 1885, 1889 ; Percy Groom, 1895 *a* and *b*.
[3] See Heinricher, E., 1896, 1897, 1901-3 ; Wettstein, R. v., 1902.

to hold firmly to their supports, but also by suitable internal structure to solve the problem of conduction of material, as well as other problems that they owe to the length and slenderness of their stems.[1] Lianes display very different degrees of adaptation to the climbing habit. The lowest stage is represented by :—

1. '*Semi-lianes*' (Warming) and *scramblers* (Schenck), which occur particularly at the margins of forests, in hedges, and in bushlands.

More elaborate and specialized in adaptation are :—

2. *Root-climbers*, which can clamber up thick tree-trunks and rocks.

3. *Twining plants.*

4. Lianes equipped with *tendrils* or other *irritable organs*, and capable of embracing slender parts of plants. These were termed by Darwin ' hook-climbers ', ' tendril-bearers,' and ' leaf-climbers '.

It is characteristic of the majority of lianes for the leaf to be broad, more or less cordate, and long-stalked.[2] (But to this rule exceptions are provided by those Papilionaceae and other lianes that climb by means of tendrils on the ends of their leaves.) In structure of leaf and stem some lianes recall xerophytes ; it seems quite natural that lianes should be exposed to the possibility of losing more water by transpiration than can be balanced by supplies from the root, and hence should be structurally adapted to provide for this contingency.[3] Certain species, for instance, of the genus Ficus occur both as lianes and as epiphytes.[4] The liane-form is a consequence of the congregation of plants to form communities, but lianes are partially independent of their fellows, since inanimate supports will sometimes serve them quite as well as living ones. Lianes belong especially to certain families, including Ampelidaceae, Asclepiadaceae, Apocynaceae, Bignoniaceae, Cucurbitaceae, Papilionaceae, Sapindaceae, Dioscoraceae, and others.

CHAPTER XXVI. COMMENSALISM. PLANT-COMMUNITIES

IN the preceding chapter we have dealt with the different bonds that may link plants together, and one individual with another : parasite with host, master with slave (helotism of lichens) ; we then discussed mutualists, epiphytes, and finally species associated with the whole plant-community. We have yet to consider the great, highly complex plant-communities which form the essential foundations of oecological phytogeography.

The term ' community ' implies a diversity but at the same time a certain organized uniformity in the units. The units are the many individual plants that occur in every community, whether this be a beech-forest, a meadow, or a heath. Uniformity is established when certain atmospheric, terrestrial, and any of the other factors discussed in Section I are co-operating, and appears either because a certain, defined economy makes its impress on the community as a whole, or because a number of

[1] For further details see C. Darwin, 1875 ; Schenck, 1892, 1893 ; Warming, 1892 ; A. F. W. Schimper, 1898.   [2] Lindman, 1899 ; Warming, 1901.
[3] Warming, 1892.   [4] Schenck, 1892, 1893 ; Whitford, 1906.

different growth-forms are combined to form a single aggregate which has a definite and constant guise.

The analysis of a plant-community usually reveals one or more of the kinds of symbiosis as illustrated by parasites, saprophytes, epiphytes, and the like. There is scarce a forest or a bushland where examples of these forms of symbiosis are lacking; if, for instance, we investigate the tropical rain-forest we are certain to find in it all conceivable kinds of symbiosis. But the majority of individuals of a plant-community are linked by bonds other than those mentioned—bonds that are best described as *commensal*. The term *commensalism* is due to Van Beneden who wrote ' Le commensal est simplement un compagnon de table ': but we employ it in a somewhat different sense to denote the relationship subsisting between species which share with one another the supply of food-material contained in soil and air, and thus feed at the same table.

More detailed analysis of the plant-community reveals very considerable distinctions among commensals. Some relationships are considered in the succeeding paragraphs.

LIKE COMMENSALS

When a plant-community consists solely of individuals belonging to one species—for example, solely of beech, ling, or Aira flexuosa—then we have the purest example of like commensals. These all make the same demands as regards nutriment, soil, light, and other like conditions; as each species requires a certain amount of space and as there is scarcely ever sufficient nutriment for all the offspring, a struggle for food arises among the plants so soon as the space is occupied by the definite numbers of individuals which, according to the species, can develop thereon. The individuals lodged in unfavourable places and the weaklings are vanquished and exterminated. This competitive struggle takes place in all plant-communities, with perhaps the sole exceptions of sub-glacial communities and in deserts. In these *open communities* the soil is very often or always so open and so irregularly clothed, that there is space for many more individuals than are actually present; the cause for this is obviously to be sought in the climatically unfavourable conditions of life, which either prevent plants from producing seed and other propagative bodies in sufficient numbers to clothe the ground, or prevent the development of seedlings. On such soil one can scarcely speak of a competitive struggle for existence; in this case a struggle takes place between the plant and inanimate nature, but to little or no extent between plant and plant.

That a congregation of individuals belonging to one species into one community may be profitable to the species, is evident; it may obviously in several ways aid in maintaining the existence of the species, for instance, by facilitating abundant and certain fertilization (especially in anemophilous plants) and maturation of seeds; in addition, the social mode of existence may confer other less-known advantages. But on the other hand it brings with it greater danger of serious damage and devastation wrought by parasites.

The bonds that hold like individuals to a like habitat are, as already indicated, identical demands as regards existence, and these demands are satisfied in their precise habitat to such an extent that the species can maintain itself here against rivals. Natural unmixed associations of forest-trees are the result of struggles with other species. But there are differences as regards the ease with which a community can arise and establish itself. Some species are more social than others, that is to say, better fitted to form communities. The causes for this are biological, in that some species, like Phragmites, Scirpus lacustris, Psamma (Ammophila) arenaria, Tussilago Farfara, and Asperula odorata, multiply very readily by means of stolons ; or others, such as Cirsium arvense, and Sonchus arvensis, produce buds from their roots ; or yet others produce numerous seeds which are easily dispersed and may remain for a long time capable of germinating, as is the case with Calluna, Picea excelsa, and Pinus ; or still other species, such as beech and spruce, have the power of enduring shade or even suppressing other species by the shade they cast.[1] A number of species, such as Pteris aquilina, Acorus Calamus, Lemna minor, and Hypnúm Schreberi, which are social, and likewise very widely distributed, multiply nearly exclusively by vegetative means, rarely or never producing fruit. On the contrary, certain species, for example many orchids and Umbelliferae, nearly always grow singly.

In the case of many species certain geological conditions have favoured their grouping together into pure communities. The forests of northern Europe are composed of few species, and are not mixed in the same sense as are those in the tropics, or even those in Austria and other southern parts of Europe : the cause for this may be that the soil is geologically very recent, inasmuch as the time that has elapsed since the Glacial Epoch swept it clear has been too short to permit the immigration of many competitive species.[2]

UNLIKE COMMENSALS

The case of a community consisting of individuals belonging to one species is, strictly speaking, scarcely ever met with ; but the dominant individuals of a community may belong to a single species, as in the case of a beech-forest, spruce-forest, or ling-heath—and only thus far does the case proceed. In general, many species grow side by side, and many different growth-forms and types of symbiosis, in the extended sense, are found collected in a community. For even when one species occupies an area as completely as the nature of the soil will permit, other species can find room and can grow between its individuals ; in fact, if the soil is to be completely covered the vegetation must necessarily always be heterogeneous. The greatest aggregate of existence arises where the greatest diversity prevails.[3] The kind of communal life resulting will depend upon the nature of the demands made by the species in regard to conditions of life. As in human communities so in this case, *the struggle between the like* is *the most severe*, that is, between the species making more or less the same demands and wanting the same dishes from the common table. In a tropical mixed forest there are hundreds of species of trees growing together in such profuse variety that the eye can scarce see at one time two individuals of the same species,[4] yet all of them

---

[1] See Chap. XCVII.   [2] See Warming, 1899 *b*.   [3] See Darwin, 1859.
[4] See Warming, 1892, 1899 *b*.

undoubtedly represent tolerable uniformity in the demands they make as regards conditions of life, and in so far they are alike. And among them a severe competition for food must be taking place. In those cases in which certain species readily grow in each other's company—and cases of this kind are familiar to florists—when, for instance, Isoetes, Lobelia Dortmanna and Litorella lacustris occur together—the common demands made as regards external conditions obviously form the bond that unites them. Between such species a competitive struggle must take place. Which of the species shall be represented by the greatest number of individuals certainly often depends upon casual conditions, a slight change in one direction or the other doubtless often playing a decisive rôle ; but apart from this it appears that morphological and biological features, for example development at a different season, may change the nature of the competition.

Yet there are in every plant-community numerous species which *differ widely* in the demands they make for light, heat, nutriment, and so on. Between such species there is less competition the greater the disparity in their wants ; the case is quite conceivable in which the *one species should require exactly what the other would avoid ;* the two species would then be complementary to one another in their occupation and utilization of the same soil.

There are also obvious cases in which different species are of service to each other. The carpet of moss in a pine-forest, for example, protects the soil from desiccation and is thus useful to the pine, yet, on the other hand, it profits from the shade cast by the latter.

As a rule, a limited number of definite species are the most potent, and, like absolute monarchs, can hold sway over the whole area ; while other species, though possibly present in far greater numbers than these, are subordinate or even dependent on them. This is the case where subordinate species only flourish in the shade or among the fallen fragments of dominant species. Such is obviously the relationship between trees and many plants growing on the ground of high forest, such as mosses, fungi and other saprophytes, ferns, Oxalis Acetosella, and their associates.[1] In this case, then, there is a commensalism in which individuals feed at the same table but on different fare. An additional factor steps in when species do not absorb their nutriment at the same season of the year. Many spring-plants—for instance, Galanthus nivalis, Corydalis solida and C. cava—have withered before the summer-plants commence properly to develop. Certain species of animals are likewise confined to certain plant-communities. But one and the same tall plant may, in different places or soils, have different species of lowly plants as companions ; the companion-plants of high beech-forest depend, for instance, upon climate and upon the nature of the forest soil[2] ; Pinus nigra, according to von Beck, can maintain under it in the different parts of Europe a Pontic, a Central-European, or a Baltic vegetation.

There are certain points of resemblance between communities of plants and those of human beings or animals ; one of these is the competition for food which takes place between similar individuals and causes the weaker to be more or less suppressed. But far greater are

---

[1] See Höck, 1892, 1893.
[2] P. E. Müller, 1887.

the distinctions.   The plant-community is the lowest form; it is merely a congregation of units among which there is no co-operation for the common weal but rather a ceaseless struggle of all against all.   Only in a loose sense can we speak of certain individuals protecting others, as for example, when the outermost and most exposed individuals of scrub serve to shelter from the wind others, which consequently become taller and finer[1]; for they do not afford protection from any special motive, such as is met with in some animal communities, nor are they in any way specially adapted to act as guardians against a common foe.   In the plant-community egoism reigns supreme.   The plant-community has no higher units or personages in the sense employed in connexion with human communities, which have their own organizations and their members co-operating, as prescribed by law, for the common good.   In plant-communities there is, it is true, often (or always) a certain natural dependence or reciprocal influence of many species upon one another ; they give rise to definite organized units of a higher order[2]; but there is no thorough or organized division of labour such as is met with in human and animal communities, where certain individuals or groups of individuals work as organs, in the wide sense of the term, for the benefit of the whole community.

Woodhead[3] has suggested the term *complementary association* to denote a community of species that live together in harmony, because their rhizomes occupy different depths in the soil ; for example, he described an ' association ' in which Holcus mollis is the ' surface plant ', Pteris aquilina has deeper-seated rhizomes, and Scilla festalis buries its bulbs at the greatest depth.   The photophilous parts of these plants are ' seasonably complementary '.   The opposite extreme is provided by *competitive associations*, composed of species that are battling with each other.

The classification and nomenclature of plant-communities are fully discussed in Chapter XXXV.

[1] See Chapter VIII, p. 37.   [2] See, for instance, Grevillius. 1894.
[3] Woodhead, 1906, p. 345.

SECTION III

ADAPTATIONS OF AQUATIC AND TERRESTRIAL PLANTS. OECOLOGICAL CLASSIFICATION

CHAPTER XXVII.   AQUATIC AND TERRESTRIAL PLANTS

In the Introduction [1] brief mention was made of the fact that there are plant-communities which are characterized by a definite physiognomy, definite constituent growth-forms, and a definite economy: this is a consequence of the circumstance that those species which make approximately the same demands in regard to the nature of the environment, or which are associated for other reasons, congregate naturally to constitute a kind of single entity.

It will now be our task to inquire as to what communities there are, and on what principles they are to be most naturally defined and ranged into some sort of system, or, in other words, which of the factors mentioned in Section II are of the greatest significance in this matter, and which of them play only a subordinate rôle, also what part in the establishment of this classification is taken by the growth-forms already discussed in Section I.   Before we can deal with the individual communities we must consider oecological classification as a whole.

The grouping of the classes of communities here adopted is based in the first place upon *the plant's dependence upon and relation to water*.[2]   Pindar's aphorism, ἄριστον μὲν ὕδωρ, is wholly true of plant-life ; water is the condition of life that exercises the greatest influence in bringing into being external and internal differences among plants ; it is likewise water that plays the leading part in determining the creation of plant-communities and their distribution over the soil.

It is quite true that the special attributes of a habitat result from the co-operation of the most diverse factors, edaphic and climatic, not one of which can be omitted without modifying those special attributes and consequently the vegetation. For instance, vegetation is greatly affected by fertility of the soil, and on a sterile soil there occur only communities of feeble productive power.[3]   But it is beyond doubt that water occupied the foremost position as a factor bringing about the greatest distinctions in vegetation and structure.   We may say :

The supply of water to the plant and the regulation of transpiration are the factors that evoke the greatest differences in *plant-form and plant-life.*

In this connexion we at once meet with two extremes :

There are many plants that pass the whole or the greater part of their

---

[1] Page 12.   [2] See Chaps. VII, VIII, XII.
[3] Gräbner (1898, 1901, 1908), indeed, would have it that nutriment in the soil is the paramount factor.

lives *submerged* in the water, which envelops them completely or, at most, leaves definite floating parts of them uncovered at its surface : these are water-plants (aquatic).[1] On the other hand there is a still greater number of plants that expose at least their assimilating organs to the air and hence to transpiration : these are land-plants (terrestrial), and amongst them we include marsh-plants.

To submerged water-plants transpiration is an impossibility ; in land-plants transpiration takes place, and it is incumbent upon them to maintain a balance between the intake and output of water, that is to say, they must *regulate transpiration*. If evaporation be greater than the supply of water the plant withers, and this has the gravest effect upon vital processes, even when it does not go so far as to kill the plant.

Here reference may also be made to the part played by water in the general economy of nature, in its promotion of putrefaction and of the production of humus, as the micro-organisms responsible for these processes need water. The significance of water in relation to the distribution of plants is demonstrated most distinctly in flat countries such as the western parts of Denmark ; a marked *zonal* arrangement of the vegetation reveals itself here, not only in water but also on land, round every lake or pool. Differences of a few centimetres in the level of the water-table suffice to evoke wide distinctions in the vegetation.[2]

The story of man also indicates the importance of water to the plant. History has shown to what an extent the prosperity of countries (density and wealth of the population) is dependent upon water. In Asia, for example, civilization was confined to those lands where a well-watered soil ensured the existence of man. In Algiers the density of population runs almost parallel with the amount of rainfall.[3] Lack of water is that factor in plant-life in the face of which man is most helpless.

The effect of their environment on water-plants is not only marked by the absence of transpiration, but is also impressed upon them by the other peculiar conditions belonging to water, such as its absorption and consequent weakening of light, its dissolved air, its movements, its buoyancy, and other characters.

In the succeeding chapters the general structural relationships of the water-plant and land-plant respectively will be considered.

CHAPTER XXVIII.   ADAPTATIONS OF WATER-PLANTS (HYDROPHYTES)

VARIOUS structural features and phenomena exhibited by submerged parts are to be regarded as adaptations to the peculiar physical qualities of water. Some of these are dealt with in the succeeding paragraphs.

1. *Roots and analogous organs.* Since nutriment may be absorbed by the whole permeable surface of all submerged parts, there is in submerged plants a reduction in those organs which normally extract mineral food-material from the soil, that is to say, in the roots, or in analogous

---

[1] Hydrophyta of Schouw, 1822.
[2] See Raunkiär, 1889 ; Warming, 1907; Massart, 1893, 1908, and others.
[3] Dehérain, 1892.

organs among Cryptogamia. Some vascular plants, such as Salvinia, Wolffia, Ceratophyllum, Utricularia vulgaris, Aldrovanda, and Genlisea, are entirely rootless ; in others, such as Azolla, Lemna, Hydrocharis, Pontederia, and Pistia, the roots soon cease to grow, do not branch, and may shed their root-caps. Root-hairs are absent in Lemna minor and L. trisulca, Myriophyllum, Butomus umbellatus, Caltha palustris, and Hippuris vulgaris ('except at the collar '), Nymphaea alba, and others.[1]

Roots and root-hairs in many cases are merely anchoring-organs.[2]

2. *Water-carrying tubes* are for the same reasons in less demand ; wood-vessels and the whole xylem are consequently reduced in vascular plants. Phloem, as the tissue conducting protein bodies, undergoes no reduction. The conducting tissues are always congregated more towards the centre of the organ, so that they finally constitute a central bundle. Van Tieghem establishes four types of degenerate roots.[3] The ramification and number of veins in the foliage leaves is less than in land-plants.

3. *Mechanical tissue* is either reduced or undeveloped because the buoyancy of water is greater than that of air. In particular, those structural designs adapted to resist bending are not developed. In order to resist stretching, due to movements of the water, large fixed water-plants living in very troubled water have their mechanical tissue collected as closely as possible to the centre of the stem so as to form a design adapted to resist tensile stresses [4] ; while certain algae have strengthening rhizoids at the base of the thallus, as Wille [5] has demonstrated in detail. *Lignification* occurs to little (in the wood-vessels) or to no extent. Among submerged water-plants there are no woody plants.

4. *Air-containing spaces* are very abundant and large in submerged water-plants and marsh-plants, and serve partly to decrease their specific gravity (as a flotation-device), and partly to facilitate gaseous interchange and especially respiration. In a number of the larger algae such as Fucus vesiculosus, Ascophyllum nodosum, Halidrys siliquosus, Sargassum, Macrocystis and other Laminariaceae, well-developed flotation-devices occur. Exceptions to these statements are provided by nearly all lithophilous hydrophytes, including the vast majority of algae, mosses, Podostemaceae, as well as by some small Spermophyta such as Bulliarda aquatica.

5. *Secondary growth in thickness* takes place only exceptionally in the axial organs of water-plants. This is correlated with the matters discussed in the three preceding paragraphs. On the contrary, submerged parts of stems and leaves of Spermophyta are far longer and thinner than are the corresponding members when developed in contact with air, and owing to weakness of illumination they almost acquire the appearance of etiolated parts.

6. The *epidermis* or the external cell-wall in contact with water is thin, and the cuticle very thin or wanting. Coatings of wax and cork are absent. In contact with air hydrophytes wither and dry up with extreme rapidity. Hairs are lacking from the assimilating organs of nearly all submerged Spermophyta, and when present may serve either

---

[1] F. Schwarz, 1888.   [2] See Henslow, 1895 ; Warming, 1881–1901.
[3] Van Tieghem, 1870-1.   [4] Schwendener, 1874.   [5] Wille, 1885.

to produce mucilage, or to promote assimilation or respiration ; litho-philous species of algae and Podostemaceae provide examples of the last two cases.

7. The epidermis or external layer of cells often contains *chlorophyll*, and in algae is actually the tissue richest in chlorophyll.[1] This must be causally connected with the weakness of light, also with the lack of any necessity for the epidermis to function as aqueous tissue.

8. *Excretion of water* is not excluded from submerged water-plants, but when occurrent assumes the form of guttation (the excretion of liquid water) induced by internal activity. At the leaf-tips of many species water-pores occur, or the tips are detached and the ends of the vascular bundles are thus brought into direct contact with the water, as has been shown by the investigations of Sauvageau, Wieler, Weinrowsky, and Minden.[2]

Transpiration in the strict sense is of course excluded, and correlated with this is the usual absence of stomata. Where these do occur as exceptions, they may be regarded as functionless vestiges. (In floating plants stomata of abnormal structure occur.)[3]

9. *Mucilage* is excreted by many water-plants, often in great quanti-ties, sometimes, as in many bacteria and algae, from the general surface of the body, sometimes from special organs, such as hairs in the higher plants, and at other times into internal passages. The function of mucilage has not been definitely ascertained, and possibly may vary widely. In many bacteria, algae, and buds, it perhaps serves as a protection against injury due to rapid physical or chemical change in the surrounding water, and according to the views of Göbel and others [4] accomplishes this by obstructing the passage of water ; according to Stahl [5] it acts as a defence against animal attack ; while Hunger [6] suggests that it acts as a lubricant facilitating plant-movement. Littoral algae living on the shore and lying dry at the ebb, as well as other algae occa-sionally exposed to drying, are protected from desiccation by mucilage ; but in those algae that grow on a rocky shore and are also exposed to the violence of the water, mucilage may serve as a defence against the force of the waves.[7] Regarding the production of mucilage, reference should be made to the works of Hunger, Schilling,[8] Göbel,[9] B. Schröder,[10] and other authors cited by these.

10. The *chlorenchyma* of water-plants is very slightly differentiated ; there is little or no indication of a distinction into palisade and spongy parenchyma in the submerged leaves of Phanerogamia. The leaves are therefore isolateral. This is possibly correlated with weakness of light. Dorsi-ventral structure reveals itself in floating-leaves.

11. The properties of water bring forth *leaf-shapes* entirely different from those of land-plants, as will be described in Section IV, which deals with water-plants. The submerged leaves of those plants that also possess aerial leaves are entirely different from the latter as regards both

---

[1] Wille, 1885.
[2] Sauvageau, 1889, 1890, 1891, 1894 ; Wieler, 1892 ; Weinrowsky, 1898 ; Minden, 1899 ; see Burgerstein, 1904, p. 246.
[3] Haberlandt, 1904, p. 413.   [4] Göbel, 1898, 1901.
[5] Stahl, 1904 *a*.   [6] Hunger, 1899.   [7] Wille, 1885.
[8] Schilling, 1894   [9] Göbel, 1898–1901.   [10] B. Schröder, 1903.

shape and anatomy. This, according to MacCulland, is directly due to the arrest of transpiration and to the cells being overcharged with water.

12. *Duration of life.* The vast majority of water-plants, at least among spermophyta, are *perennial;* this is in accordance with the favourable environment, which is but slightly affected by seasonal change. Exceptions are, however, provided by many Cryptogamia and some vascular plants, such as Salvinia, Naias, and Subularia. Vegetative propagation far exceeds sexual reproduction in many water-plants : this may go to such a length that production of fruit is entirely eliminated. Certain species, such as Elodea canadensis (at least in Europe, where only the female plant occurs), many species of Lemna, and others, multiply exclusively in a vegetative way. It is a general biological phenomenon that humidity opposes the production of sexual organs, whereas aridity promotes it.

The peculiarities of water-plants that have been mentioned here are to be interpreted in general as examples of degeneration, and of morphological and anatomical retrogression, if we compare water-plants with land-plants ; this retrogression we may, with Henslow, regard as adaptive.[1]

CHAPTER XXIX. ADAPTATIONS OF LAND-PLANTS

THE land-plant contrasts most strongly with the water-plant as regards external and internal construction, in that parts in contact with air—the assimilatory organs in particular—are exposed to transpiration, and must therefore be adapted in an entirely distinct manner.

Transpiration is a physiological process determined by factors of two kinds : *external* or environmental, and *internal* or those dependent on the precise structure or temporary condition of the plant.

The *external* (climatic and edaphic) factors were mentioned in Section I : they are more particularly isolation, temperature, saturation-deficit, and movements of the atmosphere. The supply of water depends upon the nature of the soil, including quantity of water, temperature, acidity, amount and concentration of the salts in the soil. Correctly speaking, as Clements[2] insists, climatic factors to a great extent affect conditions in the soil.

As regards *internal* factors, transpiration depends on the dimensions of the evaporating surface, and inasmuch as it is the foliage-leaves through which evaporation mainly takes place, it is likewise the size, disposition, and thickness of the foliage-leaves, as well as the whole development of the aerial shoot, that above all determine the amount of transpiration ; this is also influenced by the nature of the epidermis (cuticle, wax, cork, hairs, and stomata). The *foliaged shoot* gives the clearest indication of the conditions under which the plant has developed. An additional determinant factor is the nature of the *root-system ;* the *larger* the absorbing surface is, the more water can there be absorbed in the same time ; and the *deeper* the penetration of the root, the greater is the certainty that the supply of water will not be cut off by drought.

The regulation of the amount of water within the plant is accomplished

[1] Henslow, 1895.   [2] Clements, 1904.

by means of water-excreting organs, to which Haberlandt [1] has given the name of *hydathodes*. They are possessed by land-plants as well as by some water-plants, and cause water to be discharged by exudation-pressure in the form of drops. Not only in the tropical rain-forest, but also in the temperate countries, are there many plants, especially herbs, exhibiting the phenomenon of *guttation*. When transpiration is depressed by saturation of the atmosphere, there comes the danger that the plant, on account of continued and powerful root-pressure, may take up an excess of water from the wet soil, and thus attain a condition of maximal turgescence when the air is expelled from the intercellular spaces and completely replaced by water. The danger is averted by the hydathodes.

These organs mainly belong to the following types :—

1. Epidermal cells, sometimes of remarkable structure, or peculiar hairs which are unicellular or multicellular, and in the latter case often assume the form of glandular hairs. As these organs occur on both leaf-faces, but particularly on the lower face, drops of water excreted over the leaves simulate dew-drops.

2. In some ferns hydathodes assume the form of peculiar glandular spots on the lamina.

3. The familiar water-pores, which are constructed like stomata, occur on the upper face of leaf-teeth above a small-celled, thin-walled, usually colourless tissue (epithema) in which the vascular bundles terminate.

It must, however, be noted further that water may be excreted through the epidermis by means of pores opening outwards, without the co-operation of hydathodes ; and also that water may be excreted without the aid of living cells, for example in grasses : this contrasts with the preceding cases in which living cells are the essential and functionally active organs.

The adaptation of the land-plant to its existence in contact with air proceeds on the following lines :—

1. Control of the outgo of water, i.e. regulation of transpiration.

2. Increase of the intake of water, i.e. development of special mechanisms for absorption.

3. Arrangement for storage of water, i.e. development of water-reservoirs.

In the following three chapters these adaptations will be discussed, and in a succeeding one we shall deal with some structural characters and growth-forms of land-plants, the utility of which to the plant is obscure, although their connexion with its existence in a dry environment is beyond question.

At the outset it may be noted that the degree of adaptation of land-plants to their life in contact with air varies widely, according as the external conditions are more or less extreme. Those species that are adapted to meet the conditions of strongest transpiration and most precarious water-supply are termed *xerophytes :* [2] the remainder are termed *mesophytes :* [3] between these two classes there is of course no strict boundary.

---

[1] Haberlandt, 1894–5, 1904.
[2] Xerophyta, Schouw, 1822 (ξηρός, dry ; φυτόν, plant).
[3] Mesophyta, Warming, 1895 (μέσος, middle).

CHAPTER XXX.   REGULATION OF TRANSPIRATION IN
LAND-PLANTS

THE regulation of the transpiration, and the checking of it at critical times is, especially in xerophytes, effected by the following methods :—

1. *Anatomical structure* controlling transpiration.

2. *Diminution of the evaporating surface,* either by movements, or by reduction in the surface of leaves or of shoots which become irreversibly and characteristically adapted to the prevailing conditions.

3. *Regulation of illumination* of the assimilating organs, either by their assumption of a temporary profile-lie (accomplished by photometric movements dependent on intensity of illumination), or of a permanent profile-lie (as in compass-plants).

4. *Investing organs, such as hairs, leaves, and the like,* which weaken the light as well as directly decrease transpiration.

5. *Ablation of rain-water from the leaves.*

I. ANATOMICAL STRUCTURE REGULATING TRANSPIRATION

In this respect there is a fundamental distinction between land-plants and water-plants. It is clear that a great difference must exist between the surface of a plant that is permanently enveloped in water or moist air, and one that is surrounded by dry air and engaged in intense transpiration. But the difference concerns not only the construction of the integument and the aerating system, including stomata and intercellular spaces, but also the chlorenchyma.

A. Cuticular Transpiration.

Transpiration is either cuticular or stomatal. We shall first consider *cuticular transpiration,* which takes place through the external cell-walls of the plant, or in most of the higher plants through the epidermis. In connexion with the regulation of transpiration there occur the following devices :—

Cuticle is the first important regulator of transpiration ; its thickness is adjusted in accordance with the need on the part of the plant to limit transpiration ; yet other conditions seem to play a part, for Bergen [1] found that the cuticle of young leaves is more impermeable to water than is that of old leaves. The cuticle of hydrophytes is as a rule very thin and permeable, but that of xerophytes is thicker and often completely impermeable. The outer walls of the epidermis may be strongly thickened and cutinized, and in some cases may even include crystals of *calcic oxalate* or *silica.* The leaves, owing to the nature of the epidermis, are often *leathery* and *glossy,* and this is a frequent and striking feature in tropical (sclerophyllous) trees, but is also met with in temperate climes in leaves of evergreen plants, such as Ilex Aquifolium, some Coniferae, and Vinca. The polished surface reflects a portion of the incident light from the leaves, and may thus be of use.[2] Cuticle is often provided with fine processes, especially when the external wall is convex. Vesque [3]

---

[1] Bergen, 1904 *b*.   [2] Wiesner, 1876 *b*.   [3] Vesque, 1882.

and Henslow[1] suggest that the arrangement serves to scatter and weaken the incident rays of light. Haberlandt,[2] on the contrary, regards these lenticular cells as organs for the perception of light.

Wax may be excreted over the surface and depress transpiration, as has been experimentally established by Tschirch[3] and Haberlandt. Usually the coating produced is only a thin one ; but, to take an opposite case, Capparis spinosa, at the commencement of the dry season in the Egyptian desert, produces over the whole leaf-surface a very thick layer of wax that completely prevents transpiration.[4] The coating of wax may be very thick, more than one millimetre in Sarcocaulon in South Africa, and up to five millimetres in Wax Palms. Incrustations of wax cause plant-members to have a dull, matt, bluish surface, which is then said to be covered with ' bloom '. Such ' bloom '-covered leaves usually have at their margins no sharp teeth, and possess, at most, rounded teeth provided with hydathodes. Wax prevents water from wetting leaves, so that it protects ombrophilous foliage from rain.[5]

Incrustations of salt are produced on the surface of some desert-plants, which thereby acquire a grey tint and are perhaps protected against excessive transpiration ; at night the incrustations deliquesce as they absorb moisture from the atmosphere.[6] In the Plumbaginaceae and certain species of Saxifraga the hydathodes which excrete calcic carbonate may possibly serve to check transpiration, but their main function would seem to be the excretion of injurious salts.

Varnished leaves. Resins or similar bodies are excreted by hairs on the surface of many, particularly austral, xerophytes. The leaves are thus rendered viscid and appear as if lacquered, since they acquire a glossy, vitreous investment ; the epidermal walls are thin and feebly cutinized.[7]

The creosote bush (Larrea tridentata) in the North-American deserts has leaves which, when unfolded, are thin, but which gradually become coated with shellac.[8]

Mucilage, or a mixture of gum, resin, and other bodies, is sometimes excreted by hairs (colleters[9]), in the buds of Polygonaceae and others : it may possibly aid in the absorption of water and perhaps check transpiration during the flushing of the foliage.

The *contents of epidermal cells* may be designed to depress transpiration. The epidermis is perhaps a water-reservoir[10] when, as is usually the case with land-plants, it is colourless. In virtue of various substances contained by it the epidermis may become less permeable to water-vapour. *Tannin* is often markedly present in the epidermis of evergreen leaves during winter,[11] and appears in connexion with the aqueous tissue of desert-plants and steppe-plants such as Alhagi, Monsonia, Astragalus, Tamarix[12] ; but the functional significance of these facts is obscure.

Anthocyan is a red pigment present in many plants, and particularly in the epidermis ; according to Engelmann and Stahl[13] it acts as a heat-

[1] Henslow, 1894.   [2] Haberlandt, 1905.   [3] Tschirch, 1882.
[4] Volkens, 1887.   [5] Burgerstein, 1904, p. 207.   [6] Compare pp. 31-2.
[7] Volkens, 1890.   [8] Coville, 1893, p. 51.   [9] Hanstein, 1868.
[10] Westermaier, 1882.   [11] Warming, 1884.
[12] Volkens, 1887 ; Henslow, 1894.   [13] Engelmann, 1887 ; Stahl, 1896.

absorbing substance and thus promotes transpiration, but according to others it does precisely the reverse.[1]

An *epidermis with mucilaginous inner walls*, which are gelatinous, occurs in many land-plants, and specially in woody plants such as Empetrum, Arbutus Unedo, and other Ericaceae. The mucilage possibly serves to depress transpiration ;[2] but it may perhaps function rather as a water-reservoir.[3]

A fact of perhaps supreme importance, and one that is the cause of some of the relations already mentioned in connexion with the epidermis, is that *wettable* plant-parts wither much more rapidly than *unwettable* parts. Wiesner regards the increase of transpiration in the former as due to a peculiar swelling of the cell-walls, which consequently oppose less resistance to evaporation. Many of the devices mentioned as decreasing transpiration also serve to prevent the plant-parts from being wetted, and in this way, too, prevent rapid transpiration.

Cork, in virtue of its air-filled cavities and its other characters, depresses transpiration, as has been proved by experiment. Its thickness is sometimes obviously and directly correlated with dryness of climate, as is illustrated by the difference between the trees of the Brazilian *campos* and of the adjoining forest. The desiccating action of fires occurring in the *campos* appears to stimulate the development of cork, and thus to provide an example of self-regulation.[4] Very thick investments of cork occur in a number of desert-plants, for instance, Dioscorea (Testudinaria) Elephantipes in South Africa, and Cocculus Leaeba in Egypt.

In the aerial roots of some Orchidaceae and Araceae a mechanism designed for absorbing water assumes the form of a *velamen*, which clothes the root with an envelope of cells, usually of several layers in thickness : the cells resemble the water-absorbing cells of Sphagnum ; they are thin-walled, with annular, spiral, or reticulate thickenings. When these cells are filled with air the velamen is white ; but when they are occupied by water the chlorophyll-containing tissue of the root becomes more or less visible. Liquid water is rapidly sucked up by the velamen, and can be transported to the conducting tissue. It is possible that water in the form of vapour may also be absorbed by the velamen. Wehmer, however, expresses another view, namely, that the velamen acts as a protection against transpiration. It is possible that both views are correct.[5]

Here it may be mentioned that the roots of many xerophytic land-plants produce a very strong endodermis, which probably protects against desiccation.

B.   Stomatal Transpiration and the Aerating System.

Intercellular spaces are also seats of transpiration ; the transpiring surface of a plant is constituted not only by the surface exposed to the external atmosphere, but also by the cell-walls bounding all intercellular spaces ; it may therefore be anticipated that the air-containing intercellular spaces of land-plants, and especially of xerophytes, will sharply

[1] See Chapter V, pp. 20–1.   [2] Volkens, 1890.
[3] Pfitzer, 1870–2 ; Radlkofer, 1875, p. 100 ; Vesque, 1884 ; Walliczeck, 1893 ; Westermaier, 1880 ; H. E. Petersen, 1908.
[4] Warming, 1892.   [5] Burgerstein, 1904, p. 69.

contrast with, and be narrower than, those of water-plants in which they are usually very large.[1] At the same time there is an extreme difference in regard to the stomata present.

(a) **Stomata.**

Stomata, as Leitgeb and Schwendener have proved, are adapted by their mobility and structure to regulate transpiration. They close when excessive transpiration is threatening, when leaves are withering because of lack of moisture in the soil, also while the leaves of many plants are resting during winter ; and they re-open when there is no further danger. The guard-cells of certain desert-plants are mobile only in young leaves; but in the old leaves they become immobile owing to strong thickening of their walls, and the stomata may become blocked with wax or resin.[2] Floating-leaves of water-plants have stomata on the surface exposed to the air, but these assume a peculiar form and soon lose their power of movement.[3]

The *number of stomata* depends upon the nature of the environment. As a general rule, the drier a habitat is the fewer are the stomata, as may be seen best when comparison is made between closely allied species.[4]

The *distribution of stomata* is most intimately connected with transpiration and with the conditions of moisture. Meadow-grasses and other mesophilous land-plants, as a rule, have stomata on both faces of the leaf ; steppe-grasses only on the furrowed upper face[5] ; other xerophytes usually only on the lower face, where the stomata are often concealed in such a way as to render transpiration more difficult.

Stomata of land-plants are frequently *sunk* beneath the general surface in pits, furrows, and the like, which are often lined with hairs. There is thus formed a space containing air, which escapes only with difficulty, is screened from the wind, and becomes charged with aqueous vapour ; the result is that transpiration is retarded. Various arrangements[6] of this kind are described in the succeeding paragraphs.

The simplest device is the production outside a *single stoma* of a saucer-, urn-, or funnel-shaped cavity, which is formed either by cuticular processes giving rise to the outer stomatal cavity, or by the adjoining epidermal cells projecting above the stoma, which is sunk within an external respiratory chamber,[6] as in Pinus sylvestris and some Proteaceae. In Euphorbia Paralias,[7] also in various grasses and sedges,[8] the stoma is surrounded by low papillae.

Groups of stomata in pits, whose narrow apertures are almost occluded by hairs, occur on the lower face of the leaves of Nerium, Banksia, and other xerophytes.

In numerous plants stomata are lodged in *longitudinal furrows*, and then are usually confined to the furrows, whose margins are often more or less beset with hairs. Many stems, especially switch-shaped ones, have deep furrows to which the stomata are limited, as is the case in Casuarina,

---

[1] See p. 98.   [2] Wilhelm, 1883 ; Volkens, 1890 ; Gilg, 1891.
[3] Haberlandt, 1904, p. 412.
[4] Pfitzer, 1870–2 ; Zingeler, 1873 ; Czech, 1869 ; Tschirch, 1881 ; Volkens, 1881 ; Altenkirch, 1894.
[5] Pfitzer, 1870–2.   [6] Tschirch, 1881, 1882.   [7] Giltay, 1886.
[8] Volkens, 1890 ; Kihlman, 1890 ; Raunkiär, 1895–9.

Ephedra, Acanthosicyos horrida, and species of Genista. The furrows occur on the *upper face* of the leaf in many steppe-grasses, and others, such as Weingaertneria canescens, Festuca ovina, Psamma (Ammophila) arenaria, Aristida, Stipa, Sporobolus spicatus, Cynodon Dactylon ; in such cases the furrows may be narrowed above, and the stomata more completely enclosed by the rolling up of the leaf.[1] Furrows or broader channels clothed with hairs occur on the *lower face* of the leaf in many plants, such as Empetrum, Phyllodoce caerulea, Calluna, species of Erica, Loiseleuria procumbens, Ledum palustre, Cassiope tetragona,[2] and Dilleniaceae.[3] To this category may be added leaves, such as those of Dryas octopetala, with margins revolute to a smaller extent and with stomata on their hairy under-surfaces.

If leaves are permanently and steeply directed upwards as well as appressed, so that the lower face is the more strongly illuminated, then the lower side may be differentiated like the normal upper side and possess palisade tissue ; in such cases the stomatiferous furrow occurs on the upper face, as in Passerina filiformis, Ozothamnus, and Lepidophyllum.[4] In these plants, therefore, the easy egress of aqueous vapour is checked in more than one way.

That these features stand in direct relation to dryness of climate is shown by species such as Ledum palustre and Andromeda polifolia, whose leaves are smaller and more revolute the more they are exposed to wind and drought.[5]

The cases last mentioned form transitions to flat, broad leaves in which the sole screen over the stomata is formed by a dense investment of felted or peltate hairs, as in Olea, Rhododendron, and Elaeagnaceae, or some other kind of tomentum on the lower face of the leaf.[6] Sometimes leaves of this kind have veins strongly projecting on the lower face, and as the stomata lie in the meshes of the network of veins, they are to a certain extent sunk below the general surface, as for example in the West Indian Lantana involucrata.

When stomata are in secluded cavities containing much aqueous vapour, or lie under a dense tomentum, they are usually raised above the adjoining surface, just as in the leaves of plants that live as a whole in contact with moist air.

It may also be noted that stomata are enclosed in cavities, or sheltered under tomenta, in order that they may be protected from occlusion by water.[7]

(b) Intercellular spaces.

Respiratory cavities may be structurally fitted to regulate transpiration ; they may have cuticularized walls, or be surrounded by special cells as in the Restiaceae,[8] or may be very small. In many instances cuticle extends from the outer face of the epidermis, through the stoma, and down over the walls of the respiratory cavity.[9]

The width of intercellular spaces varies with the external conditions ;

[1] See p. 267.
[2] Warming, 1889 ; Gruber, 1882 ; Ljungström, 1883 ; H. E. Petersen, 1908.
[3] Steppuhn, 1895.   [4] Lazniewski, 1896 ; Göbel, 1891.
[5] Warming, 1887, p. 110.   [6] See p. 254.
[7] Kerner, 1887.   [8] Pfitzer, 1870-2.   [9] A. de Bary, 1877, p. 79.

intercellular spaces are larger in shade-leaves (sciophylls) than in sun-leaves (heliophylls), better developed in moist than dry air. But for reasons already given it is generally true that in xerophytes growing on land the air-containing intercellular spaces are *very narrow* ; in this respect Altenkirch's[1] measurements of respiratory cavities may be mentioned. But exceptions occur ; for instance, in Restiaceae there are wide air-spaces, which possibly play a part in the assimilation of carbon dioxide, in addition to very narrow ' girdle-canals '.

' Girdle-canals ' also occur in the Australian desert-plant Hakea suaveolens, as well as in Olea europaea, Kingia,[2] and in some arenicolous grasses such as Festuca rubra and Triticum acutum.[3] They are narrow intercellular spaces running round the palisade cells parallel to the leaf-surface ; by this tortuous course the escape of water-vapour is rendered more difficult. Certain desert-plants, such as Cynodon Dactylon, and Sporobolus spicatus, have a maze of extremely fine meandering inter-cellular canals,[4] but it is not certain that these various forms of intercellular spaces are designed to depress transpiration.

C. Chlorenchyma.

It is characteristic of land-plants as opposed to submerged water-plants to possess dorsi-ventral leaves, and in particular palisade tissue. The latter is greatly developed in xerophytes by an increase in either the number of layers or in the height of the cells, or by both means. It has already been mentioned [5] that there is a difference of opinion as to the significance and cause of this structural feature, and the suggestion has been made that it is most closely correlated with dryness of the atmosphere and with transpiration. Light doubtless also plays a part in the matter, for the oblique orientation of palisade cells must be due to illumination.[6] In halophytes growing on land the height of the palisade tissue is increased by the salts contained in the soil, as has been proved by Lesage.

D. Other Means of Regulating Transpiration.

Ethereal oils occur especially in xerophytes ; the *garigues* and *maquis* of Mediterranean countries,[7] and the *campos* of Brazil are scented with Cistus, Labiatae, Verbenaceae, Compositae, and Myrtaceae, just as are European downs with wild thyme, and Asiatic steppes with Artemisia. Neither the origin nor the significance of the correlation between dryness of climate or of soil and the occurrence of ethereal oil has yet been explained. These oils evaporate more readily than water, and surround the plant with aromatic air. According to Tyndall, air rich in ethereal oil is less diathermanous, that is to say, permits the passage of radiant heat to a less extent than does pure air ; according to this view, ethereal oils diminish insolation and consequently transpiration.[8]

It is possible that ethereal oils are of utility in other directions ; they

[1] Altenkirch, 1894.   [2] According to Tschirch.
[3] According to Giltay, 1886.   [4] Volkens, 1887.   [5] See p. 21.
[6] See Warming, 1897.   [7] See Beck von Mannagetta, 1901, and others.
[8] Volkens, loc. cit., and others.

may, for instance, protect plants against herbivorous animals, as Stahl[1] suggests. Detto[2] doubts the correctness of the view that ethereal oils serve as a screen against transpiration, and takes the same view as Stahl.

The significance of *latex* is not definitely established. According to Haberlandt, Schullerus, Pirotta, and others, laticiferous tubes are conducting channels, but according to Kerner they form a defence against animals at least in Cichoriaceae.[3]

By the various agencies described above, the transpiration of leaves is brought into harmony with the different environments. But it must not be concluded that xerophytic leaf-structure is inconsistent with capacity for vigorous transpiration : indeed Bergen[4] found that the absolute amount of transpiration (the amount of water lost in a unit of time) was scarcely less in the sun-leaves (heliophylls) of certain evergreens, including Olea europaea and Quercus Ilex, than in Ulmus campestris and Pisum sativum.

II. DIMINUTION OF THE EVAPORATING SURFACE

The extent of the transpiring surface plays an important part in determining the amount of transpiration : other relations being constant, the larger the surface the greater the transpiration. As *foliage-leaves* are essentially the organs of transpiration, it is their *size* and *number* which regulate this function and which therefore vary in the different species in accordance with climatic conditions. Divers means adopted to depress transpiration are treated in the succeeding paragraphs.

A. Temporary Diminution of Surface.

The most decisive method by which a plant can diminish its transpiring surface is the shedding of all strongly transpiring parts before the commencement of the dry season. This takes place, first, in all *annuals* which die after the seed has ripened : all seeds are very efficiently protected from desiccation. In harmony with this is the very high percentage of ephemeral species in deserts and similar places ; within the short rainy season, sometimes only from one to two months in length, these plants complete their whole life-cycle, germinating, flowering, setting seed, and dying, so that they pass through the dry season in the form of embryos enclosed in seeds ; Odontospermum (Asteriscus) pygmaeum, the ' Rose of Jericho ', is such a plant.[5]

Similar behaviour characterizes all *bulbous* and *tuberous* plants, as well as other ' renascent ' herbs whose subterranean shoots serve as reservoirs of food and water during the dry season : the epigeous shoots, with their extensive transpiring surfaces, are dispensed with while drought prevails, and the latent vitality is confined to the underground shoots. When moisture is once more supplied these species hasten to thrust new shoots and flowers into the light.[6] In fact, the rapid onset of spring after the first few showers of rain in deserts, steppes, and similar places, has often been mentioned with surprise by travellers.

In like manner behave *woody plants* which shed their foliage before

[1] But see Burgerstein, 1904, pp. 133, 214.   [2] Detto, 1903.
[3] See also p. 125.   [4] Bergen, 1904.   [5] Volkens, 1878.
[6] See p. 8.

or during the dry or cold season, and remain leafless for a long time ; such species are described as *deciduous*, in opposition to *evergreen*. In these plants during the unfavourable season all parts above ground are usually protected from transpiration by means of cork or bud-scales, the latter being covered with cork or other bodies that check evaporation.

In all these plants the structure of the *foliage-leaf* is usually not at all or only slightly xerophytic, but is mesophytic, if the vegetative season be sufficiently moist. In Egypt[1] and in the lowland of Madeira,[2] where the atmospheric humidity is small even in winter, annual herbs growing on uncultivated land adopt protective measures against drought quite different from those employed by weeds in the irrigated fields. Protection against drought is more pronounced the more a species prolongs its vegetative season beyond the commencement of the dry season. According to Kerner[3] the foliage of deciduous trees on the Austrian coast is very hairy on the under-side, because the summer is exceedingly dry.

The transpiring surface is reduced in quite a different manner in other plants, for example, in grasses whose leaves in dry weather become *rolled up*, so that they form tubes, and thus appear filiform or bristle-like. Such is the case with Psamma (Ammophila) arenaria, Weingaert-neria (Corynephorus) canescens, species of Festuca, and many other grasses inhabiting dune or heath ; and, in Mediterranean countries, with species of Stipa, Lygeum, Aristida[4] ; rolled-up leaves are particularly characteristic of steppe-grasses, and are also met with on saline soil in Triticum junceum and other grasses. As the air becomes drier the leaf rolls up so that the transpiring upper surface, where the stomata mainly or solely occur, is less exposed to transpiration ; the stomata thus become enclosed in a space in which the air is more or less motion-less. In moist weather the leaf unfurls. Among Cyperaceae similar though less considerable movements are exhibited. In these movements a part is played by the hinge-cells lying in furrows on the upper face of the leaves of grasses ; these cells are deeper than the other epidermal cells, and their cellulose walls are easily folded as the leaf curls. The motive force would appear to reside in the bast-tissue, which is usually near the under-face of the leaf, and either absorbs or gives out water, thus swelling or contracting. But the turgor of the mesophyll seems to play an important part, at least in some cases.[5]

Similar movements are exhibited by a number of Dicotyledones, including Hieracium Pilosella, Antennaria dioica, Crepis tectorum,[6] West-Indian species of Croton,[7] and Euphorbia Paralias,[8] which grows on the dunes of western and southern Europe. The leaves of Erica Tetralix,[9] and Ledum palustre are less rolled on moist than on dry soil. Among Cryptogamia may be mentioned some ferns[10] and mosses, including species of Rhacomitrium, Tortula, and Polytrichum. The leaves of Rhacomitrium canescens and Tortula ruralis in dry weather are folded together, and the shoots quite grey with densely set, long hairs ; but when the weather or soil is humid they are extended in a

[1] Volkens, 1887.   [2] Vahl, 1907 b.   [3] Kerner, 1886.
[4] See Duval-Jouve, 1875 ; Tschirch, 1882 ; Warming, 1891.
[5] Duval-Jouve, 1875 ; Tschirch, 1882.   [6] Wille, 1887.
[7] Warming, 1899 b.   [8] Giltay, 1886.   [9] Gräbner, 1895.
[10] See Wittrock, 1891.

stellate manner. Polytrichum can lay the marginal portion of the leaf over the thin-walled assimilatory cells clothing the more central portion.[1]

B. Permanent Reduction of Form of Leaf and Shoot.

In very many xerophytes the transpiring organs, the *foliage-leaves*, are extremely and unalterably reduced in size and surface, and there result a number of specialized types of xerophytic shoots. The size of the foliage-leaf, and of the foliaged shoot as a whole, shows a certain dependence upon the amount of food-material and water available to the plant at the time of its development. Lack of water may induce *nanism ;* for instance, in dry sandy places many species are dwarfed ; again, one and the same species may be small-leaved on dry soil, and large-leaved on moist soil, as is the case with Urtica dioica, Viola canina, Erodium cicutarium and many others ; a number of desert - plants, including Zilla and Alhagi, produce at the commencement of the rainy season large leaves, but later on much smaller ones or none at all. The smallness of the leaf is a *direct* result of dryness.[2] Lack of water has apparently also contributed to the evolution of a *series of definite fixed and constant types*, which are characterized by their relatively low assimilatory power and consequent slowness of growth.[3] These types are described in the succeeding paragraphs.

(a) Forms of Leaf.

1. The *pinoid or acicular leaf* is met with in Coniferae, Proteaceae, Ulex europaeus, and others. It is long, linear, pointed, and often has a more or less radial structure. The relations of this leaf to transpiration result from the fact that its surface in proportion to its volume is much less than in the case of a flat leaf, hence its evaporating surface is relatively less. This is also true of the forms of leaves described in the succeeding paragraphs.

2. The *ericoid* leaf is a rolled leaf, in other words, its margins are curled downwards, or upwards (much more rarely, as in Passerina) ; there thus arises a furrow in which the stomata are secluded from movements of the air.[4]   Ericoid leaves are short and linear ; they occur in Erica, Calluna, Cassiope tetragona and other Ericaceae, Epacridaceae, Myrtaceae, Berberis empetrifolia from Chile, South-African Thymelaeaceae, Compositae, Rubiaceae, and among other families in species growing on *maquis*, heaths, or other places where transpiration is strong.

3. The *cupressoid* (lepidophyllous, lepidoid) leaf is broad and short, appressed, apically directed, and sometimes decurrent ; it is met with in many Cupressaceae, in some Scrophulariaceae (in Veronica thuyoides and V. cupressoides, which are alpine in New Zealand), Santalaceae, Tamaricaceae, Compositae, Umbelliferae (in Azorella at alpine altitudes in South America, and in Antarctic lands).[5]

4. The *setaceous* or *filiform* leaf occurs in very many grass-like Monocotyledones ; it is usually furrowed or channelled on its upper face, and

---

[1] Kerner, 1887.   [2] Henslow, 1894; Scott-Elliot, 1905 ; Percy Groom, 1893.
[3] See Chap. XCIX.   [4] See p. 105.   [5] Göbel, 1891 ; Lazniewski, 1896.

conceals its stomata in hairy furrows. Movements associated with changes in humidity are met with—for instance, in Festuca ovina, Corynephorus canescens, many grasses growing in deserts, steppes, or on high mountains.[1] Dissected leaves, such as those of Artemisia campestris, often possess very similar small, terete segments.

5. The *juncoid* leaf is long, terete, devoid of furrows, and is seen in species of Juncus, a number of Cyperaceae, and some alpine Umbelliferae in South America. This form of leaf is mostly met with on wet, cold, acid soil that is exposed to wind.[2]

6. The *succulent* leaf may be mentioned here because, apart from its thickness, it is often more or less terete, linear, oblong, or spathulate, devoid of teeth or other indentations ; examples are provided by Sedum acre, Sempervivum tectorum and other Crassulaceae, species of Mesembryanthemum, Batis maritima, and other halophytes, and some Orchidaceae.[3] This form of leaf is characterized by the relative smallness of the surface in comparison with the volume. Henslow's view that succulence is due to the direct action of environment is probably correct.[4]

7. The *sclerophyllous* or *myrtoid* leaf. There are many other forms of leaves not belonging to any of the preceding types, yet adapted to resist excessive transpiration ; among these may be specially noted the leaves of plants that Schimper describes as being ' stiff-leaved ' or sclerophyllous.[5] The leaves may be small (as in Loiseleuria procumbens and Diapensia) ; or narrow and stiff, more or less revolute (as in Lavandula, Hyssopus, and other Mediterranean species) ; or broader (as in Myrtus communis Nerium, Olea, Rhododendron), obovate, oblong, elliptical, lanceolate, or of some other simple form, devoid of teeth or other indentations. They are flat, coriaceous, and stiff, largely owing to the thick-walled epidermis, and are evergreen. To protect themselves against excessive transpiration, such leaves usually have additional contrivances which will be described hereafter.[6] Here we may include, too, the leaf-like cladodes of Ruscus aculeatus and other species, also of Semele androgyna.

The shoots possessing leaves of the forms enumerated, and especially the pinoid, ericoid, and cupressoid, are usually extremely rich in leaves. By increasing the number of leaves the plant strives to compensate for the decreased assimilation that is caused by reduction in their size. Furthermore, possibly the close aggregation of the leaves on shoots with short segments may itself retard transpiration.

(b) Forms of Shoots.

Shoots with greatly reduced or caducous leaves occur in connexion with many xerophytes. The foliage-leaf has vanished, and its functions have been taken over by the stem, which produces palisade parenchyma. The epidermis of the stem in this case functions for a number of years. Such aphyllous shoots show several forms :—

1. The *winged* shoot is often aphyllous, and the light strikes its assimilatory tissue at acute angles.[7]

2. The *switch*, or *Spartium* form of shoot, is switch-shaped, erect,

---

[1] See p. 109.   [2] See Göbel, 1889–93, Bd. II.   [3] See Warming, 1897.
[4] See p. 371.   [5] Schimper, 1898.
[6] For further information see papers by Vesque, Volkens, Göbel, Warming, Henslow, and Schimper.   [7] See pp. 19, 113, 114.

slender, and often copiously branched; the leaves of some species, such as Genista tinctoria and Spartium junceum, are of relatively considerable size, but fall off early when once they have performed their assimilatory functions; but the leaves of other species are from the outset very reduced in form and function. The stem is terete, or deeply furrowed with stomata and palisade in the furrows and mechanical tissue in the ridges. This form of shoot is very common in Mediterranean Leguminosae, particularly in Genista, Retama, Cytisus, and Genisteae in general, also in Casuarina, Ephedra, a number of Chenopodiaceae, for example in Anabasis (which, however, is mainly halophytic), in Capparis aphylla, Periploca aphylla, and Polygonum equisetiforme.[1]

3. The *juncoid* shoot, as represented by many species of Juncus and Cyperaceae, is tall, terete, aphyllous, and unbranched, being in form similar to the leaf of some of the same species. The relative proportions of surface and volume in such a shoot have already been explained. This form of shoot also occurs in numerous marsh-plants, such as Scirpus lacustris and S. palustris, Junci genuini, and others belonging to the same families.[2] The Restiaceae also include shoots belonging to this type.

4. *Acicular cladodes* of Asparagus.

5. The *flattened* shoot is an aphyllous form which is upright, or exposes its profile: as examples may be cited Muehlenbeckia platyclada, Phyllocladus, and Carmichaelia australis. The stem in some cases (Ruscus, Semele) is so leaf-like that it is preferably included under the category of sclerophylls, a course that has been adopted above.

6. The *spinose shoot* of Colletia and others.

7. The *salicornioid* shoot, as represented by Salicornia, Arthrocnemum, and other Chenopodiaceae.

8. The *cactiform* shoot is met with under various forms in Cactaceae, Euphorbia, and Stapelia. It will be referred to subsequently in connexion with succulent-stemmed plants.[3]

III. REGULATION OF ILLUMINATION

As light has a heating effect upon the plant and thus promotes transpiration, and as intense light is injurious to chlorophyll, many land-plants possess devices by the aid of which the assimilating organs avoid too intense illumination. These devices are temporary or permanent in nature:—

A. Movements by which Illumination is regulated.

Many plants have an extremely delicate power of appreciating the intensity of light, and can regulate the amount falling upon them by movements of leaflets, which *place their blades at definite angles to the incident rays*, the intensity of which determines the precise angle. When the light is moderate, as in the early morning, the blades are exposed as fully as possible to the light, whose rays strike them approximately at right angles. But when the light becomes stronger the leaves more and more assume a profile-lie, so that the angle of incidence becomes

---

[1] See the cited works of Pick, 1881 ; Volkens, 1887 ; Schube, 1885 ; Ross, 1887 ; Nilsson, 1887 ; Kerner, 1887 ; Schimper, 1898.
[2] See Chaps. XLIII, XLV.   [3] See p. 123.

increasingly acute ; they are thus relatively less illuminated and heated, and transpiration is inevitably decreased. These movements are executed by the compound leaves of numerous plants, and particularly of some growing in tropical, dry bushlands ; among such are many species of Acacia, and other Mimoseae, Papilionaceae, Oxalidaceae (including Oxalis Acetosella), Zygophyllaceae ; but simple leaves of some plants, including Hura crepitans, likewise execute movements dependent on the intensity of light.[1] The leaves of the plants in question are wont to be more or less mesophytic in structure, for example, the leaflets of species of Acacia in the West Indies ; acacias endowed with the power of movement in response to intensity of light are often or always thin, and have a smooth, thin epidermis.[2]

Temporary vertical positions that must be of utility to the organs concerned are met with in connexion with many or most young developing leaves ; for as they shoot from the bud they are vertical, and sometimes remarkably so.

B.   Fixed Lie in Relation to Light.

All foliage leaves as they unfold execute movements as a result of which they assume a favourable fixed lie, in regard to which Wiesner has conducted investigations for many years.[3] They place their blades *perpendicular* to the strongest *diffuse light*. In unusual circumstances, when more intense light prevails, they assume a *profile*-lie. A diminution in the action of the sunlight, and consequently in transpiration, will result from a permanent profile-pose or other similar arrangements on the part of assimilating surfaces, which thus receive the intense midday light at *acute* angles. This is the case with the so-called ' compass-plants ', represented in northern Europe by Lactuca Scariola, the leaves of which, in places exposed to strong sunlight, place themselves in the meridian with their faces vertical[4] ; as a North American ' compass-plant ', Silphium laciniatum may be mentioned.

Leaves exposing their edges to the light are met with in many other plants, for instance, in some species of Eucalyptus, phyllodinous species of Acacia, and Proteaceae, in Australia ; species of Statice in South Africa ; Laguncularia racemosa in the West Indies ; and Bupleurum verticale in Spain.

Leaf-blades that are *erect*, or directed sharply upwards, are common among xerophytes growing in intense sunlight : for instance, in the West Indian Coccoloba uvifera,[5] in many grasses (including Brachypodium ramosum, Festuca ovina), in Calluna, Peucedanum Cervaria,[6] and Helichrysum arenarium ; among marsh- and moor-plants may be named Iris Pseudacorus, Narthecium ossifragum, and Tofieldia ; among halophytes, Rhizophora and other mangrove plants.[7] More rarely, blades hanging vertically downwards are met with. The switch-plants may again be mentioned as being designed upon a very similar plan.

Corrugations and folds in the lamina may play the same part, and become more frequent the drier the climate is : as examples may be

---

[1] See C. Darwin, 1880.   [2] Warming, 1899 b.   [3] Wiesner, 1876.
[4] Stahl, 1880–81.   [5] Illustrations in Börgesen and O. Paulsen, 1900.
[6] According to Altenkirch, 1894.   [7] Illustrations in Joh. Schmidt, 1903.

cited Myrtus bullata in New Zealand, Lippia involucrata and Plumeria alba in the West Indies[1], Salvia, Stachys aegyptiaca, Pulicaria and Urginea undulata in the Egyptian desert,[2] also Vicia Cracca in Europe.

The lie in all the cases mentioned above is attained by means of torsions, folds, and curvatures that take place only during the development of the individual ; hence, in all kinds of plants which assume the shapes here described, the lie of the leaves varies with the nature of the habitat. Exposed to sunlight, drought, or wind, the leaves become more erect, or have their faces more vertical, or become more crumpled, than when either in the shade or in a humid habitat where the air is moist. This is the case with Calluna, Juniperus communis, Lycopodium Selago and L. alpinum.[3] The leaves of Tilia argentea on which hot sunlight falls expose their edges to the light, but the remaining leaves present their flat faces.[4]

Hereditary profile-lie is met with in Australian phyllodinous acacias, the blade-like petioles of which have their faces vertical, but bear no lamina ; also in many plants with *flattened or winged stems, or with decurrent leaves*, as in Baccharis triptera in Brazil, Genista sagittalis, Muehlen-beckia platyclada, Carmichaelia australis, and species of Colletia. These forms of shoots are usually aphyllous ; stem replaces the leaves. In this connexion must be mentioned the ensiform leaves of Iridaceae, Tofieldia and Narthecium.

IV. INVESTING ORGANS IN THE CONTROL OF TRANSPIRATION

It is evident that transpiration will be very materially reduced when the transpiring surface is clothed by air-containing bodies, in and between which the air is so firmly lodged that its circulation is obstructed.[5] This method is adopted in various ways by many land-plants.[6]

A. Investing Hairs.

In regard to hairiness the contrast between hydrophytes and xerophytes is especially marked : the former are glabrous, the latter often clothed with grey or white cottony and woolly hairs, or by glistening silky hairs ; the optical effects are connected with the presence of air in and between the hairs. Only *dead air-containing hairs* are fitted to perform the function in question. In form, these hairs are extremely diversified.[7] It has long been known that species which are elsewhere glabrous become hairy in dry places, and that hairy species become more hairy in dry than in moist sites, as is exemplified by Ranunculus bulbosus, Polygonum Persicaria, Mentha arvensis, and Stachys palustris ; moreover, etiolated potato-shoots are nearly glabrous in moist, but hairy in dry air.[8] Woolly hairs coat many plants growing on rocks in Mediterranean countries (Corsica,[9] for instance), in the dry bushlands of the West Indies, in the desert, steppe, or alpine situations.[10] The most thickly felted plant perhaps is Espeletia,[11] one of the Compositae, living on high mountains

---

[1] Johow, 1884.   [2] Volkens, 1887.

[3] See illustrations in Warming, 1887.   [4] Kerner, 1887-91.

[5] See Haberlandt, 1904, p. 111.   [6] See Burgerstein, 1904, p. 208.

[7] See illustrations in Kerner, 1887-91.   [8] Vesque et Viet, 1881.

[9] Rikli, 1903.   [10] See Lazniewski, 1896; Göbel, 1889-93, Bd. ii.

[11] See illustrations in Göbel, loc. cit.

in South America. The woolly coat acts as a sunshade, and serves to moderate changes in temperature, also to reduce transpiration. As a particular form of hair may be mentioned the scaly hair, which when abundant lends a metallic lustre to plants such as the Elaeagnaceae, also some species of Croton and Styrax. The coating of hairs is almost invariably densest on the lower face of the leaf where the stomata occur. Young stems and leaves are frequently densely coated with hair, more densely than when older, as at the former stage they have greater need of protection against intense transpiration. Sometimes, in dry parts of tropical countries, the leaves produced just after the dry season and those developed later on differ widely in appearance, the former being more hairy, and the latter larger and greener.[1] But one group of xerophytes, namely, succulent plants such as Cacteae, species of Aloë and Agave, are usually smooth and quite devoid of any coating of hairs ; in this case other protective measures are adopted.

The production of hairs, like all other self-regulatory devices of the plant, is possibly a *direct* adaptation to external conditions. According to Vesque, hairiness and dryness of the atmosphere increase side by side. Following Mer, Henslow[2] attributes the production of hairs to local supply of nutriment, which is correlated with suppression of paren-chyma ; according to him, the more the parenchyma is checked the greater is the compensatory production of hairs. But even if this hypothesis be correct it does not carry us much nearer to the compre-hension of the correlation between hairiness and dryness.

B.   Investing Leaves.

The young parts of the shoot are usually protected against intense transpiration and intense light by older leaves. It is a quite general phenomenon for the youngest foliage-leaves to be protected by older ones in so-called ' open ' buds ; on the other hand, there are numerous buds that are provided with thick *bud-scales*, such as are met with in deciduous woody plants, not only in temperate and frigid countries, but also, though less frequently, in the tropics.[3] By the production of cork, hairs, resin, and the like, they are adapted not only to protect the young leaves resting within the bud from transpiration, but also during foliation to guard the buds against change of temperature.[4] In certain climates bud-scales are rare ; for instance, Coville[5] writes in reference to the Death Valley, ' scaly buds are almost unknown in the desert shrubs '. The same is true of Mediterranean countries, where the rainfall takes place in winter, and of the tropical rain-forest.[6] The young bud-parts of many xerophytic mosses are protected by white hairs that clothe the tips of the old leaves.[7]

Stipules and *leaf-sheaths* (the latter, for instance, sheltering the young inflorescences of dune-grasses) may perform the same service, though they are not strictly included in the category of bud-scales ;[8] in this way the

[1] H. Schinz, 1893.   [2] Henslow, 1894, 1895.
[3] Illustrations by Warming, 1892.
[4] Grüss, 1892 ; Feist; Cadura; Percy Groom, 1893.
[5] Coville, 1893, p. 53.   [6] Schimper, 1898 (1903), pp. 329–51).
[7] See pp. 109–10.   [8] See illustrations in Warming, 1907–9.

membranous stipules of species of Paronychia, Herniaria, and other plants, clothe young parts of the shoot with a dense silvery investment. *Old leaves and remnants of leaves* in many cases act in the same manner. 'Tunic-grasses' is the term employed by Hackel[1] to designate those grasses in which the lower parts of the leaves remain attached long after their upper parts have died, persisting either as coherent, firmly closed sheaths, or in a macerated condition. These tunics depress transpiration, and store water; they occur in grasses growing on dune, steppe, or desert, for example, in Nardus stricta, Andropogon villosus, Scirpus paradoxus, S. Warmingii, species of Aristida.[2] A similar relationship exists in the Velloziaceae living on the mountain-tops and high plateaux of Brazil.[3] In certain South African species of Oxalis the bulbs are invested by peculiarly constructed leaves[4]; the dead bulb-scales of Tulipa praecox bear a dense felt of hairs. Here we may mention the compact clumps, such as those of the 'cushion-plants' Raoulia and Azorella, consisting of closely-packed shoots and remains of shoots; they are met with in subglacial vegetation, and especially in South America, and are often so hard that it is difficult to cut or break them; in this case one shoot protects another—the old leaves protect the young.

Many other methods are adopted to protect the youngest part of the stem and leaves, and are described in the papers cited.[5]

The roots of many epiphytes, including the asclepiadaceous Conchophyllum imbricatum, are screened from excessive transpiration by leaves, which cover them closely, and keep them surrounded with moist air.[6] The roots of some grasses living in the Egyptian desert, for instance, species of Aristida, Andropogon, Elionurus, Panicum, and Sporobolus, are surrounded throughout their length by a sheath of sand, the grains of which are glued together by an adhesive substance excreted by the root-hairs. To a less marked extent the same is true of dune-grasses in northern Europe.[7] Volkens[8] interprets this as a device for checking evaporation.

V. THE ABLATION OF RAIN-WATER

It is of importance to the land-plant that its leaves shall not remain too long wet with rain-water; it is necessary for their surfaces to dry quickly, if transpiration is to be resumed.[9] And there seem to be adaptations serving to carry rain-water rapidly away. It is in the tropical rain-forest that such devices are especially met with, though they are perhaps not entirely lacking in temperate countries. Jungner, working in the rainy Kamerun, and subsequently Stahl in Java, arrived at essentially the same conclusions. They regard as adaptations:—

1. A smooth *cuticle* that cannot be wet; this device is very widespread.

2. *Drip-tip*, as Stahl terms the long leaf-tip terminating a blade that often becomes suddenly narrow; it is typically represented in

[1] Hackel, 1890.
[2] Hackel, 1890.   See also illustrations in Warming, 1892.   Henslow, 1894.
[3] Warming, 1893.   [4] Hildebrand, 1884.
[5] Also see Lubbock, 1899.   [6] Göbel, 1899.   [7] Warming, 1907–9.
[8] Volkens, 1887.   [9] Concerning ombrophobous and ombrophilous plants, see p. 32.

Ficus religiosa, but also in the most diverse plants (ferns, Monocotyledones, Dicotyledones), both in simple and compound leaves. The
drip-tip serves rapidly to conduct the rain off leaves that are capable
of being wet. Drip-tips are downwardly directed ; and the longer
the tip, the more rapidly does the leaf rid its surface of water. The
sabre-like tip leads water away most rapidly, apparently at times in
an almost continuous jet. Drip-tips are found neither on leaves that
are incapable of being wet, nor among xerophytes.

3. *Furrowed nerves* that conduct superficial water to the leaf-tip
are common. The arcuate course of the nerves in Melastomaceae and
others is thus of additional use.

4. *Velvety leaves* are especially encountered among herbaceous species
growing on the ground in forest, also among species forming the lower
storey of the forest, where shade and moisture are at their greatest. The
epidermal cells project in the form of countless short papillae, which give
to the leaf a velvet-like appearance, and produce a fine capillary system
in which the water spreads over the whole blade as a thin film. The
consequence is that water can evaporate much more rapidly than if
it were not spread out in this manner. But it has been suggested that
these papillae also serve to supply the leaf with an increased amount
of light, or act as light-perceiving organs.[1]

CHAPTER XXXI. ABSORPTION OF WATER BY LAND-PLANTS

SUBMERGED water-plants, or their overwhelming majority, exhibit
no organs specially adapted for the absorption of water, whereas the
opposite is the case with land-plants. These possess adaptations that
are described in the succeeding paragraphs.

i. Hypogeous Organs that Absorb Water.

Subterranean organs in the form of roots, rhizoids, and mycelia are
designed for the absorption of water ; so likewise are some rhizomes
that have absorbing hairs, like those of Corallorrhiza, Epipogum, Equisetum, Psilotum, and Hymenophyllaceae. In xerophilous land-plants
only few deviations from the normal type occur. Many xerophytes
possess deeply-descending roots, which aid them in securing water at
great depths during dry periods. This has been observed in the desert
of Afghanistan in species of Astragalus[2] ; in the Egyptian desert in the
colocynth, whose thin leaves would wither rapidly were it not for the
deep roots, in Calligonum comosum, and in Monsonia nivea. Volkens[3]
observed roots in the Egyptian desert that were twenty times as long
as the epigeous organs. The same features are to be seen in plants of
the European dunes—for example, in Eryngium maritimum and Carex
arenaria ; the latter has two kinds of roots—those that are very slender,
branched, and lie near the surface, and others that are less branched,

---

[1] See Haberlandt, 1905.   [2] Aitchison, 1887.   [3] Volkens, 1887.

and descend to a great depth.[1] Some plants in Hereroland, Acanthosicyos[2] for example, possess a specially large root-system which serves to raise the subterranean water that lies very deep. Tall perennial herbs of the Hungarian steppe are extraordinarily deep-rooted.

A peculiar device for the absorption of water is met with in the North African halfa-grass, Stipa tenacissima, the rhizome of which has peculiar epidermal cells, whose function is to absorb water.[3]

ii. Epigeous Organs that absorb Water.

In general, the epigeous parts of land-plants are not fitted to supply the plant with water by absorption, since the more or less impermeable cuticle of the epidermis will permit them to take in water only to a slight extent.[4] In ordinary phanerogams the amount of water that can be taken in thus is insufficient to compensate for the loss by transpiration. Exceptions to this are provided by lichens, mosses, and other thallophytes, which can endure prolonged desiccation and can rapidly absorb liquid water by their whole surface and may store it up ;[5] even aqueous vapour can be withdrawn from the air by many of them.

To many other land-plants living exposed to periodically extreme drought it is of great importance that they should be able to seize the moment, often fleeting, when water is available ; and, as a matter of fact, there are devices enabling epigeous parts to absorb water with ease and rapidity.

Hairs that absorb water were shown by Volkens[6] to occur on certain desert-plants, such as Diplotaxis Harra, Stachys aegyptiaca, and Convolvulus lanatus ; and by Schimper[7] in certain epiphytes, including Tillandsia and other Bromeliaceae. These hairs are not cuticularized at their base, and it is at this point that the water enters. The subject has been investigated by Mez,[8] who regards some Bromeliaceae as adapted to absorb dew, and others to absorb rain. The numerous white hairs of cacti may subserve the same function.[9]

The same rôle has been ascribed to *salt-glands*, which Volkens[10] discovered in the form of characteristic hairs on the leaves of various desert-plants, including Reaumuria hirtella, Tamarix, Cressa cretica, Frankenia pulverulenta, and Statice aphylla. These glands excrete solutions of hygroscopic salts (chlorides of sodium, calcium, and magnesium), which solidify during the day and impart to the plant-parts a white or grey appearance ; at night-time the salt deliquesces because of the increase in atmospheric humidity, and the parts concerned again become green and dotted with numerous drops of solution, even though there may have been no deposit of dew. Volkens expresses the opinion that the plants are thus enabled to absorb water. But Marloth[11] regards the

[1] Warming, 1891, 1907–09 (see illustration).
[2] Schinz, 1893.   [3] Trabut, 1888.
[4] Ganong, 1894 ; Wille, 1887 ; see Chapter VII.
[5] The case of Sphagnum is discussed in connexion with bogs. See Chapter XLIX.
[6] Volkens, 1887.   [7] Schimper, 1884.   [8] Mez, 1904 *a*.
[9] As regards hairs which in temperate Europe are reputed to absorb water reference should be made to Lundström, 1884 ; Wille, 1887, and Henslow ; and, as regards the structure and function of hydathodes to Haberlandt, 1904; see p. 101.
[10] Volkens, 1887.   [11] Marloth, 1887 *a*.

incrustation of salts as a coating that decreases transpiration, and suggests that in this way the plants rid themselves of a portion of the salts absorbed; and this view is adopted by Haberlandt.[1]

The *velamen* of Orchidaceae and Araceae has already been discussed.[2] In sundry epiphytic ferns and Araceae the aerial roots remain short, grow more or less vertically, and collect among themselves humus, and consequently water.[3]

Felted investments formed by roots or remnants of leaves, or both, occur in ferns such as Dicksonia antarctica, species of Alsophila, as well as in Velloziaceae, and palms. Some of the plants concerned are obviously xerophytes, and the investment serves not only as a means of protection against transpiration, but also as a device for collecting and storing water.[4] According to Buchenau the same is true of the juncaceous Prionium serratum (P. Palmita), which grows in the periodically dry river-beds of South Africa. In this category must also be placed the grasses that Hackel [5] terms tunic-grasses, which retain water between the macerated or scale-like persistent leaf-sheaths.

To this group of devices for the obtaining of water may be added the felted mass of *rhizoids* of many mosses. Many arenicolous xerophytes, especially arenicolous grasses, form dense *tussocks* or *cushions*, which certainly benefit them by collecting and retaining water.

Other organs, *leaves* for example, may likewise be designed to take up rain and dew. In such cases the leaves are usually more or less trough-like or, as in Umbelliferae, provided with large sheaths ; as marked examples may be cited the majority of Bromeliaceae, Pandanaceae, and the sugar-cane ; a specially remarkable form is Tillandsia bulbosa, whose narrow trough-like leaves easily obtain water and convey it to the cavities between the inflated leaf-bases.[6]

Particular forms of leaves fitted to take up and hold water are possessed by many epiphytic liverworts. Of these Göbel [7] distinguishes three types, according as the under-lobes of the leaves alone or with the co-operation of the upper-lobes form water-reservoirs, or as the leaves form peculiar bowl-like *water-sacs*.

CHAPTER XXXII.   STORAGE OF WATER BY LAND-PLANTS.
WATER-RESERVOIRS

A VERY important and common device, by which the land-plant is enabled to endure dryness of air and soil, is the construction of tissues or organs capable of conserving water for use when it shall be required for purposes of assimilation and other functions. This type of device is frequent among xerophytes, but completely lacking in hydrophytes.

---

[1] Haberlandt, 1904; see also Joh. Schmidt, 1903.   [2] See p. 104.
[3] Göbel, 1891–2 ; Karsten, 1894.   [4] Warming, 1893.
[5] Hackel, 1890 ; also see figure in Warming, 1892.
[6] Schimper, 1884, and 1888 a.   [7] Göbel, 1898–1901, vol. ii, p. 58.

CELL-CONTENTS

There are plants and parts of plants, including various Thallophyta and spores of Cryptogamia, that can be killed by desiccation only with very great difficulty, although they are quite devoid of any particular morphological means of protection. This faculty is clearly correlated with the nature of their habitat.[1] The more striking features in this connexion are treated in the succeeding paragraphs.

Mucilage is common in mucilaginous cell-walls, or in the sap of cells which are frequently large ; it absorbs water readily, but parts with it very slowly, and is therefore manufactured by xerophytes in various organs, including hairs, foliage-leaves, stems, subterranean tubers and bulbs. There is a correlation between the production of mucilage-cells inside the plant, and the development of the tegumentary tissue. Cactaceae, such as Echinocactus, that have a well-developed hypoderma, possess no mucilage-cells. The mucilage-cells of the Cactaceae are often situated in the edges, the bosses, or other protuberant parts, which are most exposed to drying.[2] Possibly acting in the same way as mucilage, there are other substances, such as—

Acids—for instance, malic acid in Crassulaceae[3] ;
Tannin, which abounds in certain desert-plants[4] ;
Salts, in halophytes ;
Latex probably plays the same part.[5]

WATER-TISSUE

Land-plants, particularly those exposed to strong transpiration, develop specialized water-storing tissues.

True water-tissue is thin-walled, contains water but no chlorophyll, is devoid of intercellular spaces (as no gaseous interchange occurs in it), and its cells are usually very large. It is capable of collapsing when water is abstracted, and of expanding when the cells once more absorb water. Water-storing tissue may be *peripheral* (*epidermal* or *hypodermal*) or *internal*.

Peripheral Water-tissue.

The epidermis is the outermost layer acting (except in plants growing in water or shade) as a water-tissue, as was indicated first by Pfitzer[6] and subsequently by Vesque[7] and Westermaier.[8] This view is supported by the facts that the epidermis usually contains no chlorophyll, that it forms a continuous layer which in certain cases is very deep and is directly connected with internal water-tissue, for instance in Velloziaceae.[9] The epidermis is specially differentiated in the Graminaceae, Cyperaceae, and Velloziaceae, which have *hinge-cells*[10] along definite lines on the upper face of the leaf, and especially in the furrows of the upper face ; these cells are larger and much deeper than the other epidermal cells ; they

---

[1] See G. Schroeder, 1886 ; V. B. Wittrock, 1891.   [2] Lauterbach, 1889.
[3] G. Kraus, 1906 a.   [4] Jönsson, 1902 ; Henslow, 1894.
[5] See p. 125.   [6] Pfitzer, 1872.   [7] Vesque et Viet, 1881.
[8] Westermaier, 1884.   [9] For illustrations see Warming, 1893.
[10] See p. 109.

play some part in the rolling and unrolling of the leaf, and probably may function as water-reservoirs.[1]

Mucilage is present in the epidermis of not a few desert-plants, including Cassia obovata, Malva parviflora, Peganum Harmala, Zizyphus Spina-Christi, and other plants in the Egyptian desert[2]; in many plants mucilage arises in all the epidermal cells, but in others only in some of these. The mode of origin of the mucilage is not known in all cases ; often it arises from the inner epidermal walls. These swell to such an extent in some xerophytes that the lumen of the cell seems not to be more than about half the volume of the wall or, at least, not so large as the latter ; this is the case in Empetrum, a number of Ericaceae, Loiseleuria procumbens,[3] Egyptian species of Acacia and Heseda, and species of Rosa.[4]

Hairs functioning as water-reservoirs form *water-bladders*. They occur in a number of African desert-plants, including Mesembryanthemum crystallinum, Malcolmia aegyptiaca, Heliotropium arbainense, Hyoscyamus muticus, Aizoon, some Resedaceae,[5] in many Chenopodiaceae, including Atriplex coriacea, A. Halimus,[6] A. (Halimus) pedunculata and A. portulacoides,[7] also as mealy hairs in other Chenopodiaceae, possibly also in Tetragonia expansa,[8] and Rochea falcata,[9] and others. In their typical form they are large, clear, watery vesicles, which project above the epidermis and glisten in the sunlight ; as their contents are gradually consumed they dry up ; in Atriplex Halimus and some other Chenopodiaceae, as well as in Oxalis carnosa,[10] the shrivelled hairs form an air-containing covering over the lamina. Whether or no all the hairs mentioned function to the same extent as water-reservoirs requires investigation.

Hairs of a most remarkable form occur, according to Haberlandt,[11] on the roots of an epiphytic fern, Drymoglossum nummulariaefolium. These hairs shrivel during the dry season ; the protoplasm withdraws to the base of the hair and shuts itself off from the dry part by a cell-wall ; when rain falls, the hairs grow out in a few hours and once more become filled with water.

Voluminous peripheral water-tissue may arise either by tangential division of the epidermis or by the formation of a hypoderma. It is situated mainly on the upper face of the leaf, and if present on the lower face it is less developed. It checks the penetration of heat-rays rather than of luminous rays, and thereby retards transpiration in addition to acting as a water-reservoir.

A multilamellar epidermis occurs frequently among xerophytes, but particularly among lithophytes and epiphytes ; there may thus arise a voluminous tissue, the thickness of which far exceeds that of the chlorenchyma, as in species of Peperomia, Begonia, Ficus, and Gesneraceae.[12]

Hypodermal water-tissue occurs in other xerophytes. It is composed of one layer of cells in certain Genisteae,[13] Velloziaceae,[14] and Orchidaceae[15];

[1] Duval-Jouve, 1875 ; Tschirch, 1882 *b* ; Volkens, 1887.
[2] Pfitzer, 1870, 1872; Volkens, 1887.   [3] Gruber, 1882 ; E. Petersen, 1908.
[4] Vesque, 1882 *a, b,* 1889–92.   [5] Volkens, 1887 ; Henslow, 1894 ; Schinz, 1893.
[6] Volkens, 1887.   [7] Warming, 1891, see illustrations, 1906.
[8] W. Benecke, 1901.   [9] F. Areschoug, 1878.   [10] Meigen, 1894.
[11] Haberlandt, 1893.   [12] Vesque, loc. cit. ; Pfitzer, 1870, 1872.
[13] Schube, 1885.   [14] Warming, 1893.   [15] Krüger, 1883.

of two to three layers in Nerium ; and is very large in certain other plants belonging to the Commelinaceae, Scitamineae, Bromeliaceae, and Rhizophoraceae.[1] A collenchymatous hypoderma functioning as a water-storing tissue is met with in a number of cacti ; it is traversed by narrow intercellular spaces leading from the chlorenchyma to the stomata.

Mucilaginous cork may here be mentioned as a cork-tissue discovered by Jönsson [2] in a number of Asiatic desert-plants.

Internal Water-tissue.

Water-tissue may occur among xerophytes in various other forms, as is indicated in the succeeding paragraphs.

(*a*) *Longitudinal bands of water-tissue* extending through the whole thickness of the leaf from the upper to lower faces occur in some desert-grasses,[3] and in Phormium tenax. Strips of chlorenchyma, in which the veins are embedded, in this case alternate with the bands of aqueous tissue. In the Velloziaceae similar longitudinal bands connect the epidermis on the upper face of the leaf with the water-containing cells that form a sheath round the vascular bundles.[4]

(*b*) *Central water-tissue*, occupying the centre of the leaf and surrounded by a thin layer of chlorenchyma, is met with in many xerophytes, including Aloë, Agave, Bulbine, Mesembryanthemum, Salsola,[5] Atriplex, Halogeton, and Zygophyllum. In aphyllous stems the aqueous tissue may be distributed in this same manner, as is the case in Salicornia and Haloxylon.[6]

Water-tissue and chlorenchyma may either be sharply delimited from each other, or may gradually merge, owing to the cells in the interior of the leaf containing but little chlorophyll, as in many Crassulaceae and Cactaceae. Water-storing idioblasts appear in the chlorenchyma of various desert-plants and halophytes.[7]

SUCCULENT PLANTS

Succulent plants are thick and fleshy forms which are provided with a water-tissue and parenchyma that contains abundant mucilage ; they are xerophytes which have especially pronounced water-storing tissue. They are commonly plump in form, and, like herbs, usually possess green stems which exhibit but feeble production of cork and of lignified tissue—lignification and succulence are, in a sense, opposed to one another. They are perennials, and often very long-lived. The cell-sap is rich in mucilage, the epidermis strongly cutinized as a rule, and the stomata are sunken. Succulent plants can store a large amount of water, which they give up extremely slowly, and they therefore dry only with great difficulty. The hottest and driest countries with a regular periodicity of climate are generally their homes.[8]

We can distinguish two main types of succulent plants, *succulent-stemmed* and *succulent-leaved*.[9]

---

[1] Warming, 1883 ; O. G. Petersen, 1893; Areschoug, 1902.
[2] Jönsson, 1902 ; also see Haberlandt, 1904, p. 363.
[3] Volkens, 1887.   [4] Warming, 1893.   [5] See figure in Areschoug, 1878.
[6] Volkens, 1887 ; Warming, 1897 *b*.   [7] See figure in Volkens, 1887.
[8] In regard to their adaptive features, consult Burgerstein, 1904, pp. 44, 205.
[9] Göbel, 1889–93.

i. Succulent-stemmed (Chylocaulous) Plants.

In these plants the stem is fleshy and juicy. The leaves are suppressed in the most marked types, or they are reduced to scales or thorns ; the stem has assumed the assimilatory functions of foliage, and the transpiring surface is thereby greatly reduced.

The most common and extreme types are Cactaceae in America, Stapelia in South Africa, and species of Euphorbia which occur mainly in Africa. To these may be added the geraniaceous Sarcocaulon in South Africa.

In the various genera there occur a series of shapes whose efficiency has been demonstrated by Göbel,[1] Noll,[2] and others. Frequent among such shapes are those like the sphere, prism, or cylinder, that combine smallness of surface with largeness of volume. This is advantageous in relation to storage of water, but disadvantageous in regard to assimilation. One stage towards increase of surface and therefore of assimilation is represented by the production of ridges, processes, bosses and the like, in Mammillaria, Echinopsis, and other Cactaceae. These protuberances are set vertically, that is, in a manner which does not render them so easily heated by the sun's rays—and this is of advantage as their internal temperature is often high.[3]

Here may be included pseudo-bulbs which occur mainly in epiphytic orchids ; they are tuberous green stems, consisting of one or more segments (thus bearing one or more leaves) ; they persist long after the leaves have fallen, serving as water-reservoirs and often containing mucilaginous sap.

ii. Succulent-leaved (Chylophyllous) Plants.

In plants with succulent leaves the stem is normal in form, except that its internodes are often short and its leaves consequently arranged in rosettes. The leaves are thick, stumpy, sessile, usually elongate and narrow, often cylindrical (if we except the sphere, a prism or cylinder has the smallest surface in relation to volume) ; they often are continued at the margins or apex into thorns, but apart from this are usually undivided and entire. As examples of plants having their leaves in rosettes we may mention Agave, Aloë, Sempervivum, Echeveria, species of Mesembryanthemum, and some epiphytic orchids ; elongated internodes are developed by Sedum, Bryophyllum, Portulaca, and Senecio (Kleinia).

Succulent-stemmed and succulent-leaved plants are both represented among halophytes.

Succulent plants deviate from other chlorophyll-possessing plants in both *respiration* and *assimilation*. The divers structural features that obstruct transpiration at the same time constitute an obstacle to the assimilation of carbon dioxide ; at night-time, during respiration, there is produced only little carbon dioxide but much malic or other organic acid, which is utilized in the manufacture of carbohydrates on the following day.[4]

---

[1] Göbel, 1889–93.   [2] Noll, 1893.
[3] Concerning the morphology of Cactaceae, see Vöchting, 1874, 1894 ; Göbel, loc. cit.   [4] Aubert, 1892 ; Jost, 1903. Lecture 15.

Succulent plants owe their origin, according to Vesque,[1] to two direct causes :—

1. Heating of the soil, which increases the osmotic power of the roots (succulent plants can endure without injury very high temperatures, and grow especially on warm rocks).

2. The supply of nutriment in alternately strong and weak solutions.

Between succulent plants and xerophytes that are poor in water there are distinctions in appearance quite apart from those in thickness and the like. The former are as a rule of a fresher green (because they are glabrous); the latter, on the contrary, white-haired or grey-haired.[2] Still there are some hairy succulent plants, Sedum villosum for example. In consequence of the production of wax, glaucous species occur in both groups of xerophytes.

BULBOUS AND TUBEROUS PLANTS

These must be considered in connexion with succulent plants. They are adapted in a different fashion to endure prolonged periods of drought. In many cases it is not only reserve food, such as starch, but also mucilage-cells or mucilage-tissue which contribute to their fleshiness, and function partly as food-materials for the production of new shoots[3] and partly as means of storage of water to provide against desiccation. Bulbous and tuberous plants belonging to the Liliaceae, Iridaceae, Amaryllidaceae, and other families, therefore occur especially in dry countries, and more particularly in South Africa ; also on the steppes of Asia, where they are among the species that develop rapidly after the commencement of spring or of the rainy season. Poa bulbosa is, according to Aitchison,[4] the commonest grass on the great plains of Baluchistan, and is certainly enabled to exist there by the aid of the thick leaf-sheaths that form a kind of bulb. Marloth[5] suggests that South African bulbous plants are designed to resist the enormous pressure to which they are subjected by the drying soil ; for some of them, namely Cape species of Oxalis, are protected by a hard coat, others by numerous, superimposed, soft, finely-fibrous layers whose bundles of bast persist as rigid bristles. In South Africa there are many remarkable, partly epigeous tubers (certain stem-tubers) which in their leafless condition are distinguishable only with difficulty from the stones among which they grow; as an example may be mentioned Dioscorea (Testudinaria) Elephantipes, which is protected from desiccation by huge cork structures. Likewise belonging to the category of epigeous tubers are the tuberous or at least swollen trunks of certain South American trees, occurring in the *Caa-tinga* forests, and including Chorisia crispiflora and Cavanillesia arborea (Bombaceae), Jaracatia dodecaphylla (Caricaceae),[6] also Jatropha podagrica (Euphorbiaceae).

Many tubers consist of root and stem combined, and thus lead the way to those consisting of root alone ; such is the nature of the lignified tuber ('xylopodium'[7]) in many herbs and small shrubs in South American savannahs.[6] In Crocus and other Iridaceae one sometimes sees clear,

---

[1] Vesque, 1883.   [2] See p. 114.
[3] Tubers of this kind occur even in aquatic plants such as Sagittaria sagittaefolia.
[4] Aitchison, 1887.   [5] Marloth, 1887 ; see also Hildebrand, 1884.
[6] See figure in Warming, 1892.   [7] Lindman, 1900.

fusiform, juicy roots radiating from the tuber[1]; these are also found on the bulbs of certain species of Oxalis,[2] and in the cactaceous Cereus tuberosus, whose shoots cannot store much water but whose roots are tuberous, juicy, and enveloped by a sheath of cork. South African xerophytes have upon their long roots many fusiform or spherical tubers which are water-reservoirs encased by cork ; Elephantorrhiza has close beneath the surface of the soil a water-reservoir of this kind, which weighs up to ten kilogrammes, although the stem is scarcely a foot in height ; while a species of Bauhinia produces tubers weighing fifty kilogrammes.[3] In Egypt there are species of Erodium with root-tubers which, according to Volkens,[4] serve to store water. Spondias venulosa has gigantic subterranean tubers. In temperate Europe, Sedum maximum possesses thick fleshy roots.

In some plants there have been discovered dwarf-roots that have been, correctly or incorrectly, regarded as water-reservoirs ; among such plants are Aesculus and some allies,[5] some Australian conifers,[6] and Sedum.[7]

Dimensions of water-reservoirs vary greatly according to the part they have to play in the life-history of different species ; in some cases they necessarily function for months or even years without intermission, but in others—for instance, leaves of trees in tropical rain-forest—only for a few hours of the day ; some resign water rapidly, others slowly. The structural features must necessarily harmonize with these differences.

Combinations of xerophilous characters—for instance, anatomical with morphological—are universal ; indeed, some characters demand the pre-existence of others before they can arise.

Correlations. One character often entails another. For example, with peripheral water-tissue there appear accessory cells in connexion with the stomata, so that the latter may be protected when the plant-member shrivels as it dries.[8]

LATICIFEROUS PLANTS

Up to the present we have dealt only with watery or slimy cell-sap in which various salts may be dissolved. But mention must be made of those plants that contain ' latex ', which is usually white and is contained in tubular organs (laticiferous cells and syncytes). The functional significance of latex is unknown, indeed it is probably multiple, and one function may be the protection of the plant against desiccation. In favour of this view is the fact that laticiferous organs are frequent in the tropics, particularly in hot, dry districts, and often precisely in plants which are thin-leaved and apparently lack any other means of replacing the water lost by transpiration.[9] The occurrence of latex in subterranean bulbs, as in Crinum pratense,[10] harmonizes with this view when these bulbs grow in clay which becomes fissured during the dry season.

ISOLATED WATER-STORING CELLS. NERVE-ENDS

The succulent plants hitherto discussed possess coherent *water-tissue*, an arrangement that seems to be the most efficient ; laticiferous plants

[1] Raunkiär, 1895.   Illustrations, 1905, 1907.   [2] Hildebrand, 1884.
[3] Schinz, 1893.   [4] Volkens, 1887.   [5] J. Klein, 1880.
[6] Berggren, 1887.   [7] Warming, 1891 ; see illustrations, 1907-09.
[8] W. Benecke. 1892.   [9] Warming, 1892.   [10] Lagerheim, 1892.

have long, tubular, branched receptacles.   But there are still other kinds of water-reservoirs.   Certain plants contain, scattered about in their general chlorenchyma, *solitary* or *grouped cells* which are larger than the other cells, have thin walls and clear contents.   Such occur in Nitraria retusa, Salsola longifolia, Halogeton, Zygophyllum, and other plants belonging to the Egypto-Arabian desert,[1] also in Barbacenia growing on mountains in Brazil,[2] as well as in loranthaceous parasites.[3]   In certain cases it has been established that if a slice of a leaf be allowed to dry these cells collapse, but expand again when water is supplied.

Lignified idioblasts (*tracheids*[4]) with spiral or reticulate thickenings occur in numerous other species, and are usually scattered in the same manner ;   they resemble the cells of the velamen of orchids[5] and the porous cells of Sphagnum,[6] being short, tolerably thick-walled, but not perforated ;   they fill with air when the contained water passes out. They show two types of distribution, being either grouped at the ends of vascular bundles or dissociated from these.   The latter is the case in the leaves of many tropical orchids,[7] species of Crinum,[8] Nepenthes,[9] Sansevieria, Capparis and Reaumuria,[10] Salicornia,[11] and Centaurea.[12] They occur in other xerophytes and halophytes close to the ends of vascular bundles ;   in this position, especially in desert-plants, they assume the form of huge, irregular tracheids with slit-like or elongated pits, and, as they stand above the delicate blind ends of vascular bundles in the leaf, they are often difficult to distinguish from tracheids apper- taining to the bundles :   arrangements of this kind occur in species of Capparis and Caryophyllaceae.[13]   These water-storing tracheids seem to play the same rôle as do wood-vessels in vascular bundles, since they fill with water and give it up again without collapsing.

The *parenchyma-sheath* surrounding vascular bundles may perhaps function as a water-reservoir in some Egyptian desert-plants [14] and in Restiaceae.[15]

Translocation of water.   Meschajeff [16] seems to have been the first to point out that the older leaves of succulent and semi-succulent plants often serve as water-reservoirs for the benefit of younger leaves ;   for in times of prolonged drought water is transferred to the younger parts of the shoot from the older leaves, which shrivel and die.

---

[1] Volkens, 1887.   [2] Warming, 1893.   [3] Marktanner-Turneretscher, 1885.
   [4] ' Reservoirs vasiformes ' of Vesque, 1882 ; ' spiral cells ', ' storage-tracheids ' of Heinricher, 1885.   [5] See p. 104.
   [6] See Chap. XLIX.   [7] Trecul, 1855 ; Krüger, 1883.
   [8] Trecul, 1855 ; Lagerheim, 1892.   [9] Kny u. Zimmerman, 1885.
   [10] Vesque, 1882 a, b.   [11] Duval-Jouve, 1868.   [12] Heinricher, 1885.
   [13] See Vesque, 1882 a and b ; Heinricher, 1885 ; Kohl, 1886 ; Volkens, 1887 ; Schimper, 1898 ; Haberlandt, 1904.
   [14] Volkens, 1887.   [15] Gilg, 1891.
   [16] Meschajeff, 1883 ; see Burgerstein, 1904, p. 228.

CHAPTER XXXIII

OTHER STRUCTURAL CHARACTERS AND GROWTH-FORMS OF LAND-PLANTS, AND ESPECIALLY OF XEROPHYTES

It has already been pointed out that some of the structural features of the growth-forms of land-plants are of such a nature that, while no one can doubt their connexion with a dry environment, their utility to the plant is not obvious. As a case in point we may cite the histology of the sun-leaf (heliophyll),[1] where, as we know, there is a greater depth and number of layers of the palisade cells in the sunlight than in shade, and in dry air than in moist air, with the correlative greater thickness and smaller intercellular spaces of the spongy parenchyma, less-sinuous walls of the epidermal cells, and other characters. Among features of this problematic nature may be mentioned lignification, which is so common among land-plants and so restricted in aquatic plants.

LIGNIFICATION

Lignification is a mechanical utility because it increases the plant's power of resisting mechanical force. In many plants, including trees, it is of service in connexion with the storage of water. The birch and the Common spruce are splint-wood trees with shallow horizontal roots ; it therefore suggests itself that the supply of water is controlled by the splint-wood.

It is worthy of note that lignification stands in direct relation to environment, for it becomes more extensive the drier the habitat (except in succulent plants). Families such as the Umbelliferae, Caryophyllaceae, Linaceae, Labiatae, Rubiaceae, and Dipsaceae, as well as genera which in temperate countries are rich in herbaceous species, become far richer in woody plants in tropical, warm-temperate, or even in Mediterranean countries.

Lignification is particularly extensive in xerophytes that contain but little sap. In these the *wood* is dense and hard, but at the same time often brittle. It resembles the so-called ' autumn-wood ', because the lumina of the vessels and cells are narrow, and the reason for resemblance is presumably that the conditions of development are identical ; the narrowness of the constituents is correlated with weakness of transpiration, which is due to great reduction in the leaves and unfavourable conditions of growth.[2] According to Cannon[3] ' the branches of irrigated plants in the desert about Tucson are poorer in conductive tissue than branches of the same diameter of non-irrigated plants '. The explanation of this must be sought in the difference ' in the length and character of the growing season of the two classes of plants '. What benefit desert-plants derive from this structural character of the wood is not clear. It must, however, be remembered that lignified parts withstand extreme temperatures better than watery thin-walled parts can, and that trees endure great fluctuations in humidity better than herbs do.

Mechanical tissue is developed in the form of bundles of *bast* above and

[1] See p. 21.　　[2] Henslow enumerates the peculiarities of desert-plants, 1894.
[3] Cannon, 1905.

beneath the veins of leaves in land-plants, below or in the epidermis, and is more extensive the drier the climate, or the more transpiration is favoured by the environment.   Parts of the fundamental tissue are developed as mechanical tissue in stems of some plants, including Restiaceae.[1]   *Stone-cells* and *mechanical cells* are differentiated in the chlorenchyma more or less as idioblasts of various forms, which Vesque[2] distinguishes by such terms as ' proteoid ', ' oleoid ', and the like ;   they occur in leaves of Proteaceae,[3] Rhizophora,[4] Restiaceae, Olea europaea (as long, sinous, sclerenchyma-cells, which interweave and run both parallel and perpendicular to the surface), Thea, and others.   In some cases the utility of these cells with thick, lignified walls is obvious ;   it is to prevent the shrivelling, collapse, or distortion of the vitally important chlorophyl-containing tissue that would otherwise take place when the plant-member withered.   The strong construction of the epidermis in sclerophyllous plants also performs the same service.

In the production of *thorns* xerophytes show also their tendency towards lignification.   It has been recognized that plants living in deserts and similar places are very thorny, and possess stiff, spiny or prickly leaves, thorny stems, and the like.   Such plants are characteristic of the scrub in Australia, of stony steppes and high plateaux in Asia (the ' Phrygana-vegetation ' of Theophrastus), the Kalahari, the deserts of Egypt and North America, and others.   Thorns vary widely in their morphological nature, and may represent complete leaves or portions of these, or emergences and prickles, or lignified stems which are vegetative axes or flower-stalks ; in accordance with these features various growth-forms—for   example   Grisebach's   ' thorn-shrub '   and   Reiter's ' thistle-form '—have been defined by different authors.

Thorns, according to Lothélier's[5] researches, are evoked by dryness of atmosphere ; species such as Berberis and Crataegus, which are thorny in dry air, become thornless in moist air.   It has long been known that spiny plants often lose their spines under cultivation on improved soil.[6]

Nearly all those who have investigated the subject express the view that, while thorns play no direct part in assimilation, they can hardly be regarded as useless organs, since they presumably serve to protect the plant against animals.[7]   Wallace[8] points out that thorny shrubs are especially abundant in those parts of Africa, Arabia, and Central Asia where large herbivora abound.   It seems to be beyond doubt that this view is correct in certain cases, and that, for instance, the long spines of Acacia horrida, A. giraffae, and other species in the dry tracts of South Africa serve as defences against numerous wandering herds of ungulates ; Marloth[9] calls attention to specialized adaptation exhibited by certain species in that the longest and strongest spines occur on young individuals or on root-shoots, which are most accessible to animals, while the branches subsequently produced on tall trees are quite devoid of spines.   A similar phenomenon is witnessed in connexion with Ilex Aquifolium, the upper leaves of which are usually not prickly when once the plant has grown

[1] Gilg, 1891.   [2] Vesque, 1882.   [3] Jönsson, 1880.
[4] Warming, 1883.   [5] Lothélier, 1890.
[6] See Henslow, 1894, p. 223 ; Vesque et Viet, 1881.
[7] Delbrouck, 1875 ; Marloth, 1887.
[8] Wallace, 1891.   [9] Marloth, 1887.

into a tall tree. It is evident that spiny plants by reason of their armed nature may defeat unarmed species and become more widely distributed ; but for all this we are not entitled to assume either that thorns are a *direct* adaptation to animals, or that they could arise by natural selection in a country rich in herbivorous animals. For example, against what animals did the Cactaceae and Agaves of Mexico and the West Indies require to defend themselves when they were evolved ? Would heredity have preserved these useless characters throughout the vast periods of time that may have elapsed since ungulates, which have recently been re-introduced, abounded in these lands ? (It is incontestible that spiny structures are now of use to Mexican succulent plants in protecting them from ungulates during the prolonged dry season). Kerner[1] assumes that the Mediterranean region is rich in thorny plants because animals also abound, and that on high mountains the absence of thorny vegetation is associated with the greater poverty of animal life. But in arctic countries there are many herbivorous animals, including large ones such as the reindeer and musk-ox, which roam about in great herds, yet no thorns occur, obviously because the conditions of humidity prevailing here and on high mountains do not conduce to the production of thorns.[2]

In the north-temperate moist climate there occur many thorny growths the significance of which is at present obscure. This is likewise true of the strong spines of many palms, including Astrocaryum and Bactris, growing in Amazonian forests.

There are other thorns whose definite utility can be demonstrated, and such is the case with those on stems of certain lianes.

The physiological reason for the strong development of lignified constituents is still somewhat obscure. But intense *light* and vigorous *transpiration* seem to be the causes. Vesque and Viet,[3] and subsequently Kohl,[4] and Lothélier[5] experimentally proved that mechanical tissue increases when transpiration is greater. Cockayne[6] found that in the rhamnaceous Discaria Toumatou thorns are not developed in moist air. Stahl,[7] Dufour,[8] and Lothélier[9] found that mechanical tissue is more strongly developed in light than in darkness : etiolated plants are very weak-stemmed. On the other hand, experiment showed that with an increased supply of water there was a diminished production of wood in the oak and Robinia, and a reduction in the development of mechanical constituents among monocotyledons.[10]

STUNTED GROWTH. SCRUB. CUSHION-PLANTS

It has already been mentioned[11] that lack of water, and strong transpiration, induce stunted growth. Wind, deficiency of water, and other conditions unfavourable to growth, bring into existence elfin-wood scrub, heath-scrub, ericaceous shrubs with bowed branches, and such growths as that of malformed and stunted Pinus sylvestris as it occurs in north-eastern Germany. Dry soil and strong transpiration impart to these

[1] Kerner, 1869.   [2] See Warming, 1892 ; Henslow, 1894 ; Cockayne, 1905.
[3] Vesque et Viet, 1881.   [4] Kohl, 1886.
[5] Lothélier, 1890.   [6] Cockayne, 1905.   [7] Stahl, 1883.
[8] Dufour, 1887.   [9] Lothélier, loc. cit.   [10] Gräbner, 1895.
[11] See pp. 29, 37.

plants their characteristic habit, by inducing short, curved, and crooked shoots and stems, with short internodes, and with a feeble or irregular production of buds : abundant moisture causes shoots to be long and possessed of long internodes. In Mediterranean and other subtropical countries with winter-rain, many species assume the form of shrubs of medium height, but in moist valleys they vary from this form up to that of a tall tree. In scrub or in the desert, the branches and leaves are often closely packed, the ramification being extraordinarily dense, and the plant as a whole being compact and rounded in form (hemispherical, or cushion-like); as examples may be cited Achillea fragrantissima, Artemisia Herba-alba, and Cleome arabica,[1] all in the North African desert ; the globular bushes of Astragalus and Genista[2] in Corsica ; Draba alpina,[3] Silene acaulis,[4] species of Saxifraga, and many mosses of cushion-growth in arctic countries[5] ; Androsace helvetica and others in the Alps.[6] The high mountains of South America and of all other lands display many examples of cushion-like shrubs or herbs which appear as if cleanly bitten or clipped ; for they are rounded off, dense in growth or even solid, and have their numerous shoots, leaves, and remnants of these closely packed together : as examples may be cited the umbelliferous Azorella and Laretia, species of Oxalis, and Cactaceae in South America. One of the most remarkable cushion-plants is Raoulia mammillaria living in New Zealand.[7]

Everywhere the cause is the same—dryness, occasioned by one or another factor. Dense ramification and tufted growth confer a benefit upon the plants, in that their young shoots are thereby better shielded from transpiration ; they protect each other and are in turn protected by older shoots from the desiccating action of wind in arctic countries.

ROSETTE-PLANTS

Many xerophytes have their leaves arranged in rosettes on shoots which resemble the first year's growth of biennial dicotyledons : rosette-plants are encountered in arctic countries, at alpine altitudes, on steppes and deserts, among epiphytes and tropical lithophytes.[8] The brevity of the internodes and the consequent arrangement of the leaves cannot perhaps always be explained in the same manner, nor is the utility always identical. In many Bromeliaceae the rosette serves to collect and retain water ; in other plants, such as Agave, it may be that the leaves forming the rosette are better screened from the sun and from excessive transpiration. In arctic and alpine plants the low rosette-shoot may benefit because the leaves spreading over the soil are not so much exposed to desiccating winds, also because these leaves are situated in warmer air and are better able to obtain heat from the soil. It is probable that in the desert they can utilize to good advantage the dew deposited by night. Meigen[9] also remarks that the leaves of many rosette-plants by overlapping one another produce niches screened from the wind, and thus reduce their transpiration. Rosette-plants thrive among open and low

---

[1] Volkens, 1887.   [2] Massart, 1898 ; Rikli, 1903.
[3] Figured in Kjellman, 1884, p. 474.
[4] Figured in *The Botany of the Färöes* (Copenhagen, 1901–8), p. 993.
[5] Andersson and Hesselman, 1900.   [6] Schröter, 1904–8.
[7] See Section IX.   [8] See p. 27.   [9] Meigen, 1894.

vegetation ; the grasslands of northern and central Europe and of the Alps, and similar types of vegetation, are very rich in low, perennial rosette-herbs such as Plantago major, Taraxacum officinale, Achillea Millefolium, and Pimpinella Saxifraga, species of Primula, Draba, Saxifraga. Bonnier [1] showed that certain species which in the plains have shoots with long internodes, when grown at alpine heights produce rosettes.

PROSTRATE SHOOTS

Many species growing on dry, warm, sandy soil have prostrate shoots, at least so far as these are vegetative. As was shown on p. 26 this is to be attributed to the thermal relations prevailing in the soil.

CHAPTER XXXIV.   OECOLOGICAL CLASSIFICATION [2]

THE foregoing chapters have made it clear that the distinctions between water-plants and land-plants are deep-seated, and concern the external form as well as the internal structure. Plant-communities must therefore be grouped in the first place into aquatic and terrestrial ; but between these there is no sharp boundary, for there is a group of plants, marsh-plants (*helophytes*), which, like water-plants, develop their lower parts (roots, rhizomes, and, to some extent, leaves) in water or at least in soaking soil, but have their assimilatory organs mainly adapted to existence in air, as is the case with land-plants to which they are closely allied. Helophytes give rise to special forms of communities. Yet we must include among water-plants all those plants that, like Nymphaeaceae, approximate to land-plants in so far as they have floating-leaves, which are more or less adapted to existence in air, but are nevertheless mainly designed for existence upon water.

It has already been shown that land-plants exhibit many grades of adaptation to their mode of life in contact with air, and that those which encounter the greatest difficulties in regard to securing water are termed *xerophytes* ; while others are described as *mesophytes* because in some respects they stand midway between the two extremes, hydrophytes and xerophytes. The differentiation of the land-plant in one or the other direction is decided by the oecological factors, edaphic and climatic, that prevail in the *station* or *habitat*. But edaphic and climatic factors cannot be regarded separately : the plant-community is always the product of both together. The nature of a soil is also influenced by climate, and it is incontestible that climate (rainfall) calls forth the wide differences between, say, desert and tropical rain-forest. But it is far from being true that climate alone calls into existence the different communities of plants which will hereafter be defined as *formations*. Characters of

[1] Bonnier, 1890, 1894.
[2] In the classification of plant-communities these are grouped into successively smaller subdivisions that are, only to some extent, analogous with systematic families, genera, species, and varieties. The most comprehensive group is termed in German a ' Vereinsklasse ' or ' Formationsklasse ', which we propose to translate as ' oecological class ', or when the context permits, as ' class '.

the soil are of supreme importance in determining the production of formations, and they must therefore be the foundation of oecological classification. Clements,[1] with reason, has objected to Schimper's scheme of distinguishing between climatic and edaphic formations, if indeed it was Schimper's meaning that a sharp distinction is throughout possible, and that both groups of factors are of equal potency.

The importance of soil in determining the development of definite plant-communities is clearly revealed in the topographical distribution of these ; there is not a single community of land-plants that extends without interruption over great stretches of land ; all are discontinuous and, according to the nature of the soil, interrupted by other communities, however uniform the climate may be. On the other hand, it is often the case that one and the same formation, either of water-plants or of land-plants, is developed in very different climates.

The differences existing in the climate of various parts of such a country as Denmark are quite inadequate to account for the great differences in vegetation. A prominent part is played by chemical differences in soils (amount of common salt), also by their fertility or amount of nutrient salts ; for instance, soil poor in food-material favours the preponderance of heath. Gräbner[2] regards the percentage of nutritive salts dissolved in the soil-water as the factor controlling the character of vegetation ; he therefore divides formations into three great groups according as the water is rich or poor in mineral matter, or contains common salt :

1. Formation, where the water is rich in mineral salts ;
2. Formation, where the water is poor in mineral salts ;
3. Formation, where the water is saline.

The same opinion is expressed by A. Nilsson[3], but this view seems to be based chiefly upon observations in the open air, and too little tested by soil-analyses. Gräbner apparently exaggerates the importance of the factor in question, for his scheme would, on the one hand, unnaturally separate formations, such as those of sour-meadow, heath-bog, and ling-heath, which belong together, while, on the other hand, it would unnaturally group together formations, such as those of sand-field and peat vegetation, which are not allied. It would appear *that the most potent and decisive factor is the amount of water in soil ;* and this, in turn, depends upon the depth of the water-table and upon the physical characters of soil (its water-capacity, the amount of air-content, the plants and animals living in it, the production of humus, and the like).

We must, however, admit that climate may favour a certain formation by causing this to become less exacting as regards its edaphic requirements, and consequently enabling it to be distributed over a large area on very diverse soils ; whereas in another climate this formation will be vanquished by communities better adapted to localities where certain special soils occur, and will be more broken up and restricted in its dis-

[1] Clements, 1899, 1904. Clements (p. 27) writes : ' From the above it follows that Schimper's so-called climatic formations, forest, grassland, and desert, are merely a somewhat incomplete expression of water-content association. As to the validity of his division of all formations into climatic and edaphic, there is also room for grave doubt . . . all plants . . . are primarily influenced by soil, i.e. they are edaphic.'
[2] Gräbner, 1898, 1909.   [3] A. Nilsson, 1902 b.

tribution. In this sense we must interpret Schimper's division of formations into climatic and edaphic. As an example, it may be mentioned that Whitford and Cowles [1] interpret coniferous forest in the eastern United States as an edaphic, xerophytic formation occurring in the district where deciduous forest prevails ; but in the entirely different climate near the Pacific coast of the United States coniferous forest is dominant, while the deciduous trees give rise to edaphic, mesophilous forest lining the watercourses.

The occurrence of the same species in different formations renders explanation difficult ; Cowles, no doubt correctly, states that a species ' in general can grow in the largest number of formations at its centre of distribution, since there the climatic conditions favour it most highly. In other regions, especially near its areal limits, it can grow only in those formations which resemble most closely in an edaphic way the climatic features at the distribution centre.' Cowles also contends that in many cases a species can occupy very different soils (for instance, clay or dune-sand) only when the atmospheric conditions are the same, and that conversely on one and the same soil very different vegetation may prevail when the atmospheric conditions are changed. An exception is provided by humus soil, for there are many species to which humus is absolutely essential. As an example showing that a species may inhabit entirely different kinds of soil, we may quote C. Schröter's [2] statement in reference to the mountain-pine. He writes : ' How fundamentally diverse are the habitats of the mountain-pine which can grow on the dry, loose, calcareous talus of a hot southern slope, and on high-moors, which are mainly composed of bog-mosses and are subalpine bogs dripping with water but poor in mineral matter. The former soil is poor in humus, rich in mineral matter, and dry ; the latter is a substratum rich in humus, poor in mineral matter, and always saturated with water. Common to both is only one character—poverty in assimilable nitrogen.' [3]

When endeavouring to arrange all land-plants, omitting marsh-plants, into comprehensive groups, we meet with, first, some communities that are evidently influenced in the main by the physical and chemical characters of soil which determine the amount of water therein ; secondly, other communities in which extreme climatic conditions and fluctuations, seasonal distribution of rain, and the like, decide the amount of water in soil and character of vegetation. In accordance with these facts, land-plants may be ranged into groups, though in a very uncertain manner. The prevailing vagueness in this grouping is due to the fact *that oecology is only in its infancy*, and that very *few detailed* investigations of plant-communities have been conducted, the published descriptions of vegetation being nearly always one-sided and floristic, as well as very incomplete and unsatisfactory from an oecological standpoint. It is to be hoped that Clements's remarks, in his *Research Methods in Oecology* will stimulate detailed research and will banish the ignorance to which he alludes in the following terms : ' Our knowledge of soil-factors is in an

[1] Cowles, 1901 *a* ; Whitford, 1905.
[2] Schröter, 1904. See also P. E. Müller, 1887, 1871.
[3] Also may be added the character of dryness—physical in the first, physiological in the second. See p. 134.

extremely elementary condition,' and 'We have no exact understanding whatsoever of the sum of physical factors which we term climatic.'

Communities of xerophytes must be set into different groups according as they are due to nature of soil or to nature of climate. The distinctions in this respect are recounted in the succeeding paragraphs.

Soil may, to employ Schimper's [1] terms, be physically or physiologically dry :

Physical Drought.   Soil is *physically dry* when it contains very little free water ; this is the case with—

1. The surface of rocks or stones occupied by plants which compose the *lithophilous* formations.

2. Sandy soil which lies so high above permanent subterranean water that this does not affect it, and which is very parched during dry seasons owing to rapid drainage and desiccation ; upon it are *psammophilous* formations, which are allied to those on rubble, where soil is formed of gravel and stones.

Here too must be placed epiphytes, which nearly all have definite adaptations for securing water (see p. 87).

Physiological Drought.   Soil is *physiologically dry* when it contains a considerable amount of water which, nevertheless, is available to the plant only to a slight extent or can be absorbed only with difficulty, either because the soil holds firmly to a large quantity of water or because the osmotic force of the root is inadequate to overcome that of the concentrated salt solution in the soil.   This may be the case when—

1. Soil is *rich in free humous acids,* or in chemical bodies that by their peculiar action on the plant evoke *xerophily* [2] ; there result those formations that grow on sour (acid) soil.

2. Soil rich in *soluble salts,* usually common salt, which brings into existence that form of *xerophily* which we see in *halophilous* formations. A halophyte is in fact a special form of xerophyte, as Clements repeatedly urges, and Wiesner [3] and Schimper [4] recognized.

In addition to the xerophytic formations thus grouped according to characters of the soil, which is dry, or dries frequently or rapidly even in a moist climate, there is another series of formations to which the physical and chemical qualities of the soil are of subordinate import in comparison with the extreme climate. The soil is neither too acid, too saline, nor too poor in nutriment, and may be sufficiently moist to sustain luxuriant vegetation, yet the climate is so extreme that the soil is either too *cold* (as in the case of formations on subglacial tracts) or periodically so *dry* for a long season that only xerophilous formations can thrive on it, excepting in situations, such as marshes or river-banks, where the soil contains sufficient moisture throughout the year ; hence in this case, also, topographical features play a part. The vegetation of savannahs (*campos*) in the interior of Brazil is a formation evoked by a dry season ; yet it is everywhere confined to the higher ground in the hilly country, while forest always occurs along the watercourses and on the mountains, where greater humidity prevails in the soil ; there can be no doubt that were the climate moist throughout the year the *campos* would be

---

[1] Schimper, 1898.   [2] See Livingston, 1904.
[3] Wiesner, 1889.   [4] Schimper, 1891, 1898.

clothed with forest.[1]   Among the formations belonging to this type
must be reckoned those forming steppes and savannahs, also certain
sclerophyllous formations.

Mesophytes grow on soil which is of an intermediate character, and
is neither specially acid, cold, nor saline, but is moderately moist, usually
well-ventilated, also rich in nutriment and in alkaline humus or in other
organic constituents. Mesophytic communities occur in very diverse
climates, near the Poles or on the equator, yet they can never be exposed
to the danger of prolonged drought. Adapted to such conditions are
plants that show a relatively weak development of the above-mentioned
arrangements for regulating transpiration ; in this respect these plants
stand *midway* between hydrophytes and xerophytes. The leaves are
large, and far more varied in form than in xerophytes ; teeth and other
incisions of the margin are common, as are compound or richly divided
leaves ; hydathodes seem to be frequent ; the vegetative organs are of
a fresh green, and devoid of thick grey coatings of hair or bluish incrustations of wax ; the leaves are usually dorsi-ventral in structure. Stomata
are numerous, often occurring also on the upper face of the leaf ; anatomical peculiarities, such as aqueous tissues, are very rare, or at least not
extremely developed.

The greatest differences among mesophytes depend upon whether
the leaves persist for only a few months in the favourable season, or for
a year or more. Ilex Aquifolium is indubitably a mesophyte growing
as underwood in forests of northern Europe ; but its leaves persist for
up to two years, and, like sclerophyllous leaves, are xerophytic in structure
because they are exposed to the harsh conditions of winter—that is, to
cold (physiologically dry) soil and possibly concurrent rapid transpiration
caused by dry, cold winds ; the same is true of the spruce (Picea excelsa)
and most other evergreen woody plants in cold-temperate countries.
In deciduous plants within the same countries the leaves are thinner, of
a paler green, and more flexible ; the cuticle is thinner, and so on ; they
are, in short, typically mesophytic in structure.

In the tropical rain-forest, which may be regarded as a mesophytic
community, there are many species possessing leaves that are xerophytic
in structure because they persist for more than one year, and must
consequently be adapted to endure all the changes during that period.
It is also difficult to regard all conifers as xerophytes, even when their
leaves are perennial.[2]

Pound and Clements [3] divide mesophytes into three groups : hylophytes (woody
plants), po-ophytes (meadow plants), and aletophytes (ruderal plants). But it
must be noted that there are forests and grasslands among xerophytic formations.

Between the different groups there are *very gradual transitional stages*.
Moreover, the peculiarities of a formation may be evoked by a combination of diverse factors of varied strengths ; for instance, there seem
to be formations for the origin of which the co-operation of coldness and
acidity of soil is responsible.

Tropophytes. Schimper [4] has introduced the term 'tropophyte', by which he
designates land-plants which, in opposition to hygrophytes and xerophytes, have

---

[1] Warming, 1892, 1899.   [2] See Section XV.
[3] Pound and Clements, 1898 ; Clements, 1904, p. 22.   [4] Schimper, 1898.

deciduous leaves, and 'whose conditions of life are, according to the season of the year, alternately those of hygrophytes and of xerophytes'; the structure of their perennial parts is xerophilous, and that of their parts that are present only in the wet season is hygrophilous. Tropophytes as a whole are included in mesophytes (a term also employed by Schimper), as are Schimper's hygrophytes. The cold-temperate flora is mainly tropophilous. The term tropophilous is a very serviceable one as applied to those plants that shed their assimilating organs during the unfavourable season, but there is no group of tropophytes contrasting with those of hydrophytes and xerophytes; for there are tropophilous hygrophytes and tropophilous xerophytes, as Schimper's own words in various passages indicate.

In accordance with the considerations given in the present and previous chapters the following oecological classes[1] may be distinguished :—

A. **The soil (in the widest sense) is very wet,** and the abundant water is available to the plant (at least in Class 1), the formations are therefore more or less hydrophilous :—

Class 1. **Hydrophytes** (of formations in water). Section IV.

Class 2. **Helophytes** (of formations in marsh). Section V.

B. **The soil is physiologically dry,** i. e. contains water which is available to the plant only to a slight extent ; the formations are therefore essentially composed of xerophilous species :—

Class 3. **Oxylophytes** (of formations on sour (acid) soil). Section VI.

Class 4. **Psychrophytes** (of formations on cold soil). Section IX.

Class 5. **Halophytes** (of formations on saline soil). Section VII.

C. **The soil is physically dry,** and its slight power of retaining water determines the vegetation, the climate being of secondary import ; the formations are therefore likewise xerophilous :—

Class 6. **Lithophytes** (of formations on rocks). Section VIII.

Class 7. **Psammophytes** (of formations on sand and gravel). Sect. X.

Class 8. **Chersophytes** (of formations on waste land). Sect. XII.

D. **The climate is very dry** and decides the character of the vegetation ; the properties of the soil are dominated by climate ; the formations are also xerophilous :—

Class 9. **Eremophytes** (of formations on desert and steppe). Sect. XI.

Class 10. **Psilophytes** (of formations on savannah). Section XIII.

Class 11. **Sclerophyllous** formations (bush and forest). Sect. XIV.

E. **The soil is physically or physically dry :**

Class 12. **Coniferous** formations (forest). Section XV.

F. **Soil and climate favour the** development of mesophilous formations :—

Class 13. **Mesophytes.** Section XVI.

[1] Concerning the classification of plant-communities, reference should be made to Clements, 1904 *a* ; and regarding nomenclature to Clements, 1902 *a* and *b*.

CHAPTER XXXV

PHYSIOGNOMY OF VEGETATION. FORMATIONS
ASSOCIATIONS. VARIETIES OF ASSOCIATIONS

THE large oecological classes indicated in the preceding chapter may be subdivided into various less comprehensive types of communities. Popular distinctions have for a long time been drawn among divers types of vegetation, to which have been allotted certain names (forest, bush, meadow, moor, heath, steppe, savannah, *maqui*, and so forth) that have been adopted as scientific terms. The leading features upon which the pertinent distinctions depend are physiognomic, and thus dependent upon biological relationships. For the *physiognomy of vegetation* is not only of aesthetic, but also of scientific significance : vegetation often essentially determines the physiognomy of landscape, and in this respect plays a part very different from that played by animals.[1] Physiognomy must therefore be scientifically considered.

A. PHYSIOGNOMY OF VEGETATION

The chief circumstances that determine the physiognomy of vegetation are :—

1. **Dominant growth-forms :** trees, shrubs, and herbs, of varied appearance, size and shape of foliage ; mosses, lichens, and other types.[2] Thus arise the oecological types : forest, bush, heath, meadow, steppe, and other kinds of herbaceous vegetation, moss-tundra, lichen-tundra, and so forth, modifications being introduced by lianes and epiphytes.

2. **Density of vegetation (number of individuals).** This depends upon the struggles of plants with inanimate Nature, and upon the biological peculiarities of growth-forms. In some communities the soil is densely covered, as in the case of meadow, but in others the vegetable covering is so open that the colour of the soil imparts to the landscape its hue. A distinction must therefore be drawn between—

(*a*) *open formations* evoked by shifting soil (seashore, dunes), extreme character of soil (rock), dryness of climate (in desert and steppe), or by extreme cold (in Polar countries), and—

(*b*) *closed formations* composed of species that grow in company[3] for some reason or other, whether it be that, like some species of trees, they can suppress all competition by their shade, or, like Phragmites, can form dense associations by means of richly branched horizontal rhizomes. A number of different species can together form one closed formation.[4]

3. **Height of vegetation.** Comparisons may be instituted between forest, bush, and heath, all composed of woody plants, or between the tall grass of a lowland meadow and the low sward of an alpine one, or between forest and tundra. Many formations exhibit *strata* or *storeys* of growth-forms : the greatest number of storeys will probably be found in well-lighted, therefore thinly wooded, forest.

---

[1] Darwin writes : ' A traveller should be a botanist, for in all views plants form the chief embellishment.'

[2] See Chapter XI.   [3] Plantes associées, of Humboldt, 1807.

[4] See Drude, 1905.

Finnish botanists have adopted a kind of nomenclature to denote the different strata or storeys :

i. Tree-stratum

ii. Bush-stratum	. . .	90 centimetres to about 4–5 metres
iii. Tallest field-stratum	. . .	45 ,, to 80–90 centimetres
iv. Middle field-stratum	. . .	10 ,, to 45 ,,
v. Lower field-stratum	. . .	5 ,, to 10 ,,
vi. Ground stratum	. . .	immediately on the surface of the soil up to 5 centimetres (mosses, lichens, algae).

It may, however, suffice to consider merely four strata :—

i. *Ground-stratum :* immediately above the soil : mainly mosses, lichens, and algae.

ii. *Field-stratum :* formed by grass and herbs, as well as dwarf shrubs of approximately the same stature.

iii. *Shrub-stratum :* formed of taller shrubs.

iv. *Tree-stratum.*[1]

4. **Colour of vegetation.** We may compare brown heath with green meadow. Here, too, may be mentioned the colours of flowers, and the contrast between entomophily and anemophily.

5. **Seasonal relationships.** This involves the duration of the resting period, and other phases of vegetation (foliation, flowering, and defoliation). We may compare evergreen forests with those that shed their foliage for winter, or for the dry season ; steppe which is green for a few months, but yellow-brown and bare for much longer ; the vegetation of north-temperate Europe in winter, and in summer.

6. **Duration of life of species:** duration of epigeous parts, the rôle played by annual and renascent species, and by woody plants.[2] Only rarely do we find an assemblage of plants consisting solely of annual species, but they do occur, as in the case of Salicornia herbacea, and of certain weeds extending over limited areas.

7. **The number of species** present gives some indication of the struggles for space between species ; this struggle may be greatly interfered with, and is, in fact, interrupted by man. Sometimes—as in spruce-forest, beech-forest, and ling-heath—a single definite species dominates ; at other times there is an extraordinary admixture of species. Rich in species are communities in warm countries, such as tropical forests,[3] the *maquis* of Cape Colony ; poor in species are the communities of northern Europe. It is evident that favourable conditions of life call into existence a more complex flora ; but geological factors have often played some part.

In the vicinity of Lagoa Santa in Brazil, on an area of about three geographical square miles, there grow about 3,000 species of Vascular plants (since more than 2,600 species have been determined, and there must be at least 400 species that have not been collected). Of these there are 1,600 species (*circa*) in the forest and 800 species (*circa*) in the *campos*, of which 400 and 90 respectively are trees; such is the case, despite of the fact that the area of forest is much smaller than the area of the *campos*, and is essentially confined to the valleys where it bounds all the watercourses. The reason for the greater richness of this forest-flora is certainly to be sought in the prevailing physical conditions (more abundant humidity and food-material, especially humus). But possibly geological causes have played a part; for probably the forest-flora is the older, and the flora of the *campos* gradually arose later, as South America raised itself more and more above the sea, and Brazil consequently acquired a more continental climate.[4]

---

[1] See A. Nilsson, 1902 *a*.   [2] See p. 8.
[3] See Humboldt, 1807.   [4] See Warming, 1892, 1899.

As the number of associated species increases the number of growth-forms as a rule does so likewise ; in this respect the premier place is taken by the warm, moist tropical forest, which perhaps owes its boundless wealth to the circumstance that it has been able to develop for vast periods of time without interruption.

The number of species depends,[1] *inter alia*, upon the means of competition possessed by the several species. Some species readily form dense masses of vegetation composed of many individuals ; others are universally represented by isolated individuals. Many species can occur in several kinds of formations, because the demands they make are bounded by wide limits, and because the more habitats they can occupy the wider are these limits. The hardiest and most accommodating species can seize upon most kinds of habitat, nevertheless they are often found only in a few, because they have been crowded out of the better ones. The more peculiar and extreme a habitat is the more uniform does its vegetation tend to be, because, as a rule, only few species are so specialized in their adaptation as to be capable of existing in such a place.

In studying the vegetation of a certain area from a floristic and geographical standpoint, it is necessary to define the relative numbers of the various species. Every community consists of *dominant* and *sub-dominant* species, as well as of others that are more or less dependent upon these and occur only here and there. Drude employs the following terms [2] :—

social : dominant species whose individuals give the main character to the vegetation ;

gregarious : species whose individuals occur in small groups so as to form small unmixed collections in the main vegetation ;

copious (with abbreviations cop.[3], cop.[2], cop.[1], to denote decreasing frequency), species represented by individuals scattered in smaller numbers among those previously mentioned ;

sparse : species having only isolated individuals occurring here and there ;

solitary : species of which individuals occur in extreme isolation.

These terms may be used in combination ; for instance, *solitary gregarious* would mean a single clump composed of one species. The relative shares normally taken by the various plants constituting a community ought to be capable of numerical expression.[3]

B. FORMATIONS

It is not sufficient merely to distinguish among the broader, physiognomical types, for between these there occur differences that demand the establishment of subdivisions. In attempting to define these many difficulties are encountered.[4]

The term 'formation', or 'vegetative formation', was introduced in 1838 by Grisebach in the form ' phytogeographical formation ', which subsequently gave place to ' vegetative formation '. Grisebach wrote,[5] ' I give the name *phytogeographical formation* to a group of plants, such as a meadow or a forest, that has a fixed physiognomic character. It is characterized sometimes by a single social species, sometimes by a complex of dominant species belonging to one family, or it exhibits an aggregation of species which, though diversified in organization, yet have some feature in common, as is the case with alpine meadow-wastes consisting nearly exclusively of perennial herbs.'

Other writers have, however, attached a narrower signification to the

---

[1] See p. 93 and Chap. XCVIII.   [2] Drude, 1888, 1889, 1895.
[3] Compare Clements, 1905.   [4] See Chap. III.   [5] Grisebach, 1838 ; 1880, p. 2.

term 'formation'. Hult, in his excellent papers on the phytogeography of Finland[1] establishes about fifty 'formations' as existent in North Finland—an Empetrum-formation, Phyllodoce-formation, Azalea-formation, Betula nana-formation, Juncus trifidus-formation, Carex rupestris-formation, Nardus-formation, Scirpus caespitosus-formation, and others. Similarly Kjellman[2] divides the algal vegetation into numerous 'formations' named after predominant species; and the same use of the term is made by Stebler and Schröter in connexion with the types of Swiss meadows that they recognized.

In like fashion we must distinguish as different formations beech-forest, oak-forest, birch-forest, and other dicotylous forest; ling-heath, Empetrum-heath, Erica-heath; or in fresh water, Scirpus lacustris-formation, Phragmites-formation, Equisetum limosum-formation, and so forth.

This means a subdivision of the vegetation based upon the local domination of certain species, and may consequently result in matters of wide and general import escaping notice, and communities naturally belonging together, as evidenced by their pursuance of identical economy, may not be recognized as such. The Empetrum-formation, Azalea-formation, and Phyllodoce-formation are oecologically essentially alike, and may be regarded as members of a more comprehensive natural community, the dwarf-shrub heath; the Scirpus-formation and Phragmites-formation are likewise members of one community, and it often evidently depends upon mere accident whether the one or the other of these 'formations' prevails at a given spot.

Obviously, such special communities play a part, and must be distinguished in any detailed description of the vegetation of a definite area; but it is better to speak of them as *associations*, and to follow Grisebach in the use of the term 'formation', since he was the first to introduce and define it, although he laid less stress on its oecological meaning.[3]

A **formation** may then be defined as a community of species, all belonging to definite growth-forms, which have become associated together by definite external (edaphic or climatic) characters of the habitat to which they are adapted. Consequently, so long as the external conditions remain the same, or nearly so, a formation appears with a certain determined uniformity and physiognomy, even in different parts of the world, and even when the constituent species are very different and possibly belong to different genera or families. Therefore—

A formation is an expression of certain defined conditions of life, and is not concerned with floristic differences.

The majority of growth-forms can by themselves compose formations or can occur as dominant members in a formation. Hence, in subdividing the groups of hydrophilous, xerophilous, and mesophilous plants, it will be natural to employ the *chief types of growth-forms as the prime basis of classification*, or, in other words, to depend on the distinctions

[1] Hult, 1881, 1887.   [2] Kjellman, 1878.
[3] The term 'formation' is so employed also by Ascherson, 1883, p. 728; Kerner, 1891, p. 830; C. Schröter und Kirchner, 1902; Kearney, 1900; Ganong, 1902; Clements, 1902 a. Adamovicz, 1898, and Cowles, 1901, adopt at least approximately the same course. On the other hand, Flahault and W. G. Smith employ the word 'association' in this sense.

between trees, shrubs, dwarf-shrubs, undershrubs, herbs, mosses, and the like.[1]

Upon this basis the following types of formations must be distinguished :—

1. **Microphyte-formation or thallophyte-formations,** in which the community is composed exclusively, or mainly, of lichens and algae. Here, except in the case of sea-weeds, there can scarcely be any question of more than one stratum (storey) of plants.

2. **Moss-formation.** Here algae may form a lower stratum (storey).

3. **Herb-formations,** such as meadow, prairie, grass-steppe, and others. Here may occur two or several strata (storeys) ; namely, a humbler vegetation of thallophytes or mosses, and a taller vegetation of herbs ; the herbs may in turn be ranged into storeys of different heights.

4. **Dwarf-shrub formations and undershrub-formations** also include admixture of herbs, which sometimes overtop the dwarf-shrubs, and undershrubs. These longer-lived constituents preponderate, however, and among them there may occur several storeys of vegetation belonging to the types described under 1, 2, 3.

5. **Bush-wood or Shrub-wood** is composed of taller, lignified, many-stemmed plants. Compared with the previous ones the community gradually has become richer in growth-forms ; there may occur epiphytes and lianes, and below the highest story, all the forms of vegetation of 1, 2, 3, 4 ; for instance, mesophilous herbs growing in the shade. Yet the oecological conditions prevailing in bush are not the most favourable to plant-life, and the ground-vegetation is often very scanty, because bush-wood may be so dense as to permit the passage of even less light than does forest.

6. **Forest** is the tallest type, and exhibits the greatest multiplicity of growth-forms as well as the largest number of storeys :

High forest, composed of trees—light-demanding and shade trees ; [2] in tropical forest more than one storey.

Underwood, composed of shrubs, dwarf-shrubs, and undershrubs.

Forest-floor vegetation, composed of herbs, mosses, thallophytes, and many saprophytes, and depending upon the light, which is more or less enfeebled by the tree-crowns, upon moisture in the soil and air, upon the humus, and other factors. Beneath such species as beech, spruce, and silver fir, which grow close together and cast a deep shade, there is a very meagre vegetation, but below light-demanding trees, in harmony with their need for light, there is a richer vegetation. The flora at the edge of a forest may differ materially from that of the interior, because the conditions of illumination in the former position permit the development of many species that are excluded from the latter. Grevillius [3] found that the tall herbs in thinly wooded and therefore well-lighted Scandinavian forests can be ranged into different types, which differ from one another in the arrangement of the inflorescences, the form and position of the assimilatory organs, innovation, time of flowering, and distribution in the different storeys of the general plant-community— all these relationships deserve closer attention and investigation.

---

[1] See Kerner, 1891.　　　　[2] See p. 18.　　　　[3] Grevillius, 1894.

It may be asked—why arrange the various types of vegetation in the classes named on p. 136 ? Why not use each growth-form as a foundation upon which to build a special class ? The following classes could then be distinguished : that of forest-formations, of bush-formations, of shrub-formations, of dwarf-shrub formations, of perennial-herb formations, of moss-formations, and of alga-formations.[1] Within each such class one would further be able to distinguish hygrophilous, mesophilous, and xerophilous formations, and to define them. From a morphological standpoint this would possess a certain interest, but from a phytogeographical one it must be dismissed, because it would involve the separation of formations that are oecologically closely allied. It is nature of locality that must be represented by formations ; and naturally allied localities may include different collections of growth-forms, yet they must be grouped in the same class.

According to the isolation or combination of the growth-forms in a formation we have to distinguish *simple* formations and *compound* formations.

Simple formations :—Formations consisting solely of one type of growth-form are few. An example is the phyto-plankton formation. The species composing this microphytic free-floating flora belong to different families or even different systematic classes, but they may all be grouped together as belonging to one growth-form adapted to the free-floating mode of existence in water ; plankton is a purely edaphic oecological formation.

Compound formations :—Usually many growth-forms, and often, to some extent, different formations, are combined to form a single whole. For instance, as will be explained hereafter, the reed-formation is one dominated by divers monocotylous herbs, which are social and perennial in habit and varied in stature ; but on the ground, also in the water between and beneath the reeds, there sometimes flourish other growth-forms comprising what may be termed *subordinate communities :* thus there may be communities composed of Schizophyceae, plankton, pleuston, and limnaea,[2] which are more or less influenced in their composition by the dominant community. Again, in forest the different lower storeys are constituted of growth-forms which, for the most part, are able in themselves to give rise to distinct formations (bush, grassland, moss-formation, and others), but in the two cases the species occurring would usually be different. For the shade of forest or of tall vegetation affects not only the conditions of illumination, but also the humidity and temperature of air and soil. As an example of a species capable by itself of giving rise to one independent formation, also of occurring as a subordinate member in another, we may cite Calluna vulgaris. This species as a dominant plant forms a widely distributed community belonging to the type of dwarf-shrub heath ; but it can occur as low vegetation in thin pine-forest ; to what extent the latter differs from the open ling-heath has not been adequately investigated.

A formation consisting of several storeys may therefore be composed of both xerophytes and mesophytes : there are forests—for instance, sclerophyllous forest—in which markedly xerophytic species form the uppermost storey, but mesophytic species the lower storeys.

---

[1] This approximates to the method adopted by Kerner, 1891, p. 821, who distinguished nine kinds of societies.
[2] See Chap. XLIX.

Mixed vegetation : Quite different from the complex, several-storeyed formations just mentioned is mixed vegetation—which consists of small patches of different formations, occurring close together, but each retaining its own individuality as a pure formation. This is specially the case where the terrain varies suddenly and greatly, and thus causes the oecological conditions to do likewise. In many mountainous districts, rocks, small bushlands, grasslands, perhaps pools of water, and the like, succeed one another within a limited area, without one formation perceptibly affecting another, and without any principle, other than chance, revealing itself in the admixture. The more level and uniform does soil remain over a wide area, the larger and more uniform are the formations ; the more uneven and variable is it, the more mixed is the vegetation. It is impossible to draw a sharp distinction between a vegetation consisting of several pure formations mixed together, and one consisting of a single complex formation ; for the smaller the patch of a mixed vegetation is the more are the species influenced in their biotic conditions by other adjoining ones. One condition especially responsible for change of this kind is human intervention.

By the mixing of formations, especially if these be extensive, there come into existence various physiognomical types of landscape which differ according to the constituents.

Fixity of formations. 'Fixed physiognomic character' is a part of Grisebach's definition of a plant-formation. Recent writers, like Beck [1], and Drude [2], emphasize the fact that 'fixity' is a character essential to the concept of a formation. We may paraphrase Drude as follows :—

A vegetative formation is an independent community of first rank, which consists of like growth-forms or of such as are necessarily associated, and is confined by its natural boundary to a site determined by the prevalence of identical conditions of existence ; it is thus assumed that without external interference no real change in the nature of any community occupying the site can set in—the community is 'fixed'.

This fixity can be regarded as being only relative, as Drude distinctly agrees ; for, as the external conditions change, every formation will be able to undergo modification, and will always do so in the course of time [3] : there are formations that may have remained, and perhaps will remain, unchanged for thousands of years—tropical rain-forest, for example. There are others that will soon be exterminated in the places they now occupy, and driven to other sites.

Secondary formations. By the term 'secondary formations' we indicate formations which have arisen through human interference.[4] There are various formations of this kind that are changed only in their flora : such may be termed *semi-cultivated* [5] formations, and are exemplified by North European heaths, which have been modified by the browsing of animals, and by other agencies. But there are entirely new formations resulting from man's activity in destroying forest, in farming, and in otherwise utilizing the soil ; of such a nature are the ' sibljak ' in Servia [6],

---

[1] Beck von Mannagetta, 1902.   [2] Drude, 1896, p. 286.
[3] Compare Section XVII.
[4] Warming, 1892. They are the 'Substitute Associations' of W. G. Smith, 1905, p. 62.   [5] Krause, 1892 *a*; see also Gräbner, 1909.   [6] According to Adamovicz, 1902.

oak-scrub in Jutland, and the Tristegia glutinosa-grasslands in Brazil[1] : these formations arise and can only be preserved by cultivation, and must be described as *secondary formations*.

Sub-formations. There are various formations which are so immense in extent and range, and display certain minor oecological distinctions, that it is advisable and convenient to subdivide them into sub-formations ; for example, plankton-formation, coniferous forest, and dicotylous forest, can thus be subdivided.

In dealing with forest one character for consideration is its deciduous or evergreen nature, another concerns the question as to whether the ground-vegetation consists of the same or different growth-forms ; the differences in the ground-vegetation in forests may be so great that they give an entirely different appearance to these, which must therefore be accounted as sub-formations. The existence of sub-formations must, however, be *justified on oecological grounds* (by the depth, water-content, or kind of soil, or by other factors). Increase in our knowledge of oecology will shed light upon these questions.[2]

C. ASSOCIATIONS[3]

The same formation, in different districts and localities, or even at different seasons of the year, may be composed of different species, and is perhaps mostly so. Plankton-formation in certain months may be composed of species different from those in other months ; reed-formation in various places has Phragmites communis as the dominant species, in others Scirpus lacustris or Typha, yet it remains the same formation. Beech-forest and oak-forest are specifically different forms of the same formation—the summer-green (deciduous) dicotylous forest in the temperate climate. Coniferous forest in one place may consist of Pinus sylvestris, at another of P. montana or P. halepensis, or in North America of entirely different species of Pinus; or again, it may be formed by species of Picea or Abies; furthermore, it may be an admixture of coniferous species. A cornfield is a ' cultivated formation ' of annual or biennial species ; but the maize-field, rye-field, wheat-field, and buckwheat-field are different forms of it.[4]

Such smaller, often more-localized subdivisions or kinds of the formation may be distinguished most correctly by the general term *association*, which has been employed in this sense by various botanists.[5]

[1] Warming, 1892.

[2] It is a matter of great difficulty to find suitable names for the various kinds of formations. Some of the many words in common use have been utilized as scientific terms, such as steppe, prairie, tundra, *Caa Tinga*, alang-alang, savannah, and others (see Warburg, 1900), but many popular words are unsuitable. Other scientific terms, such as plankton (Hensen), and *garide* (Chodat) are of recent introduction.

[3] Humboldt's ' plantes associées '. See p. 137.

[4] In regard to Woodhead's ' complementary ' and ' competitive ' associations, see p. 95.

[5] Including C. Schröter und Kirchner, Ganong, Kearnley, Cajander, Adamovicz, and partly W. G. Smith. Hult's, 1885, 'formations' (see p. 140) are associations, and so also apparently are Drude's ' unit formations '. Discussion of the different significations attached to the term ' association ' by authors is impossible here.

An **association** is a community of *definite floristic composition* within a formation ; it is, so to speak, *a floristic species of a formation which is an oecological genus*.

While the formations in different floras may be the same, associations are dependent upon the character of the flora of the country concerned.

Terminology. Associations may conveniently be denoted according to the plan suggested by Schouw,[1] by the addition of the suffix *-etum* to the name of the characteristic species or genus ; for instance, in the reed-formation there are phragmiteta, scirpeta, typheta, and the like ; similarly there are saliceta, pineta, and so on ; in order to indicate the species concerned the specific name has to be added, thus, for example, scirpetum Scirpi lacustris, salicetum Salicis albae, pinetum Pini sylvestris, pinetum Pini montanae ; or more briefly, scirpetum lacustris, salicetum albae, and the like.[2] This terminology corresponds somewhat to popular usage, in which two words are combined and one of them indicates the species, as is the case with the terms beech-forest, oak-forest, birch-forest, or Kerner's terms, gold-beard[3] meadow, feather-grass[4] meadow.

When an association is composed, not of a single prominent species, but of several in *equal shares*, it may be denoted by a compound term, for instance, scirpo-phragmitetum, fago-quercetum, or in some other way—for instance, by its popular name.[5]

In many cases associations are brought into existence within the formation by minor distinctions in the soil, because different species react in a slightly different manner. In other cases accident seems to decide the question : the species which first colonized the spot will subsequently be able to maintain itself against others, as in the instance of the reed-associations clothing the banks of rivers and lakes in north-temperate Europe.

Associations may occur irregularly as *patches* in the formation ; or may exhibit a *zonal* arrangement. The latter is always the case when the formation grows in water (in which case depth decides the matter), or on the banks of rivers, lakes and pools, where subterranean water can play its part ; with increasing height of soil above the water-table vegetation displays a zonal change.[6] Plants react to infinitely small differences in the water-content of soil. In very many cases these belts of plants at the same time represent developmental series, inasmuch as each association is successively expelled by the next one.

Associations may also be dependent upon shade and other oecological factors. An alpine meadow has a different floristic composition according as it clothes the northern or southern slope of a mountain, yet it remains a typical meadow, and thus unchanged as a formation ; grassy surfaces lining a railway differ floristically according to the aspect,[7] but they are merely different species, or, in other words, different associations, of the same formation.

D VARIETIES OF ASSOCIATION

As in oecology a formation may be regarded as a genus, and an association as a species, so we may also recognize oecological varieties

---

[1] Schouw, 1822, who referred to ericeta, rhododendreta, arundineta, pineta, fageta, and the like. [2] See Cajander, 1903.
 [3] [Chrysopogon Gryllus]. [4] [Stipa]. [5] See C. Schröter und Kirchner, 1902.
 [6] See Raunkiär, 1889 ; Warming, 1890, 1906 ; MacMillan, 1896 ; Magnin, 1893, 1894 ; Pieters, 1894 ; Clements, 1905 ; Shantz, 1905.
 [7] Stebler und Volkart, 1904 ; Stenström, 1905.

dependent upon minor differences in an association.[1] A beech-forest, fagetum Fagi sylvaticae, is not everywhere and on every terrain absolutely uniform and constant in composition, but owes its modification, *inter alia*, to differences in soil ; in one beech-forest the ground-vegetation may be dominated by Asperula odorata, which denotes alkaline, loose, well-aerated humus, while in others Aira flexuosa or Vaccinium Myrtillus may play a dominant part and thus indicate acid humus. There are also beech-forests in which still other species characterize the flora on the ground. Accordingly, beech-forests may be described as fageta asperulosa, fageta myrtillosa, and so on. Similarly there are betuleta hylocomiosa, betuleta cladinosa, and others, according as the ground is clothed with mosses of the genus Hylocomium, or lichens of the genus Cladina. The ground-flora in forests of Pinus sylvestris differs widely according to various conditions, including the character of the soil ; in Baltic forest there occur together many species that in the Hercynian hills and lower highlands are distributed in different formations.[2]

Gräbner[3] distinguishes in the heath-formation various associations : namely, ling-heath (callunetum), with several varieties, according as Pulsatilla, Genistae, or Solidago, or others predominate as perennial herbs; a tetralix-heath (ericetum), which also has several varieties ; an Empetrum-heath, and others.

Certain species are so accommodating that they can grow on very different soils, and can therefore give rise to different associations. Woodhead[4] has pointed out that this is the case with Pteris aquilina ; and he therefore distinguishes between two varieties of associations, which he terms *meso-pteridetum* (an association of Pteris with Holcus lanatus and Scilla festalis), and *xero-pteridetum* (an association of Pteris with Calluna, Vaccinium Myrtillus, Aira flexuosa, and others).

Edaphic varieties. Varieties in an association may be distributed in patches, or zones, according to the conditions prevailing in the soil ; in a meadow we can see both these forms of arrangement, corresponding to the shapes and extent of the depressions in the ground. Here we are dealing with *edaphic varieties*.

Geographical varieties. But varieties may have arisen through historic or·climatic causes; for example, according to Beck von Mannagetta[5] Pinus Laricio is distributed as a high-forest tree over so wide an area that its ground-vegetation belongs to three different floral districts : the Pontic, Baltic, and Mediterranean. Here, then, we have three *geographical varieties*. Höck's investigations show that the ' companion-plants ' of the respective forest-trees in different parts of Europe may be widely different, even if the limiting boundaries of the trees and of their companion-plants coincide to some extent.

In the succeeding Sections an attempt will be made to give a survey

---

[1] Cowles, 1899, p. 111.   These varieties are sometimes spoken of under the name ' facies '.   The term ' facies ' seems to have been employed by Lorenz, 1863, first to denote small local differences in a formation, but subsequently to denote an association, and in a somewhat similar or slightly different sense it is employed by others.   [The term ' facies ' has already the recognized meaning in English biological science of ' general aspect or appearance ' (N.E.D.), and its use in a restricted sense as that of a variety of an association is barred.   P.G.]

[2] Drude, 1902.   [3] Gräbner, 1901.
[4] Woodhead, 1906, pp. 114, 363.   [5] Beck von Mannagetta, 1902.

of the most important formations existing on the Earth. But to associations and their varieties less attention can be devoted, partly because the subject has not been sufficiently investigated, and partly because it would involve too much detail; for such detail reference should be made to special papers. It must be said that there still prevails great confusion in the use of the terms 'formation', 'subformation', and the like, and this places difficulties in the way of a clear comprehension of the subject.

In the following scheme an endeavour is made to arrange the formations within the respective classes according to definite principles in a definite order of succession. The scheme commences, so far as is practically possible, with the simplest and shortest types, which form a low stratum of vegetation corresponding to the growth-forms described in Chapter II and on page 141; they can not only form separate communities, but also occur as subordinate constituents in formations composed of taller forms. For example, the formations on acid humus soil are, as a whole, arranged as follows: First, the low formations composed of thallophytes and mosses; secondly, those composed of two storeys, dwarf-shrubs, and those forms belonging to the preceding type; and, thirdly, still taller formations composed of several storeys, shrubs and trees, among which representatives of the first two types of formations are usually found and may, indeed, have developed as the real foundation of the whole formation. An attempt has been made to show between the different formations a logical and biological connexion, which in many cases is also developmental in significance. Formations commence often as open, not fixed, communities that are composed of short and lowly organized species; with the passage of time their development is continued by the immigration of species that are taller and more successful in the struggle for existence, until a final stage is reached.[1]

Cowles [2] strongly insists 'that the plant societies must be grouped according to origins and relationships, and the idea of constant change must be strongly emphasized'. A. Nilsson [3] followed the same train of thought in his arrangement of Swedish formations into four series—the heath-series, meadow-series, marsh-series, and moor-series—as he particularly took cognizance of the amount of water and nutriment in the soil. Many of the formations brought into his series are closely allied and can readily develop into one another, but there are others that are certainly not allied, for instance, littoral meadow, alpine meadow, wood-meadow,[4] meadow spruce-forest, and meadow oak-forest.[5]

It is true that changes in the physical relationships of soil are everywhere and always taking place, and that in close correlation with this plant-communities also undergo modification.[6] This will be discussed in the final Section of this book. In the future it will be an interesting and important problem to trace out in each country or district the development of the vegetation, as regards not only flora but also plant-

---

[1] Warming, 1890, 1906; Moss, 1906, 1907; A. Ernst, 1907.
[2] Cowles, 1901 b.   [3] A. Nilsson, 1902.
[4] [Very open, thinly wooded forest, with an abundance of grass and other herbs representing meadow.]
[5] In regard to the genesis of plant-communities consult Engler, 1899, p. 179.
[6] Warming, 1899.

communities as a whole.[1] But it does not seem possible to use development as the fundamental basis of classification of plant-communities : for developmental changes are too dependent upon local conditions ; a formation does not develop merely in a single definite direction, but will modify in one direction at one place and in another at another place, according to the prevailing conditions.[2]

[1] As regards Denmark see Warming, 1904, 1906, 1907–9.

[2] Concerning the oecological nomenclature and classification see : G. Beck von Mannagetta, 1902 ; Brockmann-Jerosch, 1907 ; Cajander, 1903 ; F. Clements, 1902, 1904, 1905 ; Cockayne, loc. cit., 1905 ; Cowles, 1899, 1901 ; Drude, 1896, 1905 ; Flahault, 1900, 1901 ; Ganong, 1902 ; Gräbner, 1905, 1898 a, 1909 ; Harshberger, 1900, &c. ; Kearney, 1900 ; Kerner, 1891 ; A. Nilsson, 1902 ; Kirchner und Schröter, 1896–1902 ; Shantz, 1905 ; R. and W. G. Smith, 1898, 1899, &c. ; Stebler and Volkart, 1904 ; Woodhead, 1906.

SECTION IV

CLASS I. HYDROPHYTES. FORMATIONS OF AQUATIC PLANTS

CHAPTER XXXVI. OECOLOGICAL FACTORS

BEFORE considering the various communities of aquatic plants it is necessary to discuss the general characters of water and its oecological factors, in so far as these affect the distribution and existence of plants confined to water.[1]

Air in water. Air occurs dissolved in water in variable amounts. In the atmosphere[2] and in water the gases present are the same, but their relative proportions are different; the gas absorbed by ordinary water contains more oxygen and much more carbon dioxide in proportion to nitrogen than does the atmosphere. Just as in the case of land-plants, these two gases are the only important ones, the former in respiration, and the latter in assimilation. Only certain bacteria can dispense with oxygen. Air can reach parts that are submerged in water with much greater difficulty than it can reach parts situated in the atmosphere or in ordinary soil; indeed, stagnant water may become so poor in oxygen as to almost exclude the existence of higher plants and animals. Apparently in consequence of this, certain species occur particularly in places where the water is very troubled or has a rapid flow, and where there is a constantly fresh supply of water; possibly for the same reason many submerged parts (leaves) or whole plants (algae) are divided into capillary segments—compare the construction of gills—by which means the surface in contact with water is greater than if the organ presented a single surface; and perhaps for the same reason many algae and Podostemaceae bear long hairs that serve as probably respiratory organs or enlarge the assimilatory surface. The obstructed supply of air is also a reason, and perhaps the most important one, for the large air-containing spaces[3] which occur in many aquatic plants (sometimes occupying more than seventy per cent. of the whole volume of a plant); these spaces serve, *inter alia*, to convey air, and especially oxygen, from parts in the air to those in the water or mud. Certain swamp-plants, and especially some living in mangrove swamps, possess special respiratory organs which will be described hereafter.[4]

When access of air is prevented and water becomes poor in oxygen there are formed in soil *humous acids*, which are characteristic of moor and peat soils, and which render soil physiologically dry.[5]

As the temperature rises the power of water to dissolve gases decreases, and this is perhaps the main reason for the disappearance of certain

---

[1] See upon this subject Oltmanns, 1905.   [2] See p. 14.   [3] See p. 98.
[4] See Chapter LX.   [5] See Chapter XVI and Section VI.

aquatic plants in summer, at which season the temperature is higher and the light more intense.

Light in water. We must assume that every aquatic plant has its own minimum, optimum, and maximum intensity of light. Illumination is of profound importance in relation to the distribution of algae, also to the abundance of their species at different seasons—but on this latter point very little is known. The farther apart lie the maximum and minimum the more extensive will be the area of distribution of the species concerned.

Light plays in assimilation the same part as in the case of land-plants ; yet there are some peculiar relationships to be considered. Light is *weakened*, partly by reflection from water, partly by absorption in water, and partly by floating particles ; the weakening is therefore more considerable the dirtier the water. Submerged plants, for this reason and also because there is no transpiration, acquire as a whole the pattern of a shade-leaf ; they become lanky, like etiolated plants, thin, and their assimilatory tissue shows little differentiation.

Light penetrates downwards only to a certain distance, and constantly *weakens* with increasing depth, so that the assimilatory energy varies greatly with the depth[1] ; consequently, except in the case of bacteria, plants cannot be active at great depths. A '*regional*' *distribution* of the vegetation results from the different powers possessed by plants of living in different intensities of light. Distinctions have been drawn between :

1. *Euphotic* vegetation, which receives an abundance of light.
2. *Dysphotic* vegetation, which lives in weakened light.
3. *Aphotic* vegetation, which lives in very weak light or darkness.

Spermophyta descend at most to thirty metres (Zostera, for instance, to twelve or fourteen metres in Denmark) ; algae to forty metres, but living algae have been found 120–150 metres below the surface (in clear alpine lakes in Switzerland Characeae descend to 25–30 metres, but in Baltic lakes only to 6–8 metres) ; in Lake Geneva, according to Forel, a moss, Thamnium alopecurum var. Lemani, has been found at a depth of 60 metres ; the extreme depth to which light apparently penetrates is 400–500 metres. The presence of the protococcaceous Halosphaera viridis at 2,200 metres below the surface of the sea is certainly to be explained as a result of sea-currents or as a periodic sinking.

As the rays of different *colours* are unequally absorbed plants descend to different depths. Red rays are absorbed in the upper layers of water ; the green, blue, and ultra-violet not before the lower layers. Ultra-violet rays can be detected by means of photographic plates at a depth of 400 metres. Correlated with these facts is the '*regional*' *distribution of Algae, according to depth.*[2] Green Algae assimilate best in red light, Brown Algae in yellow light, but Red Algae are most active in green and blue light ; consequently, the first named occur only in the upper layers of water, while the last are especially in the deeper layers. Against this theory maintained by Engelmann the objection is urged by Oltmanns that with algae it is only a question of the intensity of light, and that the colour of sea-water merely acts as a screen. Recently, Gaidukow[3] has shown that when Oscillatorieae are cultivated in coloured light they

---

[1] Proved by B. Jönsson, 1903.   [2] Börgesen, 1905.   [3] Gaidukow, 1903.

change in colour and assume that which is complementary to the light acting on them, and by this means assimilate more vigorously.

Temperature of water. Submerged aquatic plants are exposed to far smaller extremes and fluctuations of temperature, both diurnal and annual, than are land-plants, because water has a high specific heat and is a bad conductor of heat : annual changes of temperature descend only to relatively small depths except in shallow water. Many aquatic plants hibernate in their green state, because no considerable degree of cold reaches them, and most of them are perennials. Their optimum for growth is generally low ; certain species, including Hydrurus (an alga belonging to the Phaeoflagellata), thrive only in very cold water. The disappearance of many algae in summer may be due to their optimum temperature having been exceeded. Arctic fresh-water lakes are usually very poor in organisms—a circumstance that may possibly be ascribed to the temperature of the water. Algae are frequently very sensitive to sudden changes in temperature,[1] in salinity or in other conditions. Each species has its own peculiarities.

High temperatures are encountered only in hot springs, where the plants growing are almost exclusively Oscillarieae and other Cyanophyceae, which may be representatives of the vegetation that was the first to appear on Earth.

The temperature of water decreases as the distance below the surface increases, but at different rates in fresh and salt water. In standing fresh water, at the bottom of lakes which are so deep that the annual variations of temperature in the upper strata cannot affect the strata at the bottom, the temperature is about 4°C., because fresh water attains its maximum density at this point. Strata of water lying above this may thus be much colder in winter. The temperature at the bed throughout the year in Swiss lakes is about 5° C., but is always below 4° C. in Baltic lakes, which are frozen at the surface during winter. In the sea, on the contrary, strata of water are colder the deeper they lie ; moreover, warm, or cold and salt, currents may be intercalated between them. The influence of temperature upon the distribution of aquatic Spermophyta is proved by an observation made by Magnin,[2] who discovered that they descend to a depth of 11 metres in the warmer Jura lakes, but only down to 6 metres in deep and cold lakes.

Temperature affects the amount of gas dissolved in water ; the colder this is the richer is it in oxygen and carbonic acid, and the more favourable may be the conditions for nutrition and consequently for growth. This is possibly the cause of the luxuriant development of algal vegetation in arctic seas.

Nutritive and other substances dissolved in water. Water contains in solution many substances which vary according to the kinds of rock or soils with which it has entered into contact. Calcic carbonate is commonly present, being dissolved by the contained carbonic acid ; and, as many aquatic plants (Characeae, species of Potamogeton, and mosses) seize upon the carbon dioxide contained in the calcium hydrogen carbonate, incrustations of chalk are excreted on their surfaces, and may lead to deposits of lime on lake-beds.[3]

[1] Oltmanns, 1892.   [2] Magnin, 1894.   [3] C. Wesenberg-Lund, 1901.

Brandt[1] asserts that the sea is rich in nitrogen, which is constantly being supplied to it, and is reduced by denitrifying bacteria. This process is more active in tropical than in temperate seas, and partly for this reason the ocean in sub-tropical and tropical regions is relatively poor in organisms, while it is rich in them in cool and cold regions.

Many waters contain organic compounds which, by consuming the oxygen, render water unsuitable for the existence of autophytes.

The nutritive substances most important to plants, such as potassium, phosphoric acid, ammonia, and sulphur, occur in all water, but only in small quantities and in an extreme condition of dilution ; in fresh water compounds of potassium and nitrogen are present in far greater proportions than are the other compounds in question. But we have no knowledge that these conditions have any distinct effect upon the distribution of aquatic plants. Certain desmids and diatoms are stated to prefer lime, others silica ; similar minor differences are attributed to other plants. *Common salt*, on the contrary, is of profound importance. In sea-water, among the numerous salts, including chlorides of sodium, magnesium, and potassium, also sulphates of magnesium and calcium, the first named is of far the greatest significance. The amount of common salt in the ocean fluctuates within only narrow limits, but that in smaller seas varies greatly, not only in different sites, but also in the same place at different times. The following are the approximate percentages :—

Red Sea 4, Mediterranean Sea 3·5–3·9, Pacific Ocean 3·5, Skager-Rak 3, Kattegat 1·5–3, the Great Belt 1·27, the Öresund 0·92 (in the last two, very variable according to the currents), Gulf of Bothnia 0·1–0·5, Gulf of Finland 0·3–0·7. These statistics relate to superficial water ; but in parts of the sea round Denmark at a greater depth there is a saline undercurrent from the North Sea.

The great difference between the floras of salt and fresh water will be discussed later in this Section.

Although not a few fresh-water algae, especially lowly organized forms, can adapt themselves to the presence of common salt, which causes an enlargement of their cells and other formal changes, yet there are no plants other than certain diatoms that are common to fresh and slightly saline water ; nevertheless, in the brackish water of the Baltic Sea there grow some Characeae, Enteromorpha intestinalis, and Potamogeton pectinatus, which also occur in fresh water.

The cyanophyceous communities occurring in special places will be dealt with later.

Specific gravity of water. Salt water and fresh water differ greatly in specific gravity, and consequently in buoyancy, which plays a great part in connexion with plankton-organisms ; for fresh water, as is well known, has a smaller power of keeping bodies from sinking. The regular seasonal changes in temperature of fresh water bring in their train corresponding ones in the specific gravity and viscosity of the water. Many plankton-organisms undergo periodic changes of shape, which all seem to be in the direction of increasing the resistance-surface, and which synchronize with the changes of temperature. It therefore seems highly probable that these seasonal variations in shape are to be regarded as responses on

[1] Brandt, 1904.

the part of the organism to periodic changes in the buoyancy of fresh water.[1]

Colour of water. Water in a pure condition is blue. Any other colour may be caused by organisms,[2] by suspended particles of clay and the like, or, especially in fresh water, by humous acids ; yellow or brown water often contains many humous acids and is acid in reaction, whereas alkaline (hard) water is clear (blue).

Movements of water. Movements of water are of great importance to vegetation. They assume the form of waves (breakers) or currents, and lead to a fresh supply of *oxygen ;* in streaming water assimilation is more active.[3] Still water is very inimical to vegetation ; and for this reason many species are absent from stationary depths over large areas, or from enclosed calm inlets. In addition, moving water conveys additional *nutriment ;* for instance, sea-water contains but little iodine and calcium, yet large quantities of these are stored by many Algae. Movements of water are all the more essential inasmuch as many aquatic plants, and particularly algae, as a rule have no far-reaching roots (in a physiological sense). Finally, movements of water act mechanically, in that they stretch and bend plants with a force that varies with their strength. In larger plants mechanical tissue is developed[4] ; calcareous incrustations may also contribute to the stability of sea-weeds, but it is worthy of note that chalk-forming algae and many crustaceous algae grow, some in deep, and others in still water. The general shape of aquatic plants is adapted to the surroundings in divers ways ; thus in rapid currents we find very elongated plant-members (ribbon-like foliage, and long filiform algae).

A distinction must be drawn between currents and wave-movements, as many species withstand the former but not the latter. Very many species flourish only in tranquil water.

Movements of water also favour the dispersal of propagative organs, such as detached vegetative parts, spores, or seeds.[5]

Aquatic species of plants throughout have a very *wide geographical distribution.* This is partly because the conditions of life are uniform, or only slightly different, over wide areas, and the transport of marine plants over great distances is very easy, and partly because many species are carried far by water-haunting birds and insects, or are transported by air-currents, as is the case with the smallest, mostly microscopic, species.

Differences in the aquatic flora associated with geographical situation show themselves in some respects more marked in the sea than in other waters ; this may be due to greater physical differences, and to the universally greater constancy in amount of salt, in temperature, and other characters of sea-water.

The general morphological and anatomical adaptation of submerged organs and aquatic plants has already been considered in Chapter XXVII.

---

[1] Ostwald, 1903 *a* ; Wesenberg-Lund, 1900, 1908.   [2] See Chap. XXXVIII.
[3] F. Darwin and D. Pertz, 1896, p. 296.   [4] Wille, 1885.
[5] Compare Hemsley, 1885; Schimper, 1891 ; Sernander, 1901 ; Rosenvinge, 1905 ; Kjellman, 1906.

CHAPTER XXXVII. FORMATIONS OF AQUATIC PLANTS

THE different oecological factors just dealt with yield many distinctions in the environments of aquatic plants, according to the modes in which they are combined. But another factor of equal importance to the production of formations, and one that lies at the very base of this, is edaphic in kind : it is the *nature of the substratum*, which brings with it many differences, including those in the mode of fixation.

The first formation to be founded is that of *plankton*, which is constituted of those microphytes that float free in water and are adapted to this mode of life.

Closely allied to plankton, but of a subsidiary and less important nature, is the glacial community forming the *cryophyte*-formation, which is composed of microphytes that are periodically exposed to ice-cold water.

A third type is the *hydrocharid*-formation or *pleuston*,[1] which is constituted of macrophytes floating on water, or, more rarely, floating in it, and adapted to this mode of existence.

Forming a sharp contrast to the preceding is *benthos*,[2] which includes aquatic plants that are fixed to the substratum, or, like some diatoms, creep over it. In opposition to free-living species, these display many types of construction associated with mechanical rigidity, and various other structural features. The formations under this head may be referred to two great and widely different groups dependent upon the *nature of the substratum*, according as this consists of solid rock or loose material.

Accordingly the fourth group of formations is constituted of *lithophilous hydrophytes*, which are water-plants living attached to stones : this group may be subdivided into several formations.

The remaining plant-communities associated with a *loose* substratum are to be separated into those composed solely or mainly of microphytes, and those in which vascular plants or larger algae play the chief part ; and after this a further subdivision may be adopted according as the plants live in saline or fresh water. In addition the kind of soil is of influence, according as it mainly consists of inorganic bodies or organic fragments.[3] Thus there result the following formations : *microphyte*-formation ; *enhalid*-formation (in the sea) ; and *limnaea*-formation (in fresh water) : the last two are closely allied.

Hydrophytic formations and their associations are often distributed in a definite manner, usually *in zones*, according to depth of water and the conditions involved in this : there are deep-water associations and littoral associations. In the deepest parts of the sea and of fresh water there is often mud, in which only saprophilous communities of microphytes (*abyssal* vegetation) can exist. In less deep parts there is a much richer *littoral* vegetation, which consists of more highly organized algae and cormophytes, and is arranged zonally according to depth, that is, according to illumination. The arrangement of these associations is greatly influenced by differences of soil and movements of the water. Very

---

[1] (C. Schröter und) Kirchner, 1896, p. 14; ' macroplankton ' (Chodat).
[2] Haeckel, 1890.   [3] See Chapter XVI.

disturbed sand is utterly devoid of plants, as is the case with vast tracts of the North Sea bed; Heligoland lies like an oasis in a desert whose sandy surface is ceaselessly set in motion by breakers or by tides, and is therefore absolutely unfitted for the germination of algal spores.

On the shore, hydrophytic formations often very gradually pass over into the marsh-plant formations: this is especially true of communities living on loose soil. Marsh-plants are also zonally arranged.[1]

Subjoined is a general synopsis of the oecological class constituted of hydrophytes, and grouped into formations, which will be treated in the succeeding chapters.

I. FREE-FLOATING OR FREE-SWIMMING (PLANKTON AND PLEUSTON)—

1. **Plankton-formation.**
2. **Cryophyte-formation.**
3. **Hydrocharid-formation (Pleuston).**

II. FIXED (BENTHOS)—

A. To rocks or stones.

1. **Lithophilous spermophytic formation.**
2. **Lithophilous algal formation (Nereid).**
   a. **Halophilous nereid-formation.**
   b. **Limnophilous nereid-formation.**

B. To loose soil.

1. **Microphyte-formation.**
2. **Enhalid-formation.**
3. **Limnaea-formation.**

CHAPTER XXXVIII. PLANKTON-FORMATION

THE term 'plankton' was introduced in 1887 by Hensen,[2] to denote bodies, dead or living, plants or animals, that float passively in water and are conveyed about by wind or current. Here we are concerned only with *phytoplankton*, which always consists of minute plants (*microphytes*); some of these are autophytes, which can manufacture organic substances from inorganic material, while a far smaller number are bacteria and fungi living on the autophytes or on their products.

FLORA

Phytoplankton-organisms are all minute; they are solitary or colonial, and unicellular or multicellular. They belong to widely different systematic groups of low organization, to wit :—

1. **Cyanophyceae**[3] may occur in masses and colour the water bluish-green, sap-green, grey-green, or red. In the *sea* there occur species of

---

[1] Warming, 1906. [2] Hensen, 1887. [3] See Wille, 1904.

Trichodesmium, including T. erythraeum (which colours the water red, and occurs in the Red Sea, also in other seas, but especially near the coast); Nodularia spumigena (which causes a greenish-grey colour, and is common and even extremely abundant in the Baltic Sea); Heliotrichum (in tropical parts of the Atlantic Ocean); in *fresh water* there are species of Anabaena and Polycystis, Aphanizomenon flos aquae, Oscillatoria rubescens (especially in mountain-lakes), Coelosphaerium kuetzingianum, Gloeotrichia echinulata, and others that give to water a sap-green or bluish-green tint, and a peculiar aroma. The majority of them are genuine plankton-organisms which are found floating below the surface, but when the water is quite undisturbed swim in numbers on its surface, just like cream on milk. Klebahn and Strodtmann [1] have minutely investigated certain small irregular bodies in their cells, and express the opinion that these are air-containing spaces within the protoplasm and enable the Algae to ascend; on the contrary, H. Molisch and R. Fischer [2] maintain that the corpuscles are not air-vacuoles, but are composed of a peculiar substance of undetermined nature. Fischer and Brand [3] come to the conclusion that they have nothing whatever to do with the power of flotation. Ripe spores contain no air-vacuoles, and they sink.

2. **Schizomycetes** may be mentioned after the Cyanophyceae. They are found in the ocean even far from land down to depths of 800–1100 metres, and in considerable numbers at depths of 200–400 metres. One and the same species shows great variability in form and size. The Schizomycetes concerned are motile, and most of them are of the spiral type, while some are luminous. In lakes the number of bacteria present shows the widest differences, as it varies between a few and many thousands of individuals to each cubic centimetre. The pelagic region of most lakes has the fewest.[4] In Zurich Lake there were at a depth of 80 metres 28–30 per cubic centimetre, in Lake Constance at a depth of 60–65 metres 31–146. The number present is smallest at the surface, and larger in the somewhat deeper strata. In the opinion of some investigators the bacteria at the surface are killed by light, but according to others the larger number of bacteria in the deeper strata of water is due to the greater abundance of decaying organic matter (dead plankton).

Among Schizomycetes nitrifying and denitrifying organisms are of special significance in regard to metabolism in water, as they oxidize ammonia to nitric acid, or reduce any excess of the latter to nitrogen. Brandt [5] has put forward the hypothesis [6] that the greater activity of denitrifying bacteria in warm seas is the cause of plankton here being poorer than in colder seas.

3. **Diatomaceae** occasion brownish or greenish [7] tints in water, especially in arctic seas, where they occur in huge masses composed of countless individuals, particularly belonging to the genera Thalassiosira, Chaetoceras, Rhizosolenia, Coscinodiscus, and Thalassiothrix.[8] In fresh water there occur Melosira (especially in lowland lakes), Cyclotella

---

[1] Klebahn, 1895; Strodtmann, 1895.   [2] R. Fischer, 1904, 1905.
[3] Brand, 1905.   [4] Forel, 1878, 1901; Schröter, 1897.
[5] K. Brandt, 1904.   [6] See also H. H. Gran, 1905, and Reinke, 1904.
[7] In regard to greenish water in the North-Atlantic Ocean, see K. J. V. Steenstrup, 1877.   [8] See Gran, 1905.

(especially in Alpine lakes), Fragilaria, Asterionella, and others.[1] The genera Rhizosolenia and Attheya, in fresh water are represented respectively by a few and by one species, which are widely distributed ; the remaining species of these genera are marine. Some are solitary, but many live combined in chains of various kinds. They are all true plankton-organisms which are incapable of forming masses swimming at the surface of the water. Some are enveloped in mucilage.

4. **Peridineae** (Dinoflagellata) occur especially in salt water. They are found in greater quantities, but in smaller numbers of species in temperate seas; while the individuals are fewer, but the species more numerous and diversified in warmer seas[2] ; the genera Ceratium and Peridinium are particularly common. The numerous forms (geographical races) of Ceratium tripos play the greatest part in the sea ; while C. hirundinella occurs in large quantities in fresh water. They are provided with two flagella and are motile. Some marine forms are luminiferous, and in the autumn, when they are most numerous in the North Sea, Skager-Rak, and western Baltic, cause the sea to be phosphorescent.

Nearly all plankton-organisms are included in the four groups named above. But both in the sea and fresh water there are other species of Chrysomonadineae (Phaeocystis, Dinobryon, and others).

Only in the sea are found Silicoflagellata, Coccolithophoridae, and other Flagellata. The Coccolithophoridae are present in vast numbers, but, on account of their minuteness, escape through the dredging-net, so that their true abundance has not been appreciated until recently.[3] Phaeocystis Poucheti is a flagellate which seems to be common to the coasts of Norway, the Faroe Isles, and Iceland, and was discovered by Pouchet in huge quantities by the Lofoden Isles. The protococcaceous Halosphaera viridis, which has the form of a sphere one millimetre in diameter, commonly occurs in the temperate and warmer parts of the Atlantic Ocean at the surface, or down to a depth of 200 metres, and has been found at the depth of 2,200 metres. At considerable depths (100–300 metres below the surface) of the tropical Atlantic Ocean, Halosphaera and the diatomaceous Planktoniella represent a kind of shade-flora.

In fresh water there live many Chlorophyceae, among which some can be described as true plankton-organisms ; among these may be mentioned Sphaerocystis Schroeteri, Dictyosphaerium, Oocystis, Botryococcus, and Golenkinia radiata.[4] Occasionally Desmidiaceae, or Scenedesmus, Pediastrum, and other forms are intermingled with plankton in fresh water.

ADAPTATIONS AND DISTRIBUTION

The power of flotation shown by plankton-organisms has recently been investigated by Wolfgang Ostwald.[5] Power of flotation constitutes the main difference between plankton and all other communities of organisms. This flotation merely denotes a tendency to sink exceedingly slowly. If a body is to sink in a liquid, its weight must exceed that of an equal volume of the liquid. The rate of sinking depends partly on

[1] (Schröter und) Kirchner, 1896.   [2] Schütt, 1893.   [3] See Lohmann, 1902.
[4] See Chodat, 1898 ; (Schröter und) Kirchner, loc. cit.
[5] See Wesenberg-Lund, 1908.

the extent of its surface and its precise shape, which constitute the factor known as ' form-resistance ', and partly on the buoyancy (specific gravity and viscosity) of the surrounding liquid.   Ostwald concludes that—

$$\frac{\text{The rate of sinking}}{} = \frac{\text{Excess of weight (of the organism over that of an equal volume of water)}}{\text{Form-resistance} \times \text{viscosity (of the liquid)}}.$$

If therefore a body is to float, or, in other words, if its rate of sinking is to be reduced to a minimum, the numerator (excess of weight) must be decreased, and the denominator increased as much as possible ; so as to make the quotient approximate to zero.   Hence, in order to decrease their rate of sinking, plankton-organisms in the first place endeavour to decrease the excess of weight.   This is effected by the nature of the cell-contents (products of metabolism, such as fat and gases, play a part), also by the thickness of the cell-wall (which is always extremely thin), and it must necessarily be different in species belonging to salt water and fresh water respectively.   Plankton-diatoms have thinner walls than have those living on the bed of the water.   It should be specially noted that fatty oil supplies an effective means of flotation, and that diatoms manufacture oil ;   this fact sheds light upon the universal and great part played by diatoms in plankton.[1]   Fat is also manufactured by other plankton-organisms, such as Flagellata.

Flotation-devices.   Again, the organism strives to increase the ' form-resistance ' as much as possible, by relative increase of surface due to decrease in volume, and an absolute increase in surface due to deviation from the spherical form.   To obtain a shape presenting the maximum vertical projection (transverse section), the flotation-apparatus appears always to be placed horizontally, and therefore at right angles to the direction of sinking.   Schütt [2] has demonstrated several arrangements designed to enlarge the surface of microscopic plankton-organisms, which thus acquire an increased power of flotation and of evading too sudden ascents or descents.   Such sudden movements tend to be caused by changes in the physical characters of sea-water, and may endanger the lives of the organisms.   Plankton-organisms, and particularly diatoms, are, relatively speaking, extraordinarily large in surface :   in some diatoms and Peridineae the surface is increased by a *flotation-apparatus* in the form of wing-like extensions, threads, bristles, and spines ; or the *body itself* is as a whole filiform, sometimes curved or spirally coiled, as in some diatoms ;   others are helmet-shaped, or parachute-like, or possess sail-like or annular processes ;   still others are combined into threads, or into gelatinous masses.   All these features remain inexplicable, except on the theory here given, though in some cases spines may serve as means of defence against foes.   Temporary changes in the form of the flotation-apparatus are known to occur.[3]

The preceding argument is strengthened by a comparison between plankton-diatoms and ground-diatoms.   The latter are fixed or creep about, and possess in their shells sutures through which the protoplasm projects, so that they can move about, seek the most favourably illuminated spots, and fix themselves there.   The majority of plankton-diatoms

[1] Beijerinck, 1895.   [2] Schütt, 1893.
[3] Wesenberg-Lund, 1908, and others.

have no sutures. Ground-diatoms, on the contrary, do not show the various outgrowths described above.

Quantity of plankton. Their power of rapid division seems primarily responsible for the frequently enormous multiplication and great abundance of plankton-organisms. Their quantity, however, varies with time and place. ' Pure blue is the desert-colour of the high sea. With the green of meadows we may compare the colour of the vegetation of arctic waters ; yet the colour of the most vigorous vegetation, of the greatest abundance of plant-life, is the dirty greenish yellow of the shallow Baltic Sea.' [1] Hensen [2] invented and employed methods for estimating the quantity of plankton. [3] His object was to determine the amount of organic matter produced in the sea at a definite time and place; and this object is of profound importance, since all marine animal-life, both lowly and highly organized, is dependent upon those plankton organisms that are plants, or at least assimilate carbon dioxide : *plankton is the ultimate source of food*, which it perhaps supplies largely in the form of fat, manufactured by diatoms and by Peridineae, from which originate the great quantities of oil in marine animals (sea-gulls, whales, and all zoo-plankton).

It is of no slight interest to note that it is not, as in the land-flora, starch, with its greater specific gravity, but oil, which is the main product of assimilation in the floating plankton-community.

The first to appreciate the significance of the microscopic plant-world of the sea as the ultimate source of food-supply for animals was the Danish botanist A. S. Œrsted, who came to this conclusion as early as 1845-8, when on his journey to Central America. [4]

Composition of plankton. A distinction may be made between homogeneous and heterogeneous plankton. Plankton is sometimes extremely rich in species, but at other times, particularly when the organisms are so abundant as to colour the water, it is dominated by one or a few species, as in the peridinia-plankton in western parts of the Baltic, and in the diatom-district of arctic seas. It is especially the Diatomaceae, Peridineae, and Cyanophyceae that lend a colour to water. The colour of Baltic lakes is mainly determined by those of the chromatophores of the dominant plankton-organisms. In harmony with the periodicity of the fresh-water plants, the colour of Baltic lakes undergoes regular change. The season when plankton-organisms determine the colour of the lake only to a slight degree is usually comprised of the early days of June, when diatoms have vanished and Cyanophyceae not yet appeared. The colour of alpine lakes in arctic regions is only slightly tinted by plankton, because this is present only in small amount.

Seasonal changes. Here it may be mentioned that plankton-associations change with the *season*, just as does vegetation on land, because they are dependent on the temperature, illumination, and chemical properties of the water. For instance, in Skager-Rak and Kattegat during February and March there occurs a rich diatomaceous plankton composed of species that later (in April and May) appear on the coasts

---

[1] Schütt, 1893.   [2] Hensen, 1887.
[3] See also Haeckel, 1890, 1891.   [4] See Wille, 1904 *b*.

of Iceland and Greenland ; in April and May there is another rich diatomaceous plankton, whose species demand somewhat higher temperatures ; in June and July, a less rich and more uniform diatomaceous
plankton (with Rhizosolenia alata) ; from August to November, the
warmest season, there is a rich plankton of Peridineae, often with an
admixture of a diatomaceous one, which is rich in species and likewise
occurs on the southern coasts of the North Sea ; finally, in December
and January there is a poorer plankton, consisting of remnants of the
preceding. Most investigators assume that in the case of coast-forms
the spores, which are mainly produced at the conclusion of the vegetative
periods of the respective species, sink to the ground, and rest there until
the commencement of the favourable season.

In fresh water lakes the seasonal changes are not less remarkable
than in seas : this is shown, for instance, by the illustrations in Wesenberg-Lund's great work (1905).[1]

Associations of the open sea are more stable, as the voluminous
masses of water in the ocean do not experience such considerable changes.
In general the amount of plankton is small ; yet, on the one hand, plankton sometimes occurs in such abundance as to cloud the water, while on
the other hand, it is meagre where different kinds of currents impinge.

False plankton (Tycholimnetic plankton) is formed in fresh water
by various Chlorophyceae, including Zygnema, Mougeotia, Spirogyra,
and others, which at first are fixed, but subsequently break loose, and,
owing to the entanglement among their threads of gas that they have
evolved, they rise to the surface of the water and float there. Other
plants, such as Sargassum in the sea, that are normally fixed, but
become detached by constant movement of the water, may also be
placed in this category.

Geographical distribution. Of the geographical distribution of plankton-
organisms little is yet known. In this respect we are best acquainted
with the temperate Atlantic Ocean, North and Baltic seas, and the sea
surrounding Iceland. Diatomaceae play the greatest part in the colder
seas, Cyanophyceae in the tropics, and Peridineae in temperate and
tropical regions. The species throughout have a wide area of distribution,
because the external conditions may be the same over such vast expanses,
and in widely separated spots ; yet each region has a characteristic
flora.[2]

SUB-FORMATIONS

Plankton may be classified into the following three sub-formations :—
    Haloplankton, salt-water plankton.
    Limnoplankton, fresh-water plankton.
    Saproplankton, foul-water plankton.

A. Haloplankton. Salt-water Plankton.

The plankton of salt water may be subdivided into *neritic* and *oceanic*
haloplankton.

Neritic plankton is confined to the coast ; in the tropics it consists
of Diatomaceae, Cyanophyceae, and Peridineae, but very little is known

---

[1] See also Kofoid, 1903, 1908.
[2] The recent literature is cited in papers by G. Karsten, 1898, 1905–6, 1907.

concerning the matter; in temperate regions the associations during the cold season are identical with those occurring during summer in the arctic region, but in the warm season they are different. Diatomaceae dominate in the temperate region, except in autumn (the warmest season, with a temperature of about 20° C.) when Peridineae or, in brackish water, Cyanophyceae (Nodularia) may be the characteristic plants. As regards the arctic neritic plankton it is known that during spring-time Diatomaceae congregate in numbers on the under-surface of the ice,[1] but that when the ice disappears pelagic diatoms and yellow Flagel-lata dominate.

Oceanic or *pelagic* plankton, i.e. plankton of the open sea, mainly consists of Peridineae and Coccolithophoridae, but also of Cyanophyceae, Diatomaceae (relatively few species), and Halosphaera. It includes a large number of species of Peridineae and Trichodesmium in the tropics, and of Peridineae and Diatomaceae in temperate and arctic regions.

These terms 'neritic' and 'pelagic' or 'oceanic' plankton approximately correspond to Haeckel's 'neroplankton' and 'holoplankton' respectively; as most species belonging to neritic plankton spend only a part of their life as pelagic organisms, and during the other part are associated with the soil as spores and the like; whereas pelagic forms always occur as organisms floating free in the water.

Among oceanic species resting spores are unknown, but some species, including Rhizosolenia styliformis, produce microspores that are apparently allied to auxospores.[2]

The depth to which phytoplankton descends in the sea differs according to various conditions, such as transparence of the water and the like; in general it may be asserted that phytoplankton occurs to a depth of 200 metres, but that even at 100 metres its amount is small[3]: diurnal migrations are unknown among phytoplankton-organisms.

In recent years investigators[4] have established a number of *associations* in phytoplankton, which are brought into existence by difference in temperature, and salinity of the water; but very little is known of their oecology. Cleve has given names to them in accordance with their dominant genera or species, thus : tricho-plankton (after Thalassiothrix), styli-plankton (after Rhizosolenia styliformis), chaeto-plankton (after Chaetoceras), sira-plankton (after Thalassiosira), tripos-plankton (after Ceratium tripos).

B. Limnoplankton. Fresh-water Plankton.

This form of plankton in fresh water is constituted of autophytic species. In large lakes it may be differentiated into two zonal subdivisions : the *pelagic* in the open water, and the *neritic* near the shore.

Other subdivisions may be recognized, such as *potamoplankton*, *heloplankton*, and probably several more.

Limnoplankton appears to be one of the most cosmopolitan of formations. Concerning that in arctic and tropical countries we are almost entirely ignorant, and it is only in regard to temperate Europe and

---

[1] Vanhoeffen, 1897.   [2] Gran, 1902; G. Karsten, 1905-7; P. Bergen, 1907.
[3] Compare previous remarks concerning shade-flora. See G. Karsten 1905.
[4] P. Cleve, 1897, 1901; Gran, 1900, 1902; Ostenfeld, 1898-1900.

temperate North America that we are well informed. Apparently great similarity prevails everywhere. In the northern part of the temperate zone and in the more southern alpine lakes diatoms dominate during most of the year. In the flat countries of Central Europe it has been observed that nearly everywhere there is a regular alternation of diatom-plankton during the cold months, and Cyanophycea-plankton during the summer. In not a few lakes an almost monotonous plankton is formed during spring and summer by Flagellata, especially by Ceratium hirundinella, and Dinobryon.[1] The maxima of the respective associations depend upon the temperature of the water, and other consequent changes in it ; moreover, light and the amount of certain nutritive substances in the water have their influence.[2]

C. Saproplankton. Foul-water Plankton

In this sub-formation is included vegetation consisting of Flagellata such as Euglena viridis and E. sanguinea, of species like the colourless Polytoma uvella, also of various Cyanophyceae and Schizomycetes. Such vegetation generally occurs in small pools of stagnant water that is rich in putrefying organic matter but very poor in oxygen, as is the case with water (manure-water, puddles in roads) near human dwellings which may be distinctively coloured. The water is usually very green and sometimes stinking. The green organisms, Chlamydomonadae and others, presumably assimilate carbonic dioxide, and obtain nitrogenous compounds as well as other nutriment from organic constituents present in the water ; they are therefore probably *hemisaprophytes*. Euglena sanguinea and others cause a red colour, and are most frequently motile. Among the saprophilous organisms in such water are many Infusoria. The farther putrefaction has proceeded the better do chlorophyll-containing plants, such as Scenedesmus, Raphidium, and diatoms, flourish therein.[3] The ' self-purification ' of rivers, after the water below large towns has been polluted, is due to the activity of bacteria and other microphytes ; the process can be traced right to the stage when the water is devoid of organic substance. Sometimes the final product assumes the form of sulphide of iron, which is a constituent of black mud. Schenck[4] investigated the Rhine between Bonn and Cologne and came to the conclusion that Green Algae play no great part in this process, and that filamentous and rod-shaped Schizomycetes absorb the organic substances.[5]

[1] Wesenberg-Lund, 1904-8.   [2] Whipple, 1894, 1896.
[3] See Kolkwitz und Marsson, 1902 ; Volk, 1903 ; Marsson, 1907-8.
[4] Schenck, 1893.
[5] An immense literature dealing with plankton has sprung up within recent years : among the investigators may be mentioned Gran and Wille in Norway, Cleve in Sweden, Ostenfeld, Ove Paulsen, and C. Wesenberg-Lund in Denmark, Apstein, Hensen, Brand, Zacharias, Chun, Haeckel, G. Karsten, Kirchner, Lohmann, and Schütt in Germany, Kofoid in North America, Chodat, Bachmann, G. Huber and Schröter in Switzerland. See also literature quoted by Oltmanns, 1905.

CHAPTER XXXIX. CRYOPLANKTON.[1] VEGETATION ON ICE AND SNOW

THIS glacial vegetation is most closely allied to plankton. It has long been known that animals and plants live on the extensive snow-fields and glaciers of arctic countries, and such high mountains as the Alps, Pyrenees, and Andes; they are in the main microphytes, yet, like plankton, they can occur in such countless numbers as to colour the snow or ice. The animals specially include Poduridae (Desoria saltans, and the blue Achorutes viaticus), Tardigrada, Radiolaria, and threadworms. To Wittrock and Lagerheim[2] we largely owe our knowledge of the vegetation, which consists mainly of aquatic plants, and particularly of algae (Diatomaceae, Chlorophyceae, and Cyanophyceae), Schizomycetes, and mosses in their protonema-stage. Some Phycomycetes also occur. In 1892 Lagerheim assessed the number of species at seventy-two. According to colour we may distinguish red, brown, also green and yellow, snow.

Red snow is the commonest, and has been known for the longest time; the tint varies from blood-red to rosy red, and from brick-red to purple-brown. It is caused especially by Chlamydomonas (Sphaerella) nivalis, and C. nivalis var. lateritia. This unicellular, spherical or ovoid alga has red contents, and colours the uppermost layers of snow to a depth of a few centimetres; in the water derived from melting snow it multiplies by zoospores. In addition there occur, among other species, Gloeocapsa sanguinea, Cerasterias nivalis, Pteromonas nivalis, as well as diatoms, and, in Ecuador, species of Chlamydomonas.

Brown snow owes its colour to various species, including the desmidiaceous Ancylonema Nordenskiöldii, which contains violet cell-sap and, together with other algae and ' cryoconite ' (very fine mineral particles), has an important action on the ice in the interior of Greenland, as it absorbs the sun's rays more strongly than does the ice and thus melts deep cavities in the latter. In company with it live Pleurococcus vulgaris, Scytonema gracile, diatoms, and other algae.

Green Snow is a phenomenon due to the presence of Green Algae, for instance Desmidiaceae, also to Raphidium nivale,[3] Cyanophyceae, protonemata of mosses, and green individuals of Chlamydomonas (Sphaerella) nivalis. *Bright yellow* or *greenish yellow* snow is tinted by another alga, possibly Chlamydomonas flavivirens, which is known to occur on snow-fields of the Carpathians.

These plant-communities provide very evident examples of extraordinary powers of resistance on the part of plant-cells: except for the peculiar properties of the protoplasm they seem to be devoid of any means of protection against cold, though perhaps red colouring-matter may enable them to absorb heat.[4] During the greater part of the year they lie frozen in ice or snow and in the lasting darkness of the arctic night; when the sun's heat melts the ice and snow, they awaken into activity, carrying on their processes of nutrition and reproduction in

---

[1] Schröter, 1904–8, p. 623.   [2] Wittrock, 1883, p. 883; Lagerheim, 1892.
[3] Chodat, 1896.   [4] Wulff, 1902.

water whose temperature scarcely exceeds 0° C. In many places the water melted during the day-time freezes every night ; and thus they pass their life in ice and icy water.[1] But in yet another respect the snow-alga is remarkably resistant ; it can preserve its existence in a dry condition even though exposed for months to relatively high temperatures.[2] The same is true of certain animals living in snow.

CHAPTER XL. HYDROCHARID-FORMATION OR PLEUSTON

On the banks of stretches of fresh water, in places protected from currents or from the violence of waves (for example among swamp-plants), and in small ditches and puddles, there occurs a type of vegetation which is *not fixed* but is free-swimming, or rarely even in part free-floating like true plankton, of whose organisms it may contain an admixture ; nevertheless this type of vegetation differs so essentially from plankton that it must be regarded as a special formation (*megaplankton*). From plankton it is distinguished by two features :

1. The occurrence of quite other growth-forms, namely megaphytes, including Spermophyta, Hydropterideae, and mosses.

2. The occurrence of algae belonging to groups entirely different from those in plankton.

The growth-forms in the two cases are however very different, and for this reason the two formations are at least theoretically to be placed apart. Kirchner in 1896 employed the term ' pleuston ' to denote the real typical representatives of the hydrocharid-formation, as he laid stress on their ' sailing ' character,[3] with which is also associated their adaptation to existence in contact with air (transpiration and the like). But in 1902 Schröter adopted the course, which we follow here, of including in the hydrocharid-vegetation such rootless, free-floating, submerged Spermophyta as Ceratophyllum, Utricularia vulgaris, Lemna trisulca, and other species.

Moreover, with perfect justice he distinguishes between the constant and the temporary floating-flora. In the temporary flora he includes the masses of algae swimming at spring-time. These are more especially composed of Conjugatae, including Zygnema, Spirogyra, Mougeotia, but also of Oedogonieae, which ascend in great quantities and remain at the surface of the water, whither they are raised by bubbles of gas, as we have already explained.

FLORA

The constant representatives of the hydrocharid-formation belong to the following groups :

Bryophyta, namely Riccia (with both submerged and swimming species), Amblystegium giganteum, and others. In pools on heaths or in pockets on heath-bogs one often finds vegetation very poor in species, and consisting of floating Sphagnum which nearly fills the water.

---

[1] See p. 22. [2] Wittrock, 1883.
[3] πλεῖν, to sail ; see Schröter und Kirchner, 1896–1902.

Hydropterideae, with Azolla and Salvinia, both of which are swimming plants.

Spermophyta, which can be ranged into the subjoined groups :—
Submerged : Ceratophyllum, Utricularia, Aldrovanda, Lemna trisulca, Stratiotes aloides.

Floating by leaves or shoots : Hydrocharis, Hydromystria stolonifera (Trianea bogotensis), Lemna minor, L. polyrhiza, L. gibba, Wolffia arrhiza, Pistia, and Eichhornia crassipes.

Transitional to the rooted limnaea-types : Hottonia palustris, Jussieuea repens, and others.

Many spermophytes, such as Lemna, Pistia, and Pontederia crassipes, can choke the water with their enormous numbers.

ADAPTATIONS

The submerged species, as in the case of plankton-organisms, must be approximately of the same specific gravity as water ; normally floating species are kept at the surface of water by air-containing cavities in their leaves and stems. This finds outward expression in thickness of the shoots and great convexity of the lower face of the floating organ of Lemna gibba and Hydromystria. Special floating-devices are shown by Eichhornia crassipes, Neptunia and Jussieuea repens.[1]

As in some cases the assimilatory organs projecting into the air are necessarily adapted to transpire, this formation shows a certain transition to land-plants, just as, on the other hand, through its submerged species it is allied to plankton.

The *morphology of the shoot* varies. In the majority of submerged Spermophyta the shoots have very long internodes and very thin stems ; the usually sessile or shortly stalked leaves are often divided into filiform segments, as in Utricularia, Ceratophyllum, and Hottonia. But in floating species the shoots mostly have short internodes and are condensed ; while the leaf-blades frequently assume the shape of typical floating-leaves, being very broad, peltate-cordate or ovate-cordate, as in Riccia natans, Salvinia, Hydrocharis, Hydromystria, Lemna polyrhiza and other species, and Azolla ; somewhat differently shaped are the shoots and leaves of Pistia. One of the offices of the floating leaf is to ensure equilibrium to the plant in water ; accordingly, floating leaves or analogous balancing-organs are developed early in the seedlings of Salvinia, Lemna, and some others.[1]

That this broad distinction between submerged and floating leaves represents a true adaptation to environment is clearly shown by Salvinia, and by aquatic plants such as Ranunculus (Batrachium), Trapa, and Cabomba, that are fixed by roots ; for all these have both submerged and floating leaves which differ in form.

In free-floating submerged plants nutriment is absorbed over the whole surface, and in Vascular plants the root is accordingly either absent, as in Aldrovanda, Wolffia, Lemna trisulca, Ceratophyllum, and Utricularia vulgaris, or very reduced ; the most important part played by the root in such plants as Lemna and Hydrocharis is indubitably to secure the plant in a definite

---

[1] Göbel, 1891.

position and to prevent its being overturned—and the same function is performed by the submerged leaves of Salvinia.

Propagation. The division of the vegetative organs plays an important part in all cases. Not only algae, but also Pteridophyta like Azolla, and Spermophyta such as Lemna, Hydrocharis, Stratiotes, multiply with exceeding rapidity by division, and for this reason they are markedly social and occur in great abundance. Vegetative parts serve as convenient means of dispersal, for instance in Lemna; the small shoots of Wolffia brasiliensis are distributed by aquatic birds. Accordingly, the production of spores and seeds is in a number of cases almost unknown or, as in Lemna, very rare.

Fertilization is necessarily connected with the water in Cryptogamia; moreover, a few Spermophyta including Ceratophyllum open their flowers under water; but the flowers of the others are developed in the air, and are mainly entomophilous, as in Utricularia, Hottonia, and Hydrocharis. The fruit is in most cases ripened under water.

Hibernation and duration of life. Nearly all are *perennial*, as is true of aquatic plants in general. Salvinia and many algae are *annual*. Flowering plants often produce special bud-like winter-shoots—*hibernacula*—which sink to the bottom in autumn[1]; among such plants are Hydrocharis, Utricularia, Aldrovanda, and Ceratophyllum; or after the death of the older shoots the younger ones, which are filled with reserve-food and do not yet contain much air, sink and hibernate without further modification—such is the case with Lemna. Certain algae, Cladophora fracta for instance, show similar behaviour, as they sink to the bottom in autumn, hibernate in the form of thick-walled cells which have rich contents, and develop in spring-time into new individuals.[2]

DISTRIBUTION AND ASSOCIATIONS

As the most prominent species in this formation float about on the surface of water, they are easily transported by wind and currents to quiet spots, where they may be collected together in vast numbers, as may be seen in the case of Lemna. Huber[3] gives some information in regard to the floating islands in calm inlets of the Amazon; these are often very extensive, and are formed partly of pleuston—for example, of Eichhornia azurea—but also of half-floating grasses which do not belong to this formation, also of other marsh-plants that have broken loose. In like manner immense multitudes of Eichhornia crassipes occur in North American rivers.

This formation seems to be strictly confined to fresh water.

Various *associations* may be distinguished according to the dominant species (pontederietum, lemnetum, pistietum, and the like). Schröter[4] separates the *emersed* hydrocharids (for example, lemnetum) from *submerged* types (including the associations, ceratophylletum, scenedesmetum, and zygnemetum) as separate formations.

---

[1] See Schenck, 1886 *b*; Raunkiär, 1895–99.
[2] Wille, communicated by letter. [3] J. Huber, 1906. [4] Schröter, 1902.

CHAPTER XLI. LITHOPHILOUS BENTHOS

THIS type of vegetation is confined to rocks, loose stones, mollusc shells, and similar solid substrata near shores and banks. Many of the species growing on these substrata also live as epiphytes.

FLORA

The *salt-water* communities are solely composed of algae, which here reach their highest and richest development in all four colours (blue-green, pure green, brown, and red), and exhibit extraordinary variety of form.

The *fresh-water* communities, though much poorer, consist of algae (nearly entirely Chlorophyceae, Cyanophyceae and Diatomaceae), of mosses (Fontinalis, Dichelyma, Cinclidotus, and others), and of Spermophyta, especially Podostemaceae.

The chemical nature of the substratum plays a part that, so far as is known, is slight and solely concerns the presence of calcium. Some algae flourish only on lime, which is perforated and corroded in furrows by their hypha-like filaments.[1] Most of the others grow equally well on stones, piles, shells, or on other algae. In addition, the inclination, illumination, and physical nature of the substratum (rock, stones, or calcareous shells) has its influence on the distribution of species.[2] According to Wille a substratum of shells is distinguished by special algal associations, for instance, by Tilopteridaceae.

ADAPTATIONS

The general peculiarities of submerged hydrophytes, such as reduction or absence of stomata, of lignified constituents, and of wood-vessels, the production of assimilating chromatophores in the outermost layer of cells, and so forth, have already been dealt with.[3] The assimilatory tissue extends to the surface ; but beyond this many algae possess an internal assimilatory tissue, which undoubtedly utilizes the carbon dioxide produced by respiration.[4]

Specialized adaptation is revealed in the undermentioned directions :

Solidity of the substratum necessitates the possession of *haptera*,[5] which in connexion with algae are sometimes described as roots though they are widely different from true roots. They occur in two forms ; as circular disks, in Fucus vesiculosus and Laminaria solidungula and others, or as branched digitate or coralloid structures, in Laminaria saccharina and Agarum Turneri and others. Here, too, may be placed the tufted rhizoids of Fontinalis and of other aquatic mosses. The adaptive means of fixation have been discussed by Wille.[6]

Haptera in some cases have the structure of root-hairs, but in others they are solid, multicellular bodies. The firmest attachment is exhibited by such crustaceous algae as Lithothamnium, Lithophyllum, Hildenbrandtia, and Lithoderma, which form an incrustation on rock. Diatomaceae and Desmidiaceae that are fixed to other bodies by mucilage belong to a special type.

[1] Chodat, 1902; Lagerheim, 1892 ; Cohn, 1893 ; Nadson, 1900 ; M. le Roux, 1907 ; P. Boysen-Jensèn, 1909.   [2] See Börgesen, 1905.
[3] See Chap. XXVIII.   [4] Wille, 1885.
[5] Warming, 1881–1901.   [6] Wille, 1885, and others: see Oltmanns, 1905.

Creeping (migrating) lithophilous species are uncommon, yet are met with among Florideae, Caulerpa, and Podostemaceae. The last named have creeping roots, but agree with algae in their mode of attachment, since the roots only indirectly play a part in fixation by bearing haptera; they have no organs specially set apart to absorb nutriment.

Air-containing intercellular spaces are entirely lacking, or at most are very small and contain scarcely any air. Exceptions to this rule occur in the sub-aerial inflorescences of Podostemaceae, and in the floating-apparatus of certain algae, such as Fucus vesiculosus, Halidrys siliquosus, Chorda filum, and Ascophyllum nodosum, which live in the littoral belt or in shallow water. Through this character, lithophilous vegetation stands in sharp contrast with other types of aquatic vegetation. The reason for it is presumably that all the lithophilous plants in question live in troubled water, where they secure rich supplies of air : Podostemaceae mostly find their homes in cascades.

The necessary power of *resistance to rupture* is structurally provided in various ways by mechanical, mainly collenchymatous, tissue.[1]

The *excretion of calcium carbonate* within cell-walls takes place in a number of algae, and with this must be mentioned the *silica* occurring in Podostemaceae. These excretions in some cases play a mechanical rôle, while in others they seem to increase the longevity; certain incrusted algae are perennial, whereas their non-incrusted allies are annual.

Heavy *production of mucilage* takes place especially in species growing in the littoral belt, and perhaps serve to prevent desiccation while the tide is down. It is calculated also to diminish friction between water and plants, and thus to preserve the latter from the violence of breakers.[2]

The **Plant-shapes** are extremely varied; and by no means all seem capable of interpretation as adaptations to environment :—[3]

Crustaceous type. There are among algae and Podostemaceae (Erythro-lichen, Lawia, Hydrobryum) crustaceous forms, which are well-suited for existence in very perturbed water; but many crustaceous Algae grow in deep, and therefore in but slightly disturbed, water.

Myriophylloid type. There are species among algae and Podostemaceae that are structurally analogous to *gills* of animals, as they are cut up into many capillary segments, by which the surface and assimilatory activity are both increased.

Muscoid type. There are moss-shaped forms among mosses (Fontinalis) and Podostemaceae (Tristicha hypnoides, species of Mniopsis and Podoste-mon).

Filiform type. There occur species shaped like an unbranched thread, which experience passive undulatory movements in the water; such are Chorda filum, many fresh-water algae, and the podostemaceous Dicraea elongata.

Phylloid type. Leaf-like forms are presented by Porphyra, Laminaria, Ulva, and Monostroma, as well as by the podostemaceous Marathrum, Oenone, and Mourera.

Special stress must be laid on the parallelism between the shapes of the marine algae and Podostemaceae, as demonstrating that these shapes are adaptive. In the sea at the southern point of South America there occur peculiar algae (Macrocystis and Durvillea) which have ' floating fronds '.[4]

---

[1] Wille, 1885.
[3] See Oltmanns, 1905.
[2] Wille, 1885 ; see p. 99.
[4] Compare Hooker, 1847 *a*.

FORMATIONS

As the lithophilous Spermophyta are biologically very different from the algae, which are always submerged and are provided with quite other methods of reproduction, lithophilous vegetation must be divided into at least these two formations :

1. **Lithophilous Spermophyta.**
2. **Algae (nereid formation).**

It may seem subsequently correct to establish several additional formations, for instance, those of mosses and of diatoms.

1. Formation of lithophilous Spermophyta.

Two remarkable families only, the Podostemaceae and Hydrostachydaceae (which were formerly combined into one) are represented in this formation. They are adapted for existence on submerged rock, in powerfully disturbed water (cascades) and have consequently evolved a number of structural features, affecting their external morphology as well as their internal structure, which are unique in the plant-world.[1]

They are almost confined to the tropics ; they occur in America from Uruguay to the southern United States, in Africa (where many most interesting forms are found), in Madagascar, and extend from India to Java; towards the east they become rarer, so that only a single species seems to live in Australia.   Mosses and algae may be intermingled with these Spermophyta.

2. Formation of Algae (nereid-formation).

This must be divided into at least two sub-formations : (a) *Fresh-water* (*Limno-nereid*) ; (b) *Marine* (*Halo-nereid*).

(a) **Limno-nereid communities** are poorer in species, individuals, and shapes than are the marine ones.   In comparison with the latter they display far less luxuriance and variety of form.   Nearly all belong to the Chlorophyceae and Cyanophyceae ; but there also occur diatoms, a very few Phaeophyceae (including Pleurocladia lacustris) and Florideae (Lemanea, Batrachospermum, for example).   In accordance with the variety displayed in environment, many *associations* (and possibly sub-formations) must be distinguished.   For example :—

Icy mountain streams have a quite peculiar flora, including Hydrurus, Prasiola fluviatilis, Tetraspora cylindrica, and others.[2]

On stones in shallow water along lake-shores there is an entirely different flora, with species of Cladophora, Rivularia, and Diatomaceae.

Sometimes associated with the algae are various mosses, among others, Fontinalis.

Schröter and Kirchner [3] recognize in Lake Constance an *encyonemetum* with different varieties, namely, spirogyretum, tolypotrichetum, and schizotrichetum.

Stones near the shores of fresh-water lakes are often encrusted by lime-producing algae.[4]   According to Wesenberg-Lund, incrustations of lime are produced mainly by Cyanophyceae (Schizothrix, Rivularia), but also by diatoms, Chlorophyceae (Cladophora), and the phaeophyceous Pleurocladia lacustris.   These incrustations occur especially on stones where the

---

[1] See Warming, 1881–1901.   [2] Lagerheim, 1892.
[3] Schröter und Kirchner, 1896, 1902.
[4] Chodat, 1902 ; Forel, 1901 ; Schröter und Kirchner, 1896.

shore is flat ; only rarely are they met with at a depth of one metre below the water-surface. In summer, when the water sinks and many stones are uncovered, the incrustations crack and fall off; while in winter they are rubbed off by ice. Thus it is that algae can contribute to the deposit of lime in lakes.

(b) **Halo-nereid communities** are those of salt-water. There are wide distinctions between the *floras* of different seas ; but even on a single coast there are many geographical features due to differences in the oecology of the various species, as is briefly indicated in the succeeding paragraphs.

The *oecological* distinctions depend largely upon differences in temperature, salinity, movements, and illumination of the water, as well as upon fluctuations in these ; one important consideration is whether or no the algae are periodically laid high and dry owing to tides, and another is the height up the rocks to which breakers reach.[1] A prominent part is played by differences in the soil (solid rock and its mineralogical nature, stones, rubble).

The temperature of the sea is of importance. The most luxuriant ' forests ' of Brown sea-weeds are developed in the coldest seas (frigid seas, North Atlantic Ocean, coasts of Tierra del Fuego, southern point of Africa), possibly because cold water is richest in air. In the southern seas named, some individuals (Macrocystis, Lessonia) are hundreds of feet in length ; in the North, species of Laminaria attain considerable dimensions—for instance, near Greenland Laminaria longicruris attains a length of 25 metres, and Nereocystis $13\frac{1}{2}$ metres in the Pacific Ocean. In tropical seas species are smaller throughout. At Spitzbergen at the depth where vegetation is richest the mean temperature of the water may not exceed at any season of the year o° C.[2]

The *seasonal* phases of species, according to Rosenvinge and others,[3] are strongly marked, and a number of species present utterly different appearances at different seasons. Some, including Chorda, Nereocystis, and a few other Laminariaceae, are annuals ; but in other species larger or smaller parts— for instance, haptera and the inferior portions of the thallus—perennate. Rhodomela subfusca in the Baltic Sea during April and May bears a richly branched shoot-system with reproductive organs which are subsequently shed. Desmarestia aculeata likewise varies greatly in appearance with the season. Some—Delesseria sanguinea, for example—fructify only in winter. The cold water is richer than warm water in oxygen and carbon dioxide and hence provides far better nutrition.[4] Kjellman's noteworthy account of algal life in extremely northern seas has already received attention in this work.[5]

Salinity of the water is another profoundly important factor affecting the composition and appearance of vegetation. The farther we proceed from the North Sea up the Baltic the less saline becomes the water,[6] and the poorer and more reduced the vegetation. The Siberian region of the Arctic Ocean is likewise poor in species, partly because the sea-bed is mainly sand or clay, and partly because of the volume of fresh water pouring out of Siberia. To fluctuations in salinity and temperature many species are exceedingly sensitive. Some species endure a slight decrease in salinity, others can adjust themselves to circumstances.

Movements of the water (wave-violence, currents) and consequent increased

---

[1] Börgesen, 1905.   [2] Kjellman, 1875.   [3] Rosenvinge, 1898; see Oltmanns, 1905.
[4] See p. 151.   [5] See p. 22.   [6] See p. 152.

supply of oxygen and food-material influence the distribution of associations. The algal vegetation on exposed coasts as a rule differs considerably from that found on sheltered coasts. In this connexion reference should be made to Hansteen's [1] investigations of the flora outside and inside the Norwegian rocky shoals and Börgesen's [2] on the Faroe Isles.

Hedwig Lovén [3] investigated the respiration and the gas contained in the vesicles of algae, and came to the following conclusions :—

The gas inside fucaceous vesicles is different in composition from that of the air in the water ;

The amount of oxygen is at the maximum at midday, and at a minimum during night-time ;

Algae can extract from water every trace of oxygen, but they live for a tolerably long time in water devoid of oxygen, and excrete into it a considerable quantity of carbon dioxide ; if oxygen be lacking in the water then they can completely exhaust the oxygen contained in the vesicles.

Light. In the first place *intensity of light* is of significance ; the Green Algae are those most photophilous, and, according to Kjellman, it may be for this reason that they are enfeebled as well as scanty in the north Arctic Ocean (though they are luxuriantly developed along rocky coasts of Greenland.) The deeper one goes the more light is absorbed, and the fewer become the species until at last they are absent. According to Berthold, Florideae are generally shade-loving plants. Berthold [4] found at Naples a luxuriant algal vegetation at a depth of 120–130 metres, whereas in the Arctic and North Atlantic Oceans only a poor vegetation subsists at a depth of even 50–60 metres.[5] The differences in their demands for light cause algae to be distributed in zones according to depth.[6]

Then the *colour of light* changes with the depth,[7] and correlated with this change are the colour of the algae and their distribution in ' regions '. Lyngbye in 1836 established the ' regions ' of the Ulvaceae, Florideae, and Laminariaceae ; Agardh in 1836 and Örsted in 1844, established those of the Green, Brown, and Red Algae. But Örsted in 1844 was the first to assume the connexion between colour of light and depth of the strata in which algae occur ; in the Öre-Sund he recognized the following ' regions ' commencing at the surface and descending :—

1. Regio algarum viridium S. Chlorospermearum.
   Sub-regio Oscillatorinearum.
   Sub-regio Ulvacearum.
2. Regio algarum olivacearum S. Melanospermearum.
   Sub-regio Fucoidearum et Zosterae marinae.
   Sub-regio Laminariarum.
3. Regio algarum purpurearum S. Rhodospermearum.

Kjellman has divided the algal vegetation off the Swedish part of the Murman coast, and in other seas, into three ' regions ', which run parallel with the coast, and are in turn composed of a great number of small ' formations ' (i. e. associations), according as one or another species predominates. The three regions are the following :—

[1] Hansteen, 1892.   [2] Börgesen, 1905.   [3] Lovén, 1891.
[4] Berthold, 1882 ; see Oltmanns, 1905.   [5] Rosenvinge, 1898 ; Börgesen, 1905.
[6] Concerning algal vegetation in caves of the Faroe Isles, see Börgesen, loc. cit.
[7] See p. 150. According to Gaidukow, 1904, the colours of algae are to be regarded as adaptations to the *quality* of the light present.

1. *Littoral 'region' :—*
Stretches between the high-tide and the low-tide marks, and includes many Green Algae, Brown Algae, and some Red Algae. At low tide these lie uncovered; many may be described as nearly amphibious, for they become dry on bright sunny days. In extreme arctic seas the rubbing action of the masses of ice prevents any strong development of this 'region'

2. *Sub-littoral 'region' :—*
Ranges from below low-tide mark down to a depth of twenty fathoms (40 metres); here algae of all colours are represented, but Green Algae cease and Red Algae become more numerous with increased depth.

3. *Elittoral 'region' :—*
Is below the preceding and descends as deep as light; it is poorer in species and individuals; moreover, the latter are smaller and distorted, as Lyngbye has already noted.

Hansteen and Gran on the whole approve of the preceding scheme; but Reinke, basing his conclusions on investigation of the Baltic Sea, from which 'region 3' is absent, suggests a further partition of regions 1 and 2 into two sub-divisions each. At a depth of 4 metres many species find their lowest limit. Rosenvinge and Börgesen [1] indicate that the littoral 'region' 'extends far beyond the highest tide-mark, in the Färöes even in some places about 25–30 metres beyond', and they set the 'lowest limit of the sub-littoral region above the lowest ebb-marks'. Rosenvinge and Börgesen do not recognize any elittoral 'region'.

Some algal hairs are assimilatory, as in Desmarestia aculeata and Chorda tomentosa, but others are colourless, especially among Red Algae. Those of the latter kind are strongly developed when the light is more intense, and Berthold has made the hardly probable suggestion that their function is to regulate the illumination; they would seem rather to be respiratory or absorbing organs.[2]

The factors already enumerated affect the vegetation notably as a whole but also in detail, and, possibly with the co-operation of other factors (kind of rock, and other topographical features), they aid in evoking a number of miniature topographical distinctions and a number of associations, which may be profusely intermingled and may owe their appearance mainly to one or a few species that compose their main mass.[3] (Phycologists often apply the term 'formations' to these miniature purely floristic groups; but it must be noted that 'formations' should be oecologically founded upon the forms of the algae, and that true 'formations' may here be quite out of the question. These groups then are, at least tentatively, to be regarded as associations.) Within extensive communities of large algae—for example, among Laminaria stems—many epiphytes and many humbler forms find suitable homes; thus an underlying vegetation of plants requiring less light arises, just as in a forest.

As the various factors enumerated act with unequal intensity at different seasons of the year, there arises a *differentiation in time* of the development of the nutritive and reproductive organs. Each species of marine alga seems to have its definite season of development, which may differ with the latitude : species that in Denmark disappear with

---

[1] Rosenvinge, 1898 ; Börgesen, 1905.
[2] See p. 168, and Wille, 1885 ; Rosenvinge, 1903.
[3] Kjellman, 1878 ; Hansteen, 1892 ; Börgesen, 1905

the commencement of summer, may in arctic seas persist throughout summer.[1] In Danish waters the marine algal vegetation in summer differs widely from that in winter,[2] and even in the more southern latitudes of Naples the same has been noticed.[3] In the latter place illumination and breakers are the determinants, but in higher latitudes temperature certainly plays a more important part.

A peculiar group, that of the *Diatomaceae*, merits special notice, as it includes forms deviating from all other plants. Among them are the ground-diatoms showing various growth-forms, including motile forms which creep over the substratum (stones or other algae), and stalked non-motile forms, which especially inhabit the marginal zones of salt-water, are easily detached and then can intermingle with plankton.[4] These communities of diatoms may perhaps be regarded as constituting a separate formation which includes many associations.

Moist rocks on sea and land may bear a vegetation which is transitional between submerged rock-vegetation and land-vegetation, and may give rise to a special formation. Inland nereid-communities are dependent upon great atmospheric humidity and trickling water, and they therefore develop luxuriantly only near waterfalls, whose spray habitually wets the rocks, and in countries where atmospheric precipitations are heavy and fall throughout the year (as in Java), and in the cloud-belt of mountains. On rocks that are wet by fresh water there may be formed a spongy felted carpet of algae (including Trentepohlia, Rhodochorton islandicum, R. purpureum and others), mosses, ferns, and other herbs; and there may even be found small shrubs that are always wet or dripping with water. On rocky coasts foam of the breakers may be carried especially high, and in these same places marine algae, such as species of Prasiola, Ulothrix and Enteromorpha, Calothrix scopulorum, also Bangia fusco-purpurea and Hildenbrandia rosea, may occur far above the high-tide mark.[5] Thus arises a kind of *supra-littoral* ' region '. Various crustaceous lichens, including Verrucaria maura, are intermingled therein. The oecology of this community differs from that in water in so far as the constituent species must be fitted to endure greater dryness than in the case of submerged types.

CHAPTER XLII. BENTHOS OF LOOSE SOIL

THE structure of soil has already been described,[6] but in the soil now under discussion the *interstices are filled with water*, and air occurs in extremely small quantities or not at all.

The texture of the soil may vary, being *mud, clay,* or pure *sand*, which is mostly quartz-sand or, in the tropics, coral-sand, and may be mixed with marine shells and stones, more or less small according to the violence of the waves. These differences in texture cause floristic distinctions, though hardly anatomical or morphological ones. But nothing further is known in regard to this matter. An exceptional kind of soil is mud that consists of dead organic matter.

[1] Rosenvinge, 1898.
[2] Kjellman, Rosenvinge, and others; as cited by Oltmanns, 1905.
[3] Berthold, see Oltmanns, 1905. [4] Schütt, see Oltmanns, 1905; compare p. 156.
[5] Rosenvinge, 1903, see Oltmanns, 1905; Börgesen, 1905. [6] See Chap. X.

Movements of the water, on the contrary, are of great morphological and floristic significance.

Salinity of the water is of even greater import. Vegetation in the sea is morphologically and oecologically very different from that in the majority of fresh waters (rivers, lakes, and pools).

Loose soil, in contrast to a stony substratum, entertains very few algae, but mainly Spermophyta.

ADAPTATIONS

Of the modifications of vegetation in such environment we may note—

Roots, or root-like organs branching in the soil, serve to attach the plant and absorb nutriment ; apart from these, special organs of fixation are lacking. Roots do not attain the dimensions or degree of branching displayed by land-plants, and some are devoid of root-hairs,[1] for instance, Hippuris (excepting at the 'collar'), Elodea, Hottonia and some others. Some tropical algae (species of Udotea, Halimeda, Penicillus[2]) on sandy soil and mud are fixed to the loose soil, and obtain nutriment therefrom by means of hypha-like hairs attached to the lower parts of the thallus that penetrate the mud[1,2] ; the same is true of Characeae.

Horizontal rhizomes, or their analogues (in Caulerpa, for instance), creeping on or, more usually, in the soil are *very common*, and bring into existence a dense vegetation composed of numerous social individuals, such as is exemplified by the submarine 'meadows' of Zostera and other 'grass-wracks'. This mode of growth clearly harmonizes with the loose texture of the soil.[3]

Gaseous interchange on the part of submerged plants is aided by the large *air-containing intercellular spaces* which are peculiar to aquatic plants.[4]

FORMATIONS AND ASSOCIATIONS

Three formations may be distinguished : i. *microphyte-formation ;* ii. *enhalid-formation ;* iii. *limnaea-formation*.

i. Microphyte-formations.

Pure associations of microphytes, particularly of Cyanophyceae appear in extreme circumstances, chiefly in hot springs, and in the shallow beds of seas and fresh waters where organic constituents abound ; also not infrequently in richly humous shallow waters of heaths. These microphytic communities differ so widely from the limnaea-formation that they must be regarded as constituting a distinct formation, or perhaps two—the *autophytic* and the *saprophytic*—in which several sub-formations may be distinguished.

A. AUTOPHYTIC MICROPHYTE-FORMATION.

1. **Sub-formation living in hot springs.** This occurs in various parts of the Earth. The temperature varies widely in these springs ; at relatively low temperatures Phanerogamia still occur in them, but high temperatures exclude all plants save Cyanophyceae (Beggiatoa, Lyngbya, Oscillaria, Hypheotrix, and others). Many of these species are ubiquitous. They form green, yellow, white, red or brown mucila-

---

[1] See p. 97.   [2] Börgesen, 1900.   [3] See p. 42.   [4] See p. 98.

ginous or filiform masses, which are often several centimetres thick and
sometimes present the appearance of almost structureless jelly.

In European hot springs we know of Anabaena thermalis (in water
at a temperature up to 57° C.), species of Leptothrix (in Karlsbad at
55·7° C.), Beggiatoa, Oscillatoria (44°–51° C.), Hypheothrix (in Iceland),
Tolypothrix lanata (in Greenland at Unartok, 40° C.), as well as others.
Lyngbya thermalis is known to occur in Iceland, in Italian mud-volcanoes,
and in hot springs in Greenland.

Many Schizomycetes occur, including true thermophilous species
(of which some dozen have been described), sulphur-bacteria, iron-
bacteria, and others. Moreover, many diatoms and other more highly
organized algae occur.[1]

The highest temperatures so far noted in this connexion are 81°–85° C.
in Ischia, 90° C. in the Azores,[2] and even 93° C. (200° F.) in California.[3]
At Las Trincheras in Venezuela there is a warm spring whose temperature
at its source is 85°–93° C.; the algae in this grow in water whose temperature exceeds 80° C.

The water of many hot springs, which mainly occur in volcanic regions,
contains sulphur, calcium, or other mineral bodies, without changing
the composition of the vegetation.

A special form of activity is displayed by certain of these algae,
which excrete crystalline concretions of calcium carbonate or of siliceous
sinter ; in the Arno travertine is deposited by Cyanophyceae ; and in
the hot springs at Karlsbad large masses of calcareous sinter are excreted.
In North America numerous hot springs and geysers occur in Yellowstone Park, and Weed[4] describes the remarkable stone-forming activities
here exhibited by algae ; these grow in waters at temperatures of 30°–
85° C. and lend to the water their various rainbow tints, ranging between
red, orange, white, and green, according to the temperature. Cohn[5]
expresses the opinion that these algae have a peculiar faculty of storing
up calcium carbonate.

Is it not possible that these thermophilous communities of the most
lowly organized algae living in hot springs present us with a picture of the
oldest vegetation on Earth? Perhaps the Cyanophyceae can even
assimilate free nitrogen.

2. Sub-formation of Sand-algae—the aestuarium of tidal shores. On
the sandy shores of seas of Northern Europe and of lakes (for example,
Michigan, Furesö in Denmark[6]), there lie, within reach of the tide-water,
communities of algae, which form a thin film on or under the surface of
the sand, and give to the shore a distinctive colour if they be present
in large quantities. On the coasts of Denmark there are various
associations of such : chlamydomonadeta, composed of species of Chlamydomonas and Diatomaceae, which lie loose on the sand which is not
cemented together : phycochromaceta formed by Blue-green Algae and
diatoms, which, by means of their mucilage, glue together the grains of
sand so as to form a thin, firm, more or less crustaceous layer, which is

---

[1] Respecting North America, see Harshberger, 1897 ; Josephine Tilden, 1898 ;
respecting Japan see Miyoshi, 1897 ; and respecting Hungary, see Istvanffi, 1905.
 [2] Moseley, 1875.   [3] Brewer, 1864.
 [4] See the literature quoted by Weed, 1887–9.
 [5] Cohn, 1892.   [6] Cowles, 1899.

usually visible immediately beneath the actual surface of the sand. With this plant-community is combined a remarkably peculiar community of animals. This association has an enormously wide distribution along shores of the North Sea, on places where high-sands and flats (aestuaria) are miles in width.

Closely allied to this subformation is a community of Cyanophyceae and diatoms which is likewise placed within reach of the tide-water, as it occupies mud-flats and thrives on marshy soil that is flooded by sea-water during the spring tides (aestuarium).[1]

We treat of these sand-algae here as essentially belonging to the benthos of loose soil, but in our classification they have no less claim to be considered amongst both the halophytic and psammophilous communities.

B. SAPROPHYTIC MICROPHYTE-FORMATION.

The vegetation on dead organic accumulations at the bottom of calm water consists of Oscillatorieae, Beggiatoeae, and Schizomycetes, but sometimes also of additional algae whose true home hardly lies here. The accumulations usually lie loose on decaying soil ; Beggiatoa lives in chalky white flocculent masses (beggiatoeta) ; while Clathrocystis roseo-persicina, as well as Bacterium sulphuratum, B. Okeni, and other sulphur-bacteria coloured with bacterio-purpurin, are in red masses. It is in calm inlets with shallow brackish water and accumulations of Fucaceae and other algae, that they specially occur in large quantities and form these associations which are rich in individuals and species.[2] The sulphur-bacteria here, as in hot springs, deposit sulphur within their cells[3] ; for by the reciprocal action of the dead organic matter and water there is produced sulphuretted hydrogen, which is absorbed and oxidized by the bacteria, thus giving rise to water and sulphur. According to Sickenberger red sulphur-bacteria play an essential part in the production of soda in the Egyptian saline lakes.

Abyssal saprophytic associations. At great depths in seas and lakes where the water is untroubled, the light is often feeble, and the temperature often low, there is frequently a collection of black mud which is filled with decaying bodies, alive with animals (thread-worms), but allows the growth of no highly organized autophytes. Probably there occurs here a rich saprophytic vegetation composed of species of Beggiatoa and other, perhaps mainly, anaerobic bacteria. But we know practically nothing of this vegetation. As an example of a place where there is probably a rich bacterial vegetation we may mention the Black Sea. According to Andrussow,[4] in this sea, at a depth of 100–600 fathoms and more, one encounters vast masses of mud containing sub-fossilized remains of mollusca, which belong to brackish water and arose in the not distant past when the Black Sea was a brackish lake, but which were exterminated when the Mediterranean Sea forced its way in. The currents are so constituted as to cause defective aeration of the deep water, which is therefore poor in oxygen though rich in sulphuretted hydrogen. It

---

[1] See Warming und Wesenberg-Lund, 1904; Warming, 1906.
[2] Warming, 1875.   [3] As was first shown by Cohn.
[4] Andrussow, 1893.

is assumed that not an animal lives in this, and that the organic constituents of the mud are not consumed by animals; but it is highly probable
that a rich anaerobic vegetation of bacteria occurs here.

In like manner in north European seas and fiords there are muddy
spots where saprophytic vegetation may lurk. The *black mud* so extremely
common in lakes, on sea-coasts, as well as in the depths of the sea, is
usually very rich in iron sulphide. According to Beijerinck,[1] and van
Delden,[2] the reduction of sulphates in water on ferruginous soil is accomplished by definite anaerobic bacteria, Microspira desulfuricans and M.
Aestuarii.[3]

Also in north European fresh-water lakes in deep situations one
encounters but scanty vegetation of an elaborate type; here many lowly
organized animals, including worms and larvae, thrive, and the eel waxes
fat by preying on them. Here, we anticipate (but do not know) there will
be found vegetation consisting of saprophytic Schizomycetes. According to Forel[4], in Lake Geneva at a depth of a hundred metres there is
a brownish layer of lowly organized algae, mainly Schizophyceae and
Diatomaceae, which thus form an organized carpet.

ii. Enhalid-formations.

To this formation belong all communities of Spermophyta, and larger
algae growing on *loose soil* in *salt water*.

FLORA

Of **Algae** there are very few : in tropical seas species of Caulerpa[5]
and Penicillus, and in European waters (especially if these be brackish)
Characeae, all of which send capillary root-like organs into the soil. (The
algae which casually occur here are attached to stones, and belong to
the lithophilous formation.)

Spermophyta, all of which are herbaceous, preponderate in number
and dimensions, though the number of species present is small (twenty-
seven). They belong to two families : *Potamogetonaceae*, with Zostera,
Phyllospadix, Posidonia, Cymodocea, Halodule, Althenia, also in brackish
water Ruppia and Zannichellia ; and *Hydrocharidaceae*, with Halophila,
Enhalus, and Thalassia. Divers epiphytic Algae occur.

ADAPTATIONS

The grass-wracks, though belonging to two different families, are
externally so alike that mistakes have often been made in the identifica-
tion of flowerless specimens.

The typical form is well illustrated by Zostera ; like this, all the representatives are submerged ; true floating-leaves are wanting, probably because
of the violence of the waves ; the leaves, except in Halophila, are ribbon-
like, rounded or blunt at the tip, and entire. The *ribbon-like* or *zosteroid*
form of leaf is well fitted for existence, not only in currents, but also in water
that is deep and therefore ill-lighted ; accordingly, it reappears in similar

[1] Beijerinck, 1895.
[2] van Delden, 1903.
[3] See also Warming, 1904.
[4] Forel, 1891.
[5] Svedelius, 1906 ; Börgesen, 1907.

circumstances in species belonging to the limnaea-formation. The breadth of the ribbon-like leaf of Zostera marina is obviously adjusted to the depth of water ; the shallower this is the narrower the leaf (variety, *angustifolia*) ; whereas in deeper water the plant becomes more vigorous and broader-leaved. The narrow-leaved species, Zostera nana, also species of Ruppia and Zannichellia, occur in shallow water not far from the shore.

In consequence of the far-stretching *rhizomes* a social mode of growth results, so that dense, grass-green, submarine ' meadows ', often extending for miles, are formed.

The *flowers* are very reduced, and inconspicuous ;[1] flowering takes place on or under the water, and with its aid ; the pollen grains in species with submerged flowers are filiform, as in Zostera and Cymodocea, or linked together in long threads, as in Halophila,[2] obviously in order that they may be more easily captured by the long stigma, when conveyed thither by the water, with which they agree in specific gravity. The stalks of the female flowers of some species, including Enhalus and Ruppia spiralis, are long and spirally coiled, and they contract after pollination. The pollination of Enhalus is dependent upon the ebb-tide.[3]

GEOGRAPHICAL DISTRIBUTION

In arctic seas this type of vegetation seems to be almost wanting, possibly because the ice will not permit of its development. Several associations of grass-wracks and the like may be distinguished.[4] In north-temperate seas of Europe, Zostera marina and Z. nana occur, while in the Mediterranean added to these are Cymodocea nodosa and Posidonia oceanica.

Zostera marina forms a shallow zone along coasts ; in Danish seas its lower limit is eleven metres [5] ; the depth attained necessarily depends upon intensity of the light, and thus upon clearness of the water. It demands a soil that is to some extent protected. Other plants also are partly confined to submarine meadows of this kind ; among such are certain algae, some of which are epiphyllous, for example, diatoms, Desmotrichum undulatum, species of Ectocarpus, Ceramium, Polysiphonia, often in great masses, while others, including Fucus, Laminaria, Cladophora gracilis, and Fastigiaria furcellata,[6] grow among the rhizomes. This is the case particularly where the soil includes stones. In addition there occur a series of more or less modified forms of various species of algae, which have been brought hither by currents, and are detained among the Zostera plants, where, remaining in a sterile condition, they undergo more or less modification in their subsequent growth ; such species are Ascophyllum nodosum *var.* scorpioides, Phyllophora Bangii which is a metamorphosed form of P. rubens, also forms of Phyllophora Brodiaei, Anfeltia plicata, Cladostephus verticillata, and others. In the shallows of the sea off Schleswig, Cyanophyceae, as a microphyte-formation, also mingle with Zostera.[7] In Danish waters and the western Baltic, close to land there is usually a shore-zone of Ruppia and Zostera nana, while in deeper water there is one of Zostera marina.

[1] Schenck, 1886 *b*.　　[2] Balfour, 1878; Theo. Holm, 1885.　　[3] Svedelius, 1904.
[4] Ascherson, 1871.　　　　　　　　　[5] Ostenfeld, 1905, 1908 *a*.
[6] Communicated by Rosenvinge.　　　　[7] Warming, 1904, 1906.

In lagoons on the coasts of the Danish West Indies, according to Börgesen,[1] there are species of Thalassia, Cymodocea, Halophila, and Halodule, creeping algae, such as Caulerpa, and non-rhizomatous Algae, such as Penicillus, Udotea, and Halimeda.

The *vegetation of brackish water* on many coasts is closely allied to that just described, but it also includes other more slender species and other genera, such as Chara, Zannichellia, Batrachium, Naias, Potamogeton pectinatus, and Myriophyllum, which reappear more bountifully in fresh water. Several of the species mentioned grow only in shallow water, down to a depth of two metres at most.

Grass-wracks play an important part in the biology of the sea as homes for marine animals, in connexion with oviposition by fish, and as food-material in the case of Thalassia testudinum, which is eaten by turtles.

Many saline waters near salt-works have a peculiar algal vegetation often intermingled with Ruppia and Zannichellia.

iii. Limnaea-formations.

To these formations belong all those *fresh-water* communities of auto-phytic spermophyta, and other plants of considerable size, the individuals of which live on *loose soil*, whether sand, clay, or mud, and are *completely submerged, or at most possess floating leaves.* (Their flowers, however, in nearly all cases rise out of the water). They are therein distinguished from marsh-plants, whose assimilatory organs for the most part project above water. But there is no sharp limit between these formations.

FLORA

The flora is composed of—

Green Algae, namely Characeae, which occur particularly on marl soil, and clothe this with a dense, peculiar-smelling carpet (characetum).

Musci, including Fontinalis, Hypnum, and Sphagnum.

Pteridophyta, including Marsilia and Pilularia among Hydropterideae, as well as Isoëtes.

Spermophyta, including more numerous species of Potamogetonaceae than occur in the sea, of Potamogeton for instance, also Elodea, Vallisneria and Hydrilla representing the Hydrocharidaceae, Sparganium minimum, S. affine, and other species, in addition to many Dicotyledones such as Nymphaeceae, Cabombaceae, most species of Batrachium, Myriophyllum, Helosciadum, Callitriche, Subularia, Elatine, Montia, Limosella, and others.

Epiphytes, among which are many Diatomaceae and Cyanophyceae which are often enveloped in mucilage.

ADAPTATIONS

The diversity of form among Spermophyta, in contrast to that in the enhalid-formation, is extreme. This is to be attributed to the great variety of environment, which includes not only powerfully streaming water but very often very calm water, such as is never met with in the

[1] Börgesen, 1900.

N 2

sea. The chief distinction in shape concerns the occurrence not only of completely submerged types, but also of *species with floating leaves, or with shoots floating on the surface.*

All the species are *herbaceous,* and nearly all are *perennial.*

The Shoot. The construction of the shoot varies widely. In agreement with the loose nature of the soil, the vast majority possess creeping axes, and therefore display *social* growth : such is the case with Potamogeton, Hippuris, Nymphaea, Nuphar, all of which have subterranean horizontal stems ; and Myriophyllum, Ranunculus, Callitriche, which have epigeous creeping stems ; the Characeae also belong here *mutatis mutandis.* Others, such as Littorella and Vallisneria, emit long runners, and on these, at certain distances from the mother-plant, they produce rosette-shoots which become firmly rooted. All such species can give rise to extensive and dense associations which clothe the lake-bed, being rich in individuals though poor in species. A minority of species, including Isoëtes and Lobelia, have also vertical rhizomes with the leaves separated by short internodes and ranged in a rosette : but these are devoid of the above-mentioned methods of migration, and are rather represented by isolated individuals.

Finally, there is a small number of *annual* species, such as Subularia, Naias, and Trapa, which grow socially only when numerous seeds are strewn over the same ground.

There are three essentially different forms of assimilatory shoots, namely :—

A. *Rosette-type.* The shoots are vertical, short, unbranched, with short internodes ; the leaves are in rosettes, sessile, and mostly submerged. Such is the case in Vallisneria with ribbon-like leaves, Isoëtes, Lobelia Dortmanna, Subularia, and Littorella uniflora with more terete leaves.

B. *Nymphaea-type.* The shoots assume the form described for the rosette-type, or are horizontal rhizomes creeping in the soil ; there are, however, long-stalked floating leaves. This type is represented by Nymphaeaceae.

C. *Long-stemmed type.* From a rhizome or a stem creeping over the ground there arise under water erect stems that have long internodes, and are slender and branched ; the main and lateral axes, as a rule, are of equal thickness, showing no secondary thickening, as is precisely the case with certain plants described in Chapter XL. The stems, which are often very long and slender, are extremely flexible and can yield to movements of the water. Their length depends upon the depth and flow of the water. Terrestrial forms of the same species have shorter internodes. These shoots are of two kinds :

a. *Completely submerged,* as in Potamogeton pectinatus, P. perfoliatus, some species of Batrachium, Myriophyllum, Zannichellia, Callitriche autumnalis, Elodea and Naias (annual). The leaves are linear or oblong (only rarely broad), and in some are very finely segmented.

b. Possessed of *floating leaves* in addition to submerged leaves, the former leaves sometimes being arranged more or less in a rosette at the end of a condensed shoot, and then sometimes having tolerably short petioles. As examples may be cited Callitriche verna, Trapa (annual) the majority of species of Batrachium, Potamogeton natans, and Elisma natans.

The Leaf. The dependence of the *leaf-shape* (and partly shoot-form) *on the medium* is particularly striking. The following are the five types of leaf-shape :—

1. The floating.
2. The submerged.
   a. The zosteroid.
   b. The elodioid.
   c. The isoëtoid.
   d. The myriophylloid.

The submerged types fall into two groups : that including very finely segmented leaves occurring mainly in Dicotyledones (type 2 d), and that including essentially long and linear leaves (types 2 a, 2 b, 2 c).[1]

1. The *floating leaf* has already been mentioned as occurring in the hydrocharid-formation.[2] It is found especially in associations thriving in calm inlets or under shelter of reed-swamps. This type is possessed by Nymphaea, Nuphar, Cabomba, Brasenia, Limnanthemum, Hydrocleys, Elisma, Batrachium, Trapa, Callitriche, species of Potamogeton (P. natans), Polygonum amphibium, and other genera with the same general form. The leaf is broad (orbicular, ovate, cordate, reniform, rhombic or elliptic ; rarely lanceolate), undivided and entire, rarely crenate or incised (as in Trapa and Batrachium), comparatively thick and tough (coriaceous), sometimes possessed of a mechanically strengthened or an upwardly bent margin, and excellently adapted to rest on water and resist movements of the latter ; the gigantic floating leaves of Victoria regia, Euryale ferox, and others are, in addition, strengthened by stout ribs on the under-surface. The floating leaf is necessarily adapted to transpire, and thus provides a transition to the land-plant.

Stomata occur solely or mainly on the upper face, whose epidermis contains no chlorophyll ; they are protected from being plugged with water by the deposition of wax in or on the cuticle, which is thus rendered unwettable. This it is which gives often a glossy or whitish appearance to the upper face.[3]

The *lamina of the floating leaf* is dorsi-ventral, showing palisade tissue towards the upper face, and very lacunar spongy parenchyma towards the lower face. The lower face is often coloured dark red by erythophyll, the significance of which is not yet known. Prickles on the lower face of the lamina and on the stalk are shown by Victoria and Euryale.

The *petiole of the floating leaf* has the power of adjusting itself according to the depth of water, in such a way that its growth ceases when the lamina comes in contact with the atmosphere. In the case of shoots with long internodes, the latter are likewise similarly arrested in growth, as in Trapa and Callitriche ; in such cases, the proportions as regards length between the petioles and insertion of the floating leaves is such that all the blades are accommodated with space on the water. Frank suggested that growth of the petiole is promoted by the pressure of the overlying column of water ; other investigations have shown that contact with the atmosphere and more intense illumination are responsible for the shaping of the floating leaf.[4]

---

[1] For the literature, see Schenck, 1886 b.   [2] See Chapter XL.
[3] See Jahn, 1886.   [4] Frank and others ; see literature in Schenck, 1886 b.

Various species are *heterophyllous*, possessing not only floating leaves but also submerged leaves. According to Askenasy [1] and others, the floating leaves of Batrachium and Cabomba do not appear until the plant is about to flower, so that they may serve specially to raise the blossom above water.

2. The *submerged leaf* differs both morphologically and anatomically (particularly as regards epidermis and chlorenchyma) from the floating leaf: [2]

  a. The *zosteroid, or ribbon-like leaf*, which generally occurs among grass-wracks, is less frequent in this formation, though it occurs in Vallisneria, Sparganium, species of Potamogeton, and others. Convincing evidence is forthcoming that this form of leaf is adapted to and evoked by deep or running water (both these conditions appear to act in the same direction), when we observe that it appears on certain marsh-plants, including Alisma Plantago, Sagittaria sagittifolia, Echinodorus ranunculoides, and species of Sparaganium, if these be compelled to grow in such water. Similar forms of leaf are met with under the same conditions in Scirpus lacustris and Potamogeton natans, which produces 'current-leaves' half a metre long. [3]

  b. The *elodioid leaf*, which is *narrow, linear, undivided, flat, sessile*, and *short*, is frequent, as shown by Elodea, Potamogeton densus, P. obtusifolius, P. pusillus and other species, Hippuris, Zannichellia, Callitriche autumnalis and other species, and Naias. In this category may be placed the leaves of the aquatic mosses. Broader forms of leaf are displayed by other species of Potamogeton.

  c. The *isoëtoid leaf* is *linear, undivided, terete*, often *tubular*, and sessile, and occurs in Pilularia, Isoëtes, Lobelia Dortmanna, Littorella lacustris, and others, most of which are rosette-plants. Subularia and the Characeae may most fittingly be appended to these.

It becomes clear that the two, tolerably similar, linear types of leaves just described result, at least partially, from the action of the water, when we observe the behaviour of Juncus supinus, Hippuris vulgaris, Elatine Alsinastrum, Isoëtes lacustris, Pilularia, and other plants that assume terrestrial and aquatic forms ; the submerged leaves are much longer, more flaccid than the subaerial leaves.

  d. The *myriophylloid leaf, or leaf dissected into filiform or linear segments* (like the gills of a fish) is very widespread, occurring in Myriophyllum, Heliosciadium inundatum, Batrachium, and Cabomba, as well as in certain marsh-plants, including Oenanthe Phellandrium, O. fistulosa, and Sium latifolium, when these grow in deeper water. Allied to this type of leaf is the uncommon fenestrated leaf of Ouvirandra fenestralis. Many observations show that the depth of the incisions, and the fineness of the segments are due to the influence of the medium (depth of water, strength of its flow, and so forth) : when the shoots reach the surface there appear floating leaves, as in Batrachium, or leaves with shorter, broader, thicker segments, especially when the shoots project out of the water as in the case of Myriophyllum. The physiological cause of this difference must presumably lie in the elongation due to decreased illumination, and in

---

[1] Askenasy, 1870.
[2] In reference to the different forms of Polygonum amphibium, see Massart, 1902.
[3] See Costantin, 1884, 1885, 1886; Göbel, 1889 ; Raunkiär, 1895–9; Glück, 1905, 1906.

the exclusion of transpiration. Finely divided leaves are well suited to the medium, because their surface is thus relatively increased, and consequently the absorption of food-material and of oxygen is facilitated. The movements of the water would hardly permit of larger surface.

Reproduction. The reproduction of the Cryptogamia takes place under water. But nearly all the Spermophyta thrust their flowers above water ; some, including Hottonia and Nymphaeaceae, are entomo-philous, but others, represented by Hippuris, Myriophyllum and Pota-mogeton, are pollinated by the aid of wind or water, or adopt self-pollina-tion. The pollen is conveyed by the water in Zannichellia, Callitriche, and Naias ; while semi-cleistogamy is enacted under water by Subularia aquatica, Limosella aquatica, Euryale ferox, Elisma natans, and rarely by Batrachium. Peculiar behaviour, paralleled by that of Ruppia spiralis, is displayed by Vallisneria, whose small male flowers break loose, swim on the surface and pollinate the stigmas of the female flowers which rest on the water ; nearest in this respect to Vallisneria stands Elodea.[1]

After pollination, the developing fruits of many species, including Trapa and Ranunculus, are dragged or curled down under water where they ripen. Dispersal of the seed is often accomplished by special means which are suited to the medium : the seeds or fruits of many species, owing to their peculiar structure, are lighter than the water, and are thus conveyed to other habitats.[2]

Vegetative Propagation. Propagation by vegetative means is very widespread among all aquatic plants, and very easily takes place by mere separation of fragments of the shoot ; it is of profound biological signifi-cance, and some species have almost ceased to reproduce sexually.[1] Calla palustris has special buds that easily become detached.[3] The rapid spread of Elodea and the inconceivable number of its individuals in Europe result from vegetative multiplication, as it produces no seed because the female plant rarely occurs in Europe. The great power of vegetative multiplication is to be ascribed to the production of propagative buds, to branching, and to the facile manufacture of adventitious roots.

Hibernation.[1] Most of the species hibernate as green plants at the bottom of the water, where thermal conditions are not so extreme as in air ; such is the case with Callitriche, Zannichellia, Nymphaeaceae, Vallisneria, and others. Hibernating organs of a special kind, which become detached from the decaying shoot, are represented by the cartila-ginous winter-shoots ('hibernacula') of Potamogeton crispus and other species,[4] the spherical buds of Myriophyllum and species of Utricularia which include densely packed leaves, also by the gemmae (buds) of Hydrilla and Elodea.

FORMATIONS, AND DISTRIBUTION

A great number of associations occur in the limnaea-formation, and some of these, like those composed of Cryptogamia, might perhaps be regarded as being themselves complete formations or sub-formations. The common *zonally arranged associations* are evidently identical over wide areas in temperate countries.

In the deepest parts of some lakes in Europe there occurs a zone of

[1] Schenck, 1886 *b*.   [2] Ravn, 1894.   [3] J. Erikson, 1895.
[4] Sauvageau, 1888–94 ; Raunkiär, 1895.

ground-algae, which may include vast numbers of Cladophoraceae, with which Diatomaceae and Fontinalis antipyretica may be mingled.[1] Brand [2] found a zone of this kind at a depth of 20 metres. Generally, *characeta* form the deepest zone of macrophytes; they are ranged in dense sublacustral ' meadows ', in which no plants, other than possibly mosses, may find place. In Lake Geneva they descend to 20–25 metres, according to Forel,[3] while in Lake Constance they go down to 30 metres according to Schröter and Kirchner,[4] but the most frequent depth is 8–12 metres. The association next in depth is that of Elodea, which descends to 6 metres; then succeed submerged species of Potamogeton, including P. lucens, at depths as great as 4–6 metres, and with them may be Myriophyllum. In the next higher zone appear plants with floating leaves, *nymphaeeta* and *nuphareta* down to 3–5 metres, and *batrachieta* to 2–3 metres. As a rule, Spermophyta seem to cease at a depth of 10 metres.[5]

In North America [6] there are zones completely corresponding to these European ones.

By the shores of shallow waters in Europe, and especially where the soil is sandy, there occurs a peculiar association composed of the *limnaea rosette-forms*, Lobelia, Littorella, Isoëtes, and Subularia, which are characterized by shortness of stems and by rosulate leaves that belong to type 2 *b* described on p. 182.

The distribution of the associations is determined by—

a. The conditions of *depth, light,* and *clearness of water.* Some species can descend to much greater depths than others. The limnaea-communities in larger lakes are confined to a zone of slight depth which borders the shore and entertains an abundant fauna.

b. Distinctions in *soil.* Some species prefer sand, and others mud. The Characeae are calciphilous, yet they can be found in pools on heaths and moors where the water contains but little lime.

c. Movements of the water. Some species, and particularly those with floating leaves, live only in placid water.

Limnaea-vegetation is allied to hydrocharid-vegetation. The boundary between the two is not sharp; they are often intermingled, and in both there occur genera that are the same, although represented by different species. Certain plants, including Eichhornia crassipes, Stratiotes aloides, Hydrocharis, and Pistia, that usually are free-floating, may on occasion become attached by roots; and conversely other species, of Ceratopteris [7] for example, that are normally fixed and rooted, may become free-floating. Plankton necessarily must occur in limnaea-formation: the two give rise to a combined formation. There is, of course, no sharp distinction between aquatic plants that are fixed by their roots and marsh-plants; among them are many ' amphibious ' species, including Polygonum amphibium, that can assume an aquatic or a terrestrial form. Plants such as Montia rivularis living in springs are in a sense transitional between land-plants and water-plants; they seek water that flows rapidly and is rich in oxygen and carbonic acid.[8]

[1] (Schröter und) Kirchner, 1896; G. Huber, 1905.   [2] Brand, 1896.
[3] Forel, 1891.   [4] Schröter und Kirchner, 1896–1902.   [5] Magnin, 1893, 1894.
[6] According to Coulter, Cowles, Transeau, and Pieters.   [7] Göbel, 1889–92, II.
[8] In addition to the literature already cited, attention should be directed to papers by Chatin, 1856; Früh und Schröter, 1904; Pond, 1905; Glück, 1905–6; Göbel, 1908.

SECTION V

CLASS II.   HELOPHYTES.   MARSH-PLANTS

CHAPTER XLIII.   ADAPTATIONS.   FORMATIONS

AMONG *aquatic plants* are included all plants whose assimilatory organs are submerged, or, at most, swim on the surface of water[1]; *marsh-plants* or *helophytes* include all those which normally have their roots under water or in soaking soil, but, like land-plants, raise their foliage branches above the water-surface.   It has already been pointed out[2] that there is no sharp limit between marsh-plants and land-plants. Moreover, many marsh-plants are more or less ' amphibious ' and plastic, so that they can change their structure according as they are submerged or not.[3]

Marsh-plants and bog-plants are confined to shallow, still, or gently flowing water, and to soil that contains a large amount of water (apparently more than eighty per cent.) at least for a prolonged period.   The soil is loose, often very loose and soft, also usually rich in humus in the form of peat or mud.[4]

ADAPTATIONS

1. Marsh-plants, like aquatic plants, are largely *perennials*.[5]   But in marshes that are completely dried up during the dry season annual species may prevail.

2. Many marsh-plants readily produce *adventitious roots* and possess horizontal *rhizomes* or runners ;   in Europe such is the case with Equisetum limosum, Iris Pseudacorus, Phragmites, Typha, Acorus, Butomus, Scirpus lacustris and S. (Heleocharis) palustris, Eriophorum angustifolium and E. alpinum, Sparganium, Carex acutiformis, C. rostrata, and other species, Cladium Mariscus, and other Monocotyledones, Lysimachia vulgaris and L. thyrsiflora, Ranunculus Lingua, Sium latifolium (with roots that produce buds) and S. angustifolium.

Caespitose plants with a small power of vegetative migration or devoid of such a power are exemplified by Lythrum Salicaria, Cicuta virosa, Alisma Plantago, and Rumex Hydrolapathum.   Plants of this form often grow partially on their own dead fragments by which they are gradually lifted upwards ;   one obvious reason for this is that water is raised by capillarity in the sponge-like tufts formed by the interwoven remnants of stems, leaves, and roots ;   such is the case with Eriophorum vaginatum, Carex stricta, C. paniculata, and many other species.

In addition there occur plants showing other modes of growth ;   for

---

[1] See Chap. XXVII.   [2] See p. 131.
[3] Costantin, 1897 ; Schenck, 1884, 1886; Massart, 1902.
[4] See p. 61.   [5] See p. 100.

example, those living on Sphagnum must necessarily have the power of raising themselves as their substratum grows.[1]

3. As in aquatic plants, *internal air-containing spaces* develop in stems, leaves, and roots, and are adapted to meet the scarcity of air in wet soil, whose air-content is sometimes further decreased by special conditions, such as the accumulation of organic remains, production of peat,[2] the interweaving of roots, and other means by which there is formed a covering that shuts off the supply of air. In order to secure adequate aeration for the plant the following special devices exist :—

(a) *Aerenchyma*[3] : a tissue which, like cork, has its own phellogen, but consists of thin-walled, non-suberized cells, and includes large, air-containing, intercellular spaces. The tissue shows itself as a white, spongy envelope. It occurs in Epilobium hirsutum and other species, Lythrum Salicaria, Lycopus europaeus, the mimosaceous Neptunia oleracea, and others.

(b) *Respiratory roots* (pneumatophores). In some trees and shrubs there are developed *erect roots* which thrust their tips above water and convey air to the intercellular system of parts in the mud by means of their pneumathodes, that is to say, by means of lenticels or other communications with the atmosphere.[4] They occur particularly in mangrove-swamps; also on certain palms, Taxodium distichum,[5] and possibly on Jussieuea.[6]

4. Marsh-plants commonly have *mesophytic leaves ;* but in a remarkable number of them *xerophytic structure* is encountered.[7]

5. The *seeds* and *fruit* of many marsh-plants are provided with air-containing spaces and other devices that facilitate their dispersal by water.[8]

FORMATIONS AND ASSOCIATIONS

Qualities of the soil play a great part in evoking floristic, and, more or less, oecological distinctions in the communities :

Saline swamps are not only floristically, but also anatomically and morphologically so peculiar, that they differ widely from fresh-water swamps, and they will be considered in Section VII, dealing with halophilous vegetation.

Fresh-water swamps, which alone will be dealt with here, display many differences according as the water is troubled or not, according as the soil is muddy, sandy, or gravelly, and so forth. The constituent formations have been but little investigated. These are probably several, but certainly there are two : *Reed-swamp* and *Bush-swamp*. There are also communities on the ' boundary ' zone of wet land which live an amphibious life, and are hydrophytic in adaptation, and for these Schröter and Kirchner set up a third formation consisting of *amphiphytes*. They write[9] : ' Every point on the boundary zone is annually inundated for a shorter or longer period, which is longer the nearer it stands to the lake. . . . Thus this zone represents a very gradual transition from terrestrial

---

[1] See P. E. Müller, 1894.   [2] See Chap. XVI.
[3] Schenck, 1889 b; Göbel, 1889–92, ii.
[4] Göbel, 1886; Wilson, loc. cit.; Schenck, 1889 a; Schimper, 1891; G. Karsten, 1891.   [5] Kearney, 1901.   [6] Göbel, 1889.   [7] See Chap. XLVI.
[8] Guppy, 1891–3 ; Ravn, 1894.   [9] Schröter (und Kirchner), 1902, p. 42.

to lacustral conditions. It therefore exhibits a zonal arrangement of
its occupants according to the degree of adaptation of these to the lacustral conditions. . . . The geographically defined " boundary zone "
must be classified biologically into three main subdivisions : (a) Meadow-
swamp, lying nearest the dry land and inundated only for a short time;
(b) zone that is being converted into land . . . : (c) strips of gravel or
sand poor in vegetation.' Only the last two concern us here. In these there
grow land-forms of plants derived from the lake-flora, also typical occupants of the ' boundary zone ', as well as plants that have advanced to
this point though belonging to swamp meadows and ditches. The above-
named botanists distinguish, in connexion with Lake Constance, two
associations : a heleocharetum (with Heleocharis acicularis, Littorella,
Ranunculus reptans, Myosotis palustris var. caespititia, Agrostis alba,
and others) and a polygonetum. They furthermore distinguish a fourth
formation, namely that of *alluvial plants,* including a tamaricetum-
association (with Tamarix germanica and Hippophaë and others), which
occupies the ' boundary zone ' where this takes the form of a gravelly or
sandy shore, and including also plants from the riparian alluvia as well
as alpine plants.

Swamp vegetation is also allied to that which is described by Drude
under the heading of *Formations near springs and brooks,* and includes
tall herbs like Ulmaria, Geranium palustre, Impatiens noli-me-tangere,
Equisetum Telmateia, and small herbs, mosses, and algae.

As these last formations have been oecologically investigated only
to a slight extent, we shall here consider in detail only—

1. **Reed-swamp Formation.**
2. **Bush-swamp Formation.**

CHAPTER XLIV. REED-SWAMP OR REED-FORMATION

REED-SWAMPS occur either in fresh flowing water, or in stagnant,
more or less acid, water. Many species, such as Phragmites communis,
seem to be not in the least exacting as regards choice of soil. The swamps
of Arundinaria macrosperma are on acid peat soil, and the vegetation
thus provides a transition to oxylophytes.

The vegetation is mainly composed of tall monocotylous perennial
herbs, grows in more or less deep, usually standing, water, and seems
to be most nearly allied to the communities of fresh-water plants ;
between the individual shoots and leaves one sees everywhere clear water,
which entertains representatives of the plankton-, hydrocharid-, and even
limnaea-formations.

ASSOCIATIONS IN TEMPERATE ZONES

Among the various genera and species to be found are Phragmites
communis, Scirpus lacustris, Typha, Butomus umbellatus, Glyceria
spectabilis, and other species, Phalaris arundinacea, Iris Pseudacorus,
Cladium Mariscus, Carex paniculata, C. rostrata, C. gracilis, C. filiformis,
C. acutiformis, C. stricta, C. riparia, C. vesicaria, and other species,

Alisma Plantago, Sagittaria, Sparganium ramosum, S. simplex, Acorus Calamus, and Calla palustris, which together represent the most important monocotylous representatives of this oecological class in temperate Europe ; to these may be added Equisetum limosum, and, among Dicotyledones, Senecio paludosus, Sonchus palustris, Menyanthes trifoliata, Lythrum Salicaria, Epilobium hirsutum, Rumex Hydrolapathum, Lysimachia vulgaris, L. thyrsiflora, Ranunculus Lingua, Oenanthe fistulosa, O. aquatica, Sium latifolium, S. angustifolium, Cicuta virosa, and many others. Many diatoms and Green Algae occur as epiphytes.

Especially on the banks of water-stretches is found this vegetation, which advances as a pioneer of land-vegetation, acts as a check to wave violence, and adds to the land.[1]

According to depth of water and other conditions dependent thereon (light, temperature, and water-movement) this formation is likewise arranged in *zones*, which in Denmark and over the greater part of Europe are identical, and may be pure associations, such as phragmiteta, scirpeta, and so forth. Magnin[2] and Schröter[3] have observed, in Lakes Jura and Constance respectively, the following zones : *scirpeta* (Scirpus lacustris down to 3·5 metres in Lake Constance), *phragmiteta* (Phragmites communis down to 2 metres), after which succeed *heleochareta* (Heleocharis palustris), and *cariceta* (in Denmark, C. aquatilis, C. rostrata, and others). All these plant-communities work in the direction of filling up collections of water and draining them dry. There also occur *typheta*, *equiseteta*, and others.

Precisely the same zones present themselves in North Europe and North America. According to Transeau,[4] in the Lakes of Michigan, after the aquatic societies containing Potamogeton and Nymphaea, there succeeds the ' cat-tail-Dulichium association ' with Typha, Phragmites, and Dulichium. Further inland follow the ' Cassandra society ', ' shrub-and-young-tree ' association, and forest. Cowles[5] finds in the vicinity of Chicago the following zones : (1) Chara ; (2) Nymphaea ; (3) Carex and Scirpus ; (4) Cassandra calyculata and other shrubs ; (5) Forest. In other places the cariceta are succeeded by grass meadows.[6]

Adaptations. *Vigorous creeping stems* fasten the plants to the loose soil, causing *social growth* and dense pure associations of Phragmites, Scirpus lacustris, Equisetum limosum, Typha, and many others; also, on the Nile, Cyperus Papyrus. The production of shoots by roots, which is so frequent in dry spots, rarely occurs in the vegetation of reed-swamps, but is exhibited by Sium latifolium. Caespitose species are likewise rare.

The *vegetative shoots* vary in construction, but mainly belong to one of three types :—

(*a*) Leafless, bare stem consisting of one internode, which may be either one or two metres in length, and is capped by the inflorescence ; as examples, we may mention Scirpus lacustris, S. Tabernaemontani, which are as much as 1–2 metres in height, also Heleocharis palustris and species of Juncus which are shorter.

---

[1] See Section XVI.   [2] Magnin, 1893, 1894.
[3] Schröter und Kirchner, 1896–1902.   [4] Transeau, 1903, 1905.   [5] Cowles, 1901.
[6] See also Pieters, 1894, 1901 ; Hitchcock, 1898 ; V. Borbas (Bernátsky), 1907 ; Früh und Schröter, 1904.

(b) In addition to long linear leaves rising from a rhizome or radiating from the base of the flowering axis, there are tall culms bearing inflorescences, as may be seen in Typha, Acorus, Butomus, and others.

(c) Tall haulms with long distichous, spreading leaves, as in Phragmites and other grasses.

The *character common* to all is that the dominant, mainly monocotylous, plants which give the stamp to the vegetation, are *tall, slender, upright,* and *unbranched.* Even in Ranunculus Lingua and in Rumex Hydrolapathum there is a recurrence of the same habit which thus suggests an adaptation of obscure significance. It may, however, be pointed out that these tall slender shoots easily bend to breeze or current, and elastically recover ; this is specially true of the unbranched stems of plants such as Scirpus palustris, or the tall, long leaves projecting above the water from the stems of Typha and Sparganium.

Nearly all the species are *perennial* herbs, or, like Ranunculus sceleratus, biennials. Special hibernating and propagative organs are produced by Sagittaria, in the form of stem-tubers on runners. An occasional woody plant, such as Salix cinerea or Alnus glutinosa, may also occur.

Among the plants of reed-swamps many are almost devoid of protection against desiccation—they are mesophytes ; but others are xeromorphic, for one finds among them leaves with a profile-lie in Iris, rolled leaves in Cladium, juncoid shoots in Juncus and Scirpus, dense hairiness in Epilobium hirsutum, and so forth. Aerenchyma is very widely represented in plants of the reed-swamp.

Geographical Distribution. Associations of this same stamp are ubiquitous on the Earth. Phragmites, the reed, is extraordinarily widespread ; over an area of many square miles it forms impenetrable associations (phragmiteta) in the Danube delta, in deltas of the Caspian Sea, and Lake Aral, and even in Australia ; on the Syr-daria it attains a height of 6 metres and endures salt water quite well, while in Lusatia in Germany its variety *pseudo-donax* attains nearly 10 metres.[1] It can grow in water 3 metres in depth. In Mediterranean countries this reed forms communities, and is sometimes accompanied by Arundo Donax and Erianthus Ravennae, two grasses which are often even 5 and 6 metres in height. As an illustration of its capacity of adjusting itself to external conditions, we may mention that on the shores of the North Sea and of many inland lakes it produces long epigeous runners.[2]

Extensive reed-swamps of Glyceria spectabilis on the saline soil of Neusiedler Lake assume the form of veritable ' grass-forests ' 2 metres in height ; the same is true of Cyperus syriacus in Sicily, and in an exaggerated degree of C. Papyrus on the Upper Nile, where this, together with Panicum pyramidale, Phragmites communis, Typha australis, and others, give rise to a ' sudd-formation '[3] ; the shores of Valencia Lake in Venezuela are fringed with dense reed-swamps of Typha domingensis, which exceeds man's height, and the same is true of the cyperaceous Malacochaete Tatora on the shores of Lake Titicaca.

In Virginia, according to Kearney,[4] there occur similar reed-swamps,

[1] According to Ascherson und Gräbner 1898-9.
[2] Illustrations in Warming, 1906 ; Schröter und Kirchner, 1902.
[3] Broun, 1905.   [4] Kearney, 1901.

in which he distinguishes associations of Typha-Sagittaria along the rivers, and of Scirpus-Erianthus at the edge of the swamp-forest. Here also occurs Arundinaria macrosperma - association, which clothes large expanses of the Dismal Swamps. Along the rivers of Pennsylvania, according to Harshberger,[1] there occur extensive reed-swamps, in which he distinguishes various associations, including those of Zizania, of Typha, of Sagittaria latifolia, and of Ambrosia trifida. In swamps with slowly flowing water one finds other associations, including those of Symplocarpus (with Spathyema foetida, and species of Osmunda), of Iris-Typha-Acorus, and of Heracleum-Veratrum-Eupatorium. Everywhere the physiognomy, and to some extent the genera, are the same as in Europe.

ASSOCIATIONS IN TROPICAL ZONES

Other species and families occupying similar habitats play the same part in nature within the tropics, but, as their associations include entirely different forms, they present a different physiognomy. They have been studied only to a slight extent. Many species of Araceae are swamp-plants—just as are Calla and Acorus in Europe—and usually have sagittate or cordate leaves ; dense associations, often several metres in height, are formed by them, for instance by Montrichardia arborescens in Trinidad and the adjacent parts of South America, also by Caladium.[2] Among Scitamineae, species of Heliconia likewise occur in tropical America, and gigantic Amaryllidaceae, represented by Crinum, fringe the rivers of Guiana. These types of vegetation are never absolutely pure ; other, possibly many other, species mingle with those that have been mentioned as giving the tone to the vegetation. In the tropics larger numbers of woody plants occur and affect the appearance of reed-swamps. But there are also swamps with grasses and perennial herbs very like those of temperate countries, for example in Brazil.[3]

CHAPTER XLV. BUSH-SWAMP AND FOREST-SWAMP OF FRESH WATER

IN reed-swamp and low-moor there often occur some *woody plants*, but in other places these become so numerous as to create *bushland* and *forest* (woodland-swamp, paludal forest, foliage-moor). In North Europe the first steps towards these are shown in collections of alders, birches, and willows growing on the banks bounding fresh water.

Alneta may occur on mud on which only few plants can thrive as undergrowth ; these latter include ferns, mosses, Calla, Lythrum, Spiraea Ulmaria, Menyanthes, Carex, and others, which particularly grow on the drier spots at the foot of the alders. Intermingled with the alders there may be species of Salix, Viburnum Opulus, and Rhamnus Frangula. In other spots Urtica dioica forms impenetrable thickets.[4]

Saliceta at other places in North Europe form the riparian bush-

---

[1] Harshberger, 1904.   [2] Martius, 1840–7.   [3] Warming, 1892.
[4] See Domin, 1904.

lands, mainly composed of species of Salix, S. alba, S. fragilis, S. cinerea, and S. pentandra, between which there grow dicotylous perennial herbs, including Lysimachia vulgaris, Epilobium hirsutum, species of Valeriana, and Spiraea Ulmaria, as well as grasses such as Calamagrostis lanceolata and Phragmites. The lianes in these bush-swamps are Solanum Dulcamara, Convolvulus sepium, and Humulus Lupulus.

Betuleta and **Pineta,** according to Fleroff,[1] occur on bog-lands in Russia.

More extensive bush-swamp and forest-swamp occur in the southern part of the United States, where they give rise to large forests on wet, peaty soil. In Virginia there are two main associations—juniper-swamp and black-gum swamp—with several (subordinate) associations.[2]

Juniper-swamp is formed of Chamaecyparis thyoides, and may be pure. The soil is a very acid peat, which even in summer is covered by water 30–60 centimetres deep.

Black-gum swamp is composed of Nyssa biflora and Taxodium distichum. On the horizontal roots of the latter there arise conical roots which attain a height of a metre, are similar to those of Bruguiera in mangrove-swamps, and in like manner serve as respiratory organs. On the mud they form the solitary firm spots over which man can walk. Many pseud-epiphytes occur on the trees.[3] In the water between the trees grow Azolla, Wolffiella, and others. The soil is acid, but not so peaty and dry as in juniper-swamp. The water as a rule forms a sheet 30–100 centimetres deep lying above the soil. Nyssa and Taxodium are deciduous. And the same is mainly true of the subordinate species in Virginia. Farther south there appear a number of evergreen shrubs, and several short palms, including Sabal and Chamaerops; near the tropics Tillandsia usneoides and other epiphytes show themselves on the tree-crowns. Nearly all woody plants growing in the American forest-swamp are protected against rapid transpiration. Stomata in nearly all species occur exclusively on the lower face of the leaf, and are sunken in some species. In addition, the following features are present: coating of hairs or of wax, thick cuticle, and thick outer wall to epidermis, conversion of epidermal cells into mucilage, hypoderma, multiplication of the palisade-layers. This strong development of measures guarding against desiccation is a consequence of acidity of soil, which abounds in organic remains.

Bamboo-forest (bambusetum). Tropical bamboo-forest must apparently be regarded as one type of association belonging to swamp-forest. Tropical rivers are often fringed with bamboo-brake, which forms most impenetrable vegetation. Humboldt mentions that, along the river Magdalena, there are uninterrupted forests of bamboo and banana-leaved species of Heliconia.

Nipetum. Under this heading is included the eastern Asiatic and Australian vegetation composed of Nipa fruticans. This palm is all but stemless, yet it possesses immense pinnate leaves, which may be six metres in length, and its growth may be so dense that one requires an axe to cleave a way through the vegetation, in which other species, including Chrysodium aureum, also occur. Nipetum lies on the land-

---

[1] Fleroff, 1907.   [2] Kearney, 1901.   [3] Theo. Holm, in letter.

ward side of mangrove-swamp, from whose species it derives subordinate constituents; it also appears in connexion with lagoons and swamps, but mostly on less saline soil. According to J. Schmidt [1] it does not truly belong to the saline mangrove-swamp, next to which Schimper places it, but rather belongs to fresh-water swamps along rivers.

In the tropics several other, but little known, forms of forest-swamp and bush-swamp occur. A small fan-palm, a species of Bactris, clothes large swampy tracts along the river Caroni on the plains of the island of Trinidad. Another palm, Phoenix paludosa, is encountered in eastern Asiatic swamps. While, according to Kurz, in Burma there are swamp-forests that are leafless during the rainy season. Koorders [2] gives an interesting account of a forest-swamp in the interior of Sumatra, in which respiratory roots (in Calophyllum, Eugenia, and others), prop-roots, plank-roots, and remarkable besom-like aerial roots ($1-1\frac{1}{2}$ metres in length) occur. The physiological dryness and these peculiarities of construction are to be attributed to lack of oxygen in the soil.

It is impossible to establish any sharp distinction between swamp-forests and forests on dry land; this is shown by the semi-aquatic varieties ('facies') of the primaeval forests that line the Amazon, which are annually inundated, and are locally known as 'igapo'.[3]

Bush-swamp and forest-swamp occur not only in tropical and temperate, but also in arctic lands; for instance, in Kanin peninsula, on the White Sea,[4] Picea excelsa, Betula pubescens, and Pinus sylvestris, occur as dwarfed trees on insufficiently drained soil, and are accompanied by genuine marsh-herbs such as Eriophorum vaginatum.

Again, in the rainy zone of the territory near the Magellan Straits the forests composed of evergreen species of Nothofagus (N. betuloides and others) are regarded by Dusén [5] as hydrophytic communities; here, extraordinary humidity brings forth a swampy soil nearly wholly carpeted with mosses. In this, as in all other cases, sharp distinctions are lacking. But it must be insisted that in these bush-swamps the soil is always more or less sour (rich in humous acids), and that consequently this formation is allied to the one about to be described.[6]

[1] J. Schmidt, 1903.   [2] Koorders, 1907.   [3] J. Huber, 1906.
[4] Pohle, 1903.   [5] Dusén, 1905.
[6] W. G. Smith, 1903; N. Walker, 1905; see also, on the subject of this chapter, Coulter, 1903; Hitchcock, 1898; Kearney, 1901.

SECTION VI

CLASS III.   OXYLOPHYTES.   FORMATIONS ON SOUR (ACID) SOIL

CHAPTER XLVI.   XEROMORPHY.   FORMATIONS

On a soil that contains an abundance of free humous acids, and is more or less peat-like, there occurs a group of closely-related formations ; these rise in stature from humble communities of mosses and lichens, to dwarf-scrub and bushland, and finally to forest.

All these communities share the character of choosing soil that is poor in nutriment and particularly in easily assimilable nitrogen—they are *oligotrophic* ; furthermore, they are not *calciphilous*. They often clothe sterile sand, on which they themselves soon produce raw humus, and usually give rise to dense, exclusive, and extensive communities.

Sour soil is intimately associated with a moist, cold or temperate, climate.

These communities all exhibit *xeromorphy*, that is to say, they are protected from desiccation by certain devices, of which the most important are :—

1. *Well-developed coating of hairs :* Hairs on the lower face of the leaf form a felt in Ledum, Salix repens, S. lanata, and S. glauca, but are scale-like in Cassandra calyculata, and in the North American Nyssa uniflora, Persea pubescens, and Magnolia virginiana, which grow in swamps.[1] The essential function of the hairs may be to prevent water from occluding the stomata, but the hairs also depress transpiration. In this connexion it may be noted that Salix Myrsinites, growing in the swamps of Lapland, retains its faded leaves which serve to protect the year's shoot.[2]

2. *Papillae* project over the stomata of various Gramineae and Cyperaceae, such as Carex limosa, C. panicea, and C. rariflora ; also of Lysimachia thyrsiflora, Polygonum amphibium.[3] They also may prevent the stomata from being blocked by water.

3. *Wax* forms incrustations over the whole leaf, as in Vaccinium uliginosum, or only over the stomatiferous lower face, as in Andromeda polifolia, Vaccinium Oxycoccos, Primula farinosa, Salix groenlandica, Carex panicea, and in the North American swamp-plants,[4] Acer rubrum, Persea pubescens, and others.

4. *Thick cuticle* is shown by various leaves, and by the stems of Scirpus caespitosus and others.

---

[1] Kearney, 1901 ; see p. 191.   [2] Kihlman, 1890.
[3] Volkens, 1890 ; Kihlman, 1890 ; Raunkiär, 1895-9, 1901.
[4] Kearney, 1901.

5. *Sclerophylly* is frequent and is due to thickness of the epidermal wall, as in Andromeda polifolia, Vaccinium Oxycoccos, V. Vitis-Idaea, and Ledum palustre, and is perhaps correlated with the perennial character of the leaves concerned.[1]

6. *Mucilage* occurs, for instance, in the epidermal cells of Berchemia scandens[2]; and a continuous hypoderma is present beneath the epidermis on the upper leaf-face of Pieris nitida.

7. *Ericoid leaves.* Many moor-plants have flat broad leaves; but some species possess *ericoid or filiform leaves* whose stomata are enclosed in secluded spaces, so that water-vapour escapes with difficulty—such is the case in Erica Tetralix, Empetrum, Calluna vulgaris, and other species to be mentioned in the next group.

8. *Terete leaves, aphyllous stems.* The assimilatory organs of many moor-plants and marsh-plants assume the form of *erect, terete leaves* or *aphyllous stems*, as in Equisetum limosum, species of Junci genuini, and other species of Juncus to a less extent, Scirpus palustris, S. caespitosus, S. (Heleocharis) lacustris, and other species, Eriophorum vaginatum, Carex microglochin, C. dioica, C. chordorrhiza, and C. pauciflora.

9. *Bilateral leaves.* Leaves exposing their edges (profile-lie) are met with in Iris, Narthecium, Acorus, and Xyris. The leaves are flat, broad, but likewise erect or upwardly directed, long, and undivided, in Alisma Plantago, Sagittaria, and other Alismaceae, Butomus, Typha, Sparganium, Ranunculus Lingua, and Lathyrus Nissolia.

10. *Closure of leaves.* Broad-leaved Cyperaceae can close their leaves together (always?), and distinctly so in Carex Goodenowii; yet the stomata are not confined to the upper face.

This xeromorphy of plants growing on wet moor-soil occurs all the world over; it is known not only in Europe,[3] but also in America[4] and New Zealand.[5]

It is evident that there must be a causal connexion between the soil and the xeromorphic structure which has been described, but it would never be anticipated. The soil must be physiologically dry. In genera which include not only paludal species, but also mesophytic species not growing in dry places, we frequently find that the latter have the broadest leaves, although the contrary might be expected. The paludal Epilobium palustre and Lysimachia thyrsiflora are the narrowest-leaved species of their genera in our country; Galium palustre and G. elongatum likewise have narrower leaves than the mesophilous species possess; and other cases might be cited.

Here attention may be directed to the remarkable fact that many species of heather-plants can grow both on extremely dry, warm soil, and on extremely cold, wet soil; such is the case with Calluna, Empetrum, several species of Pinus, Juniperus communis, Betula nana, Saxifraga Hirculus, Ledum palustre, and Vaccinium Myrtillus in Europe,[6] Pinus Taeda in the Dismal Swamps of the United States,[2] and Phormium tenax in New Zealand.[7] One would, therefore, assume that between the two

---

[1] For anatomy see H. E. Petersen, 1908.   [2] Kearney, 1901.
[3] Volkens, 1890; Kihlman, 1890; Raunkiär, 1895-9,1901; concerning the British Isles, see Miall, 1898; Rob. Smith, 1899.
[4] Kearney, 1901; Pound and Clements, 1900.   [5] Cockayne, 1901, 1904, 1905a.
[6] See Gräbner, 1895, 1901.   [7] Cockayne, 1904.

kinds of soil there is some essential agreement, and that some of the life-conditions under which marsh-plants exist compel them to deal economically with water. Several different factors may be of moment and co-operate : for instance—

Johow[1] and Kihlman[2] directed attention to the observation made by Tschaplowitz[3] indicating the existence of a *transpiration optimum*, and that therefore even marsh-plants may be forced to depress their transpiration.

Wet soil is cold, and therefore physiologically dry.[4] Consequently, on moors and swamps vegetation develops late, and flowering is late, except in certain species. Kihlman[5] and Göbel[6] point out that many plants, though growing in soaking spots, are clothed with woolly hairs (as is the case with species of Espeletia in Venezuela) or are protected from rapid transpiration in some other way, because strong winds dry the vegetation at a time when the activity of the root is checked by coldness of soil. This well accounts for the xerophytic structure of plants in the extreme north and high up on mountains, and it plays an important part in the places under discussion.

Another circumstance of potential significance is that in every wet, badly-aerated soil, which is poor in oxygen, *respiration is obstructed*, and consequently the root's functional activity is depressed.[7] According to Freyberg, roots of marsh-plants consume less oxygen than do land-plants in a given time, and in order to maintain the balance between their working and that of epigeous parts the functional activity of the latter must be depressed. The fact that many plants, such as Calluna and species of Pinus, growing on heath and on other dry warm soils can also grow on moors, is no longer incomprehensible when we remember that heath often has an extremely ill aerated, periodically soaking, raw-humus soil, which exemplifies the 'dry production of peat'.

It must also be noted that peat has a great *power of retaining water*.[8]

Many moors and heaths may become very dry in their upper layers in summer. We can often walk dry-footed not only over moors clothed with such marsh-plants as Scheuchzeria, Rhynchospora alba, and Carex limosa, but can note that the bog-mosses are so dry as to crackle as we tread on them. Many arctic swamps and moors often become completely dried up.

Livingston[9] arrives at the conclusion that at least in some bog-waters there occur *chemical substances* 'which are not in direct relation to the acidity of the water', but which 'act on the vegetation' : and 'it is suggested that these substances may play an important rôle in the inhibition, from bogs, of plants other than those exhibiting xerophytic adaptations'.

But the weightiest cause of the physiological dryness of the soil probably lies in the presence of free *humous acids* and other dissolved substances which chemically affect the roots.[10] Moor-soil is probably

[1] Johow, 1884.   [2] Kihlman, 1890.   [3] Tschaplowitz, 1883.
[4] See p. 50.   [5] Kihlman, 1890.   [6] Göbel, 1889–91.
[7] Transeau, 1903, 1905.   [8] See pp. 47, 61.   [9] Livingston, 1904.
[10] See Weber, 1902, 1903 ; Schimper, 1898 ; Cowles, 1901 ; Bruncken, 1902 *a* ; Früh und Schröter, 1904.

always acid ; humous acids depress the root's activity and render it more difficult for the plant to replace the water lost by transpiration.

In fact, various factors enter into the question, and possibly all of them play a part in evoking xeromorphy.

Finally it may be noted that there are not only moors inclining to xerophily, but also others leaning rather to hydrophily, and that, in addition to the structural types [1] and forms of leaves mentioned, there are others which apparently show no signs of xerophily and cannot be shown to be in harmony with this habitat : for example, broad, hastate, sagittate, or cordate leaves occur in many Araceae concerned, while broad, orbicular or reniform leaves are shown by Rubus Chamaemorus, Caltha palustris, and Viola palustris.

The formations growing on acid soil may be arranged according to the following scheme :—

1. **Low-moor formation**: Often represents the first stage and the one most nearly related to hydrophytic formations.

2. **Grass-heath. Tussock-formation** (Peculiar to the Southern Hemisphere).

3. **High-moor formation**: Likewise very wet.

4. **Moss-tundra formation**.

5. **Lichen-tundra formation**.

The others are drier formations in which woody plants are the dominant constituents :

6. **Dwarf-shrub heath formation**.

7. **Bushland and forest formation**: The formations 4, 5, 6 may occur as undergrowth here.[2]

CHAPTER XLVII.   LOW-MOOR FORMATION

THE vegetation of low-moor requires less water than does that of reed-swamp, with which it is often continuous on the landward side and at the expense of which it often develops. It shows less open water, which, moreover, is visible to a less extent than in reed-swamp and frequently is only to be seen at certain times. The water-table is always at a high level. The vegetation is closer, and the vegetative shoots project nearly entirely into the air. The water is still or flows but slowly ; the land flat and horizontal, though in arctic countries it may slope slightly. *Humous acids* arise in the soil, which becomes *moor-like* because of the accumulation of vegetable fragments [3] ; thick layers of peat may be produced, especially from certain species, including at times some, such as Phragmites, also present in reed-swamp. The peat *contains much nitrogen*, which, however, is not always in a form easily available to plants.

---

[1] In regard to the great variation in the anatomical characters of monocotylous marsh-plants, see Gräbner, 1895.

[2] In reference to these formations, special reference should be made to the great work by Früh and Schröter, 1904.   See also Pound and Clements, 1900; Clements, 1904 ; Weber, 1902, 1903 ; MacMillan, 1893, 1896, 1897 ; Livingston, 1904; Ramann, 1895 and 1906; Yapp, 1908.   [3] See Chap. XVI.

The water coming from low-moor is *rich in calcium and potassium*. Low-moors arise not only around reed-swamps, but also at the margin of standing or flowing water whose boundaries they constantly narrow, as the reed-vegetation gradually advances into the water. The moor nearly always commences to form on the side towards the prevailing wind ; waves caused by the wind prevent or obstruct its production on the opposite side.[1] Low-moor is also termed meadow-moor, grass-moor, sedge-moor, lowland-moor, infra-aquatic moor, or swamp-meadow.[2]

FLORA AND ASSOCIATIONS

In northern Europe Monocotyledones mostly dominate, and with them many Dicotyledones are intermingled. The following families and genera are represented : most important are the Cyperaceae which often grow in *tufts* or form a matted covering, and are particularly represented by numerous species of Carex (hence the name sedge-moor [3] or caricetum), but also by species of Eriophorum, Scirpus, Schoenus, and others ; among Gramineae, Aira caespitosa, Agrostis vulgaris ; among Equisetaceae, Equisetum limosum and E. palustre ; many Juncaceae ; Juncaginaceae, including Triglochin palustre ; Orchidaceae, including Epipactis palustris and species of Orchis ; Umbelliferae, including Peucedanum palustre, Angelica, and Archangelica ; Ranunculaceae, including Caltha, Trollius, and Ranunculus ; Rosaceae, including Comarum palustre and Geum rivale ; in addition, Menyanthes, Galium palustre, Epilobium palustre, E. parviflorum, Parnassia palustris, and many others. Mingled with these may be *shrubs*, including species of Salix, Betula, Alnus, Rhamnus Frangula, Empetrum, and Ericaceae, which grow especially on the turf and drier places. Some European marsh-plants, such as Saxifraga Hirculus and Carex chordorrhiza, are possibly relics of the Glacial Epoch.[4]

Among *mosses* there are species of Hypnum, Amblystegium, Mnium, Polytrichum, Paludella, and other genera ; but Sphagnum is scanty or absent.

In Austrian swamp-meadow, according to Günther Beck, there are thirty-four Cyperaceae, twelve Gramineae, three Juncaceae, also a number of perennial and other herbs, among which eighteen belong to the Monocotyledones.

In various places, according to the dominant plants, we can distinguish various *associations*, including *amblystegieta, cariceta* (parvo-cariceta and magno-cariceta [5]), *eriophoreta, molinieta*, and others,[6] or, according to species, *stricteta*, named after Carex stricta, and others. In Greenland, here and there are *junceta* composed more especially of Juncus arcticus [7] ; similar associations composed of J. effusus, J. compressus, and others, also occur in Denmark.

Beneath and between the latter plants, in low-moor there are usually two storeys ; for, in addition to the *humbler perennial herbs* which may occur singly, the *ground* entertains a vegetation of mosses which are an unmistakable sign that there is no circulation of air in the soil. But

---

[1] Forchhammer, in lectures about 1860; Klinge, 1890.   [2] See Früh und Schröter, 1904.   [3] 'Gräskjär,' i.e. grass-swamp. See Warming, 1897 ; [see footnote 4, p. 198.]
[4] See Section XVII, Chapter XCVI.
[5] Schröter und Kirchner, 1896, 1902 ; Brockmann Jerosch, 1909.
[6] Stebler und Schröter, 1889–92; Hult, 1881, 1887.   [7] Hartz, 1895.

the mosses by no means play the same part as in Sphagnum-moor. Lichens generally do not occur in such places in temperate Europe ; whereas in arctic moors they are sometimes admixed, often on wet spots where they occasionally form pure associations.

ADAPTATIONS

Duration of life. The constituent species are mainly perennial and herbaceous ; only a few are woody. Some are biennial ; but annuals are scanty, being represented by parasitic Rhinantheae. During winter the bog shows grey withered leaves and shoots. The spring sets in late because of cold due to the abundance of water and to evaporation, also because of cold air above depressions in the ground ; but certain early-flowering species, including Eriophorum vaginatum, provide exceptions to this rule.

The *form of shoot* varies greatly in this rich flora. Any general type of adaptation can hardly be proved to exist. Among dominant Mono-cotyledones some produce *dense, tall tufts* ; this is the case with Carex stricta, which sometimes forms a zone, a ' strictetum ', outside the reed-swamp on the landward side, and shows between its tufts open water until this is occupied by other vegetation.[1] These bogs, in contrast to reed-swamp, entertain many *caespitose plants*, such as species of Carex and of Gramineae.

Runners and *travelling-rhizomes* are possessed by an abundance of species, including Equisetum palustre, Carex Goodenowii, C. panicea, C. gracilis, C. acutiformis, and Menyanthes.[2] The runners and rhizomes above the soil or water are often woven together with numerous roots to form a coherent mat.

Meadow which consists of grasses, but whose soil is neutral in reaction, is closely allied to caricetum : it will be discussed in connexion with mesophilous communities. The amount of water most favourable to it is apparently 60–80 per cent., whereas for arable field 40–60 per cent. suffices. Caricetum contains an amount of water exceeding 80 per cent. ; the water-table of meadow in summer stands at a depth of 15–30 centimetres.

Moss-bog. Sometimes mosses, including Aulacomnium, Hypnum cuspidatum, and others, preponderate over flowering plants. In such cases there arises a dense, soft carpet of moss, with isolated flowering plants, lycopods, and lichens scattered here and there. Moss-bog of this kind occurs in Arctic and Antarctic countries[3] ; it merges into moss-tundra as well as into Sphagnum-bog, that is to say into Polytrichum-tundra and Sphagnum-tundra.

By the word ' myr ', employed in Norway and Iceland, we understand any kind of moor. And in Norway a distinction is drawn between grass-moor[4] (bog-moor) and moss-moor, the latter of which is a high-moor. The grass-moor [sedge-moor] probably nearly always has a carpet of mosses (Paludella, Amblystegium, and others) and a covering, half a metre in height, of Cyperaceae (Carex chordorrhiza or C. filiformis).

---

[1] The ' Zsombék-formation ' of Kerner, 1858. [2] See Yapp, 1908.
[3] ' Meadow-moor ' of Brotherus ; see also Warming, 1887 ; Dusén, 1905, p. 403.
[4] [Here the Norwegian, like the German, word ' Gras ' is definitely applied not only to grasses, but also to sedges. Hence the English version should be ' sedge-moor '. P.G.]

DISTRIBUTION

Low-moor occurs on the most widely-separated parts of the Earth, even in Arctic countries—for instance, on the White Sea.[1] Stages transitional between it and swamp-bushland or swamp-forest often present themselves. We may cite, for instance, a form of swamp-meadow in Servia described by Adamovicz[2] : here associations are formed by Salix pentandra, and Betula pubescens, which never attain man's height and are richly branched from the base. Between these shrubs appear smaller groups of Phragmites and Typha latifolia, and in the vegetation clothing the ground are Calamagrostis lanceolata, Avena rufescens, Cirsium palustre, Succisa pratensis, Caltha palustris, Trollius europaeus, Polemonium caeruleum, and others.

CHAPTER XLVIII. GRASS-HEATH. TUSSOCK-FORMATION

THE formations on acid humus are allied to divers plant-communities which most probably should be regarded as separate formations, but have been inadequately studied. Among such we may include those mentioned below, which show a preponderance of grasses, but apart from this exhibit great differences among themselves.

Grass-heath occurs in northern Europe, and Gräbner[3] alludes to its presence in northern Germany. In the Alps the setaceous-grass meadow represents this formation and covers vast areas. It occurs on all kinds of soil, calcareous or primitive, where the soil is dry and shallow, and where the slope is not too steep to permit of the accumulation of raw humus. The dominant species are Nardus stricta, Molinia caerulea, Carex montana, Agrostis alba, and Anthoxanthum odoratum. In addition, Cladonia and Calluna may lodge here and there.[4] Similar *nardeta* occur in Ireland[5] and Denmark.[6]

Tussock-formation. Differing widely from these grass-heaths are the austral communities about to be described :—

The *tussock-formation* of New Zealand is described by Cockayne[7] as follows : ' The surface of the ground is very wet peat, in many places being actual swamp. . . . The tussock meadow is by no means a uniform formation, but varies so much in composition and physiognomy that ' it must be divided into various sub-formations. One of these is—

(a) The *maritime tussock slope*. Here, ' From the steeply sloping bank of soft, wet, and spongy peat rises up a dense mass of tussocks about 1·5 metres in height, growing upon thick trunks so very closely together that it is not easy to walk between them. These tussocks consist chiefly of a species of grass ', probably Poa anceps. ' Mixed with this tussock are others . . . of Poa foliosa and Carex trifida.'

(b) *Flat tussock meadow*. ' Perhaps this should rather be classed as a heath . . .' ' The badly-drained soil, poor in nourishment, the abun-

[1] Pohle, 1903.   [2] Adamovicz, 1898.   [3] Gräbner, 1901.
[4] Kerner, 1863 ; Stebler und Schröter, 1889, 1892.
[5] Pethybridge and Praeger, 1905.
[6] Raunkiär, 1889; Warming, 1894, 1907-9.   [7] Cockayne, 1904.

dance of lichens and lycopods, the stunted bushes of *Coprosma,* and the semi-xerophytic ferns certainly point to its classification as a heath ; but, on the other hand, the presence of a grass as the dominant plant seems to mark the society as a meadow . . .' ' The soil of the meadow consists on the surface of rather loose brown peat . . .' ' Amongst the most frequent plants are the following : foliaceous and fruticose lichens, mosses of several species, liverworts . . .'

These quotations from Cockayne suffice to show that we are dealing with a formation belonging to acid peat soil ; but it deviates so widely from the formations described as occurring in the northern hemisphere that it cannot be grouped with any of these.

On other islands adjacent to New Zealand the tussock-formation is mainly formed of Poa foliosa, Danthonia bromoides, and Carex trifida.

In Patagonia, South Georgia, also on the Falkland Isles and other sub-antarctic islands the tussock-formation occurs. S. Birger [1] has described the tussock-vegetation of the Falkland Isles. Here it may clothe vast tracts ; the tussocks attain a height and diameter of more than 2 metres. The individual tufts are separated from one another by passages of such width that a man can traverse them. The passages are caused by sea-lions, and in the interior of the ' tussock-forest ' one finds abundant animal life. Where the tussocks are close together Poa flabellata dominates. The tussock-grass of South Georgia is likewise Poa flabellata ; it attains a height of 1·5 metres ; the large tufts show glaucous leaves, which, though 1–2 metres in length, very consistently withstand the wind ; they are mounted upon thick peaty cushions, which are 50–60 centimetres in height and are composed of a mouldering mass of rhizomes, roots, and leaves. In this island the tussock-formation continues from high-tide mark on the shore up to an altitude of approximately 300 metres, and, on the sheltered northern slopes, covers large areas without interruption. The individual cushions are isolated from one another by intervening spaces which are completely hidden by the over-arching leaves of the tussocks.[2]

CHAPTER XLIX. HIGH-MOOR FORMATION

THE moor, known as high-moor, Sphagnum-moor, sphagnetum, or heather-moor,[3] is mainly formed by *bog-moss* (Sphagnum) and arises on moist soil, which is only slightly permeable to water, but does not necessarily show open water, though very damp air hangs over it. Humid air and dew are essential to sphagnum-moor, which acquires the whole of its moisture from atmospheric precipitations.[4] It often arises *on top of old low-moor* ; it may also take origin on wet sand, and even on rocks if these be frequently wet, as on the west coast of Norway and Sweden. Bog-mosses prefer abundant atmospheric precipitations, but thrive neither at high temperature nor in dry air.

The water in high-moor contrasts with that in low-moor by being

[1] S. Birger, 1906.
[2] J. D. Hooker, 1844-7 ; Schenck, 1905 ; Skottsberg, 1906.
[3] See Früh und Schröter, 1904.
[4] See Gräbner, 1895 ; Weber, 1894, 1902; Früh und Schröter, loc. cit.

poor in lime. The peat formed is *poor in assimilable nitrogen, potassium, phosphorus,* and consequently in the most important nutrient bodies belonging to soil. According to most authorities lime in the soil opposes the production of high-moor because Sphagnum—so it is reputed—is calciphobous. But Gräbner's [1] opinion is that Sphagnum cannot endure a concentrated solution of salts, whatever be the nature of these, and can only grow where very little nutrient matter is contained in the water. Weber,[2] as well as Transeau,[3] on the contrary, regard the amount of nutrient salts in solution as being only of subordinate significance.

In regard to the production of high-moor the important work by Früh and Schröter [4] should be consulted.

ADAPTATIONS AND FLORA

Sphagnum and its growth. The construction, conditions of life, and mode of growth of Sphagnum, cause the peculiar type of vegetation of this moor. The smooth stem is densely beset with leaves and emits a branch at each fourth leaf ; in many species the branches are pendent and apply themselves more or less closely to the stem. At its periphery the stem shows 1–5 layers of large, thin-walled, hyaline, capillary cells, whose walls are often strengthened by tracheidal thickenings and contain pores. By this means and by the dense growth of the moss there are formed capillaries, which eagerly suck in water and hold it firmly. It is erroneous to suppose that Sphagnum sucks up water from the soil ; it raises water only for an inconsiderable distance. The movement of water in a Sphagnum-moor is essentially a descending one.[5] The depth at which the water-table lies is dependent upon the atmospheric precipitation and upon the permeability of the peat and of the substratum. The leaf consists of a single layer of cells ; some of these are narrow, long, and green, and together form a network, whose meshes are occupied by larger cells which, being colourless, perforated, and like the capillary cells of the stem, act in the same manner as these. The result is that where moisture is present the Sphagnum-plants are laden from top to bottom with capillary water. As the older parts die off and are converted into peat,[6] the apices constantly and vigorously elongate ; thus one generation is founded upon another. In this way the Sphagnum-moor continues to grow in height, surface, and periphery, so long as the rainfall and the dew suffices : desiccating wind is directly hostile to this type of vegetation. In this manner there arise thick, soft masses of moss, which raise themselves to a considerable height, often bulging upwards more at the centre than at the periphery, because the water in the centre has had the most prolonged access. Hence the name 'high-moor' When once a high-moor is established it constantly extends at its edges and invades areas hitherto intact. Sometimes Sphagnum-clothed tracts, kilometres in width, surround the convex margin of a high-moor and may increasingly convert more and more of the adjoining forest into bog. The forest is thus gradually exterminated.

[1] Gräbner, 1895.   [2] Weber, 1900.   [3] Transeau, 1905.
[4] Früh und Schröter, 1904.   [5] Weber, 1902.
[6] Concerning the power of this to raise water, see p. 61.

Among the species concerned are Sphagnum cymbifolium, S. fuscum, S. Austini, S. rubellum, S. teres, S. recurvum, S. medium, which produce different varieties of sphagneta and often show a zonal distribution.

On the soft loose soil formed by Sphagnum there are other plants, including some species found on low-moor ; but the flora is scarcely so rich as on that. Species with travelling-shoots are especially abundant, in accordance with the loose nature of the substratum. The constituent plants must all of them live chiefly as saprophytes. Among other mosses are species of Polytrichum, Aulacomnium, Bryum, Paludella, Dicranum, and others ; among liverworts, Aneura, Cephalozia, Jungermannia ; in some cases lichens also occur ; among Cyperaceae in North Europe, Rhyncospora alba, several species of Carex and Eriophorum (especially E. vaginatum), and Scirpus caespitosus ; among grasses, Molinia caerulea, Agrostis canina, and others ; among other Monocotyledones, Narthecium ossifragum, Scheuchzeria palustris, Triglochin palustre ; among Dicotyledones, Ericales are particularly frequent, and include Vaccinium uliginosum, V. Oxycoccos, Vaccinium Vitis-Idaea, Andromeda polifolia, Ledum palustre, Erica Tetralix, Calluna vulgaris (this last-named species, though scarcely ever absent, spreads particularly when the moor has become high and very dry, and then is so abundant that the moor is essentially a *calluna-moor*) ; in addition, there are Rubus Chamaemorus, Drosera, Pedicularis sylvatica and Cornus suecica ; among woody plants, Empetrum, Myrica Gale, Salix repens and Betula odorata. On older, higher and drier moors are found pines, including P. sylvestris and especially P. sylvestris var. turfosa, also P. Pumilio ; these are deformed and produce low woodlands, like the elfin-woodland on alps.[1]

In other countries entirely different genera and species are encountered : in North America, Kalmia, Sarracenia, Darlingtonia, and others ; one of these, Kalmia angustifolia, has become completely naturalized in Hanover. But in all oecological essentials the North American is similar to the European high-moor.[2]

Duration of life. Nearly all the constituent species are perennial. In addition to some parasitic Rhinantheae only Cicendia filiformis and a few other species are annual.

Concerning *construction of shoot* any general statement is scarcely possible.

Different associations of the moor are formed according to the prevalence of Cyperaceae, such as Eriophorum, dwarf-shrubs such as Calluna and Betula nana, or trees.

The associations are usually zonally distributed round the water-basins which may occur in the moor ; for instance, a caricetum next to the water may be succeeded by an eriophoretum, and that by a callunetum, and so forth, until the series may conclude with a cladinetum or the like.[3]

On Sphagnum-moor only those species can grow which are capable of following the growth of these mosses, just as on shifting dune-sand the only species that can succeed are those capable of permeating the over-lying sand. As the lower parts of the plants are gradually overgrown by Sphagnum and give rise to peat, their remains become buried. Peat

[1] See Chapter LIII.   [2] Ganong, 1897 ; MacMillan, 1893, 1896.
[3] A. Nilsson, 1899 ; MacMillan, 1896, 1897, 1899.

may attain a thickness of 3–4 metres, or in East Prussia even of 6–10 metres. Peat is produced not only by species of Sphagnum, but also by Polytrichum juniperinum, Scirpus caespitosus, Eriophorum vaginatum, Erica and Calluna, and others. Remains of animals and other objects may be enclosed and preserved in peat. Humous acids efficiently prevent the putrefaction of organic bodies ; the water of moor contains few or no bacteria. Parts of plants, such as leaves and fruits, and even remains of human beings and animals, may be preserved for thousands of years in the water of moor.[1]

Forest-moors in northern Europe arose thousands of years ago in small lakes or pools within forest, and were formerly surrounded and at least partially or temporarily overgrown by trees. They now contain rich stores of plant-remains, which depict the evolutionary succession of the local vegetation and flora. The substratum of such moors is often constituted of fine clay deposited from the neighbouring heights shortly after the Glacial Epoch, together with remains of Dryas, Betula nana, Salix polaris, S. reticulata, and other tundra-plants, which formed the first vegetation on the moraines when these had been deserted by ice after the Glacial Epoch.[2] The later evolution of these forest-moors has been elucidated by Steenstrup[3] in his excellent work on those of Vidnesdam and Lillemoose in Danish Seeland. According to Steenstrup the first tree-growth after the tundra period was produced by Populus tremula, which was accompanied by mosses, including species of Hypnum and Sphagnum : thus began the formation of moor. Betula also appeared early and accompanied the succeeding strata. Gradually the lakes became fringed with forest vegetation, the trees of which fell into the bog, and, together with leaves and fruits conveyed by wind, were buried therein. The first high-forest was constituted of Pinus sylvestris ; this gave way to Quercus sessiliflora and Q. pedunculata, and these to beech, which is only very scantily encountered in the uppermost layers of the moors.[4] Blytt[5] thought that he had discovered in Norway an alternation of strata of peat and of forest (tree-stems) corresponding to an alternation of moist and dry periods, and he employs this evidence to support the theory of the climatic alternation. Possibly it is merely a case of the conversion of forest into marsh under the influence of increased humidity due to local conditions. When forest has been exterminated by conversion into marsh, the soil ultimately becomes drier, and forest once more seizes upon the place. Forest-moor is richer in remains of trees than is low-moor, and contains more numerous aquatic mosses than does the latter.

According to Cajander,[6] the moors of northern Europe are undergoing regressive development ; for example, on the White Sea they are being changed into hilly land which sometimes shows immense hillocks of peat ; Cajander is of the opinion that, with increasing latitude, and possibly with increasing altitude (in the Alps), this regressive development becomes more accentuated.

[1] In regard to high-moor in Scotland, see W. G. Smith, 1902; Lewis, 1905–1907.
[2] Nathorst was the first to discover remains of this kind in Sweden in 1870.
[3] Steenstrup, 1841 ; see Chap. XCVI.
[4] Steenstrup, loc. cit. ; Vaupell, 1851–63.
[5] Blytt, 1882, 1893.   [6] Cajander, 1905.

Distinctions between low-moor and high-moor:[1]

1. Low-moor arises on a surface that is covered with water—it is *infra-aquatic* in origin.

High-moor arises in water or on moist soil or even above water, from which it emerges and rises, becoming dependent solely upon atmospheric water—it is *supra-aquatic*.

2. Low-moor has a flat surface (either horizontal or inclined).

High-moor has a convex surface like a watch-glass.

3. Low-moor is produced particularly by grass-like plants, including Cyperaceae, Gramineae, and Juncus, also by Hypneae.

High-moor owes its origin particularly to bog-mosses, Sphagnum and others, and includes many Ericaceae.

4. Low-moor water is calcareous.

High-moor water contains little or no lime.

5. Low-moor forms black amorphous peat from its vegetable remains, which are so disintegrated as to be scarcely recognizable.

High-moor preserves its plant-remains in a higher degree.

6. Low-moor peat is heavy and rich in mineral bodies (with 10–30 per cent. of ash).

High-moor peat is light in weight and poor in mineral matter (with about 5 per cent. of ash).

7. Low-moor peat is usually close in texture, and when wet is consequently 'greasy', and conducts water badly, so that it may be quite dry on top but wet and 'greasy' at a slight depth (it is almost useless for horticultural purposes).

High-moor peat is almost throughout uniformly moist or dry because it conducts water well, and is rich in air (it is frequently used by gardeners).

8. Low-moor soil is very rich in food-material for plants, so that it entertains mainly 'eutrophic' plants,[2] which are adapted to live at the expense of nutritive solutions present in the soil : fungal life is scanty.

High-moor soil, on the contrary, is very poor in nutriment, for instance, in nitrogen, so that the vegetation consists of 'oligotrophic' plants : fungal life is very abundant, so that there is a keen struggle for the nutritive salts.[3] Probably to this is due the occurrence here of plants that obtain nitrogen by means of their leaves—carnivorous plants.

9. On low-moor mycorhiza and carnivorous plants are rare.

On high-moor mycorhiza and carnivorous plants are common : all the Ericaceae, as well as Empetrum, Betula, and others have mycorhiza;[3] while, for instance, Drosera, Dionaea, Sarracenia, Darlingtonia, Cephalotus are inhabitants of high-moor.

GEOGRAPHICAL DISTRIBUTION OF HIGH-MOOR

Even more than low-moor is high-moor governed by climatic conditions, because it is dependent solely upon atmospheric precipitations and not upon water in the soil. Consequently the distribution of high-moor is more restricted than that of moors as a whole.

Within the tropics the production of peat is almost confined to

---

[1] Früh und Schröter, 1904, pp. 12–15. See also Weber, 1907, and his other papers cited in the Bibliography.

[2] C. Weber, 1907, and his other papers cited in the Bibliography. [3] Stahl, 1900.

mountains, because high temperature greatly favours the decomposition of organic bodies. On parts of the east coast of Brazil where rain-forest prevails, cushions of Sphagnum occur in moist places, but probably no peat is formed. From subtropical lands with winter rainfall high-moor is also excluded. The production of peat is most active in lands of moderate temperature and high humidity. In arctic countries it is scanty and feeble, mainly because the amount of vegetation is small ; in Greenland peat is formed from Webera nutans and Hypnum stramineum [1]; it is produced in Siberia (though not in such quantity as near the Baltic Sea),[2] in Spitzbergen,[3] in Vaigach, and on Tierra del Fuego ; on Antarctic islands extensive sphagneta occur, and there is a production of peat.

The proper domain of high-moor is the cold-temperate coniferous zones, and west parts of the dicotylous forest-zone where the climate is oceanic. The dicotylous forest-zone in the eastern part of continents (Asia, America) is unfavourable to high-moor, because of high temperature in summer and low relative humidity of the air. High-moor is rare in the eastern parts of the United States. In the steppe-regions of Russia high-moor occurs here and there in clumps of pine-forest.[4]

High-moor is especially widespread in the northern parts of the coniferous belt of Russia and Siberia.[5] South of the limit of forest the main part of the country is largely devoid of forest. The soil is frozen nearly throughout the year, and cannot be drained of water during the short summer. Only in the vicinity of river-valleys is the soil dry, because here it is possible for the water to flow away rapidly. The slopes of the valley are therefore clothed with mesophilous forest, next to which comes moor-forest with stunted spruces and birches, and this passes over into vast plains clothed with high-moor.

True high-moor does not proceed far beyond the limit of forest. In the tundra-domain, Sphagnum is characteristic of the depressions that are constantly covered with water. On the peat-ridges standing above water Sphagnum plays only a subordinate part.[6]

In cold-temperate parts of the Southern Hemisphere—for instance, in Patagonia and Tierra del Fuego—high-moor is composed of Sphagnum, Azorella, Carex, Empetrum rubrum, and others.[7] Even in northern parts of the west coast of Patagonia it becomes rare as the atmospheric precipitations in summer become less frequent.[8] New Zealand has alpine moors that are characterized by a number of Antarctic genera.[9]

CHAPTER L. MOSS-TUNDRA OR MOSS-HEATH

THE *tundra* is a large flat or gently undulating tract, devoid of forest, and occurs in the north of Russia and Siberia; the ' barren ground' in North America is perhaps mainly tundra. The original word is Finnish,

---

[1] Warming, 1887.   [2] Middendorff, 1867; Pohle, 1903.   [3] Nathorst, 1883.
[4] Kusnezow, 1898.
[5] Middendorff's ' Schwappende [soaking] Tundra ' ; see also Pohle, 1903.
[6] See Chapter L.   [7] Alboff, 1902.   [8] Dusén, 1905.
[9] Diels, 1896. Regarding the more recent literature dealing with high-moor, the cited works by Früh and Schröter, Weber, and W. G. Smith should be consulted.

and denotes any open, forestless stretch of land, and therefore also includes mountain-tops devoid of forest. In phytogeography the term refers solely to treeless moor-like plant-communities occurring within the Polar climate. The tundra arises from arctic ' fell-field ', when mosses gain the upper hand over all other plants, and form a continuous soft sward. Transitional stages between the tundra and ' fell-fields ' have been described by Porsild [1] as occurring in Greenland.

According to Middendorff [2] the tundra is always associated with wet soil and moist air ; moreover, the exceeding prevalence of marshy soil is characteristic of Polar regions. As the first cause of this we must regard the very short duration of the snowless season coupled with low temperature in summer and with frequent mist. The arctic summer is like spring in temperate zones. In winter, however, great atmospheric aridity prevails. Evaporation is small, the soil is wet. In Polar lands there is, however, no warm season during which the soil can dry. Added to these climatic factors are edaphic factors, and especially the ground-ice which prevents the water from sinking. Ground-ice is not the chief cause of the marshiness, as is shown by the fact that in the northern part of the forest-domain of North Russia the larger part of the soil is marshy, although there is no ground-ice. Here also, there is not sufficient time during the short summer for the water to flow away. On steeply inclined soil the flow of the water is facilitated, and the production of marsh is consequently checked. And this explains why it is that in mountainous country like Greenland tundra occupies only a small area, whereas in northern Siberia it seizes upon nearly the whole of the land. According to Porsild, [3] in Greenland moss-tundra occurs on rocks in depressions of the terrain, on horizontal, not-drained terraces of the basalt cliffs, on the flat headlands beneath the basalt cliffs, and on the horizontal moist moraines. The soil is cold, especially when ground-ice lies near the surface.

Coldness of soil checks the absorption of water by plants, so that, especially when the wind is strong, the plant cannot replace the water lost by transpiration, and is in danger of desiccation. The mosses are therefore xerophilous, and the Spermophyta are in various ways protected against being dried up.

The moss-tundra of the peninsulas of Kola and Kanin in northern Russia has been described by Kihlmann [4] and Pohle.[5] In these districts the tundra is differentiated into two components—peat-hillocks and puddles :

The *peat-hillocks* are large cushions of moss which form isolated hillocks or long ridges. They rise to a height of two or three metres above the general level of the surrounding surface. Pohle is of opinion that the peat-hillocks are quite normal clumps of moss, which, in the course of centuries or possibly millennia, have arisen by gradual growth.[6] The peat-hillocks during winter are devoid of snow, whereas the surrounding depressions are filled with it. Consequently there arises a distinction in the soil. Where the snow lies deep, the low temperature prevailing during winter cannot penetrate deep into the soil, which remains relatively

[1] Porsild, 1902.   [2] Middendorff, 1867.   [3] Porsild, 1902.
[4] Kihlmann, 1890.   [5] Pohle, 1903.
[6] See the remarks made upon regressive development by Cajander, 1905 b.

warm, and thaws rapidly after the snow has melted, so that in such places there is no ground-ice. On the contrary, where the covering of snow is thin or wanting, the low temperature descends deep into the soil, and there is formed ground-ice which cannot be melted in the deeper layers during the cool summer. Ground-ice melts with the greatest difficulty in peat-soil, and consequently persists during the summer at a depth of only a few centimetres.

The vegetation of the peat-hillocks is definitely xerophilous. Species of Sphagnum occur scantily ; whereas species of Polytrichum abound in the form of tall mosses, whose erect stems are closely crowded and produce a low, soft carpet. Even when the soil on which they grow is very wet with melted snow below the surface, it may be superficially dried by the summer sun and consequently become hard. While in winter, when great dryness of air prevails over the northern tundra, it is to wind that plants owe their desiccation. The Polytricha can hold water in their dense, matted tufts, but despite this they show xerophilous structure, as some species have peculiar leaves, which when dry can close their margins over the assimilatory tissue. In addition to this genus, Dicranum elongatum, D. tenuinerve, and other species of Dicranum, form similar dense, firm tufts ; and intermingled in this air-excluding carpet of *mosses that produce raw humus* are species of Hylocomium, Hypnum, Rhacomitrium, Jungermannia and other Bryophyta, lichens, dwarf-shrubs, including Empetrum, Betula nana, and Vaccinium Myrtillus, also herbs belonging to the same species as on ' fell-field '.

Mosses are enabled to colonize the inhospitable region in virtue, not only of their capacity of drying up and reviving after a fresh supply of moisture, but also of their hardy nature, and of their powers of assimilating, apparently at very low temperatures, more readily than can Spermophyta.

The *puddles* have no ground-ice. The soil is not so cold as that of the peat-hillocks, and in the dry season water accumulates in these places. Here a formation belonging to the temperate forest-domain, namely Sphagnum-moor, can invade a portion of the tundra. These Sphagnum-moors do not extend far beyond the limit of forest ; east of the Timan Mountains and in Asiatic tundras they are merely sporadic, while far north they are completely absent.

The production of peat in the tundra varies. Pohle [1] in northern Russia saw layers of peat more than six metres in thickness. According to Kihlman,[2] in the Kola Peninsula there is not so much a production of moss-peat as one of raw humus, which is permeated with living parts of plants. As a whole, the modern production of peat in the tundra does not seem to be great ; on the contrary, the tendency seems often to be retrograde,[3] as has been mentioned already. The soil must necessarily abound in humous acids, and this character must materially contribute to the xeromorphy of the constituent plants.

Moss-heath apparently occurs only in the Northern Hemisphere, and especially in Siberia and Lapland. Heuglin describes it on the Straits of Jugor ; it is also known in North America and Greenland. In the

---

[1] Pohle, 1903.   [2] Kihlman, 1890.
[3] Cajander, 1905.

Alps, Polytrichum septentrionale would appear to form carpets of moss on deserted glacier-soil; for instance, in the Oetzthal one may see large tracts which are composed of sand and talus thrown down from the mountains, and are covered by a soft carpet of Grimmia and isolated little spruces, junipers, and herbs.

In temperate zones similar plant-communities are found upon periodically inundated soil. Especially in the neighbourhood of heaths there are formed small associations that take a position intermediate between heath and moor. Danish botanists[1] have applied a special term 'Moskär' to these, and they distinguish Sphagnum-kär, Polytrichum-kär, Dicranum-kär, and Grimmia-kär. Identical associations occur in Ireland,[2] and on the Faroe Isles.[3] Even so far south as Madeira carpets of Grimmia occur on high mountains upon periodically inundated soil, yet there is no sign of any production of raw humus.[4]

CHAPTER LI. LICHEN-TUNDRA OR LICHEN-HEATH

LICHEN-HEATH or lichen-tundra is drier than moss-heath; according to Hult,[5] Cladina-formation, as defined by him, apparently has not more than 40 per cent. of moisture in the soil. Lichen-heath occurs especially in hilly, mountainous countries whose rock occurs at a slight depth. A thin layer of raw humus clothes this and supports lichen-heath. The soil is dry, but lichens are incapable of dispensing with atmospheric moisture; even when they can endure periodic desiccation caused by transpiration, they can flourish only where mist, rain, and dew are frequent. Lichens can absorb aqueous vapour from the atmosphere. The species vary in hardiness. According to Kihlman[6] there are several forms (associations) of lichen-heath exhibiting various stages of sensitiveness, in particular, to dry winds. *Fruticose lichens* thrive best where the air is still and moist, and are therefore sparse in the extreme north. *Cladina-heath*, composed of Cladonia rangiferina, C. alpestris and others, with an admixture of Sphaerophoron corallioides, is the most sensitive; it prefers places where the snow lies long, it endures no dry wind, and therefore especially seeks out depressions in the land; it is common on all extensive alpine plateaux lying inland in northern Europe and Arctic America. The different kinds of *Platysma-heath*, including Platysma cucullatum, P. nivale and others, with Cetraria crispa, C. islandica and others, are more hardy. But most hardy is *Alectoria-heath*, which consists of Alectoria ochroleuca, A. divergens, and A. nigricans, together with more abundant dwarf-shrubs. In accordance with these degrees of hardiness, the habitats of the various lichen-heaths are different.

On heaths where tall *fruticose lichens* grow close together there is formed a soft, thick carpet, which gives to the landscape a characteristic yellow-grey tint that is visible from afar. They are to be seen on the

[1] See Börgesen and C. Jensen, 1904; Mentz, 1902.
[2] Pethybridge and Praeger, 1905. [3] Ostenfeld, 1908.
[4] Vahl, 1904 b; in addition to the works already cited, reference should be made to those by Ramann (1895), Sernander (1898), and Dusén (1905).
[5] Hult, 1881. [6] Kihlman, 1890; also see Hult, loc. cit.

fells (plateaux) of Norway, for instance between Gudbrandsdalen and Œsterdal, on the Faroe Isles,[1] Iceland, and Lappmark and Siberia. In Greenland they are only scantily and feebly developed, in fact, typically only in the interior of the south, where they may cover extensive tracts, and consist especially of Stereocaulon alpinum and Cladonia rangiferina.[2] In Antarctic lands lichen-heath, with Neuropogon Taylori, appears to occur on the higher mountains of Kerguelen,[3] and occurs in South Georgia with Sphaerophorus, Stereocaulon magellanicum, Neuropogon melaxanthus, and Sticta Freycinetii.[4]

Between the lichens one finds, in the Northern Hemisphere, Empetrum, Betula nana, Loiseleuria procumbens and other Ericaceae, Juniperus communis, and other creeping low shrubs and dwarf-shrubs. Various herbs occur sparsely scattered in this carpet, just as in moss-heath, including species of Lycopodium, Carex, Hieracium, Deschampsia flexuosa, Nardus stricta, Juncus trifidus; mosses are present also. Both dwarf-shrubs and herbs are xeromorphic, and the species are the same as in the adjoining fell-fields; they usually remain humble and more or less concealed within the carpet of lichen. The causes for the xeromorphy may be the same as in the case of moss-heath—acidity and coldness of soil, also wind.

On the level or undulating tundra-plains of northern Europe, which are swept clear of snow by storms in winter, fruticose lichens succeed but ill. Here *crustaceous lichens* preponderate, and Lecanora tartarea becomes extremely widespread, for instance on heaths in Lapland, covering with a brittle incrustation the dense mat of the lichen-heath that has been killed by dry winds[5]; it also appears, though in smaller quantity, in Greenland. In temperate zones lichen-heath occurs, but to no great extent.[6]

When the dwarf-shrubs, Betula nana, Calluna vulgaris, Vaccinium Myrtillus, V. uliginosum, and a few willows, become taller, there arises a vegetation consisting of two storeys, and when the dwarf-shrubs become more numerous lichen-heath passes over into dwarf-shrub heath.

Fell-field, moss-heath, and lichen-heath (merging into dwarf-shrub heath) are distributed over most of the barren tracts in the far north of Lapland, Siberia, North America (where they are known as ' barren grounds '), Greenland, Spitzbergen, and Iceland, also at higher altitudes on high mountains, and in Antarctic countries near the Straits of Magellan. They certainly present to us a picture of the first vegetation that prevailed in the North after the Glacial Epoch. Under more favourable conditions they become associated with Sphagnum-moor and dwarf-shrub heath.[7]

[1] Ostenfeld, 1908 b.   [2] Rosenvinge, 1889 ; Hartz, 1895.
[3] Schenck, 1905 b.   [4] Skottsberg, 1905.   [5] See p. 72; Kihlman, 1890.
[6] Gräbner, 1895 ; Warming, 1907–09.   [7] For the literature see p. 258.

CHAPTER LII. DWARF-SHRUB HEATH

By the term *heath*,[1] so far as it concerns northern Europe, is meant a treeless tract that is mainly occupied by evergreen, slow-growing, small-leaved, dwarf-shrubs and creeping shrubs which are largely Ericaceae (*ericaceous heath*). The vegetation varies in height according to the temperature, humidity, and light prevailing, and often rises to thirty centimetres or more, but often only to ten or twenty ; on the one hand it may be so dense that the soil is invisible, but on the other so open that the soil is partially bare, and leaves space between the shrubs or other plants.

The stunted and xerophilous character of the vegetation is due to climate and soil, but particularly to the latter.

The vegetative season is usually dry, and transpiration may then be intense, even though the prevalence of heath demands a certain atmospheric humidity. In spring (May and June) the atmospheric humidity is at its minimum, at least in Denmark. During summer, periods of drought are frequent, and the air hangs quivering over the hot surface of the heath. During winter, in the northernmost sites, cold and drought together with storms play an important part in reference to the evergreen plants. Winds blow with great violence over the dry surface, upon which dwarf-shrub heath is wont to occur.

The nature of the soil is of far greater import than is the climate. The soil is for the most part a sterile quartz-sand,[2] which has been thoroughly elutriated and deprived of nutrient substances since the Glacial Epoch ; some of the Spermophyta present occur especially on sandy soil. This soil has its characters greatly changed by a more or less thick layer of raw humus deposited over it by the heath vegetation.[3] Calluna and Vaccinium Myrtillus are among the plants flourishing luxuriantly on raw humus in which *heather-peat* is formed by their matted roots, as well as by the rhizoids of mosses and hyphae of Cladosporium and the like. The layer of raw humus greedily takes up moisture, retains it firmly for a long time, prevents evaporation from the soil, and obstructs the entrance of air, thus occasioning the manufacture of humous acids that retard the absorption of water by roots. In times of drought this layer, owing to its dark colour, may easily be heated and robbed of moisture. If the process of drying is carried far the ling disappears and lichen-heath usually arises, for example in North Germany.

The production of raw humus must be regarded as the most characteristic peculiarity of heath. The acid soil excludes plants belonging to forest or to other formations in the temperate zone, and thus gives absolute dominance to heather. The conditions necessary for the production of raw humus are therefore those essential to the occurrence of heath. Ramann sets these conditions forth as follows :—

1. Want of nutrient substances (these are required by organisms responsible for decomposition).

2. Exclusion of air (in most cases only when the soil has been covered with water for a long time).

---

[1] See Focke, 1893 ; E. H. L. Krause, 1892 *a* ; Gräbner, 1895, 1901, 1909.
[2] See p. 59.   [3] See p. 62.   [4] Ramann, 1905, p. 159.

3. Excess of water usually combined with—
4. Low temperature.
5. Lack of water (dryness in the warmer season).

Of these, want of nutrient substances and lowness of temperature in summer are particularly important. Dwarf-shrub heath occurs only in arctic and alpine sites, also in the oceanic west coasts of the cold-temperate zone that are characterized by cool summers. The significance of the nutrient substances contained in soil has been expounded by Gräbner [1] in particular. When soil is relatively fertile, heath passes over into meadow or pasture, as is especially obvious in mountainous places. Extreme drought excludes dwarf-shrub heath; in the driest spots it is replaced by grass-fields (Weingaertneria) or by lichen-heath.

ADAPTATIONS AND FLORA

In many cases the dwarf-shrubs assume a decumbent form—such is the case with Betula nana, Salix and Juniperus [2] in arctic heath—in other cases this form of growth is normal, for instance, to species of Arctostaphylos. A heath is often a fell-field with a denser vegetation of dominant dwarf-shrubs, and with at least two storeys of vegetation; but just as in fell-field, moss-heath, and lichen-heath, one finds here also herbs, grasses, mosses, and lichens, and the last two groups often fill the interspaces beneath the dwarf-shrubs. Moss-heath and lichen-heath are thus interspersed in this heath.

The dwarf-shrubs possess bent, curved, brittle shoots. The majority of them are evergreen, and this is particularly true of the prominent forms such as Calluna, Empetrum, Juniperus, Arctostaphylos Uva-Ursi, and Lycopodium; but their tint is always dark and brownish green, and more so in winter than in summer. The leaves are closely set, very numerous, small, mostly sclerophyllous, linear and often ericoid.[3]

The shrubs in northern Europe are mainly represented by the following evergreen dwarf-shrubs:—

Ericoid type: Calluna and Empetrum, also in moister places Erica Tetralix.

Broader coriaceous, flat, entire *leaves:* Arctostaphylos Uva-Ursi (which occurs particularly in more open spots in the vegetation), Vaccinium Vitis-Idaea, Thymus Serpyllum.

Pinoid type: Juniperus communis.

By dwarf-shrubs with *thin leaves falling or fading in autumn:* Salix repens, and Vaccinium Myrtillus (which, however, occurs rather in forest-soil), also in the northernmost parts, Arctostaphylos alpina, Betula nana, Salix herbacea, S. polaris, S. reticulata, and several others, including, in more southern spots, such representations of switch-plants as Sarothamnus and species of Genista. A number of these deciduous, or at least not evergreen, plants are guarded against excessive transpiration by coatings of grey or white hairs, or of wax.

By *thorny plants:* Species of Genista and Ulex.

The dominant dwarf-shrubs Calluna, Erica, and Empetrum form tufts and cushions, and possess long-lived primary roots and prostrate shoots that emit roots.

[1] Gräbner, 1895, 1901, 1909.   [2] See pp. 26, 38.   [3] See p. 110.

Many of the shrubs, including Empetrum, Vaccinium, Arctostaphylos, have fleshy fruits that are eaten by birds.

Beneath and between the dwarf-shrubs grow *mosses* and *lichens*, which permeate the soil with their rhizoids ; among the lichens are Cladonia rangiferina, Cetraria islandica, and Sphaerophoron corallioides ; and among the mosses, species of Polytrichum, Rhacomitrium, Hypnum, and Hylocomium. In addition, there occur numerous grasses and herbs which are mainly perennial : annual and biennial species, such as Aira praecox and A. caryophyllea, maintain themselves with difficulty amidst the dense growth, and at most occur in well-lighted bare spots (to this rule parasitic Rhinantheae provide an exception). The herbs and grasses, for example, Arnica montana, Solidago Virgaurea and Campanula rotundifolia, associated with this habitat are *caespitose*, and thus are better fitted to grow in dense soil than are species possessing shoots that travel underground.

The evergreen dwarf-shrubs are distinctly xerophytic in structure, but so likewise are many of the herbs. In regard to the latter we merely note here that broad, thin, smooth leaf-blades scarcely occur, and that the grasses mostly have setaceous or filiform leaves with stomata in furrows that open or shut according to circumstances, as is exemplified by Weingaertneria canescens, Nardus stricta (a tunic-grass), and Festuca ovina.[1] Many species such as Rumex Acetosella, Campanula rotundifolia, Scleranthus, and Artemisia campestris, which are interspersed here and there, have small and narrow leaves when compared with their congeners of other habitats ; other species, such as Antennaria dioica, and Gnaphalium arenarium, are densely hairy ; while Sedum acre represents succulent plants.

ASSOCIATIONS AND DISTRIBUTION

Dwarf-shrub heath occurs in a number of countries within the temperate and cold zones of the Northern Hemisphere, and develops typically over extensive areas—for example, in West Denmark (Jutland) and in North-West Germany. Several kinds of associations of it are met with.

In *northern Europe* in most cases there is *callunetum* with Calluna vulgaris as the dominant species ; on wet soil *ericetum Ericae Tetralicis* prevails ; and in other places with moist soil is *myricetum* composed mainly of Myrica Gale, Erica, and Calluna. Among other associations may be named *myrtilletum* in which Vaccinium Myrtillus preponderates.[2]

Calluna vulgaris, the ling, which is the species that forms associations and gives the tone to heath, is a remarkable plant. It is accommodating and tenacious of life : it demands a not inconsiderable degree of atmospheric humidity, yet is not nice in its demands as regards soil. It can grow equally upon most sterile sands which may be at least temporarily somewhat dry, and upon very wet, boggy soil, which is periodically dry,

---

[1] See pp. 107, 110.
[2] In regard to heath in North-Western Europe, see Gräbner, 1901 ; Rob. Smith, 1900 ; W. G. Smith, 1902, 1905 ; W. G. Smith, with Moss and Rankin, 1903 ; Pethybridge and Praeger, 1905 ; Mentz, 1900 *b*, 1902 ; Börgesen and C. Jensen, 1904.

and it also thrives upon humus-laden soil. Though capable of growing upon moderately good soil it seldom has the opportunity, because it is expelled by other plants. These latter, which are more exacting in their demands, and are more active in anabolism, avoid the more sterile and acid soil of heath and resign it to ling. Here ling becomes a social species and reigns alone for miles. On chalk and marl it rarely grows on good soil, being excluded by competition, but sometimes occurs upon poor calcareous soil, such as on muschel-kalk.[1] In a climate of high atmospheric humidity it demands sunlight and open country (in continental sites it grows in the lowlands only within thinly wooded, well-lighted forest), and does not withstand extreme winter-cold combined with drought, for it is sensitive to extreme dryness in any form. One sometimes sees the Calluna-vegetation suddenly disappear from large tracts, probably because the plants concerned have attained their limit of age (which is certainly estimated very high at 20–30 years) and are replaced by young plants, which usually spring up in great numbers.

It has already been mentioned[2] that heath-moor often gradually passes over into ling-heath. In the moister spots of dwarf-shrub heath we often find Erica Tetralix abundantly mixed with ling. Not infrequently the former species produces associations, and with it often appears the majority of those plants characteristic of high-moor. If the climate be sufficiently moist, Sphagnum may grow luxuriantly, and the community easily changes into high-moor. On the other hand, when for any reason Sphagnum-moor becomes drier, then ling gains the upper hand. When the dryness becomes excessive ling disappears, and only a few easily satisfied mosses, lichens, and Spermophyta are dotted over the bare soil.

In other places, for instance in northern Norway, communities of this kind are encountered. Hult[3] mentions a peculiar 'formation' that is 'perfectly intermediate between heath and moor'; its vegetation consists mainly of dwarf willows—Salix reticulata, S. herbacea, and S. polaris—also numerous perennial herbs and dwarf-shrubs, such as Dryas, Arctostaphylos alpina, Loiseleuria, and Phyllodoce.

All these occurrences demonstrate that heath and heath-moor belong to the same class of formations.

On the *Alps* there are considerable dwarf-shrub heaths, in which several associations are to be distinguished, according to the dominant species : On calcareous Alps, heath formed by Erica carnea occurs both in and above the forest-belt ; in the subalpine elfin-wood region of the eastern calcareous Alps is found heath composed of Daphne striata with Polygala Chamaebuxus and Globularia nudicaulis ; at a greater altitude Azalea procumbens produces heath ; in Azalea-heath the layer of raw humus may attain a thickness of half a metre.[4]

In *south-eastern Europe*, in place of the absent Calluna, appears the ericaceous Bruckenthalia spiculifolia which gives rise to heath.[5]

In *arctic dwarf-shrub heath* Calluna and Erica play little or no part, but among dwarf-shrubs present are Cassiope tetragona, Vaccinium

---

[1] de Candolle, 1874.   [2] Chap. XLIX.   [3] Hult, 1887.
[4] See Kerner, 1863 ; Christ, 1879 ; Krasan, 1883 ; Hayek, 1907 ; Brockmann Jerosch, 1907, p. 278 ; Schröter, 1904–8 ; Engler, 1901.
[5] Adamovicz, 1898.

uliginosum var. microphyllum, V. Vitis-Idaea, Ledum palustre f. decumbens, Arctostaphylos alpina, A. Uva-Ursi, Rhododendron lapponicum, also Diapensia lapponica, Dryas octopetala, Betula nana, B. glandulosa, Juniperus communis, Salix glauca, S. herbacea, and S. polaris ; in Iceland Dryas octopetala may be represented by such numbers of individuals as to form an association, Dryas-heath, and mingled with it are Silene acaulis, Armeria maritima, Thymus Serpyllum, and others.[1] Several other species may give rise to associations ; in Finland there are Loiseleuria-association, Empetrum-association, Phyllodoce-association [2] and others, which occasionally may occur on fell-fields.

Herbs, including many evergreen species, grasses, mosses, and lichens are intermixed in greater or smaller numbers, and in many places there are gradual transitions to fell-field, moss-heath, and lichen-heath ; the species are to some extent identical, but their relative abundance is different.

Dwarf-shrub heath covers large areas in Greenland, North America, and north-eastern Asia. It has provided many arctic travellers with fuel, yet it does not extend to the northernmost land nor to extreme altitudes ; there it is replaced by the more meagre and more easily satisfied fell-field. The soil, as in Europe, is formed of raw humus.[3]

Types intermediate as regards flora between arctic and North German heath occur in Iceland, Lapland, and northern Scandinavia.[4]

Antarctic heath is formed by Acaena adscendens on Kerguelen. On northern and eastern slopes where the atmospheric humidity is high Acaena reigns almost alone on the very humous soil. The creeping main shoots cover the ground with a narrow-meshed net, from which the foliaged shoots rise vertically to a height of half a metre. The habit of the formation is different in situations where the air is less moist. For Acaena then applies itself closely to the soil, and the only erect parts are the shoot-tips and the numerous flowering shoots. Stations of this kind are also characterized by the abundance of companion-plants, such as Lomaria alpina, Azorella, Pringlea, Galium antarcticum, and Ranunculus biternatus. The heath on Kerguelen is the formation which occupies those stations that are the most favourable to plant-life, especially as regards shelter from wind.[5] On South Georgia Acaena adscendens also produces heath.[6] On the Falkland Isles heath consists of a number of dwarf-shrubs, including Empetrum rubrum, Pernettya pumila, Gaultheria microphylla, Drapetes muscosus, Vaccinium Oxycoccos, and Myrtus nummularia.

CHAPTER LIII. BUSH AND FOREST ON ACID HUMUS SOIL

On bushland allied to dwarf-shrub heath there occur not only species that elsewhere produce high-forest, but also shorter shrubs peculiar to them. The individual plants are stunted forms, such as have been described on pp. 37 and 129, with bent, contorted stems and branches.

[1] Stefansson, 1894. As to morphology and anatomy, see Warming, 1908 a, and H. E. Petersen, 1908.
[2] Hult terms these ' formations '. See also Pohle, 1903.
[3] See Warming, 1887. [4] Grönlund, 1884 ; Hult, 1887 : Brotherus, 1886.
[5] Schenck, 1905. [6] Will, 1890.

The vegetation on the ground consists of those communities that belong to the formations already described as occurring on sour soil.

As examples of such *bushland composed of true shrubs*, we may mention the following :—

i. Arctic Bushland on Acid Soil.

In the far north, for example on the tundra in Lapland, Betula nana and other kinds of birches occur as low flat bushes, and are often associated with grey-haired willows, such as Salix glauca and S. lanata. On Scandinavian mountains immediately above the tree-limit occurs the *zone of grey willows*, including Salix lanata, S. glauca, and others, whose leaves are protected from excessive transpiration by means of a coating of hairs, and also by a thick-walled epidermis, wax, and the like. In Greenland, at 70° N., one finds similar willow-bushlands which are up to one metre in height, with stems and branches interlaced, and are composed of Salix glauca and Betula nana. But these bushlands should perhaps be regarded as mesophilous.[1] On Norwegian mountains in like manner, Betula nana, willows, and juniper form extensive low bushlands which are about ½–⅔ metre in height and correspond most closely to the Rhododendron-bushlands of the Alps.

ii. Subalpine Bushland on Acid Soil.

Where, as in Central Europe, the summits of the highest mountains still lie in the cloud-belt, the production of raw humus is promoted by the moist misty air and low temperature, and there arise on acid soil bushlands that cover considerable areas above the limit of forest. Rhododendron-bushland is closely allied to dwarf-shrub heath. It occurs on the Alps and Pyrenees, also, in a taller, more forest-like form, on the Himalayas. It is produced by species of Rhododendron (sometimes accompanied by Juniperus communis), and the protection against transpiration here is provided by scale-like hairs and a coating of resin. On the calcareous Alps Rhododendron hirsutum gives rise to the formation, but in the central Alps it is R. ferrugineum. In Rhododendron-bushland there grow a number of dwarf-shrubs that occur on heaths—for example, Vaccinium and Calluna.[2]

Other subalpine associations—for instance, in Servia[3]—are composed of Juniperus communis or species of Vaccinium, or of mixtures of these.

On high, windy situations on mountains and on windy sites in *extreme northern latitudes*, there is, just as in high-moor, bushland or stunted forest composed of arboreous species that elsewhere form high-forest. The common spruce (Picea excelsa) in Lapland occurs as a creeping shrub that repeatedly strikes root; it assumes peculiar, rounded, very compactly branched, low, shrub-like shapes, and occurs singly or forms extensive carpets of bushland, as its slender branches are partly concealed in the lichen-heath.[4] It derives its protection against transpiration from the structure of its leaves. The Scots pine (Pinus sylvestris), and, in Siberia, Pinus

---

[1] See Chap. XCI.
[2] For details see Kerner, 1863 ; Hayek, 1907 ; Christ, 1870 ; Schröter, 1904–8.
[3] Adamovicz, 1898.   [4] See illustrations in Kihlman, 1890.

Cembra, form the same type of scrub. Betula pubescens, possibly represented by the sub-species B. pubescens carpathica, grows on lichen-heath and high-moor in Lapland : the individuals are often isolated, and are closely applied to the ground as deformed individuals, which require fifty to sixty years for the stem to attain a length of 2 metres and a thickness of 4 centimetres, and possess branches that do not rise above the general surface of the lichen-heath. But in more favourable spots the plant is taller and forms scrub, which is about 1–2 metres in height and may entertain large-leaved, mesophilous, perennial herbs. The birch oecologically approaches species that are xerophytic in structure ; like the conifers it attaches itself firmly to bare, sun-heated sandstone rocks of Saxon Switzerland ; it also produces bushland and forest above the coniferous belt in northern Europe. Its 'lacquered' leaves obviously provide protection against transpiration.

At *high altitudes* forest does not suddenly cease ; it dwindles to form scrub composed of short trees and shrubs, below the level of the open grassland and fell-field, which consist of herbs, lichens, mosses, and dwarf-shrubs. This scrub is composed of different species in different parts of the Earth. In high alpine situations elfin-scrub is the most widely known xerophytic bushland.[1] It is formed by varieties of Pinus montana (var. Pumilio, var. uncinata, and var. Mughus), which in more western places (Western Alps and Pyrenees) become tall trees and appear between the limit of forest and the alpine grassland. An erect stem is not developed ; the stems creep over the ground, descend slopes, are clothed with mosses and other plants, strike root, and send up bow-like strong branches ; these last may exceed man's height, and are packed together often so tightly and firmly as almost to form cushions that can bear the heaviest loads of snow. Whole mountain slopes or ridges may be clothed with dark green, interlacing masses of elfin-wood so densely as to be impenetrable, or often more easily passed over than passed through. The soft, humus-laden, often completely peat-like, soil absorbs much water. Screened from wind by the crowns of the elfin-wood bushes, there develop a number of plants which blossom earlier than those on the adjoining rocks and alpine grassland, and which vary in nature according to the amount of light, the number of fallen coniferous needles, and the like. In younger communities there are rhododendrons, juniper, roses, Daphne, Polygala Chamaebuxus, Empetrum, species of Vaccinium, Erica carnea, Calluna, and other low xerophilous shrubs, also many species of Prunella, Digitalis, and Campanula, many grasses and sedges, as well as mosses and lichens. This elfin-scrub is a xerophytic type of vegetation which is well able to withstand, on the one hand, rapid transpiration, intense sunlight, and cutting cold winds, and, on the other hand, the exceeding moisture of a wet soil, frequent and dense mists, falls of rain and of snow. The elfin-tree and ling are two parallel and readily satisfied species, which are easily driven out by other species to places where the conditions of life are most unfavourable. This alpine scrub, composed of Pinus montana, on peat soil is allied to dwarf-shrub heath. The bog-pine, Pinus montana var. uncinata, is a characteristic tree in the high-moor of Switzerland, according to Schröter ;[2] sometimes it spreads

[1] Kerner, 1863 ; Früh und Schröter, 1904 ; C. Schröter, 1904–8 ; Hayek 1907.   [2] Schröter, 1904.

as high-forest (uncinato-pinetum) and in the shade of its trees, which are 10–20 metres in height, there grows a dense underwood composed of shrubs and perennial herbs belonging to high-moor.

Similar scrub is found on all other high mountains above the limit of proper forest. For example, it is met with on high mountains in Japan at altitudes between 2,200 and 2,500 metres, and is here formed by Pinus parviflora (which is related to P. Cembra) as well as by birch, Alnus viridis, and others.

In like manner scrub appears in America and in Antarctic lands, and in the latter case is produced by species of Nothofagus. Moreover, in northern Europe the shrubs mentioned on page 197 as occurring on high-moors become so numerous that they form 'large and continuous thickets'[1], and there also occurs scrub that is formed of dicotylous trees, such as oak and beech ; the soil is raw humus and the trees are more or less elfin-trees. None of these species, however, occur exclusively in this shape ; they will be discussed in the chapters dealing with mesophytic vegetation.

[1] Yapp, 1908.

SECTION VII

CLASS V. HALOPHYTES. FORMATIONS ON SALINE SOIL

CHAPTER LIV. INTRODUCTORY GENERAL REMARKS ON HALOPHYTES

SALINE soil occurs upon the Earth in many places, namely, along the shores of all oceans and saline lakes, by salt-springs which occur in many spots (for instance in Germany),[1] on steppes where salts contained in the soil cannot be washed out or are carried down into depressions by flood of rain. According to Bunge there are nine of such great salt tracts, of which each has its own floristic peculiarities : these tracts are identical with the steppe-regions—the Australian lowland, Pampas, central part of North America, western and eastern Mediterranean regions, South Africa, regions of the Red and Caspian Seas, and Central Asia. The salts concerned are especially common salt, gypsum, and magnesium salts.

Wherever the soil is saline there appears a special type of *vegetation* which is produced by a few *definite* families, and is composed of morphologically and anatomically peculiar forms. A certain amount of soluble salts must be present before halophytic vegetation is called into existence but the nature of the salts seems to be a matter of indifference ; at least the vegetation agrees in all essentials [2] in the following situations : saline spots in Hungary where the salt is sodic carbonate, in Transylvania where it is sodic chloride, near Budapest at the bitter salt-springs, where the salts are sodic and magnesic sulphates. But the deleterious action of various salts differs widely.[3]

Halophytic vegetation is extremely hardy in relation to climate, for instance, in regard to altitude above the sea-level ; in all parts of the world, under all climates, at all altitudes, where it occurs it bears the same stamp. Certain species have a wide distribution ; for instance, this true of Salsola Kali (in many places this is not a halophyte) and of Glaux maritima, which occur not only on the coasts of north-western Europe even on the rainy shores of Norway, but also on the salt-steppes of Tibet while Salsola in North America has become a pestilent weed in cornfields.

Two other features common to halophytic communities are the *exceeding poverty of the flora* and the *very open nature of the vegetation* The action of salts in excluding plants has already been explained p. 67. But it must be added that the degree of ease with which soil dries plays a part ; for when the soil dries readily a small amount

---

[1] See Ascherson, 1859 ; Petry, 1889.   [2] Bernátsky, 1905.
[3] Kearney and Cameron, 1902.

(perhaps one per cent.) of salt may expel all plants save halophytes, whereas if the soil does not dry readily 2-3 per cent. of salt is required to act in the same manner.

The following families are *halophilous* : Chenopodiaceae, Aizoaceae, Plumbaginaceae, Portulacaceae, Tamaricaceae, Frankeniaceae, Rhizophoraceae, and Zygophyllaceae. In addition the following are often represented on saline soil : Cruciferae, Caryophyllaceae, Euphorbiaceae, Cyperaceae, Gramineae, Malvaceae, Primulaceae, Asparageae, Compositae, and many others. Some genera are nearly world-wide in distribution, though represented by different species.

Markedly halophobous are : Betulaceae, Fagaceae, Piperaceae. Urticaceae, Rosaceae, Ericaceae, and Araceae. Furthermore, mosses and lichens do not thrive well upon saline soil.

CHAPTER LV.   ADAPTATIONS OF HALOPHYTES

Duration of life. Into the composition of halophytic vegetation there enter annual and perennial herbs, as well as woody species, including shrubs and trees. The number of annual species seems to be large ; for instance, in the north of France, according to Masclef,[1] of the thirty-five species confined to saline soil twenty are polycarpic suffruticose herbs, the remainder, or nearly half, being monocarpic ; the like is true of Denmark ; and in the Spanish peninsula, according to Willkomm, about one-third of the halophytes are annuals. The reason for this preponderance is unknown ; probably it is indirectly due to the circumstance that halophytic vegetation is usually very open and thus provides space for such annual species.[2]

Anatomy. Attention has already been directed on p. 134 to the very close agreement in external and internal structure between halophytes and xerophytes : this agreement exists because *salt in the soil renders it physiologically dry ;* a halophyte, in fact, is one form of xerophyte.

Some of the morphological and anatomical characters that are shown by xerophytes and reappear among halophytes[3] are—

Succulence. The most striking feature among halophytes is that they are nearly all *succulent* plants : the leaves are thick, fleshy, and more or less translucent. This is due partly to the *abundance of cell-sap and poverty in* chlorophyll, but partly to *smallness of the intercellular spaces.* It has long been known that certain species are dimorphous, showing the maritime or halophytic form with juicy, thick leaves, and the ordinary inland form with thin leaves : among such species are Lotus corniculatus, Geranium Robertianum, Convolvulus arvensis, Matricaria inodora, Hieracium umbellatum, Solanum Dulcamara. Culture-experiments[4] have shown that certain halophytes grown on ordinary soil poor in salt acquire thinner leaves and lose other characteristics—such is the case with Cakile maritima, Cochlearia officinalis, Salicornia herbacea, Spergu-

---

[1] Masclef, 1888.
[2] The production of seed is a most effective method of protection in such halophytic areas.   [3] Warming, 1897, 1906 ; Schimper. 1891 : Kearney, 1900.
[4] Batalin, 1884 ; Lesage, 1890 ; Boodle, 1904.

laria media, Salsola Soda—whereas other species are unchanged; and conversely, that certain inland species when cultivated in a substratum treated with common salt acquire thicker leaves, as is the case with Lotus corniculatus and Plantago major. This thickness of leaf is caused by an enlargement of the mesophyll-cells, which become large and roundish, and, in the interior of the leaf, poor in chlorophyll, so that they are hyaline and form an almost true aqueous tissue. In some cases a typical *aqueous tissue* occurs and is surrounded by palisade tissue, for instance in Salsola Kali,[1] Batis maritima,[2] Salicornia.[3]

Mucilage-cells are developed, as in xerophytes. Hypodermal aqueous tissue is met with in species possessing more coriaceous leaves, for instance in mangrove-plants (which may also possess large mucilage-cells, e. g. Sonneratia), and in the grass Spinifex squarrosus.

Cell-sap. Large quantities of salt may occur in the cell-sap and cause the solution to be more concentrated than in the soil.

The *palisade tissue* of halophytes is massive. Lesage[4] has experimentally proved that the individual cells become taller, and often divided transversely; common salt acts morphologically approximately in the same manner as sunlight. According to Schimper,[5] the leaves of plants standing nearest to the sea in the Barringtonia-formation are thicker than those of plants farther inland because their palisade tissue is larger.

The *intercellular spaces* are *small*.

The majority of species are leaf-succulents (Grisebach's ' chenopodform '); some are stem-succulents with reduced foliage-leaves, as in the asclepiadaceous Caralluma.

Succulent halophytes, as a rule, show a *dark-green* colour which later on passes over into yellowish-green or red; on certain steppes near the Caspian Sea, when all else has been dried up by the sun, the solitary green patches visible to the eye are on saline soil. Lesage proved that side by side with an increase of common salt in the plant goes a decrease in the amount of chlorophyll, and that this is due to the reduction in size and number of the chloroplasts. In accordance with this seems to be the fact established by Griffon[6] that the assimilatory activity is less in the halophytic form than in the ordinary form of the same species.

Wax. Coatings of wax, causing a glaucous and mat surface, characterize many species, including Eryngium maritimum, Triticum junceum, Elymus arenarius, Crambe maritima, Mertensia maritima, Glaucium flavum, and Spinifex squarrosus.

Hair-coating. The majority of halophytes are glabrous. Yet some species possess *hairs*, though only rarely do they have soft or grey hairs, as in Kochia hirsuta, Senecio candicans, and Tournefortia gnaphalodes. Halophytes possessed of hairs are generally sand-plants; some have special water-storing hairs,[7] whose large, spherical, thin-walled, pearl-like, terminal cells filled with sap fall off or collapse to form a grey coat ('meal'), as in Atriplex, Obione, and Mesembryanthemum.

Coriaceous and *glossy leaves* occur on trees and shrubs in mangrove-swamps, and in allied vegetation, for instance in Rhizophora, Bruguiera, and Nipa fruticans, also in sandy littoral forests.

[1] Areschoug, 1878. [2] Figured in Warming, 1890, 1897.
[3] Warming, 1906, Fig. 77–84. [4] Lesage, 1890.
[5] Schimper, 1891. [6] Griffon, 1898. [7] See p. 121.

The *stomata* of true, succulent, littoral halophytic herbs, in cases so far investigated, are not sunken, but usually lie at or near the level of the epidermis. The thickness and cuticularization of the epidermal wall are small in the succulent species : this is worthy of note, and may denote that the atmosphere of the habitat is not very dry, but probably it should be correlated with the fact that protection against transpiration is attained by other means. Exceptions to this rule are provided by tall woody species in mangrove, also by the saxaul-tree (Haloxylon Ammodendron), Niederleinia juniperoides, Acantholippia Riojana,[1] and others.

Storage-tracheids. In addition to what has already been stated regarding the anatomical relations of aqueous tissue, it may be mentioned that, in some species, *storage-tracheids* apply themselves to the nerve-ends (in several genera belonging to mangrove-swamps), but, in other species (including Salicornia), are isolated in the parenchyma quite apart from the nerves.[2]

Lignification. Except more especially in the case of mangrove-vegetation, lignification occurs only to a slight extent, and in this respect halophytes deviate from other xerophytes. There are, however, several *thorny* species, in most of which the thorns belong to leaves, as in Salsola Kali, Eryngium maritimum, Echinophora spinosa, Carthamus lanatus, and others ; but these species are possibly confined to sand, which may be responsible for the production of the thorns.

Idioblasts, in the form of stone-cells, may occur in the palisade or aqueous tissue, for example in Sonneratia, Rhizophora, Carapa, and other mangrove-plants, also in Scaevola Koenigii.

External Form. According to Lesage's investigation, the height of certain species, Lepidium sativum for instance, decreases on saline soil. Halophytes, as a rule, attain neither great stature nor great circumference. According to the investigations of Stange [3] and others, concentrated salt solutions (not only of common salt, but also of potassic nitrate and glycerine) decrease growth in length, but do not always increase growth in thickness.

(*a*) **Leaf.** Among halophytes there is the same endeavour as in xerophytes to reduce the surface, as is shown by the *leaves* remaining *small*. Lesage's experiments prove that an abundance of salt in soil causes leaves to be smaller as well as thicker. The leaves may be *linear* and *semi-terete*, as in Suaeda maritima, species of Portulaca, and Salsola Kali ; *spathulate* and *oblong* shapes are very common.[4] The leaves are rarely indented, being usually undivided and entire. Some plants, such as Tamarix, are *scale-leaved* ; others are almost *aphyllous* stem-succulents, as is illustrated by Halocnemum, Arthrocnemum, and Haloxylon, or, like Ephedra and Casuarina, they remain poor in sap.

The *ericoid* [5] form of leaf, with the stomata concealed in a hairy furrow on the lower leaf-face, characterizes the frankeniaceous Niederleinia juniperoides (in the Argentine salt-steppes) and species of Frankenia.

As in many xerophytes the leaves assume an erect lie, so that when the sun is highest in the heavens its rays strike at acute angles : this, as well as narrowness of the leaves, occasions *isolateral leaf-structure :* as

---

[1] Göbel, 1891, p. 13.   [2] Duval-Jouve, 1868 ; Hultberg.
[3] Stange, 1892.   [4] As figured by Warming, 1897.   [5] See p. 110.

examples may be cited Atriplex (Obione) portulacoides, Suaeda maritima, Sesuvium Portulacastrum, and some species in mangrove.[1] In the
verbenaceous Lippia (Acantholippia) Riojana the leaves are appressed ;
and between the leaf and stem hairs occur, while on the assimilatory
outer side of the leaf there are deep, hairy furrows.

(*b*) **Stems.** The stems of halophytes are often prostrate, radiating
on all sides from a common point, which is the base of the main axis.
This may be observed in species of Atriplex, Suaeda, Salsola, and other
Chenopodiaceae, in Polygonum Persicaria and its allies, Senecio vulgaris,
and other plants on European shores. This is not caused by wind, since
the stems take no definite constant direction ; the great irregularity
prevailing points to local influence, which is certainly to be attributed to
the differential heating of the frequently stony soil.[2]

The majority of the structural peculiarities just mentioned also occur
in xerophytes. There thus exists a remarkable agreement between
halophytes and xerophytes, as Schimper was the first to point out ; both
these types of plants dry slowly when exposed to strong evaporation and
aridity ; such is the experience of every one who has tried to dry succulent
species for the herbarium. But this slowness of drying is due not only
to the devices guarding against rapid transpiration, but also, in the
case of halophytes, to the saline cell-sap which evaporates more slowly
than pure water. Moreover, in matters floristic, certain features in
common have been demonstrated—for instance, the occurrence of the
same species on the sea-shore and on mountains.

What is the reason for this remarkable agreement between xerophytes
on non-saline soil and halophytes which may even grow on soil saturated
with water, as is the case with Salicornia-vegetation on the coasts of the
North Sea at high tide, and with mangrove-plants at all times ? Schimper[3]
directs attention to the deleterious action of common salt in the cell-sap
upon assimilation and upon the life of the plant as a whole ; and this
has been experimentally proved.[4] Common salt behaves as a poison to
the plant, because it is readily absorbed in large quantities and then acts
lethally. In order to prevent an excess of chlorides from being conveyed
to the leaves by the transpiration-current and deposited in the cells, the
plant must, according to Schimper's explanation, guard against intense
transpiration, and therefore requires the many protective devices already
described. It is doubtful if this explanation be correct ; for, supposing
that transpiration were extremely slow and weak, but nevertheless
prolonged, quantities of salt would certainly accumulate in the plant
until finally an injurious amount was present ; it might then be that
the plant would possess means by which to decompose and dissipate
the absorbed salts—and such is apparently the case according to Diels.[5]
If Diels be correct (which Benecke[6] denies) the xerophytic structure is
probably designed to obstruct the free gaseous interchange between the
tissue and atmosphere, so that, as in the case of succulent plants, there
may be formed in the tissue malic and other organic acids, which in this
instance would serve to decompose the chlorides.

[1] According to Johow (1884), Karsten (1891), Warming (1897 *b*), and Schmidt
(1899, 1903).   [2] See p. 27.
[3] Schimper, 1890, 1891, 1898.   [4] Kearney, 1902.
[5] Diels, 1898.   [6] Benecke, 1901.

But more probable is another explanation offered by Schimper, namely, that the protective arrangements against intense transpiration are required owing to the plant finding it difficult to absorb water from a relatively concentrated salt solution ; this latter truth was first demonstrated by Sachs in 1859[1] ; a saline soil is therefore *physiologically dry*.

The statement made by Stahl[2] that halophytes are incapable of closing their stomata and of thus regulating their transpiration, does not correspond with fact, according to Rosenberg,[3] Benecke,[4] and Diels.[5]

Halophytic communities may be divided into those that are *lithophilous, psammophilous, pelophilous*, and *helophilous*, according as the substratum consists respectively of rock and stones, of sand, of mud, of swamp. There are communities consisting solely of thallophytes or solely of herbs, and others including shrubs, trees, and true forest. Lichens and mosses are very rare, yet some markedly halophilous species occur.

We shall deal with these plant-communities growing on saline soil in five chapters, but this is only a matter of practical convenience. We cannot speak of formations, because each chapter includes communities differing so widely that they should properly be regarded as separate formations, the growth-forms composing them being so diverse. The following is the arrangement adopted :—

Lithophilous halophytes.
Psammophilous halophytes :
   (a) Sand-Algae,
   (b) Iron-sulphur-bacteria,
   (c) Halophilous herbs,
   (d) Shingle-banks,
   (e) Halophilous forest and bushland.
Pelophilous halophytes :
   (a) Aestuaria with Zosteretum and Salicornietum,
   (b) Salt-meadow,
   (c) Salt-bushland,
   (d) Salt-steppe.
Salt-swamp, and Salt-desert.
Littoral swamp-forest (mangrove).

[1] Also see Hedgecock, 1902.
[2] Stahl, 1894.
[3] Rosenberg, 1897.
[4] Benecke, 1901.
Diels, 1898. There is an extensive literature dealing with the structure and nature of halophytes : special reference may be made to the papers by Areschoug, Benecke, Brick, Contejean, Diels, Ganong, B. Jönsson, G. Karsten, Kearney, Lesage, Ricome, Rosenberg, Schimper, J. Schmidt, Stahl, Volkens, and Warming.

CHAPTER LVI. LITHOPHILOUS HALOPHYTES

ON rocks near the sea plants acquire xerophytic structure for two reasons : first, the general one that the substratum is rock,[1] and. secondly, proximity to the sea. Spray cast up by the breakers, and particles of salt deposited by foam and wind on the plants, introduce a floristic modification into the rock-vegetation, so that there is an admixture of halophytes or it is purely halophytic. There thus come into being communities, including those composed of lichens, which display a zonal distribution. On granite rocks at Bornholm (in Denmark), at Kullen (in Sweden), and elsewhere on northern coasts, the vegetation is arranged in several belts. Lying lowest, where the rocks are very frequently wet, is Verrucaria maura, a very thin black scaly lichen divided into small pieces ; this lichen is widespread along the coasts of northern and arctic seas. Higher up, the rocks are reddish yellow with Placodium murale, which is accompanied by Xanthoria parietina. Above this follows a belt of Ramalina scopulorum ; here the action of the saltwater is reduced to little or almost nothing. In the first two lichen-belts cracks in the rock entertain halophytic Spermophyta, including Matricaria maritima, Aster Tripolium, and species of Atriplex. A little higher the action of the salt water vanishes, and the vegetation on littoral rocks is identical with that on inland rocks.[2]

Similar vegetation colonizes rocks on Adriatic shores. Here, likewise, Verrucaria maura clothes wet rocks with its ' often pitch-black crusts ', and in the clefts of the cliffs are lodged fleshy-leaved halophytes, including Crithmum maritimum, Statice cancellata, Inula crithmoides, and others.[3]

In Madeira on rocks exposed to salt water are a few succulent plants. Dotted here and there is an individual of Mesembryanthemum nodiflorum, Portulaca oleracea, Beta maritima, or Crithmum maritimum.[4] On the Canary Isles, inaccessible rocks, which are constantly wet with salt sea-spray from the breakers, are gay with numerous species of Statice, which possess large light-green rosettes of leaves, and blue, red, or white blossom, surmounting inflorescences that are about half a metre in height.[5] Rocks on the shore often entertain extremely halophytic spermophytic species that do not grow in any other habitats ; such is the case on the Mediterranean, in the West Indies, and other places.

In this, as in other cases of rock-vegetation, a distinction must be drawn between those species that are truly attached to the rock, and those (chasmophytes) that are rooted in earth contained in crevices : this distinction was noted by Schimper.[6] The former plants possibly include only lichens, algae, and perhaps some mosses.

Lithophilous vegetation on sea-shores has been the subject of but little investigation.

Compare Chapter XLI on Lithophilous Benthos of the sea-shore, into which this may pass.

---

[1] See p. 238. [2] Warming, 1906. [3] Beck von Mannagetta, 1901. [4] Vahl, 1904 b.
[5] Christ, 1885; C. Schröter, 1908. [6] Schimper, 1898; Cockayne, 1901; Öttli, 1903.

CHAPTER LVII. PSAMMOPHILOUS HALOPHYTES

THE vegetation on sand is by no means a single formation, but must be divided into several, which are determined by the nature of the soil and particularly by its humidity, and which include various growth-forms.

If we walk on the shores of northern Europe, say, in Denmark, we meet with the following formations zonally arranged, commencing nearest the sea :—

1. **Sand-Algae.**
2. **Iron-Sulphur-Bacteria.**
   These two occur nearest the sea in the aestuarium.
3. **Halophilous spermophytic herbs.**
4. **White dunes (shore dunes).** } These are to be regarded as two formations, and are by no
5. **Fixed or grey dunes (inland dunes).** means essentially halophytic. They will be considered in Section X with other psam-
6. **Sand-fields, in many places.** } mophilous vegetation.
7. **Shingle-banks**—occasionally.
8. **Halophilous forest and bushland**—frequently as a succession inland.

a. Sand-Algae.

These have been already described as part of the microphyte formation of the benthos of loose soil. (See Chapter XLII, p. 175.)

They also form vegetation beneath spermophytic communities that occur as members of the vegetation succeeding on the landward side.

b. Iron-Sulphur-Bacteria.

In the immediate vicinity of the Sand-Algae, but in deeper layers of the shore-soil, there are often black masses of sand, which owe their blackness to the reduction of sulphates dissolved in water contained in ferruginous sand ; in this process, according to Beijerinck[1] and van Delden,[2] definite anaerobic Spirilla (Microspira desulfuricans and others) play an essential part. This formation is not only met with on the sea-shore, but is also general in the mud of fresh-water lakes and pools ; it seethes not only with sulphate-reducing bacteria, but also with Bacillus subtilis and others.[3]

c. Halophilous spermophytic Herbs.

In this third zone the soil is drier, being flooded with sea-water only occasionally ; the sand is loose and therefore white, dotted with a very open scanty vegetation.   One plant stands here, another there, and yet another at a considerable distance—an arrangement seemingly due to the very un-settled state of the loose soil, which is disturbed by wind and very high tides.   The species are for the most part *annual* herbs, such as, in North

---

[1] Beijerinck, 1895.   [2] van Delden, 1903.
[3] Warming and Wesenberg-Lund, 1904; see also Chap. XLII, p. 177.

Europe, Cakile maritima, Salsola Kali, and species of Atriplex, which always find the open space they require, and are not prevented from developing by the shifting nature of the soil. In addition, perennial herbs with creeping rhizomes occur, as they accord with the often unsettled or shifting, loose soil, and easily maintain their position when once firmly rooted : among such species are Honckenya peploides and Triticum junceum. Only on less disturbed, and particularly on stony soil, raised slightly above the sea-level, does one encounter perennial species with a multicipital, deep tap-root, such as is possessed by Mertensia maritima and Crambe maritima.

As the soil is now and again flooded by sea-water, and the water in the soil lies at a slight depth below the surface and is strongly saline, and furthermore as the soil is saturated with salt brought by sea-foam and spray from the waves, the vegetation is definitely halophytic. Its xerophytic and halophytic nature is revealed in a number of characters. Fleshy leaves are possessed by most ; and a glaucous wax-coated epidermis shows itself in some types, such as Triticum, Eryngium, Crambe, Mertensia, and Glaucium flavum. Hairs clothe Kochia hirsuta and Senecio viscosus ; while thorns appear on Salsola. On certain spots in this formation may be found the cactiform Salicornia herbacea, whose home, apart from this, lies in salt water. All these plants are certainly photophilous and incapable of enduring shade.

The species often range themselves zonally into associations according to stability of the soil and other relations (including humidity of soil). In many places the succession, travelling from the sea landwards, is as follows :—[1]

1. *Salicornietum*, with S. herbacea.
2. *Atriplicetum*, with many species of Atriplex, Suaeda maritima, and others.
3. *Cakiletum*, with Cakile maritima, Salsola Kali, Honckenya, and Crambe maritima.
4. *Triticetum*, with Triticum junceum, and others.

The zones nearest the sea consist solely of annual species, but those on the landward side also include perennial species belonging to the growth-forms already mentioned.

Farther south, for instance on the shores of Holland, one meets with several other species, including Convolvulus Soldanella, which belongs to plants possessing subterranean runners, and Euphorbia Paralias. Still farther south, on the shores of France, yet other species, among them Matthiola sinuata, appear : yet the growth-forms remain the same and the formation the same.

On the shores of other seas the same formation seems to reappear, and to differ merely in flora.[2]

On the borders of fresh-water lakes the zonal distribution of vegetation is similar, as are the growth-forms, in so far as they are dependent upon the loose texture of fine-grained sand, and on the degree of repose to which the soil is subject. But only a few observations have been made regarding the sandy shores of fresh-water lakes. According to Cowles,[3] the ' lower beach ' is characterized by Sand-Algae. The ' middle beach ', lying between the high-water marks of winter-storms

[1] Warming, 1906, 1907–09.
[2] In regard to North America, see Harshberger, 1900, 1902.   [3] Cowles, 1899.

and summer-storms, is colonized by annual herbs, among which many are succulent plants that, apart from this, grow on the salt sea-shore : as examples of these may be mentioned Cakile americana, Corispermum hyssopifolium, and Euphorbia poly-gonifolia. Above the high-water mark of winter there commences a sand-field composed of species with travelling rhizomes : among such species are Agropyron dasystachyum, and Lathyrus maritimus. Added to such species are biennial and annual herbs, as well as some stunted shrubs.

Tropical sandy sea-shore has been but little investigated, but we may note—

Pes-caprae-association. Under the name ' Pes-caprae formation ', Schimper [1] dealt with that association on the tropical sea-shore in which the convolvulaceous Ipomoea Pes-caprae plays a prominent part. The large-leaved, fleshy, dark-green shoots,[2] several metres in length, and sometimes decked with large red blossom, creep *over* the sand, strike root, and form a frequently close network. In addition, there are several other species which likewise for the most part grow *over* the sand, and do not, like the North-European Carex arenaria, have long creeping rhizomes buried in the sand ; this contrast may be due partly to the circumstance that shifting sand is rarer in the tropics, as the sand is often calcareous (coral) sand composed of heavier and larger grains, and partly to the fact that the wind does not blow so violently as on the northern coasts of Europe.

Canavalia-association. In physiognomy the Pes-caprae-association is more or less similar to others, such as the Canavalia-association, which occurs on the Moluccas, according to Warburg, and in the West Indies.

Among species or genera playing a part on tropical shores may be named the fleshy Sesuvium Portulacastrum ; Alternanthera, Achy-ranthes and Philoxerus vermicularis among Amarantaceae ; Spermacoce and Hydrophylax among Rubiaceae ; Sporobolus virginicus and Cynodon Dactylon among Gramineae ; Remirea maritima and Fimbristylis sericea among Cyperaceae.[3]

On Asiatic shores the blue-green Spinifex squarrosus plays a part similar to that of Ipomoea Pes-caprae, but, like Psamma arenaria in northern Europe, it has a subterranean rhizome ; the large development of its aqueous tissue is probably to be associated with its occurrence on saline soil. The spherical inflorescences, nearly as large as the human head, are extremely light, and possess stiff, elastic, long, spike-axes radiating on all sides ; rolling before the wind and bounding with great leaps over the sand they shed their seeds in the same manner as certain steppe-plants (' wind-witches ').[4] This plant sometimes forms pure associations to the exclusion of other species. Sometimes a large quan-tity of Nostoc accompanies it. In Ceylon, on the seaward side, it reigns almost alone, but towards the land many other species mingle with it.[5]

Sand on the tropical sea-shore provides examples of *creeping perennial*

[1] Schimper, 1891.
[2] Figured in Warming, 1897 ; see also Tansley and Fritsch, 1905.
[3] The anatomy of various species is figured in Warming, 1897.
[4] Cleghorn, 1858 ; Göbel, 1889-92.
[5] Regarding the morphology of this and other Indian littoral plants, see Tansley and Fritsch, 1905.

herbs, whose frequently prostrate shoots stretch loosely over the sand on all sides, many of them without striking root : this is true, not only of Ipomoea Pes-caprae, but also of the East Indian Euphorbia thymifolia, E. pilulifera, species of Sida, Indigofera enneaphylla,[1] also of the American Euphorbia buxifolia, Heliotropium inundatum, Cakile aequalis, Portulaca pilosa, and others. All these are small-leaved and more or less fleshy.

It is difficult to draw a distinction between the true halophytic vegetation of the sea-shore, and the xerophytic vegetation of dry, warm, non-saline sand, which borders on the shore or occurs in inland dunes ; such a delimitation must be left to the future.

d. Shingle-banks.

In another direction the sea-shore is not sharply delimited, at least in parts of the Earth where the Ice Age has left numbers of erratic blocks scattered on the beach ; in such cases the vegetation of the sandy beach merges with that of rocks, or a combined vegetation representing sand and rock arises, though with characteristics appertaining to the latter.[2] A shingle-bank may also arise by weathering of the shore-rocks and the rounding-off of fragments by the action of waves. See also Chapter LXII.

e. Halophilous Forest and Bushland on Sand.

The vegetation on sandy sea-shore is frequently succeeded on the landward side by forest and bushland, which also appear inland on saline sand. But there is always a question as to whether this formation must be regarded as halophilous or as psammophilous : and this question Kearney[3] has propounded in reference to dune-plants in North America. Tentatively I regard the undermentioned types as halophilous. But in any case the distinction is of subordinate import. Psammophytes and halophytes intermingle on sandy shores near the sea. Towards land the vegetation gradually becomes purely psammophilous, in proportion as the salt is washed out of the sand ; nevertheless the undermentioned low littoral forests and bushlands on flat coasts are halophilous, since they occur only on sea-coasts, and their roots probably extend down to saline water permanently present in the soil.

Barringtonia-association. As the first of these may be mentioned one in Eastern Asia and Australia described by Schimper[4]—the Barringtonia-association—in which a part is played by the large-leaved, large-flowered myrtaceous Barringtonia racemosa and other species, also by Hibiscus tiliaceus, Casuarina, Thespesia populnea, Terminalia Catappa, Heritiera litoralis, and others. The trees in Barringtonia-forest are short, have curved trunks and branches, and large leaves that are leathery or xerophytic in some other way. The forest may be rendered wellnigh impenetrable by Caesalpinia Bonducella, species of Canavalia, and other lianes. In Eastern Asiatic littoral forests there may occur coco-nut palms, cycads, casuarinas with their Equisetum-like aphyllous switch-shoots, and peculiar types of Pandanus, including P. labyrinthicus and P. odora-

---

[1] Schimper, 1891.
[2] 'Mixed vegetation', see p. 143; for figures and details, see Warming, 1906.
[3] Kearney, 1904.   [4] Schimper, 1891.

tissimus, which resemble species of Rhizophora in growth, as their prop-roots serve to fix them in a loose soil.[1]

Coccoloba-association. Here also must be placed the West Indian Coccoloba-association, where C. uvifera dominates in the form of a small tree or shrub with large, very rigid leaves, which are directed sharply upwards [2]; it produces bushland on the shore, and may possess creeping branches that strike root. Intermingled with it are many other species of shrubs and herbs, including some belonging to species or genera that occur in Asia.

Restinga. Allied to these is the Brazilian 'restinga'-forest, which in many respects recalls the *campos serrados*, described in Section XIII as present in the interior of Brazil. These littoral forests form a transition to ordinary xerophytic forests; for the crooked stems and branches often encountered in some of the latter forests also occur here; the leaves in some species are coriaceous, stiff, thick, and possessed of hairs, without being fleshy, but in others are fleshy and glabrous. Cactaceae and Bromeliaceae play a prominent part. Restinga-forest in Brazil apparently is not strictly confined to the sea-shore, as, according to Schenck,[3] it can penetrate far inland where no saline soil occurs.

The seeds and fruits of plants belonging to these tropical, littoral forests are frequently adopted for transport by water, as was demonstrated by Hemsley and Schimper.[4]

Aphyllous halophytic forest. In addition to true forest on saline sand there is in Central Asia a special association produced by the *saxaul-tree*, Haloxylon Ammodendron, which belongs to the Chenopodiaceae. This tree attains a height of five or six metres, and its trunk a thickness of twenty centimetres; its grey, curved and twisted trunk, with its numerous, scaly, thin, short-segmented branches, resembles a 'green besom'.[5] It produces a forest which recalls Casuarina-forest in Australia, as the tree is devoid of needles or foliage, yet it is green and bears flowers. (The wood is hard, very brittle, and without annual rings.) Added to this tree are a few other species, including Calligonum persicum and Pteropyrum Aucherii; and here one also finds the root-parasite Cistanche tubulosa with dirty-violet flowers.

The amount of salt in the trunk is considerable, that in the cortex being 6·25 per cent.; even the epidermal walls are permeated by fine crystals and similar deposits. Excreted masses of hygroscopic salts serve to absorb dew by night. The cork is so constructed that certain mucilaginous swelling layers take up water; when the supply of water ceases the production of mucilage is arrested and protective cork once more appears. In addition, numerous idioblasts containing tannin occur. Assimilatory activity is maintained by the production of chlorophyll in the secondary cortex.[6]

Tamarisk-bushland. Halophilous tamarisk-bushland arises locally in Asia and in Mediterranean countries; it may attain a quite considerable

---

[1] Regarding the flora of Singalese shores, see Tansley and Fritsch, 1905.
[2] Figured by Warming in Börgesen and Paulsen, 1900.
[3] See Schenck, 1903 *a*.   [4] Hemsley, 1885; Schimper, 1891; Guppy, 1906.
[5] Basiner, 1848.
[6] In regard to the remarkable adaptive features of the saxaul-tree, reference should be made to B. Jönssen, 1902.

height and present a dull, bluish appearance ; at the blossoming season the scaly thin branches become decked with countless small, bright-red flowers.

As halophytes are so xerophytic in structure, it is not surprising that growth-forms of the one type of vegetation reappear in communities belonging to the other type. For example, in Venezuela and the West Indies, on the shore, intermingled with Batis, Sesuvium, and other true littoral plants, one finds certain Cactaceae, and Bromeliaceae that are not halophilous in the strict sense. According to Schimper, in Java alpine plants recur in saline, moist spots ; and Battandier discovered a floristic likeness between the littoral and alpine floras of Algiers.

CHAPTER LVIII. PELOPHILOUS HALOPHYTES

a. Aestuarium with Zosteretum and Salicornietum.

Zosteretum. Where the saline soil is loamy and contains clay there appear species of plants and, to some extent, formations differing from those on pure sand ; and in this case, again, the distribution of the species as well as of the associations is zonal. The North-European shores provide admirable illustrations of this. Special reference may be made to the marshy tracts on the eastern coasts of the North Sea. Here the tide twice a day brings a quantity of extremely fine, organic, and inorganic constituents, mainly clay, which are deposited in calm spots during high tide. These constituents are held fast and filtered, in the first place, by grass-wracks (Zostera), which form large, mud-collecting banks (zostereta), and, in the second place, in shallower water by Cyanophyceae and by Salicornia herbacea. The Cyanophyceae form a zone parallel to the Sand-Algae-formation on the sand of the shore ; they also appear in zosteretum, and may cover the soil of salicornietum and of true littoral meadow.[1]

Salicornietum. Salicornia herbacea produces a pure but very open association (salicornietum), which forms the outermost zone of true land-vegetation (formation of halophilous herbs) ; it clothes large stretches of the beach where this is not covered by water at low tide, and it is under water at high tide, although it is a cactus-like, succulent plant, and apparently a pronounced xerophyte. Being devoid of foliage its fleshy stem has undertaken the work of assimilation,[2] and possesses a two-layered palisade tissue sharply cut off from the internal aqueous tissue, and in addition contains tracheid-like cells that store water.[3]

b. Salt-meadow.

Glyceria maritima-association, in North Europe. When the soil has become higher and drier, owing to the deposit of mud year after year between the annual salicornias, it is occupied by Glyceria maritima-association (Festuca thalassica-association). This plant, with its narrow-leaved glaucous shoots, which bend aside just above the soil and bear

[1] Warming, 1904, 1906.
[2] Dangeard's (1888) interpretation of the succulent portion as a foliar sheath is maintained by others.   [3] See p. 126.

tufted lateral shoots, gives rise to a littoral meadow, whose sward is low, coherent, and dense, or, toward the sea, interrupted. Mingled with Glyceria are other pronounced halophytes, including Triglochin maritimum, Spergularia marina, Suaeda maritima, Plantago maritima, Aster Tripolium, Glaux maritima, Statice Limonium, also species of Atriplex and Cochlearia : all these in one manner or another show the structure of halophytes.

Agrostis alba var. stolonifera plays the same rôle as Glyceria, but especially on more sandy soil.

Blue-green Algae, diatoms, also species of Rhizoclonium and Vaucheria, are not uncommon in moist saline meadows.

Higher littoral meadow (juncetum J. Gerardi). As the above-mentioned species occur in greater numbers, and as the soil is gradually raised and becomes drier, Glyceria maritima is suppressed, and the vegetation gives way to that of more elevated littoral meadow, which differs in flora, and whose low, dense vegetation consists essentially of perennial herbs, including grasses. This meadow cannot be regarded as being of a mesophilous type, as it is associated with a clearly saline soil. It includes, among others, Juncus Gerardi, Plantago maritima, Glaux, Armeria maritima, Trifolium fragiferum, Artemisia maritima, as well as the gramineous Hordeum secalinum and Festuca rubra. Representing annual species one finds Lepturus filiformis, Bupleurum tenuissimum, species of Erythraea, and the hemiparasitic Odontites. Lichens are entirely absent, and mosses are almost or quite so. By diking littoral meadows and thus washing out the salt, and by cultivating the soil, *artificial marsh-meadows* or ' water-meadows ' are produced.

Northern salt-meadows do not always commence on clay, but may arise on sand.[1]

The vegetation surrounding salt-springs in the interior of a continent, Europe for instance, differs only slightly from littoral meadow. In Siberia, according to Cajander, near salt-springs meadows are formed by Potentilla anserina, Glaux maritima, Salicornia herbacea, and Glyceria distans. On Hungarian steppes grow many of the same species as on the shores of northern Europe.[2]

On American coasts there also occur littoral meadows which seem to differ materially from the European only in flora. According to Ganong,[3] on the shores of the Bay of Fundy the outermost zone is a spartinetum formed by Spartina stricta ; this is succeeded on the landward side by a belt of Salicornia and Suaeda ; and this passes over into meadow (staticetum) of Statice and Spartina juncea. The salt-meadows of Nebraska are mainly composed of Distichlis spicata stricta.[4]

Here we should probably also place the *Salicornia-association* of the Upper Andes, which has been described by R. Fries [5] and appears to be widespread.·

Many species and genera growing in saline meadow have remarkably

[1] Warming 1906.
[2] Bernatzky, 1905. The Bohemian salt-meadows are mentioned by T. Domin (1905 *a*).   [3] Ganong, 1903.
[4] Pounds and Clements, 1898 (1900). In regard to North-American littoral meadows see also Harshberger (1900), Hitchcock (1898), and Kearney (1900).
[5] R. Fries, 1904.

wide distribution ; not only does, for example, the halophytic flora of North America agree throughout in many points with that of Europe, but even in New Zealand a number of genera, including plants occurring in European saline meadows, are found, and are represented by the same or by different species. Among the genera are Apium, Atriplex (A. patula) Carex, Chenopodium (C. glaucum), Eryngium, Glyceria, Lepidium, Samolus (S. litoralis), Scirpus, and Triglochin.

c. Salt-bushland

On mud and clay soils bounding the Mediterranean Sea, for example at Montpellier,[1] there occurs a dense, dark-green vegetation, which is from one-third to one-half of a metre in height, and consists of the shrubby Salicornia fruticosa[2], intermingled with Atriplex portulacoides, Statice Limonium, S. bellidifolia and other species, and Scirpus Holoschoenus. In the shade of the shrubs often grow felted masses of the cyanophyceous Lyngbya aestuarii. This community differs from those European ones already described as occurring on clay shores in containing shrubby species, and must be set apart from them as a separate formation—that of salt-bushland—which may be compared with saline steppe on clay soil.

On coasts bounding the Caribbean Sea one finds along the shores of lagoons flat expanses of clay, which maintain vegetation that is obviously nearly related oecologically to the South-European community just described. Among the shrubs present are Batis maritima (usually half a metre in height), Salicornia ambigua, Sesuvium Portulacastrum (which may be social and may clothe extensive tracts with often low, lush, glaucous vegetation), also species of Portulaca and Heliotropium, including H. curassavicum.[3]

In the Argentine Andes is found a Lepidophyllum-association[4] which should perhaps be placed under this category : it consists of shrubby Compositae, a metre in height, including species of Lepidophyllum as well as other plants.

d. Salt-steppe.

This occurs in many central, continental parts of countries, including Spain, Hungary, south-eastern Russia, Asia, North America, Argentine Pampas, and Australia.[5] As a rule it is met with in depressions into which water from the higher environs flows. In the centre often lies a small salt-lake, or a swamp that dries up completely in summer. Some salt-steppes are oecologically in close alliance with the outermost zones of clay shores, namely, with the Salicornia zone and Glyceria zone in saline meadow,[6] but others are oecologically related to the shrub vegetation already described as occurring on Mediterranean and American shores. The soil is more or less clayey, and is incompletely stocked with vegetation ; the constituent species, which are not numerous, form on the grey or white ground scattered tufts, which stand out as dark patches and

[1] Flahault et Combres, 1894.   [2] Duval-Jouve, 1868.
[3] Börgesen and Poulsen, 1900.   [4] R. Fries, 1905.
[5] See p. 219 and Chap. LXXIV.
[6] Regarding Hungarian halophilous vegetation, see Bernátzky, 1905.

are mostly either dark-green and glabrous, or grey with a coating of mealy, scaly, or felted hairs, or are blue-green with wax. Salt-steppe remains green when all the surrounding vegetation has withered. Many of the species are more or less shrubby, and have narrow, linear, or spathulate leaves, or are aphyllous.

On Euro-Asiatic salt-steppes one finds species of Anabasis, Halimocnemis, Haloxylon, also of the asclepiadaceous Brachylepis, and others.

In North America, among others, the following Chenopodiaceae occur : Sarcobatus Maximiliani (' pulpy thorn '), Atriplex canescens, Spirostachys occidentalis, Salicornia herbacea, and Suaeda, some of which are shrubs. The salt-steppes formed by these lie on the large plateaux west of the Rocky Mountains—for example, in the neighbourhood of the Great Salt Lake of Utah.

The Argentine salt-steppes (*los Salitrales*) are interspersed in the Pampas and merge with these. Among the plants occurring only on saline soil are Suaeda divaricata, Spirostachys patagonica, and S. vaginata, Halopeplis Gilliesii, Niederleinia juniperoides, and Statice brasiliensis.

Between salt-steppes and other steppes there are frequently very gradual transitional types, because steppe-soil always contains some salt. Salt-steppe also gradually passes over into salt-desert.

CHAPTER LIX.　SALT-SWAMP AND SALT-DESERT

THE communities discussed in the preceding chapter grow in soil that is, at least periodically, more or less dry. Still more laden with water is the soil in connexion with the formations about to be described.

Salt reed-swamp. Along many of the shores in northern Europe there are muddy spots in inlets where the water is calm. In such places there arise swamps entertaining either halophilous or salt-enduring species, in the form of tall perennial herbs, such as Phragmites communis, Scirpus Tabernaemontani and S. maritimus, all of which possess long, creeping rhizomes and therefore may produce dense communities. Added to these are other herbs, such as Aster Tripolium and Triglochin maritimum ; while the wet or flooded soil is covered by Cyanophyceae and bacteria.[1] This is a formation parallel with fresh-water phragmitetum and scirpetum (S. lacustris). Similar salt reed-swamps occur in Spain [2] and elsewhere. They also present themselves in connexion with steppes and deserts when water is present. According to Martjanov those in Central Asia on the Altai Mountains are surrounded by a dense palisade of Phragmites communis, which is several metres in height ; outside this, on drier soil, grow Salicornia herbacea, Suaeda maritima, Taraxacum collinum, Lactuca sibirica, Triglochin maritimum, Plantago maritima, Glaux maritima, Atriplex litoralis, Aster Tripolium, and others, most of which are familiar species belonging to the North-European flora.

Here also may be placed the *shotts* and *sebakh* of North Africa, depres-

---

[1] Warming, 1906.　　　　　[2] Wilkomm, 1852, 1896.

sions which contain salt water during the rainy season, but many of which are dry and covered with incrustations of salt in summer. In them grow Chenopodiaceae, Plumbaginaceae (Statice, Limoniastrum), Nitraria dentata, Reaumuria vermiculata, Frankenia Reuteri, and others.[1]

Similar depressions occur scattered about countries that have a marked continental, dry climate, for instance in the steppes of Hungary, in southern Russia,[2] in the interiors of Asia and North America,[3] in Australia. The salt deserts of Argentina are often like vast snow-fields, or ice-fields, but, in the rainy season, like lakes ; some of them are entirely devoid of plants.[4] Among the Chenopodiaceae present are species of Atriplex, Spirostachys, Halopeplis, Suaeda ; among grasses, Munroa, Muehlenbergia, Pappophorum, and Chloris. In addition there occur Papilionaceae, Portulacaceae, Apocynaceae, Cactaceae, and others.

Apparently almost wanting in plant-life, according to Buhse's description, is the great Persian salt-desert,[5] which is more sterile than the Sahara and covers one-thirtieth of Persia. The clay soil, which in deeper layers assumes the form of mud, retains salt, and thus crystallizes out in places so as to form incrustations as much as a foot in thickness. Of this yellowish-grey tract, extending for 115 geographical miles, the main part is sand, with which lime, oxide of iron, common salt, sodic sulphate, and clay are mixed ; on it grows not a single plant, not a grass haulm, not a moss, and not even any humbler plant : it is the desert of all deserts.

Of the oecology of salt-swamp but little is known. The growth-forms present seem to be for the most part perennial herbs and shrubs. Much more is known regarding the formation about to be described.

CHAPTER LX. LITTORAL SWAMP-FOREST. MANGROVE[6]

Among all plant-communities confined to swamps in salt or brackish water mangrove-swamp is the most extensive and interesting, as well as the best known. It occurs by all tropical seas, especially on flat, muddy shores, where the water is relatively calm, as in lagoons, inlets, estuaries, but not where rocky soil and breakers prevail ; the soil is flooded with water either permanently or at high tide. In many places mangrove-vegetation extends far inland along the banks of large rivers. The water is usually more or less brackish, as far as the tide extends.

Mangrove-vegetation mostly assumes the form of low forest or bush-land, and, viewed from the sea, reveals itself as a dark-green, dense, often impenetrable mass of low trees with countless arched aerial roots. Rhizophora Mangle in favourable sites in the West, for instance at the mouths of rivers in Venezuela, like R. mucronata in the East, grows up to produce stately high-forest.[7] The bottoms of the crowns of the trees are usually truncate, and stand a small distance above the water, and beneath them one sees, where Rhizophora-vegetation forms the outermost

[1] Massart, 1898.
[2] See Krasnoff, 1886 ; Tanfiljew, 1905 ; Basiner, 1848.
[3] See Hitchcock, 1898.　　　　　　　　[4] Brackebusch, 1893.
[5] Buhse, 1850.
[6] ' Tidal forest ' of various writers, including Tansley and Fritsch (1905).
[7] Johow, 1884 b.

fringe of vegetation, a tangle of countless brown roots more or less clothed with algae. The soil, which in places is not covered with water at low tide, is soft, deep, black mud, full of rotting, stinking, organic bodies in which bacteria abound. The water between the trees may be covered with dirty film, and bubbles of gas rising from the bottom burst at the surface.

Many Crustacea belonging to various genera live here, burrow in the ground, bury dead leaves, and play a part similar to that played by earthworms in non-saline humus soil.[1]

FLORA

The *flora*, as it consists only of about twenty-six species[2] belonging to nine families, is poor in species and tolerably uniform over the extensive tropical and subtropical regions of the Old World, from Africa to Australia; the closely allied American flora is still poorer, consisting of only four species. The species are :—

Eastern Mangrove	*Western Mangrove*
Meliaceae	
Carapa moluccensis	
„ obovata	
Rhizophoraceae	
Bruguiera caryophylloides	
„ eriopetala	
„ gymnorrhiza	
„ parviflora	
Ceriops candolleana	
„ roxburghiana	
Kandelia Rheedii	
Rhizophora conjugata	Rhizophora Mangle
„ mucronata	
Combretaceae	
Lumnitzera coccinea	Laguncularia racemosa
„ racemosa	
Lythraceae	
Sonneratia acida	
„ alba	
„ apetala	
Rubiaceae	
Scyphiphora hydrophyllacea	
Myrsinaceae	
Aegiceras majus	
Acanthaceae	
Acanthus ilicifolius	
Verbenaceae	
Avicennia officinalis	Avicennia nitida
„ „ var. alba	„ tomentosa
Palmae	
Nipa fruticans	

Of these twenty-six species only one, Acanthus ilicifolius, is a herb, the remainder being shrubs or trees.

[1] C. Keller, 1887.
[2] If we reckon Avicennia officinalis var. alba as a distinct species.

ADAPTATIONS

1. *Fixation.* The degree of softness of the soil and the variety in depth of water evoke differences, and cause the species to assume a zonal distribution into associations.

Distinctions in salinity of the different zones probably also play a part. On the landward side mangrove-plants vanish, and are replaced by other shrubs and trees.

Furthest out towards the sea are those that are best capable of fixing themselves in deep water, namely, species of Rhizophora (rhizophoretum); within these, in shallower water or on drier ground, succeed those species that have smaller capabilities in this respect, namely, species of Avicennia, Bruguiera, Aegiceras, Carapa, and others.

The species of Rhizophora are fixed by *prop-roots*, that is, by aerial roots which spring from the trunk and often emit radiating branches, which curve down to the ground in an arching fashion. These bow-like roots upon which the tree depends are very numerous; the foundation of the tree is firmer and its power of resistance to bending, which may be caused by wind or movements of the water, is greater than if the stem were the sole support. The anatomical construction of these roots agrees with the unusual nature of the demands to which they are exposed as supporting structures, and deviates from that of most other roots in that the mechanical tissue is made to assume a tubular arrangement round a large pith.[1] Similar prop-roots are possessed by Ceriops and Acanthus ilicifolius. The lower parts of these prop-roots are often beset with algae.[2]

Inasmuch as Rhizophorae form the outposts of the mangrove they entangle mud between their roots and thus add to the land.

2. *Respiratory roots.* Respiration is a matter of difficulty in the soil, which is water-logged, rich in organic bodies, and poor in oxygen. For this reason all the mangrove-plants have a strongly developed system of aeriferous spaces; the submerged parts are very spongy and soft in structure; stomata and unusually large lenticels on parts projecting above water place the atmosphere in communication with the intercellular spaces. The prop-roots of Rhizophora serve as respiratory roots. Other species possess quite peculiar respiratory roots. Avicennia has erect, unbranched ' asparagoid ' roots a foot in height; these stand in very long rows, which radiate from the tree and indicate the position of the horizontal roots from which they spring.[3] Similar respiratory roots are owned by Sonneratia acida [4] and Laguncularia. Bent knee-like roots, with the knee projecting above water, occur in Bruguiera and to a less extent in Lumnitzera; while comb-like prolongations in connexion with the root occur in Carapa. Karsten's [5] experiments confirm the view that these peculiar structures are respiratory roots. And their anatomical construction harmonizes with this function.[6]

3. *Germination and vivipary.* Several species of mangrove-plants

---

[1] Warming, 1883.　　　　　　　　　　Tansley and Fritsch, 1905.
[3] Figured by Warming in Börgesen and Paulsen, 1900.
[4] Göbel, 1886.　　　　　　　　　　[5] Karsten, 1891.
[6] For similar respiratory roots in plants occupying fresh-water swamps, see p. 186; Kearney, 1901 ; Koorders, 1907.

show the rare phenomenon of vivipary : that is to say, the embryo-plant, while still attached and nourished by the parent tree, grows into a more or less developed plant without undergoing any resting period : this behaviour, abnormal to other plants, is normal to these. The following series of stages may be noted :—

a. In Aegiceras the embryo emerges from the seed but remains inside the fruit ; it has a large stem and is green.

b. In Avicennia the endosperm and subsequently the embryo emerge from the seed and lie uncovered within the chamber of the ovary. The embryo is green and obtains food from the parent-plant by means of a long, repeatedly branched, hypha-like haustorial cell, which traverses the placenta.

c. In Rhizophora and allied genera (Bruguiera and Ceriops) the embryo grows not only out of the seed, but also out of the fruit, and projects from the latter in the form of a green seedling displaying the hypocotyl and root, which in some species exceed one-third of a metre in length ; ' like long, green pods the full-formed seedlings hang down from the branches '. The cotyledons serve as a haustorium, sucking food from the parent-plant. Finally, the seedling breaks loose from the cotyledons (Rhizophora has only one cotyledon), which remain behind inside the fruit and shrivel with it ; it falls into the water or mud ; its club-shape and pointed root-end adapt it to pierce the mud, into which it rapidly thrusts lateral roots that had previously been initiated.[1] If the seedling does not succeed in fixing itself, it floats and may strike root on some distant shore ; in this way the species is provided with means of dispersal by water. Vivipary is most pronounced in those Rhizophoraceae that grow in very deep water and very soft soil, and is obviously of great advantage to the species.

We must regard as adaptations to the environment the *green nature of the infant-plant*, and the presence of *anchoring organs* which take the form of upwardly-directed bristles on the seedling, or (in Avicennia, Aegiceras, Sonneratia and Rhizophora) of lateral roots which are already prepared and can rapidly burst out.

4. *Means of migration.* All littoral plants show very wide areas of distribution. The mangrove includes approximately the same species along all the tropical shores from Australia to East Africa. This is due in part to the fact that the medium and temperature remain uniform throughout, and in part to the efficient means of dispersal. Thanks to air-containing spaces in the integument or in other parts, and to the consequent decrease in specific gravity, fruits, seeds, and seedlings of mangrove-plants can float for a very long time, and in doing so are not robbed of their germinative power.[2] The strongest likeness exists between the mangrove of the West Indies and of West Africa, and, on the other hand, between those of East Africa and Asia.

5. *Xerophytic structure.* The constituent species of the mangrove, with one exception,[3] are shrubs and trees. Despite the circumstance that these plants grow in inundated muddy soil, their vegetative shoots[4] display a number of structural features that occur in plants adapted to withstand drought. The features in question are the following :—

[1] Warming, 1883 ; Karsten, 1891.
[2] Hemsley, 1885 ; Schimper, 1891 ; Guppy, 1906.   [3] See p. 235.
[4] Regarding their morphology, see J. Schmidt, 1903.

a. The *leaves* are thick, coriaceous or somewhat fleshy, and entire : Sonneratia, Lumnitzera, Carapa, Rhizophora, Avicennia.

b. The *epidermis* is thick-walled and strongly cutinized ; the leaves are often very glossy : Rhizophora Mangle.

c. Various organs of the leaves of mangrove-plants have been interpreted by Areschoug [1] as *hydathodes.* In the mangrove on the coast of Siam, Schmidt [2] found that glands on the upper face of the leaves of Aegiceras corniculatum excrete salt. During the night the salt-crystals absorb water from the atmosphere and deliquesce, but during the morning the water again evaporates and the salt crystallizes out. In no other species was the excretion of salt found by Schmidt.

d. *Stomata* are often sunk beneath the general surface, and urn-shaped anterior chambers are general.

e. *Aqueous tissue* is always present, and sometimes massive. In Rhizophora mucronata, the older leaves, which no longer assimilate, become thicker than they were in youth ; this is caused by an enlargement of the aqueous tissue : thus the leaf changes its function.[3]

f. The *mesophyll* is almost devoid of intercellular spaces, and the palisade tissue is the sole or main chlorenchyma : Sonneratia, Lumnitzera, and others.

g. The *nerve-ends* dilate into water-storing tracheids : Bruguiera, Avicennia, Ceriops, and others.[4]

h. Long *stone-cells* or bast-like mechanical cells are lodged between the palisade cells in Rhizophora, Sonneratia, and Carapa,[5] and in the pith of Rhizophora.[6]

i. *Mucilage-cells* occur in species of Sonneratia, Rhizophora, and some others.

j. Some species are strongly and densely clothed with *hairs* : Avicennia.

k. The leaves assume a *profile-lie* [7] and consequently acquire isolateral structure : Sonneratia, Lumnitzera, Ceriops, also Laguncularia.

The cause of this xerophytic structure is to be sought in the physiologically dry soil, and in the position of the leaves, which are raised free into the air and are exposed to intense tropical light, high temperature, and wind.

ASSOCIATIONS

Various species may build up associations :

Rhizophoretum occurs, for instance, and Acanthus unaided may form mangrove-vegetation along estuaries.[8]

According to Schmidt *nipetum* belongs rather to fresh-water swamp.[9]

---

[1] Areschoug, 1902.   [2] Schmidt, 1903.   [3] Haberlandt.
[4] See p. 126.   [5] See p. 128.   [6] Warming, 1883.
[7] See J. Schmidt, 1903.   [8] Tansley and Fritsch, 1905.
[9] Cp. Tansley and Fritsch, loc. cit. From the extensive literature dealing with mangrove, special reference may be made to Warming (1883), Johow (1884), Göbel (1886), Schenck (1889), Schimper (1891), G. Karsten (1891), Haberlandt (1893), Börgesen and O. Paulsen (1900), Areschoug (1902), Schmidt (1903), Tansley and Fritsch (1905), Holtermann (1907).

SECTION VIII

CLASS VI. LITHOPHYTES. FORMATIONS ON ROCKS

CHAPTER LXI. ROCKY COUNTRY

THE more uniform a soil, the more uniform and purely typical is the plant-formation growing on it. The soil is generally uniform where it remains flat and horizontal over a wide area. Contrasting with this is a rocky country with numerous declivities, masses of rock, and chasms succeeding one another. A single formation may be composed of many species and many different growth-forms, but it must constitute a single entity, in which one species is more or less dependent upon another or at least stands in some relation to it. In a mountainous, rock-strewn district fragments of various, perfectly independent, formations intermingle in a chaotic manner, which is the more diversified the more uneven and varied the surface.[1]

The *exposure* of walls of rock or sides of mountains varies extremely, and has a most important influence on vegetation ; neighbouring rocks may differ entirely in their exposure, and consequently may bear radically different plants, and such is the case with mountain-sides respectively exposed to and removed from the sun's rays.

Slope of the surface of rocks or of the sides of mountains varies in like manner and is of no less significance ; the steeper the general surface the greater the extent to which natural rock will come to the surface, the more flat and horizontal is a rocky tract the more does it favour the accumulation of detritus, of products of weathered rock, and of vegetable fragments, that is to say, the production of a loose covering of soil ; and hand in hand with these distinctions go corresponding ones in the vegetation.

Of further import is the *nature of the rock* (if it be primitive, or limestone, or slate, and so forth), because its hardness, tendency to split, specific heat, and other properties are of the deepest significance to vegetation)[2] ; and it is a matter of importance whether or no *water* flows away over the sides of the rock or mountain.

The vegetation of a rocky tract of country will therefore present an extremely characteristic and kaleidoscopic picture ; xerophytic and mesophytic formations are lodged among one another, usually in small, but none the less typical, parcels ; here may be a bare, vertical rock bathed in burning sunlight, there a humus-laden, sopping, shady gully ; here, a rock over which the water slowly trickles, there a completely dry slope ; here a mountain-side exposed to the prevailing wind, there a mild and sheltered dell ; and so forth. Furthermore, the vegetation in many spots is in the act of changing from one formation to another, and the most

---

[1] 'Mixed vegetation' ; see p. 143.   [2] See Chapters X and XVII.

gradual intermediate stages between the two present themselves. For the oecologist a rocky and mountainous piece of country is the most difficult of all objects to investigate.

In this Section we deal solely with vegetation on *true rock*, not with vegetation on the loose soil covering rock, even though this may entertain species that are very intimately associated with the rock. Still, to this limitation an exception must be made in favour of the vegetation growing in clefts and niches, because in such habitats it is characteristic alike in species and in growth-forms.

We shall therefore distinguish two formations :

1. **Lithophytes.**

2. **Chasmophytes—**

in accordance with the suggestion made by Schimper,[1] who wrote : ' The vegetation on the surface of rocks or stones may be termed that of *lithophytes*. Crevices in rocks, in which more finely grained components and more water accumulate than on the surface, produce a somewhat more copious vegetation, that of *chasmophytes*.'

Öttli[2] defines as ' rock-plants (petrophytes) ' ' all those plants growing on sides of rock or blocks which are able, as the first of their kind, to colonize the rock permanently, and which display in distribution or structure a more or less pronounced dependence upon rock as a substratum. Within this definition are included both lithophytes and chasmophytes. But it is not a natural scheme to co-ordinate chasmophytes and lithophytes ; in opposition to the latter should be placed those Cryptogamia and Phanerogamia which only colonize rock where detritus has accumulated, whether this be in crevices or on the general surface of the rock. Plants of this latter type we term *chomophytes* (τὸ χῶμα, an aggregation) '. He suggests the subjoined scheme :—

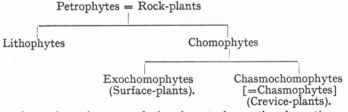

```
            Petrophytes = Rock-plants
        |                         |
   Lithophytes                Chomophytes
                           |              |
                   Exochomophytes   Chasmochomophytes
                   (Surface-plants).  [=Chasmophytes]
                                      (Crevice-plants).
```

The ' exochomophytes ' are merely immigrants from other formations such as sward, meadow, bush, heath (callunetum), forest, and the like.

Following our account of Lithophytes and Chasmophytes we shall consider in Chapter LXIV the closely allied formations on Shingle and Rubble.

CHAPTER LXII. LITHOPHYTES

LITHOPHYTES are those plants that can colonize steeply inclined and bare rock. They are exclusively cryptogamic, and include algae, lichens, and mosses. Among these, algae and lichens are perhaps in general the first colonists, and are the growth-forms that can settle upon the steepest rocks.

Algae may give the colour even to vertical rocks over large tracts ;

[1] Schimper, 1898 ; Eng. Edit. p. 178.   [2] Öttli, 1903.

far north, as well as in Scandinavia and on the Alps one sees black stripes running down rocks and indicating vegetation composed of Cyanophyceae (species of Stigonema) which follow the course of the trickling water. The algae, Trentepohlia iolithus and T. aurea, colour rocks red and yellow. The ' Black Rocks ' in Angola earn their name from investing algae, and the conical tops of the granite mountains near Rio de Janeiro owe their brown colour to a small kind of alga. In many tropical places where a considerable amount of moisture is combined with a high temperature there is a rich sub-aerial vegetation of algae, which are mainly Cyanophyceae.[1] Algae fix themselves in most cases simply by aid of a mucilaginous layer of the cell-wall.

The *lichens* concerned include crustaceous lichens (Lecanora, Lecidea, Biatora, and others), and foliose lichens (Parmelia, Xanthoria, the black species of Gyrophora, and others). On less precipitous spots, where the vegetation is older, fruticose lichens are added. Here, attention may be directed to the fact that on the littoral rocks of northern Europe halophilous lichens display a zonal distribution, as already mentioned on page 224.

The *mosses*, including species of Hypnum, the blackish - brown Andreaea, and the grey species of Grimmia, may form dense cushions on the rock, over which their protonemata spread as flat incrustations.

Though these plants on rocks are often dull in hue, being black or grey, yet some species are bright-coloured—for example, the lichens Buellia geographica and Xanthoria elegans are greenish-yellow and yellowish-red.

ADAPTATIONS.

As the *substratum is absolutely physically dry* lithophytes belong to growth-forms that are capable over their whole surface of absorbing water derived from rain, dew, melting snow, or water running down the rocks. And the plants enumerated have this power.[2] Devices for collecting water are possessed by many mosses, in the form of felted rhizoids, and by certain liverworts, in the form of peculiar, concave leaves.[3]

Lithophytes require *haptera* by which they can attach themselves to rock, unless the thallus itself adhere closely to this.

Rock is to many plants (for instance, marine algae) only a basis of support, but in other cases (for example, lichens) it is likewise a nutritive substratum into which the plants delve more or less deeply ; the rhizoids of calcicolous and silicicolous lichens, according to Bachmann,[4] penetrate the mica of granite. In the case of these plants the nature of the rock plays an important part ; the harder and freer from clefts it is, the more difficulty will plants find in attaching themselves. On Etna, according to J. F. Schouw, there are pre-historic streams of lava which up to the present sustain no vegetation ; apart from this, the lichen Pterocaulon vesuvianum may be mentioned as the first plant to settle on lava. On the other hand, soft rocks, such as many limestones, are readily colonized ; rhizoids of mosses and lichens, and filaments of algae perforate and erode them ; indeed, the whole thallus of certain endolithic lichens lies buried to a depth of several millimetres in the stone, the apothecia alone becoming

[1] Fritsch, 1907.
[2] Regarding lichens, see Zukal, 1895 ; and regarding mosses, see Oltmanns, 1885.   [3] See p. 119 ; Göbel, 1898-1901.   [4] Bachmann, 1904.

eventually visible on the outside.[1] But in the main, lithophytes are dependent for their mineral nutriment on atmospheric precipitations and on dust conveyed to them by wind.

The substratum would seem scarcely favourable to saprophytes, yet some of these seem to occur as soon as a small supply of organic matter is present. In the Bernese Oberland, on the Faulhorn, a nitric bacterium is alleged to permeate rocks and cause them to crumble.

Lithophytes must be able for a time to endure great *desiccation*, whether this be due to heat and the burning sun's rays, or to cold, dry winds. Vegetation on rocks exposed to the sun may be heated up to temperatures (50°–60° C.) which approach the normal limits of life (such, for instance, is the case with plants growing on the limestone mountains of Dalmatia)[2] ; while at night-time the temperature may sink very low, lower than in the case of plants occupying other substrata. Lithophytes are oecologically closely related to epiphytes and often identical with them. Furthermore, during frosty periods in winter lithophytes are but feebly sheltered from desiccation, since snow can remain lying only on a few spots on rocks.[3]

A rock substratum is the warmest of all soils.[4] Consequently, there are numbers of plants which on high mountains occur only in fissures of rocks, but in the plains grow on loose soil.[5]

ASSOCIATIONS.

Lowly organized though they be these plants display great differences in their oecological demands ; and different formations and associations of them can be established in accordance with the like conditions prevailing on rock. The oecological factors concerned include the following :—

Nature of the rock (with great differences according as the rock is eugeogenous or dysgeogenous) ;

Exposure (involving differences in illumination, heating, wetting by rain, exposure to wind) ;

Steepness of the rock-surface ;

Supply of flowing water ;

Geographical situation and altitude.

Kihlman's[6] investigations, for instance, prove that different species of lichens show different powers of resistance to the action of cold winds. According to Zukal,[7] species of Parmelia require more moisture than do crustaceous species of Lecanora ; the latter are less capable than the former of absorbing water-vapour from the atmosphere.

Even among lithophytes one sees competition, struggles for space, and the suppression of one species by another. In North America useful contributions to our knowledge on this matter have been made by Bruce Fink,[8] who recognized various associations of lichens, including ' Lecanora-associations of exposed boulders ', ' Lecanora calcarea contorta associations of exposed, horizontal limestone surfaces,' and others and in Sweden the competition between species has been studied by A. Nilsson.[9] One sees crustaceous lichens settling at first in patches and the patches gradually extending until they meet, either preserving

[1] Bachmann, 1904.   [2] Kerner, 1868.   [3] Öttli, 1903.   [4] See Homén, 189?
[5] Flahault, 1893, 1901 *a* ; Vahl, 1904 *b*.   [6] Kihlman, 1890; see p. 208.
[7] Zukal, 1895.   [8] Fink, 1903.   [9] A. Nilsson, 1899 *b*.

sharply circumscribed outlines or dovetailing with one another ; and the one species eventually suppresses the other.[1]

The first colonists of bare rock, namely algae and lichens, gradually produce a nutritive substratum suitable for more highly organized species, the earliest of which are mosses and fruticose lichens. While the first colonists have horizontally extended vegetative organs and are appressed to the rock, or even crustaceous, their immediate successors rise above the surface as cushions or miniature shrubs (e. g. species of Cladonia), which overgrow the algae and crustaceous and foliose lichens, retain more water than these, and produce more organic detritus. In such communities of mosses and fruticose lichens space may be provided for an under-vegetation of algae.

Means of firm attachment are always the first necessity to the invading species ; and just as the first colonists provide a favourable substratum for taller species, so these in turn produce a substratum in which still more highly organized plants, including Pteridophyta and Spermophyta, can fix and nourish themselves. In cushions of moss one notes seeds of Sedum acre, and other Spermophyta, germinating and developing ; and in the cushions of Sedum other species find place for their roots. The less inclined is the slope of a rock the more easily does this successional development take place. Gradually, in this way, a block of rock or a whole slope may become clothed with a colony of Spermophyta and mosses, and in the course of years, as humus accumulates and earth is conveyed thither by wind or water, bush or forest may develop on these spots. Such superficial vegetation is, of course, oecologically greatly influenced by the thickness of the substratum thus deposited.

In this superficial vegetation we must also include that which develops on all projecting pieces of rock that rise above the surface in rocky country ; on many rocks one sees quite small prominences surmounted by a vegetation of grasses and perennial herbs, or of Calluna, or even of shrubs if the layer of soil be sufficiently thick. These small communities of plants must not be regarded as belonging to true rock-vegetation, of which, on the contrary, chasmophytes are genuine constituents.

CHAPTER LXIII. CHASMOPHYTES

CHASMOPHYTES are plants rooted in clefts of rock that are filled with detritus. In these clefts, particles of earth conveyed by wind and water accumulate, and water collects. The amount and rate of accumulation depend upon the width and situation of the clefts. In the soil thus constituted plants settle, and their dead fragments further add to the supply of nutritive material in the clefts. Thereafter earthworms appear, as do other animals that burrow in the humus soil and improve it.[2] The flora of the clefts varies with such prevailing factors, as exposure, width of cleft, position of this, and, according to Öttli, with the presence or absence of any covering of snow during winter.

When a rock has a very steep surface and is devoid of crevices or clefts, only lithophytes develop upon it, and rarely is rock entirely bare of vegetation (e. g. algae). When, on the contrary, the rock shows cracks and clefts, chasmophytes develop in these. But in many cases, in northern

[1] Bitter, 1898.   [2] Öttli, 1903.

Europe for instance, chasmophytes are merely species that also occur in other situations ; yet these plants must be regarded as constituting a formation of their own, because their flora is affected by the character of the locality, and they form a definite society that develops in a particular habitat ; moreover, they possess certain biological peculiarities enabling them to occupy this type of habitat. But the formation has received only scant attention.[1]

Thallophyta, Musci, Pteridophyta, and Spermophyta are all to be found occupying such clefts.

ADAPTATION.

Rhizoids and roots are often constricted within narrow fissures and flattened out as thin as paper. Roots of grasses, for instance, may form a very flattened tuft filling the cleft. The tap-roots of other plants are likewise flattened, when the clefts are narrow. The roots, guided by water in the cleft, seem to penetrate very deeply ; ' their twine-like roots penetrate incredibly deep into the moist interior of rock ' writes Christ [2] in reference to chasmophytes. Contractility of the main root seems to be common, and to be of great importance to the species concerned.

The constituent species are usually *tufted* in growth, and do not spread far from the general root-system ; although means of vegetative migration are not absolutely excluded, yet migratory plants with elongated horizontal rhizomes are unusual. There rather occur creeping perennial herbs, Fragaria for instance, that produce long thin epigeous runners, which give rise to new shoots, either in another cleft or in the one in which the parent plant is established.

Rosette-plants are common ; the rosulous habit is, indeed, character-istic of plants freely exposed to light.

Xerophytic types are common, and succulent plants particularly so ; in northern lands Crassulaceae, including Rhodiola and Sedum, and Saxifragaceae, including Saxifraga Aizoon, occur ; in more southern Europe Sempervivum appears. Moreover, there are plants, such as Saxifraga oppositifolia and Silene acaulis, with small, thick, imbricate leaves, or, like Diapensia, with small, dry, leathery leaves. Among peren-nial herbs with rosettes of leaves may be mentioned Papaver nudicaule, species of Draba, and other species that occur in great numbers on fell-fields.

The warmer and drier the climate is, the more do mosses and lichens retreat into the background, and the more abundant do Spermophyta become. Many of these are xerophytes. On rocks of the calcareous Alps, among stones in the stony tracts on the mountains of Herzegovina, and in similar places, one often finds white, woolly species of Cerastium, rigid tufts of species of Arenaria, species of Veronica, Alchemilla, Saxi-fraga, and others, which form low, dense tufts, and possess small, stiff leaves, a strong epidermis, abundant hairs, and many other signs of their xerophytic nature.

Even in lower parts of the Alps the succulent types, Sempervivum and Sedum, become more common, and gradually lead us to true *tropical rocks.* Here we can still see crustaceous lichens flourishing, but succulent and other xerophytic Spermophyta have become more abundant. We find not only rosette-plants, such as Bromeliaceae, Agave, Velloziaceae

---

[1] Except from Öttli in Switzerland. [2] Christ, 1885.

and species of Yucca in America ; Aloë, Dracaena, Mesembryanthemum, Aizoon, Sempervivum, Cotyledon and other Crassulaceae, or Senecio (Kleinia) in Africa, including the Canary Isles[1] ; but also succulent-stemmed euphorbias in the Old World, and Cactaceae in the New. Besides these plants one finds grey-haired, small shrubs, such as species of Croton and fragrant species of the verbenaceous Lippia, also small fleshy-leaved plants, such as Peperomia, Pilea, Pedilanthus and, finally, Orchidaceae with pseudo-bulbs.

Many of these plants almost seem to live on air, yet they attain a considerable size ; filled with sap they hang down in all their beauty from jagged solid rock, at first sight seeming to be purely superficial, but in reality sending their roots into crevices, and abstracting the water retained there by capillarity. At certain seasons, and particularly in the brief spring-time the brown or grey rocks are dappled with gay flowers.

Annual plants are rare on rocks, as they find but few spots suitable for germination. In lands with a long, dry season they may, nevertheless, play a prominent part. Very common on rocks in Madeira are Gnaphalium luteo-album, Campanula Erinus, Gymnogramme leptophylla, two annual species of Aichryson, and Sinapidendron rupestre ; even in the lowlands annual herbs are richer in individuals than are perennial herbs, and only exceeded in this respect by undershrubs.[2]

It is by no means universal for the vegetation on rocks to be exclusively composed of xerophytic types. *Mesophytes* often form no inconsiderable part of the plants present. This may be due to difference in the crevices. Some of these receive water percolating from higher parts of the mountain, and may remain moist throughout prolonged periods of drought ; other crevices obtain their water exclusively from the strictly local rain. Some crevices contain abundant detritus and are therefore endowed with a greater power of storing water ; others are poor in detritus and allow the water to pass away. The chemical composition of the detritus also varies, as some crevices contain abundant humus, in which numerous earthworms may lurk, whereas others are poor in humus. Cracks in rocks supply an endless variety of habitats, each of which forms a special kind of environment.[3]

Plants occupying crevices of the lower parts of rocks on the Danish shores are halophytes which are mostly species belonging to the sandy beach, whereas those higher up are perennial herbs and shrubs, the majority of which are xerophytes, though some mesophytes occur.[4]

Öttli does not regard vertical rocks in the Alps as habitats poor in water. Nevertheless the majority of plants that he terms lithophytes are xerophytes, and show that they find difficulty in obtaining water, especially during winter, because the rocks retain no coating of snow. But many mesophytes also occur. On the highlands of Madeira, according to Vahl,[5] the vegetation on vertical rocks is a varied admixture of xerophytes and mesophytes ; side by side with white woolly-haired undershrubs grow ferns and liverworts. But in the hot dry lowlands of Madeira all the springs dry up in summer, and the crevices of the rocks are not sufficiently moist to permit mesophytes to replace the water lost in transpira-

[1] Christ, 1885.   [2] Vahl, 1904 b.   [3] Öttli, 1903.
[4] Warming, 1906.   [5] Vahl, 1904 b.

tion. The vertically cleft basalt rocks are well-nigh devoid of plant-life, because they are quite incapable of retaining water. The irregularly cleft basalt rocks are inhabited by a few strongly xerophytic under-shrubs and some herbs. Where strata of basalt and tuff alternate, the difference in their powers of conducting water becomes obvious. The tuff, which is a good conductor of water, is clothed with Adiantum Capillus Veneris and other plants, or, near cultivated soil, with Parietaria judaica, and stands out in the form of horizontal bands on the walls of rock. Yet during summer these mesophytes are in a partially faded condition. Rocks consisting entirely of tuff display vegetation quite as scanty as that of basalt rocks, because there is no impervious stratum lying below.

In tropical places where the air is moist or where the site is shady, for instance in forests, or in humid mountain-valleys where mist often hangs over the ground, we may encounter a vegetation of chasmophytes, including dense, green cushions of moss and many mesophytes—just as in moist temperate situations.

CHAPTER LXIV. FORMATIONS ON SHINGLE AND RUBBLE

CLOSELY allied to formations on rock are those on rubble, which develop on a soil that has arisen through the disintegration of rock by atmospheric agencies (heat and frost). This soil produced by weathering is composed of detached, angular pieces of rock, and smaller or larger stones which have fallen from the faces of rocks ; at the base of rocks in mountainous countries there are often found accumulations of this kind—*talus*. In some cases the fragments of rock are so large as to remain stationary, but in others they are smaller and produce an unstable rubble that slips down. The formations produced show corresponding differences.

In the former case, in the course of time earth accumulates between the fragments of rock in greater or smaller quantities, and thus provides a soil capable of sustaining highly organized plants and even trees. And when this is not the case many herbs may send their roots down between the stones into the subjacent soil and grow luxuriantly. Among such masses of stones one often finds delicate mesophytes, including ferns, that with this exception are confined to forest : the reason for this is that the roots under and between the stones find sufficient water to replace that lost in transpiration. There occur not only non-migratory plants, but also migratory plants, because beneath and beside the stones there is adequate space for rhizomes and runners.

On talus of this kind two formations occur intimately mingled together :—

1. Lithophytes on the stones.

2. The remaining vegetation between the stones. As time passes, the accumulation of earth and humus between the stones may become so considerable that the stones are covered and the lithophytes vanish. The vegetation thus has a developmental history, and presents an appearance that changes with its age.

On talus consisting of small-stoned, slipping rubble, the species are necessarily to some extent different from those on stationary talus. On the former, the proper fixation of the plant is a matter of difficulty; the vegetation is constantly destroyed by movements of the rubble and must start afresh. The particularly successful species are largely those possessing very long roots or long creeping rhizomes.[1] This kind of substratum is poor in plants and the vegetation remains open, until the soil is stationary or ceases to receive fresh masses of stones arising by weathering of the rocks.

These formations are closely allied with those of fell-fields, with which they merge. In regard to the number of formations and associations on rocks future research must decide.[2]

[1] See C. Schröter, 1904-8.
[2] Special reference should be made to the papers by Hitchcock (1898), Öttli (1903), Pohle (1903), Rikli (1903), Adamovicz (1898), Ostenfeld (1906), Brockmann Jerosch (1907). For the most recent treatment of the subject of formations and associations, see C. Schröter, 1904-8; also C. Flahault, 1906 b.

SECTION IX

CLASS IV. PSYCHROPHYTES. FORMATIONS ON COLD SOIL

CHAPTER LXV. CLIMATIC CONDITIONS IN SUBGLACIAL FELL-FIELDS

IMMEDIATELY below the snow-line in Polar countries, and on high mountains there develops a vegetation composed of *subglacial communities*, which though exhibiting many differences consequent on the variety of the alpine and Polar conditions, yet can be grouped together because they show many points of agreement. The natural conditions prevailing in these situations are discussed in the succeeding paragraphs.[1]

Temperature. The mean temperature of the warmest month is low and does not rise more than a few degrees above zero centigrade ; on mountains it falls by 0·6 of a degree centigrade with each increase in altitude of 100 metres. This fall in temperature brings in its train certain limits to the distribution of species according to altitude and latitude, and certain limits to the distribution of snow (local conditions, such as slope and the like, play a great part in the matter[2]). In the vegetative season sharp frosts and falls of snow occur, arresting the development of plants and modifying their forms, particularly because the soil is rendered cold.

The annual and diurnal changes in *temperature on mountains* are not essentially different from those in the plains of the same climatic zone, though the temperatures are lower and their fluctuations less. Moreover, the date of the average arrival of the lowest and highest temperatures is later on mountains than in plains. Consequently, on high tropical mountains the change of temperature is tropical in type, and there is only a small difference between the temperature of the extreme months. On Antisana in Ecuador the difference in temperature of the hottest and coldest month is only 3·2 centigrade degrees. The mean temperatures of all the months are above 0° C., but frosts and falls of snow occur in all months. Approaching the Poles the seasons become more sharply defined, just as in the plains. Even in the cold-temperate zones, on mountains temperatures almost as low as in Polar countries are met with in winter. And particularly those plants that are not covered with snow are exposed to very severe cold. The specific zero of species must lie low. Yet the minimal temperatures in winter seem to exclude certain species from only a few places, and without exercising any notable influence on the character of the vegetation. The minimal temperatures

---

[1] See Kerner, 1869 ; Schröter, 1904–8.   [2] See Chap. XXI.

in Polar countries, and on high mountains are not lower than in parts of Siberia and Canada where forest occurs. Of far greater significance to vegetation are the temperatures during summer and the length of the vegetative season.

Wind. Movements of the air are much stronger on mountains than in plains, and especially above altitudes at which a mountain begins to divide into its separate peaks. The wind is also very strong on arctic coasts during summer. *Wind has an intense drying action on plants,* and even when these grow with their roots in water the coldness of this renders it difficult to replace the water lost by transpiration : the plants must guard themselves against desiccation.

Moisture. A sufficiency of moisture is found periodically during the vegetative season both in air and soil. Rain and mist may abound, and much snow may fall ; and water derived from melting snow may moisten the soil perhaps throughout the whole vegetative season. Nevertheless the soil is *cold and the activity of the roots is depressed* — a circumstance that is of the most profound oecological import to the plant. In Polar countries the relative humidity of the air during summer is high in most places, the number of days on which atmospheric precipitations fall is very small, but mists are frequent. On the Polaris Expedition it was found that the mean relative humidity during summer was 75 per cent., but such a low figure is exceptional : in Lady Franklin Bay, however, it was 81 per cent. In winter in high latitudes great dryness of air prevails.

On mountains the rainfall increases with the altitude up to a certain height, and decreases above this ; the height in question varies according to situation and season. The zone of greatest rainfall is the inferior limit of the cloudy belt. Higher up, the rainfall decreases because as the temperature sinks there is a diminution in the absolute amount of aqueous vapour in the atmosphere ; yet the frequency of rainfall is not lessened. Where the summits of high mountains extend into the region of the upper air-currents, there is above the cloudy belt a dry region where the air is very dry, and only a slight rainfall occurs—for example, on Teyde Peak (Teneriffe). According to Meyen, on the Andes the wind is sometimes so dry that one's skin is ruptured, blood exudes, and one can travel only in woollen clothes. Evaporation on mountains is always greater than in the plains also because of the lower atmospheric pressure. Vertical currents of air among mountains are always of great import. Every ascending current of air brings with it a high atmospheric humidity, mist or rain ; every descending current brings with it great atmospheric aridity. Accordingly, aridity and saturation of the air may alternate very rapidly. The air on high mountains may become suddenly very dry, and even extremely so, on account of rarefaction of the air and intense sunlight. Thus, both air and soil may be very dry periodically, and for this reason the vegetation must be xerophytic, even where the dry periods last for only a few hours. Vegetation on high alps, according to Kerner, may at times be so soaking with moisture that water can be squeezed out of the tufts of moss, but a few hours later, after an east or south wind, it may be so dry as to crackle as one steps on it.

Duration of daylight. Vital activity commences at a time when the day is long—from fourteen to sixteen hours on the alps of temperate countries, and much longer in Polar lands. This prolonged illumination decreases growth in length.

Intensity of sunlight. The intensity of direct sunlight increases with altitude,[1] and is very great on high mountain-tops, owing to rarefaction of the air, to its absolute dryness, and to the thinness of the intervening atmospheric stratum. It is calculated that the intensity of the sun's rays falling on the summit of Mont Blanc is 26 per cent. greater than at Paris. On high mountain-tops the temperature in the sunlight may be thirty-four centigrade degrees higher than in the shade ; moreover, the temperature of the soil is much higher than that of the air. On mountains the fluctuations of the temperature of the air are much less than in the plains, but those of the soil are much greater; for the maximal soil-temperatures greatly exceed the maximal air-temperatures, though the minimal temperatures of the soil are not much lower than those of the atmosphere. Thus the excess of the mean temperature of the soil over that of the air increases considerably with the altitude. The sun's rays awaken shoots into activity and growth at a time when the soil is still very cold. In Polar countries light is less intense and temperature lower, but their action on growth is subject to less intermission ; the distinction between the temperature by day and by night becomes obscured. Intense light by day and cold by night co-operate at high altitudes in retarding growth ; prolonged, though weaker, light and lower temperatures in Polar lands have the same effect. These circumstances bring nanism in their train.

The vegetative season. In regard to the vegetative season there are two extreme types—the *arctic-temperate*, and the *tropical*.

The *arctic-temperate type.* Within the Arctic zone, as well as on high mountains within the temperate zone, the vegetative season is short, lasting, as a rule, only a few weeks. In Franz Josef Land only one or two months enjoy a mean temperature exceeding 0° C. In the Eastern Alps at the extreme altitude reached by Spermophyta (about 3,300 metres) and in unfavourable spots the vegetative season lasts only for one month. As regards humidity and illumination there are greater differences between the arctic lowlands and the alps in temperate zones. Nevertheless the general character of the vegetation and many of the species of plants are identical in both these places. Common to both are strong winds, the very brief vegetative season, and above all the frequency of frosts during the vegetative season.

The *tropical type.* This differs from the arctic-temperate in that the vegetative season endures throughout the whole year.

The *subtropical* intermediate type with a long vegetative season stands between the two extremes.

On tropical mountains the insolation is more intense than in higher latitudes, and consequently the maxima of the soil-temperature are much higher; and this gives to the vegetation an entirely different character.

---

[1] Cp. Wiesner, 1905 a, and other papers.

CHAPTER LXVI.   ADAPTATION OF SPECIES IN SUBGLACIAL FELL-FIELDS

THE climatic conditions already recounted stamp the vegetation with their impress. It is obvious that a series of extreme xerophytic types must result from the different factors that promote transpiration, and from coldness of soil, which checks the absorption of water.[1]

The adaptations of the vegetation are discussed in the succeeding paragraphs.

A. Duration and Development of the Plant.

1. *Perennation.* The vast majority of the plants concerned are *perennial* herbs or dwarf-shrubs. *Trees and tall shrubs are absent—* excluded by drought and wind. In arctic countries annuals are represented by the polygonaceous Koenigia islandica, probably also by Gentiana nivalis, G. serrata and other species, and the gentianaceous Pleurogyna ; a few others, including Draba crassifolia, are probably biennials. In the Alps there are several species of Gentiana that flower only once ; (as reputed annuals may be mentioned species of Euphrasia ; but these and similar hemi-parasites may be left out of consideration, as their conditions of life are so different). Bonnier and Flahault[2] give the following statistics in reference to the western Alps: the number of annual species at altitudes 200–600 metres above the sea-level is sixty per cent., at 600–1,800 metres is thirty-three per cent., and above 1,800 metres is only six per cent. And at this last altitude Kerner[3] gives the estimate at four per cent. in the Tyrol, but judges the annual and perennial species to be about equally numerous in the valleys below. In regard to different latitudes Vahl[4] gives the following figures concerning annual species : Portugal, 34 per cent. ; Denmark, 20 per cent. ; Iceland, 11 per cent. ; Greenland, 8 per cent. According to Warming,[5] in Greenland north of 73° N. there are no annual species, leaving out of consideration those probably introduced by man ; between 73° N. and 71° N. there is 1 per cent. ; between 71° N. and 69° N., 2 per cent. ; between 69° N. and 67° N., 3 per cent. ; between 67° N. and 62° N., 4 per cent. ; and between 62° N. and 60° N., 5 per cent. Some species are annual in the lowlands, but perennial on mountains : such is the case with Arenaria serpyllifolia and Poa annua ;[6] or annual lowland species are replaced on mountains by perennial species—for instance, in the Alps, Draba verna by D. laevigata, Viola tricolor by V. lutea, and so forth. In northern Europe some species, which south of the limit of forest are annuals, become perennial north of this limit.[7]

The cause for these phenomena is to be sought in brevity of the vegetative season and lowness of temperature.[8] Annual species blossom when the temperature is highest ; as the climate becomes colder their seeds must ripen under unfavourable circumstances and therefore are easily rendered sterile. Possibly some annual species become changed into

---

[1] Hence the term ' Psychrophytes ', that is, plants belonging to cold soil.
[2] Bonnier et Flahault, 1878.   [3] Kerner, 1869.   [4] Vahl, 1904 *b.*
[5] Warming, 1887.   [6] Kerner, loc. cit. ; Bonnier, 1884.
[7] Kjellman, 1884.   [8] See p. 25.

perennials because seed-production is arrested, and through correlation the vegetative organs consequently become more vigorous and longer-lived.

2. *Development* commences late, but proceeds with great rapidity during the vegetative season. The spring season in Polar countries is hurried through. Plants that in the plains belong to the late-flowering species blossom earlier in the Alps, though they commence to develop much later. The growing season of many species, thanks to the action of winter-cold, is on the whole much shorter than elsewhere (in high latitudes precociously ripening varieties are brought into existence—for instance, barley in northern Norway).

3. *Early flowering.* Subglacial species in general are *spring-plants*, that is to say, *they blossom very early*, before the foliage is fully developed; Soldanella, Primula acaulis, Crocus vernus, and others, even blossom under the snow ; this depends on the circumstances that the flowers have been initiated in the year preceding that of flowering, and that the flower-buds are provided with rich supplies of food stored in the surrounding bud-scales.[1] The consequences are that flowers may possibly find insects to pollinate them, and that the short vegetative season can be fully utilized for the maturing of seeds, as could otherwise scarcely be the case owing to lack of heat. Exceptions are provided by Compositae that can ripen their fruits within a few weeks, and by such species as the northern forms of Cochlearia, which can pursue the processes of blossoming and ripening of fruit, uninjured even after the most severe and prolonged periods of frost.[2]

4. *Vegetative propagation*, particularly by the aid of detached buds, plays a great part in the life of certain species, perhaps to atone for failure to set seed or produce flowers, as in Saxifraga cernua, S. stellaris var. comosa, S. flagellaris, Polygonum viviparum, and viviparous grasses. In many stations the conditions are so unfavourable to existence that the soil is not covered by plants, which are scattered about at considerable distances from one another.

B. Structural Features.

1. Most shoots are *epigeous :* in this way neither time nor food is lost in shooting out of the soil. The shoots usually live longer than one year, and produce a series of vegetative year's-shoots before they blossom and thereafter perish ; a prolonged period of manufacturing nutriment necessarily precedes the production of flowers, to which the final year of the shoot's existence is dedicated.

2. *Evergreen* shoots characterize a number of species of herbs and dwarf-shrubs : this is of advantage, in that favourable temperatures and illumination can be utilized throughout the year.[3] The hibernating foliage-leaves of some species, at least, contain abundant food, which is consumed in spring, after which the leaves wither. Kerner[4] very appropriately compares the short rosulate-leaved shoots of species of Saxifraga and the like with epigeal bulbs. The withered leaves persist for a long time.[5]

3. True *bulbous and tuberous plants* are *rare*, perhaps a more direct mode of development of shoots involves no loss of time : yet in the Alps such

---

[1] Kjellman, 1884 ; H. Jonsson, 1895 ; Middendorff, 1867 ; Warming, 1908 *a*.
[2] See p. 23.   [3] See p. 25.   [4] Kerner, 1869.   [5] See p. 75.

occur in Lloydia serotina, and Chamaeorchis alpina. On the Andine fell-fields ('*pŭnas*') and in Tibet many plants with subterranean organs storing reserve-food are however stated to occur. The vast majority of dicotylous plants possess a multicipital stout tap-root but produce no adventitious roots ; as examples, may be cited Silene acaulis, species of Arenaria, Draba, Dryas, and Saxifraga oppositifolia.

Rarer are herbs with horizontal epigeous or hypogeous shoots *that strike root*, and undershrubs, such as the small arctic willows, with subterranean shoots.

4. *Nanism.* Extremely characteristic is the prevailing nanism, which is caused by the factors mentioned in Chapter XLV as checking growth. The *foliage-leaves* are *small*, and in many species assume rounded-off, more or less entire shapes ; even among mosses the leaves are shorter and broader than in individuals of the same species in other habitats. On the other hand the leaves are linear in some of the plants, which consequently become moss-like, as is the case with certain species of Saxifraga and Sagina. The *vegetative shoots* are *short*, and *have short internodes*, and often rosulate leaves ; whereas the flowering axes are frequently more or less scape-like and bear small, bract-like leaves. Alpine species therefore differ in habit often markedly, from allied or parallel species in the lowlands : for instance, Artemisia nana from A. campestris, Aster alpinus from A. Amellus.[1] Yet the *shoots* are sometimes *elongated*, though *prostrate* and closely applied to the ground. Woody species have bent, curved, and twisted shoots, which often trail along the ground, as in Betula nana, Juniperus, Empetrum, Dryas octopetala, Loiseleuria procumbens, or are entirely subterranean so that only the short tips of the branches project, as in Salix herbacea ('turf-shrubs').[2]

5. *Size of flower.* Very often one can see that this nanism concerns only the assimilatory shoots, and that *flower and fruit are of the same size* on mountain and lowland.[3] When it is asserted that flowers are larger, this is a subjective impression (possibly aroused by the small size of the vegetative organs) and is not supported by measurements.

6. *Lie of leaves* may be different from that assumed on the same species in another habitat ; they become more erect, appressed, and concave in Juniperus and Lycopodium,[4] and are always directed upwards or even vertical in some species, such as Ottoa oenanthoides and Crantzia lineata which have juniper-like leaves.

7. *Cushion-like and tufted plants.*[5] Branching is often very close ; this is due partly to shortness of the stems and internodes, and partly to desiccating winds, which kill the youngest shoot-tips and call forth an irregular and vigorous development of branches.[6] A part in this may also be played by the fact that the flowers are terminal in many species.[7] In this way many species acquire a very low, dense, convex, often hemispherical, tufted or cushion-like form, which is characteristic, not only of Spermophyta, but also of mosses, and is very obvious when one compares subglacial species with allied or parallel species growing in lowlands. The density of the tufts is increased by the circumstance that old dead parts, including leaves, remain attached without undergoing decay.[8]

---

[1] Bonnier, 1890 *b*.   [2] C. Schröter, 1904; cp. p. 26; also see Lidforss, 1908.
[3] Bonnier, 1890 *b*.   [4] See figures in Warming, 1887. [5] See p. 129.
[6] See p. 38.   [7] Reiche, 1893.   [8] See p. 75.

Occurring in their typical form (e. g. Azorella and others) on high mountains in South America, on Kerguelen Island, and in New Zealand, these dense cushions may be protected against desiccation by their old, densely packed parts greedily absorbing and retaining water ; and when the surroundings become cold they possibly can remain warm for a considerable time because of the high specific heat of the water.[1]

8. *Lianes.* Climbing plants are lacking because tall plants are absent.

9. *Thorns* and *prickles* are almost unrepresented in subglacial plants ; the species of Rosa and Rubus concerned usually have few or no prickles. This must be attributed to the intense humidity prevailing during the spring season.

10. *Leaf-structure.* Xerophytic structure, evoked by the factors mentioned in Chapter LXV, reveals itself in the perennial, evergreen, foliaged shoots. For the leaves are *coriaceous*, rigid, and very glossy (cutinized), as in Loiseleuria procumbens and Globularia cordifolia ; or are *thick and juicy*, as in species of Saxifraga and Sempervivum ; or show a more or less dense coat of *hairs*, as in species of Rhododendron, Draba, Cerastium alpinum, Espeletia, and Culcitium. Many possess the *pinoid, juncoid,* or *cupressoid* or *rolled* type of leaf.[2]

Stomata are concealed in furrows, or under the revolute leaf-margins, or under hairs, as in Cassiope tetragona, Ledum palustre *var.* decumbens and other Ericaceae, also in Empetrum, Dryas, and others. Deciduous foliage shows this xerophytic structure to little or no extent.

The leaf-structure of alpine plants has been investigated by Lazniewski, Leist, Wagner, and Bonnier.[3] The two last-named agree in their general conclusions, which relate to alpine leaves when compared with the leaves of corresponding lowland plants, and are as follow : Alpine leaves are designed for probably more vigorous assimilation by means of a better developed palisade, and are consequently thicker (to the extent of $\frac{1}{6} - \frac{1}{5}$ or even $\frac{1}{3}$) in proportion to the surface or often absolutely, than those of lowland plants. They are always dorsiventral and of loose texture, thanks to large intercellular spaces. They possess many stomata on both faces, but especially on the upper face, where these are often more numerous than on the lower face. The guard-cells lie at the level of the general epidermal surface, excepting in the case of the evergreen leaves already discussed. Wagner is of opinion that alpine plants require greater powers of assimilation, because the absolute amount of carbon dioxide in the air is less, and the vegetative season may be shorter [4] ; added to this the intensity of light to which alpine plants are exposed may be greater, also richer in strongly refractive rays.

Bonnier [5] compared the leaves of nineteen species growing respectively in Spitzbergen or Jan Mayen Island and the Alps ; he came to the following conclusions, which certainly are too sweeping : the arctic leaf is thicker and more fleshy, has looser mesophyll, in which the palisade

---

[1] Göbel, 1889 ; Reiche, 1893 ; Meigen, 1894 ; Diels, 1896 ; G. Andersson and Hesselman, 1900.   [2] Göbel, 1891 ; Lazniewski, 1896 ; Diels, 1905.
[3] Lazniewski, loc. cit. ; Leist, 1889 ; Wagner, 1892 ; Bonnier, 1894 *a.*
[4] In this connexion reference should be made to Schröter (1904-8) and to pp. 14-15, also to the remarks in the sequel dealing with construction of arctic leaves.
[5] Bonnier, 1894 *a.*

is less differentiated and includes rounded cells ; the arctic leaf also has a thinner cuticle (in evergreen species this is scarcely the case).[1] This difference in structure is due to the circumstances, first, that the atmospheric humidity[2] in Polar countries increases with the latitude, whereas on high mountains it decreases above a certain altitude ; secondly, that alpine plants live in an atmosphere usually free from mist and are exposed to frequently alternating light, which is very intense during the day time, whereas Polar plants live almost continuously in mist or in light of low intensity. This explanation harmonizes with the researches made by Lothelier,[3] and Bonnier, on plants growing in dry and moist air respectively, and with Bonnier's investigations on plants grown in continuous (electric) light. The low intensity of the light seems to be of greater import than the mist, which at a sufficient distance from the coast is scarcely more frequent in Polar countries than in the Alps. These results obtained by Bonnier agree with older ones obtained by Theo Holm[4] and more recent ones by Börgesen.[5]

11. *Aromatic*, also bitter and resinous, *substances* are scantily formed in Polar lands, but are more frequent on high mountains. For instance, on the Andes[6] small Compositae containing these substances are more numerous than in the allied lowland-flora. This is presumably due to the more intense light. The flowers on high mountains are throughout more fragrant than those of Polar countries.

12. *Pigments, foliage-leaves*, according to Bonnier,[7] often become *deeper green* in colour with increasing altitude (and latitude ?) ; they produce more chlorophyll, whereby they acquire greater assimilatory powers and atone for their small size. Bonnier remarks that there is an optimum altitude at which leaves attain their deepest shade of green. Red cell-sap, anthocyan, frequently occurs in Polar lands,[8] also on high mountains. Anthocyan apparently is capable of converting the incident rays of light into heat-rays[9] ; and according to Tischler[10] red races of plants endure cold better than do green ones.

The *colours of flowers* become deeper and purer with increasing altitude and latitude. The full, pure colours of gentians, hare-bells, and potentillas on the Alps, and Mimulus, Lupinus, and Sida on the Andes are well known. White-flowered species show more marked reddish tinges in subglacial situations than elsewhere ; according to Blytt the flowers of Achillea Millefolium, Trientalis, and of Carum Carvi, and the bracts of Cornus suecica, on Norwegian mountains are more red in tint than in the lowland. Subjective impressions, nevertheless, play some part in this ; floral hues *seem* to be more intense on humble plants, which often grow within a sterile environment, but Bonnier and Flahault,[11] by the use of graded colours, found that the colours in reality are deeper. This must be attributed on mountains to intensity of sunlight, and in Polar lands to its long duration.

---

[1] Börgesen, 1894; H. E. Petersen, 1908.   [2] See p. 249.
[3] Lothelier, 1890, 1893.   [4] T. Holm, 1887.   [5] Börgesen, 1894.
[6] Meyen, 1836.   [7] Bonnier, 1894 b, 1895.
[8] According to Wulff, 1902.   [9] See p. 20.   [10] Tischler, 1905.
[11] Bonnier et Flahault, 1878 ; see also Bonnier, 1890, 1894, 1895.

CHAPTER LXVII. SUBGLACIAL FELL-FIELD FORMATIONS

MOST characteristic of the *fell-fields* [1] is the short stature of the plants (showing nanism) ; also the fact that the *soil is never completely covered by plants*. One individual stands here, and another there ; between them we see bare, pebbly, stony, sandy, or clayey soil, which is devoid of humus and determines the prevailing colour of the landscape. In eastern Greenland Pansch and Hartz saw places which were so bare that scarcely a moss or lichen was visible. The cause for the poverty in individuals does not lie in the soil itself, for this indubitably contains a sufficiency of nutritive substances and water, and could certainly produce luxuriant vegetation *were sufficient heat* supplied. Between climate and density of vegetation there evidently must be a certain constant relation of such a kind that no more seeds or other propagative organs germinate or develop into plants than just suffice to maintain the vegetation at the standard once established. Only under specially favourable circumstances are the seeds capable of germinating and developing into plants. Subglacial species (and lithophytes) may be regarded as the pioneers of the plant-world, since they are least dependent upon other plants or upon animals.

Another characteristic feature of the fell-field is the *abundance of Cryptogamia*. These are mainly *lichens* and *mosses*—at least in the northern and arctic fell-fields ; this is due to the ability these plants have of thriving at low temperatures. Their abundance varies with the situation, and partly according to the soil ; on slate in the North it would appear that the number of Spermophyta is greater than that of mosses and lichens, and the vegetation is more varied ; but the reverse is true on mountains composed of primitive rocks, where the vegetation passes over into the lichen-heath and moss-heath. In addition to these Cryptogamia there occur taller plants, including both herbs and dwarf-shrubs. The Spermophyta mostly assume the cushion-form or tufted-form, and possess a perennial, strong, tap-root (the root or rhizome is multicipital) ; the leaves of herbs usually form a rosette on the ground. The dwarf-shrubs are mostly evergreen.

The soil varies in nature. In general in Polar countries and on many high mountains it may be described as being one of older or younger moraine-stones : and the vegetation on soil that is very stony accordingly approximates to rock-vegetation, which, indeed, establishes itself here and there in typical form. Elsewhere, clay and sand preponderate over stones and gravel. *Various forms of vegetation* consequently develop side by side and are intermingled.

Fell-fields are found upon the highest and most sterile situations on high mountains and towards the Poles as far as the region of perpetual snow and ice.

The *flora* of fell-fields may be considerably diversified, and is mostly so because no single species dominates. The species and genera present in different parts of the Earth differ very widely.

---

[1] German, Felsen-fluren ; Danish, Fjäld-marker ; ' Gesteinsfluren ' of Schröter, 1904–8, p. 503.

We must subdivide subglacial fell-fields into several, probably many, *formations* ; but they have not yet been sufficiently investigated in low latitudes to render possible accurate scientific subdivision.

On *high mountains in Europe* and in *arctic countries* fell-fields prevail where the mean temperature of the warmest month is below 6° C. At a somewhat lower altitude or latitude fell-field is confined to edaphically unfavourable localities, while plants on more favourable sites produce a closed formation. Low temperature in summer favours the production of *raw humus* where the vegetation is somewhat richer, and in such places there arise various formations belonging to sour soil, such as moor, heath, moss-tundra, and the like, which have already been dealt with.

In tropical places the vegetation shows character essentially different from that of the arctic-alpine fell-field, and may provisionally be termed *tropical-alpine fell-field.* Here, too, may be placed the vegetation of high mountains in very dry countries that recalls steppe-vegetation in many respects. It may be termed *mountain-steppe.*

In depressions lying within the subglacial tract where snow remains for a long time, one finds characteristic, greasy mud, which sustains a vegetation of its own — Öttli's[1] *snow-patch flora.* The oecological conditions of this habitat are : (1) The characteristic soil, which owes its origin to melting of the snow ; (2) Brevity of the vegetative season ; (3) Humidity.

Arctic Fell-field.

Arctic fell-field occurs round the North Pole in northernmost North America, Siberia, northern Europe, Greenland, Iceland (where it is termed ' melur ').[2] The most important shrubs, dwarf-shrubs, and prostrate and ' turf ' shrubs, are Juniperus communis, many species of Salix, Betula nana, Empetrum ; among Ericaceae, Cassiope tetragona, Arctostaphylos alpina, Loiseleuria procumbens, Rhododendron lapponicum, Phyllodoce caerulea, Vaccinium, Ledum, and Kalmia ; and the rosaceous Dryas octopetala.[3] The most important herbaceous genera are Poa, Festuca, Trisetum, Hierochloe, Nardus, and others among grasses ; Carex, Elyna (E. Bellardi), and Kobresia (K. bipartita), among sedges ; Luzula and Juncus, among Juncaceae ; and the liliaceous Tofieldia. In addition there occur many Caryophyllaceae, including Silene acaulis and Viscaria alpina ; Compositae ; Cruciferae, including Draba, Cochlearia, Vesicaria, and Braya ; Campanula uniflora, Papaver nudicaule, Polygonum viviparum, Pyrola rotundifolia, Rhodiola rosea, also species of Ranunculus, Potentilla, Saxifraga, and Pedicularis. Moreover, there are always many mosses and lichens of various forms, including such fruticose lichens as Cetraria, Cornicularia, Sphaerophoron, and Cladonia ; these Cryptogamia in many places play the leading part or reign almost alone.

In different stations, according as the soil is more gravelly, clayey, or sandy, more warm or cold, certain species sometimes occur in greater abundance and give a special appearance to the vegetation, so that it is possible to distinguish various *associations*, such as Juncus trifidus-association, Diapensia-association, Carex rupestris-association, Dryas-

---

[1] Öttli, 1903 ; C. Schröter, 1904–8.   [2] Stefansson, 1894.
[3] Warming, 1887 *a*, 1908 *a*.

association, Silene acaulis-association ; Ranunculus glacialis-association, Dryas-Potentilla nivea-association.[1] On high Scandinavian mountains parched fell-fields that are poor in humus may take the form of dwarf-shrub heath,[2] which represents a transition to the already mentioned dwarf-shrub heath occurring on raw humus.[3]

Fell-fields of European Mountains.

On European mountains fell-fields presenting the same physiognomy occur up to the lower limit of perpetual snow and ice, and even higher up, interspersed among the snow-fields, where the sun's rays and slope of the soil bring bare places into existence. On the Alps fell-fields occur both on limestone and on primitive rocks, each of which has its own peculiar species. Particularly in limestone districts one finds accumulations of rubble (talus) entertaining a definite herbaceous vegetation ; the plants lie upon bare, periodically very dry rubble, in the form of roundish tufts widely separated from one another and developed on all sides from a single point. In the Tyrolese Alps, according to Kerner,[4] such talus is first colonized by some Cruciferae, including Arabis alpina and Hutchinsia alpina, and by species of Saxifraga, Linaria alpina, Salix retusa and S. herbacea, among which are found grasses, sedges, Dryas, and subsequently Loiseleuria procumbens and the two species of Arctostaphylos. Loiseleuria may here and there acquire mastery over the others and form an association. Mosses and lichens are of less significance than in northern Europe and in Polar lands ; nevertheless Polytrichum septentrionale plays a great part along all places that have been long covered by moraine-stones.

The soil tends to a greater extent than in Polar countries to produce xerophytes. The snow-fields on the mountains of Herzegovina, according to G. Beck,[5] in summer have many spring-plants, which—it is remarkable to note—are partly bulbous and tuberous plants, and include Scilla bifolia, Muscari botryoides, Corydalis tuberosa, Anemone nemorosa, Crocus Heuffelianus, Saxifraga, and Viola. This contrast to the fell-fields of Polar countries must be attributed to greater dryness and heat in summer.[6]

Very many genera are common to arctic countries, alps of the northern hemisphere, and the alps of Java.[7]

Tropical Fell-fields.

South America. On high mountains in South America we find extensive fell-fields, which are termed *paramos*, in countries ranging from Ecuador

[1] See Hult, 1887, where these are termed *formations*.
[2] A. Cleve, 1901.
[3] Special reference should be made to the papers by Kihlman (1890), Hult (1887), Warming (1887), Hartz (1895), T. Holm (1887), Nathorst (1883), Kjellman (1882, 1884), G. Andersson (1900, 1902), Porsild (1902), A. Cleve (1901), Sernander (1898), Pohle (1903), C. H. Ostenfeld (1908 b).
[4] Kerner, 1863.   [5] Beck von Mannagetta, 1890, 1901.
[6] Reference should be made to the papers published by Christ (1879), Kerner (1869, 1886), Beck von Mannagetta (1901), Stebler and Schröter (1889, 1892), Schröter, 1904-8, Öttli (1903), and Brockmann-Jerosch (1907).
[7] See Meyen, 1836.

to Venezuela ; these support the typical, open vegetation, the individuals of which are scattered in small tufts, and display growth-forms exactly corresponding to those in northern fell-fields ; cushion-like growth is perhaps more common. But other species and genera impart to the vegetation a distinctive appearance ; in addition to Viola, Anemone, Alchemilla, Draba, Senecio, Gentiana, Poa, Hordeum, and many other European genera, there are found Nassauvia, Chuquiraga, species of Baccharis of wondrous shapes, and other Compositae, Tropaeolum, Loasa, Blumenbachia, Verbenaceae, Cactaceae, Calceolaria, Mimulus, Melastomaceae, Krameria, Lupinus, Calyceraceae, and others. The umbelliferous genus Azorella deserves special mention.[1] The rhododendrons of Switzerland are replaced here by species of Escallonia and Bejaria.

Here, too, one finds the composite genera, Espeletia and Culcitium (termed ' frailejon '), of which Espeletia grandiflora is a remarkable plant attaining a height of 2 metres, and remaining unbranched. The stem is clothed with numerous dead remnants of leaves, that form an investment as thick as a man's body ; while higher up it bears a number of leaves enveloped in very thick coatings of wool, and inflorescences. In the highest regions these genera, together with short alpine herbs, grasses, and ferns, form the sole vegetation.[2]

Despite great humidity, frequent rain and mist, which the sun may suddenly dissipate, the vegetation is xerophytic, as Göbel's descriptions demonstrate : many plants occur with pinoid, cupressoid, juncoid, or woolly-haired leaves. On the Chilian *pŭnas*, according to Meigen,[3] mosses and lichens are greatly curtailed, lichens occur only at isolated spots in any abundance ; while mosses never form a carpet or an extensive cushion. Extreme dryness is the cause of this.

Africa. The vegetation on high African mountains is of similar character. Viewed from a distance it seems to be a continuous grassy sward, but closer inspection shows that the tufts of grass are isolated. Grasses and sedges give rise to cushions, whose size varies from that of the human fist to that of a dinner-plate ; and their haulms, about 70 centimetres in height, rise up from the axils of leaves which are erect or have fallen backwards on to the ground. In the dry season the soil is bare or is clothed with a mat of mosses and lichens. Shortly after the rains set in there shoot forth many herbs ; the first to appear are monocotylous bulbous and tuberous forms, such as Hypoxis angustifolia, Hesperantha Volkensii, and others, which are succeeded by dicotylous herbs and undershrubs, such as Wahlenbergia Oliveri, Tolpis abyssinica, and Helichrysum Meyeri-Johannis, and many others. Moreover, there are some little trees which attain heights of from 5 to 8 metres and are bent away from the wind to the south-west. Of these there are only a few species, including Agauria salicifolia, Erica arborea, and Ericinella Mannii. Their boughs are hung with lichens. Higher up the tufts of grass become scantier, while puny shrubs, especially Ericinella Mannii and Senecio Johnstonii prevail. The last-named represents here the same growth-form as Espeletia in South America. In Abyssinia it is replaced by the lobeliaceous Rhynchopetalum montanum.[4]

[1] See next page.   [2] Göbel, 1891.   [3] Meigen, 1893.   [4] Volkens, 1897.

Characteristic of tropical fell-field is the occurrence of *pygmy-trees*, which render it different from other formations belonging to this class. Tropical fell-field thus approaches tropical savannah, but differs from this in the frequent presence of cushion-like plants.

Antarctic Fell-fields.

The climate of antarctic islands shows a strong likeness to that of tropical high mountains, in that the difference of temperature in the different seasons is slight. There is a long season during which the temperature rises a few degrees above 0° C. On Kerguelen the winter has a mean temperature of 2° C., and the summer one of 6·4° C. On South Georgia only the three coldest months show mean temperatures below 0° C., yet the mean temperature of the warmest month is only 5·3° C. This explains the close resemblance between tropical and antarctic fell-fields.

In South Georgia fell-field is essentially formed by scattered tufts of Poa caespitosa (tussock-grass).[1] The tussocks are from 1 to 1½ metres in height, and at the base are surrounded by withered leaves. Between the tussocks only few other species grow. On the Falkland Isles the tussock-grass is also common. But the fell-field is much richer in forms here. There occur evergreen dwarf-shrubs, Chiliotrichum amelloideum, Pernettya empetrifolia, which often give rise to true heath. In addition we find here the peculiar, cushion-like umbelliferous Azorella caespitosa.[2] This assumes the form of a dirty-green, hemispherical cushion, which is more than a metre in height and is extraordinarily hard. The periphery is composed of numerous little shoots, which are all equally tall and are densely clothed with scale leaves. These shoots adhere so closely and firmly to the intervening old leaves and shoots that it may be difficult to remove a fragment of the plant even with the aid of a knife. Dusén describes 'Bolax-heath', as a 'heath' composed of Bolax glebaria, Comm. (Azorella caespitosa, Vahl) occurring in tracts south of Rio Grande : the cushions of Azorella fuse together nearly everywhere, thus forming an almost uninterrupted, compact, very hard vegetable covering, which extends over wide areas.[3] Lichens, mosses, e. g. Racomitrium, and other plants may spread over these cushions.

To this vegetation that of fell-field on high mountains in New Zealand is allied. Here the majority of species are xerophytic cushion-plants, which grow scattered about and seem everywhere to be deserting the rocks. Especially remarkable are the 'vegetable sheep' plants which are species of Raoulia and Haastia. Dense cushions are also produced by Celmisia viscosa, species of Veronica, Hectorella, and others.[4]

Mountain-steppe.

In drier districts on high mountains there occurs a type of fell-field that approximates to steppe in many respects.

Europe. On Etna the presence of tragacanth-shrubs is a true sign of an affinity with steppe.

---

[1] Concerning the characteristic tussock-formation, see Chapter XLVIII.
[2] See Göbel, 1891 ; Schenck, 1905 b. [3] Dusén, 1905. [4] Diels, 1896, 1905.

Africa. On the Canary Isles the peak-region lying above the clouds is desert almost devoid of plant-life. Here and there grow hemispherical shrubs of Sparto-cytisus nubigenus, which are almost the only plants that one sees.[1]

Asia. On high mountains in central Asia, above the forest in the cloud-belt, there extend alpine steppes whose flora is a remarkable admixture of steppe-plants and alpine forms. In Tibet, Rockhill found at a great altitude a vegetation consisting of scattered tufts of grass, rhubarb, and Allium senescens.

South America. The *punas* of the Andes must be regarded as mountain-steppes.[2] In summer their temperatures are higher than those of the *paramos*. At Potosi even in November the mean temperature is 14·2° C. Nevertheless in the dry rarefied air the fluctuations of temperature are so great that brooks may freeze at almost any time of the year. Snow mostly falls in summer but never remains lying for a whole day, even when it is a foot in depth. During night-time radiation from the soil is so intense that rime may be formed at a time when the temperature of the air is 6–7° C. Strong winds dry everything up. Dead animals last like mummies and do not undergo putrefaction. The vegetation is mainly formed by a grass, Stipa Ichu. Its setaceous tufts are 35–50 centimetres in diameter and nearly always inclined away from the prevailing wind. During most of the year the tufts of grass are blackened as if burnt by the sun. In addition to grasses the *punas* display numerous xerophytic perennial herbs and undershrubs, which assume the forms of rosette-plants and cushion-plants. Bulbous and tuberous plants also occur. Even the Cactaceae give rise to cushions. The *punas* are poorer in plant-life than the *paramos*. Cactaceae which are absent from the latter are common here.[3] Here, too, grow herbs with enormously long roots, often a metre in length, and with reduced assimilatory shoots, which are only a few centimetres in length and bear rosettes of woolly or glandular-haired leaves that lie flat on the ground.[4] In the Argentine Andes, Fries[5] distinguished three kinds of shrub-steppes : Hoffmanseggia-association, Cactus-association, and Azorella-association.

---

[1] Schröter, 1908. [2] See p. 259. [3] Tschudi; Göbel, 1891.
[4] Benrath ; Weberbauer, 1905. [5] Rob. Fries, 1905.

SECTION X

CLASS VII. PSAMMOPHYTES. FORMATIONS ON SAND AND GRAVEL

CHAPTER LXVIII. OECOLOGICAL FACTORS. FORMATIONS

SANDY soil and its qualities have already been discussed on p. 59. The vegetation developing on this loose soil is invariably characteristic, and owes its distinctive features to the substratum and to other physical conditions, but particularly to those concerning the temperature and the degree of moisture prevailing in the substratum. Of paramount significance in determining the nature of the vegetation on sand is the depth of the layer of sand above the water-table ; when this layer is so deep that the peculiar properties of the sand are brought into play, then *sand is physically dry and bears a definitely xerophytic type of vegetation.* Here we deal solely with cases of this kind, and not with sandy soil that is saturated with moisture because it fringes open water.

The majority of sandy soils owe their origin to water, and especially to the disintegrating and comminuting action of breakers, but in a less degree to other agencies, such as the sun's heat which splits stones into pieces. Consequently one meets with sand along many coasts, often in the form of dunes ; but one also encounters it inland (in tropical deserts) occasionally in the form either of dunes or of sand-fields.

The chemical nature of the soil [1] shows differences not only in composition of the grains, but also in the amount of salts present ; and in the latter respect sand on the sea-coast contrasts with that lying inland. Vegetation on sand bounding seas or salt-lakes is influenced by the salts contained in the adjacent water, and may largely owe its xerophytic character to these salts.[2]

In the north of Europe, near the sea and on sandy shores one may encounter the following formations, which to some extent show a zonal succession :

1. **Sand-Algae :** in the aestuarium.
2. **Iron-sulphur-Bacteria.**
3. **Psammophilous halophytes.**
4. **Shifting, or white, sand-dunes.**
5. **Stationary, or grey, dunes,**
6. **Dune-heath and dry Sand-field.**
7. **Dune-bushland.**
8. **Dune-forest.**

[1] See p. 59. [2] See Section VII.

The first three formations are halophytic communities and have been studied (see p. 225). But the fourth formation, composed of large dune-grasses, is much more psammophilous than halophilous; for the dominant and most prevalent grass, Psamma arenaria, thrives admirably at inland spots far from the sea-coast; other species, including Elymus arenarius, Eryngium maritimum and Lathyrus maritimus, are more restricted to the coast; whilst still others are not in the least confined to the coast. The influence of common salt vanishes at a short distance from the sea; dune-vegetation is by no means halophytic in constitution.[1]

CHAPTER LXIX. SHIFTING, OR WHITE, SAND-DUNES

ESSENTIAL to the building up of a dune are two main conditions— sand and wind—and to these is added another, namely, that the sand must be dry and bare, or, in other words, so free from vegetation that the wind can act on it. Dunes mainly arise along sea-coasts and river-banks; waves and tides throw up on the shore sand which is usually quartz-sand, whose grains generally average one-third of a millimetre in diameter; these are dried by the sun and transported by the wind. Shifting sand, like snow drifted by wind, is deposited wherever it finds shelter from wind—behind stones, shells, fragments of wood, and, above all, among plants. The sand accumulates in contact with the last-named, among their shoots and leaves, and behind them on the lee side arises the so-called ' tongue-like heap',[2] which is a long tongue-shaped accumulation of sand sloping gradually downwards.

Subsequently, these ' dune-embryos '[3] themselves act as obstructions to the drifting sand, and gradually change their forms until they give rise to typical dune, whose angle of inclination in typical cases is 5°–10° on the windward side, and about 30° on the lee side; the sand slips down the steep lee side, on which the grains become arranged according to size, power of cohesion, and so forth, quite uninfluenced by the wind. Dunes become most regular where there is a prevailing wind and where vegetation is quite scanty or lacking; as the form of the dune is largely determined by plant-growth.[4]

Dunes also arise far removed from water, in deserts of Central Asia, Africa, and elsewhere, where only the fundamental conditions for their genesis prevail; namely, that sand should be formed and not be completely clad with vegetation, and that a sufficiently strong wind should blow.

Dunes are described as ' white ' so long as the vegetation clothes the soil to such a limited extent that it is the colour of the sand which determines the prevailing tint and appearance of the dune: in fact, there are some dunes entirely devoid of vegetation. White dunes are usually at the same time 'shifting dunes ' which travel in the same direction as the prevailing wind.

[1] See Kearney, 1904; Warming, 1907-9.
[2] Sokolow, 1894 (' Zungenhügel '). [3] Cowles, 1899.
[4] See Sokolow, loc. cit.; Gerhardt, 1900; Cowles, loc. cit.; Cornish, 1897; Wessely, 1873; Warming, 1891, 1907; Massart, 1893, 1898 a, 1908; Wery, 1906.

FLORA

In the north of Europe the sea-marram, Psamma arenaria, is the most important dune-grass, and far excels all others in the densely caespitose arrangement of the leaves and in its faculty of collecting sand and growing up through this. In addition we may mention Elymus arenarius, Carex arenaria, and Lathyrus maritimus. Added to these are also Hippophaë rhamnoides and Sonchus oleraceus, whose long creeping roots emit numerous suckers, and some others.

ADAPTATIONS

Plants fitted for life on dune-sand assume characteristic growth-forms, which are encountered on shifting dunes all the world over. The dune is almost entirely constituted by a very loose aggregation of particles and is easily traversed by roots and rhizomes. Plants strictly limited to a confined spot in the soil are scarcely fitted to exist on dune, as the wind is incessantly bringing new supplies of sand, which cause constant change in the conformation of the dune and in its soil. Consequently, the characteristic features of the *vegetation* are that it is very *open*, and consists of *xerophytic perennial species* possessing *long and richly branched* rhizomes; while *exceedingly long* roots (often many metres in length) are found permeating the sand.

There are certain 'sand-binding' species of plants that first arrest the sand and produce dune-embryos; in addition, plants promote the growth in height of sand-dunes. Triticum junceum in the north of Europe is one of the halophilous psammophytes that begin the production of dune both on the shore and at the base of the white sand-dune; another plant like it, is Honckenya peploides : but these can only give rise to *low dunes* on the beach. Psamma arenaria and Elymus arenarius expel them and produce *high dunes* (psammeta). This is due to their faculty of enduring burial by the sand and of growing up through this; it is evident that, inasmuch as new shoots accumulate fresh stores of sand, the dune must constantly increase in height. In the fine sand composing the interior of the dune one finds a great number of old rhizomes and roots ; and if the wind breaks down an old dune these enclosures are revealed.

ASSOCIATIONS

In northern Europe we find various associations including—

Elymetum, composed of Elymus arenarius ; usually occupying only a narrow zone nearest to the sea-beach ; this species may also give rise to inland associations.

Psammetum, consisting of Psamma arenaria, sea-marram; in the vast majority of cases white dunes are produced by Psamma ; for binding the shifting sand this is the most important species, and is often artificially utilized for this purpose.

In white dunes here and there *woody species*, particularly Hippophaë rhamnoides, Salix repens, and Empetrum nigrum, appear. The first two may give rise to bushland.

In northern Europe the next developmental stage to the white sand-dune is—the *grey dune*.

CHAPTER LXX

STATIONARY, OR GREY, SAND-DUNE. SAND-FIELDS. DUNE-HEATH. DUNE-BUSHLAND. DUNE-FOREST. OTHER EXAMPLES OF PSAMMOPHILOUS VEGETATION

Grey Dune and Dry Sand-field in North Europe.

WHEN wind does not disturb the dune other plants may settle between the shoots of Psamma arenaria and Elymus arenarius ; the more effectively do these two species maintain the sand in a stationary condition the more do they prepare the soil for other species and for their own extinction. There now appear shorter plants which have less vigorous subterranean organs, are more confined to their point of fixation, are annuals or perennials, and cannot endure being buried by much sand. The vegetation continues to become closer, as mosses (Polytrichum, Ceratodon purpureus, Grimmia and others), lichens, and some Schizophyceae, establish themselves and permeate the sand with their rhizoids or thalli :[1] the soil thus becomes firmer and more densely clothed with vegetation. Perennial species of caespitose habit or possessing a multicipital primary root can at this stage establish themselves here, so that the soil ultimately is covered with a low, dense, greyish-green carpet.[2] The *grey dune* has arisen. The space thus becomes too confined for the two tall species of dune-grasses, whose long vegetative shoots carry on a prolonged struggle for life—and this is particularly true of the sea-marram—only to succumb in the end.

Next to dunes, and often in the interior of countries, for example in Jutland and North Germany, are found extensive flat or undulating expanses whose soil is dune-sand or glacial sand. On such soil ' heatherless sand-field '[3] arises. Such dry sand-fields on inland spots in northern and central Europe are, for the most part, a product of cultivation. Sand-field arises on old heath-soil poor in nutriment and is reconverted into heath, unless such a change is prevented by human agency. It includes essentially the same species as grey dune, as well as some species common to dwarf-shrub heath ; the species are accommodating and generally fitted to endure prolonged drought. As the growth-forms present seem to be like those on grey dune, sand-field may be combined in the same formation with grey dune.[4]

OECOLOGICAL FACTORS

The vegetation is xerophytic in character. The necessity for this xerophytism has already been explained on p. 59 in connexion with the account then given of sandy soil ; but additional factors contributing to the same result are the intense light and heat, and strong winds, all of which tend to accelerate transpiration.

[1] Gräbner, 1895 a.
[2] In regard to the Flora in northern Europe see Warming, 1891, 1907-9 ; Buchenau, 1889 a, 1889 b.
[3] Gräbner, 1895 a, 1901 ; Massart, 1908.
[4] Regarding sand-field : in Denmark, see Warming, 1891, 1907 ; in Bohemia, see Domin, 1904, 1905 ; in Servia, see Adamovicz, 1904.

The *soil* is very *sterile ;* only in dunes lying nearest to the sea is there a certain amount of calcic carbonate, which is derived from the shells of marine animals ; but in dunes more remote from the sea this salt is dissolved. Of nitrogen and humus extremely little is present ; the humus bodies are rapidly oxidized to carbon dioxide and water, and thus disappear.

Heat. The dune is strongly and rapidly heated by direct rays of the sun ; the temperature at the surface may rise to 50°–60° C. at midday in July[1] ; warm currents of air stream out of the soil and play upon the plants. The sun's heat dries the upper layers of sand so completely that the grains of sand lie loose, but *at a slight depth the sand is cold and moist ;* for the upper loose stratum checks evaporation. The changes of temperature within twenty-four hours may be very great.

Light is reflected from the surface of the sand and impinges upon the lower faces of the leaves. The illumination as a whole is intense.

Strong winds are wont to prevail where sandy soil and dunes occur ; and wind has a double action : it *desiccates*[2] and acts *mechanically*, in the latter case partly through the agency of the particles of sand that it transports. These blown particles by their action may even polish stones and perforate the thin broad leaves of plants that are not fitted to live in this community : such for instance is the case with poplars, which are often planted out on tracts of shifting sand.[3]

ADAPTATIONS [4]

These accord with the differences in kind of the soil and vary with them. The more shifting is the sand the more there come to the front species which possess far-reaching subterranean organs (rhizomes and roots) and exhibit a luxuriant activity in the production of shoots and adventitious roots ; for these species can endure being buried beneath a covering of sand, through which they can force their way upward. The more coherent and stationary is the soil the more prominent do other growth-forms become.

In grey dunes of northern Europe we note growth-forms showing the following characters :

(*a*) Long *creeping rhizomes*, or roots capable of producing adventitious shoots, are possessed by Carex arenaria, Galium verum, Sonchus arvensis, Helichrysum arenarium, Festuca rubra, Rumex Acetosella. Under this category we may also place mosses and the shrubs Hippophaë Salix repens, and Rosa pimpinellaefolia.

(*b*) A *tufted habit* is exhibited among grasses by Weingaertneria canescens, Festuca ovina, and Nardus stricta ; and among Dicotyledones by Ononis repens, Anthyllis Vulneraria, Eryngium maritimum, Dianthus deltoides, Artemisia campestris, and Armeria vulgaris, nearly all of which are very deep-rooted. Here, too, may be included such dwarf-shrubs as Calluna and Empetrum, and the undershrub Thymus Serpyllum. Many species have shoots that lie prostrate on the sand and closely applied to it, but do not strike root ; and they show leaves radiating from a common

---

[1] Giltay, 1886. [2] See p. 38. [3] See Massart, 1893, 1908.
[4] See Giltay, 1886 ; Buchenau, 1889 ; Warming, 1891, 1907 ; Kearney, 1900 ; Abromeit, 1900 ; Massart, loc. cit. ; Chodat, 1902 ; Coville, 1893; Cowles, 1899.

point at the upper end of the primary root ; such is the case with Artemisia campestris and Ononis.

(c) A few possess *epigeous creeping* shoots : among such are Antennaria dioica, Hieracium Pilosella, Polypodium vulgare, and in most cases Sedum acre.

(d) *Annual* (*ephemeral*) and *biennial* (*hibernating annual*) *species* are common, showing that the dune largely partakes of the nature of steppe. The latter species germinate in autumn or spring, develop, and flower in early spring, but conclude their lives before the onset of the summer's heat. It is the warm soil that occasions this acceleration of development, which is exhibited by Cerastium semidecandrum and C. tetrandrum, Trifolium arvense, Filago minima, Aira praecox, Bromus mollis, and Phleum arenarium, as well as by Jasione montana, Draba verna, and Teesdalea nudicaulis. To these types we may append Eryngium maritimum which is, at least often, perennial but is reputed to be monocarpic. This abundance of small, annual, rapidly-flowering plants is in accord with the prevailing *dryness, intense insolation,* and *sterility of the soil.* But we have further to note that :—

The perennial herbs, grasses, and shrubs as a whole, are low, small-leaved, and narrow-leaved.

Most of the grasses, including Psamma, Triticum junceum, Nardus, Festuca ovina, F. rubra f. arenaria, Koeleria glauca, have deeply *furrowed* leaves that can close themselves more or less by rolling up ; not a single grass possesses broad, lush, bright-green leaves.

Elymus arenarius has broad leaves which, however, are blue-green with *wax,* like those of Triticum junceum. *Wax* coats the leaves of Lathyrus maritimus, Eryngium maritimum, Mertensia maritima, Glaucium flavum, and Crambe maritima.

Woolly hairs clothe Salix repens, Gnaphalium, and Antennaria.

Hippophaë has *scaly hairs,*

Glandular hairs are developed on Senecio viscosus, Ononis repens, Cerastium semidecandrum, and some other plants, to such an extent that the surface of the plant is densely coated with sand.

Nardus and Koeleria glauca are *tunic-grasses.*

Not a few species depress their transpiration by having the leaves *vertical,* as in Salix repens, or *crumpled,* as in Eryngium.

The leaves of certain psammophytes are dorsiventral in structure and, although their blades are horizontal, they have palisade parenchyma on the lower side : according to Vesque[1] and Giltay[2] this must be attributed to the intense light reflected from the sand.

Thorny structures occur in Hippophaë, thus rendering its bushes almost impenetrable, also in Eryngium and Ononis.

The leaves of many plants present are closely applied to the soil, and many species spread all their shoots horizontally over the sand, presumably because of the prevailing thermal conditions.[3]

Succulent plants are represented only by a few species—including Sedum acre.

As a means of defence against the mechanical action of wind, the sea-marram has a power of turning its leaves in an arched fashion with

---

[1] Vesque, 1882.   [2] Giltay, 1886.   [3] See p. 26.

the dorsal side opposed to the wind ; this firm, smooth, and glossy dorsal side is provided with hypodermal sclerenchyma.

Efficient protection against the force of the wind is supplied by the large *leaf-sheaths* that for a long time enclose the inflorescences of the sea-marram, Elymus arenarius, and Weingaertneria and other grasses.

Deep-growing, only slightly branched, *roots*, which not only prevent the plant from being dislodged, but also raise water from the depths when the surface is dry, are possessed by many species, including Psamma arenaria, Elymus arenarius, Carex arenaria (with two kinds of roots),[1] and Eryngium.

The root-hairs function for a long time ; grains of sand cling firmly on to the roots of the sea-marram, lyme-grass, Koeleria glauca, and others, and sometimes encase the roots in firm tubes.

Dune-Heath.

On many grey dunes and on old-established sand-fields there develop various dwarf-shrubs, but particularly Empetrum nigrum and Calluna vulgaris, also Genista anglica, Rosa pimpinellaefolia, and small specimens of Salix repens. Among these Calluna vulgaris may attain an enormous distribution, so that grey dune is changed into dune-heath—*callunetum*—in which ordinary heath-plants are intermingled with dune-plants. In such heath hard pan does not seem to occur, and raw humus is scarcely developed.[2]

Dune-Bushland.

Many old dunes in northern Europe become occupied by taller shrubs, so that bushy vegetation arises on them. Hippophaë rhamnoides is especially liable to appear in these circumstances. On certain dunes on the eastern and southern coasts of the North Sea there are extensive, thorny, dense, impenetrable thickets of Hippophaë—*hippophaëta*. The xerophytic nature of this shrub is expressed in the narrow leaves and their dense coating of scaly hairs, and in the thorns. In the shade of these shrubs there flourishes a flora that somewhat partakes of the character of the ground-flora of forest.

Salicetum also occurs and is composed of Salix repens, together with rose-bushes, Sarothamnus scoparius, and Populus tremula ; but it is of limited distribution.

Dune-Forest.

In Denmark artificial forests occur on dunes, and include Pinus montana, Picea alba, and, in older plantations, Pinus sylvestris, P. austriaca, Picea excelsa, and others. But on the Baltic coasts natural pine-forests thrive on dunes.[3]

Other Examples and Distribution of Psammophilous Vegetation.

Psammophilous vegetation showing the same and different defences against transpiration occurs *in other parts of the Earth*, and includes many species, growth-forms, and formations which are unfamiliar to us and

---

[1] Buchenau, 1890 ; Warming, 1891.   [2] See Chap. LII, pp. 212-3.
[3] Concerning bushland and forest on dunes in northern Europe, see Warming, 1891, 1907-9 ; and Abromeit in Gerhardt, 1900.

have not been adequately investigated. To classify these other types of psammophilous vegetation is impossible at present.

Europe. There is considerable floristic agreement in the floras of the *coasts of northern Europe.* The dunes of similar floristic character also occur in Arctic regions, for instance in the east of Greenland, in Iceland,[1] and on the White Sea ;[2] but Psamma is absent, and is replaced by Elymus arenarius, Juncaceae, and other species.

Similar types of vegetation also occur at a distance from the coast, for instance in the interior of *North Germany.* The production of dune in Denmark and Germany takes place in heather-districts and on the coasts, where strong winds are hostile to tree-growth.

Borbas and Kerner[3] have described the similar vegetation of the sandy soil of the *Hungarian* plains. Here, in accordance with the loose texture of the soil, one finds the same long roots and rhizomes, metres in length (for instance in Festuca vaginata, which here plays the part of the sea-marram), and the same defence against transpiration ; tuberous subterranean organs are also alleged to occur here. Adamovicz[4] has given a description of dunes in *Servia.* Here the first colonists are the annual Polygonum arenarium and Veronica triphyllos ; which are succeeded by Medicago minima, species of Bromus, Viola tricolor, and others ; subsequently biennial and perennial rosette-herbs are found. In the second year the plants to appear are all perennials, and various associations (Festuca-association, Euphorbia-association, and others) develop. The sand-dune may pass over into sand-puszta.

Lacustral dunes in *Switzerland* exhibiting various types of oecological adaptation have been described by Chodat.[5]

The dunes along the *French Mediterranean coasts* are low and insignificant, and include as sand-fixing grasses Psamma arenaria, Cynodon Dactylon, and others. On old dunes the flora differs from that of northern dunes in being much richer, in species, and apparently in including more numerous grey-haired species. On the delta of the Rhone the dunes are clothed with nearly impenetrable scented *maqui*, whose bushes include Juniperus phoenicea (which attains a height of 6–8 metres), Pistacia Lentiscus, Phillyrea angustifolia, Tamarix gallica, and Ruscus aculeatus.[6] On old dunes forests of Pinus often appear.

Africa. In Africa vast tracts of sand occur both on the coast and in the interior. True shifting sand-dunes are encountered in the Sahara,[7] and thence as far as Syria. The vegetation is exposed by day to the most scorching heat, and by night to a considerable degree of cold. Here the dry season is very prolonged and the vegetative season short ; and the plants must be fitted to withstand the former, and utilize the latter.[8] Among characteristic plants may be mentioned the grass Aristida pungens, the polygonaceous Calligonum comosum, Herniaria fruticosa, Ephedra alata, species of Limoniastrum, and others, of which some reappear on sandy deserts in Asia.[9] Kotschy describes the vast brownish-yellow expanse of sand stretching east of Suez ; here the dunes owe their origin

[1] H. Jonsson, 1905.   [2] Pohle, 1903.   [3] Kerner, 1863.
[4] Adamovicz, 1904.   [5] Chodat, 1902.
[6] Flahault, 1893 ; Flahault et Combres, 1894.
[7] For illustration see Schirmer, 1893 ; Massart, 1898; Flahault, 1906 *b.*
[8] See p. 275.   [9] Massart, 1898 *a.*

partly to Nitraria tridentata. In German south-west Africa there are, here and there, impenetrable bushlands clothing the dune-hillocks of the shore with plant-types which either assume the Erica-, myrtle-, or oleander-form, but belong to entirely different families, or are thickly coated with woolly hairs, or are fitted in some other way to avoid rapid transpiration. One very remarkable shrub in the African dunes is the cucurbitaceous Acanthosicyos horrida which attains man's height. It lacks leaves ; but paired thorns are crowded so densely and in such a manner on the felted branches that impenetrable bush arises similar to that formed in Europe by Hippophaë. The roots may attain a length of 15 metres or more, and the thickness of the human arm ; they descend as far as the ground-water. Wind causes sand to accumulate round the plants, whose shoots grow with the rising sand and repeatedly emerge from this, just as is the case with the sea-marram in Europe.[1]

Asia. In regard to tracts of sand in *Asia* it may be added that on sand-dunes in the Kirghiz Steppe, Pinus, Betula, Populus, Salix, and Ulmus grow in company. The most efficient sand-fixing plants of the Transcaspian steppes are Carex physodes, Aristida pungens, and a number of shrubs. On sandy soil we find bluish-green and aphyllous species of Calligonum, Ephedra, and of the papilionaceous Ammodendron, also the remarkable saxaul-tree (Haloxylon Ammodendron) which almost gives rise to forest.[2]

North America. In *North America* the phenomena of dune-production in a forest country have been described by Cowles,[3] in connexion with his observations on dunes bounding the shores of *Lake Michigan.* Here first rank as a dune-producer is taken by Psamma arenaria, and second rank by Agropyron dasystachyum, Elymus canadensis, Calamagrostis longifolia, Salix adenophylla, S. glaucophylla, Prunus pumila, and Populus monilifera. When one of the willow-shrubs is buried by the sand, roots shoot forth from the buried branches. Populus monilifera and P. balsamifera also germinate on the shore and sometimes give rise to groups of trees sufficiently close to cause an accumulation of sand. As the dunes grow the conditions of life become unfavourable to the dune-producing plants. The dune becomes high and dry and the length of life of the plants is not indefinite, so that they die and the sand is blown away. Even after death Calamagrostis and Psamma can act as efficient sand-fixing agents, whereas other species, such as those of Populus, cannot act in this manner. On the lee side willows, poplars, Vitis cordifolia, grasses, and other dune-plants grow, whereas on the windward side these cannot establish themselves, as not one of them is able to fix the rapidly shifting dunes. Not until the distance from the shore is greater, the wind weaker, and protection is provided by other dunes, does the dune commence to be clothed with vegetation. The first plants to appear on the lee side of the slowly moving dunes are Psamma arenaria, then Asclepias Cornuti, Equisetum hiemale, and Calamagrostis longifolia. In the course of a few years the lee side becomes overgrown by the shrubs and trees, Cornus stolonifera, Salix adenophylla and S. glaucophylla, Vitis cordifolia, Prunus virginiana, and Tilia americana. Shrubs suppress the herbs ; lime-trees grow and

[1] Marloth, 1888.
[2] See the Sections on Halophytes (Chapter LXVII), Deserts, and Steppes.
[3] Cowles, 1899.

produce a forest, where poplar, oak, ash, walnut, sassafras, and other trees occur, and are festooned with numerous lianes, such as Celastrus scandens, Vitis cordifolia, Rhus Toxicodendron, Ampelopsis quinquefolia, and Smilax hispida. Shrubs are frequent in open places and at the margin of the forest. In addition to the original shrubs many other species settle. The majority of the species in the dune-forest are definitely mesophytes. On the windward side annual and biennial herbs are the first to germinate, and among them Corispermum hyssopifolium leads the way. In the course of time small shrubs arise and produce scrub[1] composed of Arctostaphylos Uva-ursi, also Juniperus Sabina and J. communis. Scrub in turn gives way to coniferous forest composed of Pinus Strobus, P. Banksiana, P. resinosa, Thuya occidentalis, Abies balsamea, and Juniperus virginiana. Beneath the trees grow shrubs belonging to the dune-scrub. In the shade many mosses thrive. On dunes lining the shores of Lake Michigan bushland composed of Quercus coccinea partly replaces coniferous forest.

In Virginia and North Carolina, according to Kearney,[2] the earliest sand-fixing species are Psamma arenaria, Uniola paniculata, Panicum amarum, and Iva imbricata, which belongs to the Compositae. Farther from the beach the dunes are covered by grassland composed of Psamma and Panicum, between which grow shrubs of Myrica carolinensis, Quercus virginiana, and Rhus copallina. The oldest dunes are clothed with pine-forest.

The dunes of Nebraska are occupied by the following sand-binding grasses : Calamovilfa longifolia, Redfieldia flexuosa, Eragrostis tenuis, Muehlenbergia pungens, and many others.[3]

In sandy parts of *northern Mexico* (in the Tularosa desert) there occur dunes in which Yucca radiosa is the most important sand-fixing plant. Its roots extend horizontally for a distance of forty feet from the shoot. Two grasses (Andropogon and Sporobolus), a few shrubs and undershrubs, and many annual herbs, comprise the remainder of the vegetation.[4] In the Tularosa desert the shifting sand is not composed of silica, but of gypsum.[5]

South America. In *South America* psammophilous vegetation occurs on the coast, but there are also enormous tracts of sand and immense dunes in the interior of Chile,[6] and of the Argentina.[7] In the latter there grow not only a number of kinds of grasses, including Cenchrus, Diachyrium, and Bouteloua, but also other plants which, for the most part, are apparently aphyllous and include the zygophyllaceous Bulnesia Retamo, a true psammophyte which often sets a limit to the advance of the sand, species of Ephedra and Cassia, Mimosa ephedroides, and the boraginaceous Cortesia cuneifolia. In the Argentine Andes Fries[8] distinguishes a Lampaya-association and a Patagonium arenicola-association.[9]

---

[1] Which Cowles terms ' heath '.   [2] Kearney, 1901.
[3] Rydberg, 1895.   [4] Coville and MacDougal, 1903.
[5] In regard to North American dunes and their oecology, special reference should be made to the works of Rydberg, 1895 ; Coville, 1893 ; Cowles, 1899 ; Kearney, 1901 ; Pound and Clements, 1898 ; Hitchcock,1904 ; and Harshberger, 1900, 1902.
[6] F. Albert, 1900.   [7] Brackebusch, 1893.   [8] R. E. Fries, 1905.
[9] Respecting the oecology of dunes in *New Zealand* the works of Diels, 1896, and Cockayne, 1904, should be consulted.

The foregoing remarks render it clear that shifting dunes in all lands show a strong general likeness, and that similar sand-fixing plants always occur. But the final result of the fixation varies very widely. In each country there arises a xerophytic *vegetation* that does not essentially differ from that occurring in other dry and sandy localities in the same country.

Dunes that have naturally been clothed with forest occur all the world over, and in many places forest must have been compelled to contend keenly against shifting sand.

Here we must recall to mind the already-mentioned [1] *tropical littoral psammophilous forest,*—Barringtonia-association in eastern Asia, Coccoloba-association in the West Indies, 'restinga' on the shores of Brazil, and other littoral forest on sandy soil. It has already been pointed out that it is difficult to decide whether the forests owe their xerophytic character mainly to the qualities of a sandy soil or to the salinity of this and to the proximity of the sea.

[1] See p. 229.

SECTION XI

CLASS IX. EREMOPHYTES. FORMATIONS ON DESERT AND STEPPE

CHAPTER LXXI. OECOLOGICAL FACTORS. FORMATIONS

IN regions *where the atmospheric precipitations are less than a certain limit,* the *vegetation acquires a xerophytic structure,* excepting in spots where the soil is rendered wet by flowing or subterranean water.   Loam is there as dry as sand or stony soil, for all kinds of soil alike are inadequately supplied with moisture by rain.   The soil varies greatly in nature, but apparently is always very rich in nutriment, as the salts that arise by weathering of the soil are only very incompletely washed out.   Yet dryness of the soil checks absorption of nutriment by the roots ; and this necessarily reacts upon the dimensions and numbers of the plants present.   The abundance of nutriment may be such that the soil is supersaturated with salt, and the amount present increases with increasing aridity of climate.   The various kinds of soil behave differently in this connexion. Salts are readily washed out of sandy soil, so that, other conditions remaining constant, sand is poorer in salts than is clay, which is relatively impervious to water.   Also the form of the terrain is responsible for distinctions.   Slopes of hills are more completely elutriated ; on the other hand water collects in depressions, and, after evaporation has taken place, leaves behind deposits of salt.   In this way steppes and deserts may display salt-impregnated spots, where incrustations of salt carpet the soil.[1]   Excepting in some grassy steppes, the *soil is utterly devoid of humus ;* for sun and drought burn it up completely.   The air is very dry, and the variations in its temperature are very great.   Frosts occur in deserts even of low latitude.   In summer the temperature may exceed 50° C.   In the interior of Arabia in the winter of 1893, according to Nolde, at night minimal temperatures of −5° to −10° C. were reached, while by day the thermometer rose to more than 25° C. The soil during daytime becomes exceedingly hot, and its temperature in the Sahara rises above 70° C.   Similar high temperatures in the soil have been observed in tropical savannah ; for instance one of 85° C. was registered at Chinchosho in West Africa.   In dry regions mountain-slopes are favoured with greater atmospheric humidity and heavier rainfall than are the plains, because they experience an alternation of an ascending wind at daytime, and a descending one at night-time.   Especially favoured in this respect are slopes exposed to the prevailing wind—here mesophytic vegetation may reign—whereas on lee-slopes and plains dry steppe prevails.   Mountains are of additional significance in relation

[1] See p. 219.

to arid lowlands, inasmuch as they are the sources of rivers, which may descend more or less into the lower-lying land and provide this with permanent ground-water, even after it has become parched nearer the surface.

According to the minimum moisture with which the vegetation can content itself, the following formations may be distinguished :—

1. **Desert** prevails in very dry places, and its soil shows very sparse vegetation.

2. **Shrub-steppe** occurs where the rainfall is very scanty, but nevertheless regular.

3. **Grass-steppe** is found where the rain is moderate in amount, but falls only in a few days in the year; grass-steppe, as a rule, can be utilized for cultivation without artificial irrigation.

A transition between grass-steppe and forest is provided by the bushland that Adamovicz [1] terms **'sibljak'.**

Vegetation growing on saline soil has already been described in connexion with halophytic vegetation (Section VII).

CHAPTER LXXII. DESERT

WHERE rain falls at quite irregular times, and is continuously lacking for a long time during the season otherwise most favourable to plant-growth, only the scantiest vegetation can exist. Here and there are dotted little greyish-green plants which stand wide apart, while intervening large stretches of soil are utterly devoid of vegetation. Only when a shower of rain casually moistens the soil do numerous annual herbs germinate, and then merely bear fruit and die a few weeks later.

On the desert of Atacama in South America rain scarcely ever falls. At Walfisch Bay on the coast of South Africa the mean annual rainfall, which descends on only six days in the year, is seven millimetres. In regard to the rainfall in several stations on the northern fringe of the Sahara statistics have been published. Here rain falls in winter; but even at this season several months may pass with only a single day's rain, or possibly without any.

In all deserts there are, however, more favoured localities, whether these be mountains, where the rainfall is heavier, or river-beds, where water flows after falls of rain and where the subsoil may remain moist for a longer period.

The soil in desert is by no means uniform.[2]

There are pure sandy regions—*vast stretches of sand and dunes*—which are gradually transformed by the wind. The vegetation of such places has been discussed in the preceding section.

In other regions of the world, such as parts of North Africa, the soil is practically solid *rock ;* the lithophilous vegetation on this has been dealt with in Section VIII.

In yet other regions the combined action of sun and wind have disintegrated rock into *stones* and *gravel.* In Egypt one encounters ' silica-

---

[1] Adamovicz, 1902.
[2] See Schirmer, 1893; Hochreutiner, 1904; Massart, 1898; Flahault, 1906 *b.*

deserts' ('serir'), where rounded, blackish-brown, resonant, siliceous pebbles clothe the essentially sandy expanses and stand out as dark objects from the reddish-yellow desert-sand; gravel-steppes[1] occur in Algeria, and extensive stony plateaux (which are termed 'hammada' by the natives), carpeted with sharp flints and calcareous stones, form the greatest part of the Sahara; again, in the upper terraces of the Karroo in Cape Colony one encounters waterless stony desert, and, in the Kalahari, 'stone-fields.' Finally, there are deserts, for instance on the Mexican Plateau, whose soil consists of a firm, reddish clay which is rich in stones: in the dry season the clay becomes as firm and hard as rock, and consequently fissured, so that it may almost be regarded as equivalent to a rock substratum.

EGYPTO-ARABIAN DESERT

As a type we may select the Egypto-Arabian desert described by Volkens.[2] This includes rocky, gravelly, and sandy deserts, where eight or nine months often elapse before a drop of rain falls. Rain descends almost exclusively in winter, from December to April. Nowhere else has the atmosphere been observed in daytime to be drier than in North Africa, where the relative humidity may be 10–25 per cent.; yet at night the temperature may sink very considerably, often below 0° C., and there may be a rich fall of dew, which is the sole atmospheric source of water during the prolonged dry season. During this season the temperature of the air may exceed 50° C., and at daytime the soil is considerably hotter than the atmosphere. In general, too, there is not a breath of wind, particularly in the valleys.

The vegetation during the dry season presents the following appearance. Most of the plants are greyish-white or dirty green stumpy shrubs, which sometimes attain a height of three feet, and are rounded and hemispherical in form: some of the plants are low, mainly prostrate, caespitose herbs: rarely do there occur herbs that twine or are provided with larger long-lived leaves.

Scarcely have the first showers of rain fallen, about the commencement of February, before the shrubs shoot forth foliage and soon burst into blossom; seeds of numerous ephemeral species (Odontospermum pygmaeum, the 'rose of Jericho', for instance) germinate, thus initiating an active life that will last for only one or two months, and some few juicy and therefore longer-lived annual species (of Mesembryanthemum for instance)[3] develop also; there emerges thereafter a crowd of bulbous plants, whose shoots and flowers had already been initiated and were only awaiting the rain to reach full development. In addition there are many other perennial herbs with hypogeous shoots, most of them possessing a multicipital primary root; many of these have rosulous shoots and spread their leaves flat over the ground.

Among the ephemeral and annual species there is but little structural adaptation to the dry climate; for the active life is passed under favourable circumstances, and the adaptive feature is brevity of existence. But in all other species structural adaptation reveals itself. The construction of succulent and bulbous plants has been described in Chapter XXXI,

---

[1] Trabut's 'Steppes rocailleux'; Flahault, 1906 b.   [2] Volkens, 1887.
[3] See p.121.

while aqueous tissue and water-storing hairs have been dealt with on p. 120. The *hammadas* almost exclusively bear small shrubs whose leaves and stems are clothed with felted hairs. The leaves of the grasses present are short, stiff, of the rolled type, and poor in sap. Many shrubs have aphyllous shoots, or shoots showing only scale-like leaves, as in Tamarix, Ephedra, Polygonum equisetiforme ; many leaves are changed into thorns.[1] Conversion of the inner part of the bark into mucilage is frequent, and in Halimodendron a like conversion even affects the pith. Assimilatory tissue persists for a long time. After the original assimilatory cells have disappeared chlorophyll appears in the secondary cortex, and may be still seen in quite old branches.[2]

SAHARA

In regard to the Sahara the works of Schirmer[3], Massart[4], Hochreutiner[5], and Flahault[6] should be consulted. Its nature is the same as that of the desert just described, as Volkens has indicated.

SOUTH AFRICA

Like North Africa, South Africa has gravelly, sandy, and other deserts, including the Kalahari, Karroo, and Great Namaqualand, which, however, are not so poor in vegetation. Here many remarkable growth-types appear. Among them is Welwitschia mirabilis (Tumboa Bainesii), which was discovered in Damaraland by Welwitsch and Baines. On a parched arid plain these naturalists found, in addition to a little grass, only this species, which spreads its two gigantic leaves over the dry ground, sends its roots down to the depths, and can vegetate without intermission throughout the year, remaining active despite cold or drought.

The coast of German South-West Africa is an almost rainless desert. On it Euphorbia-steppe may appear, like an oasis, mainly where there is a permanent supply of subterranean water.

Many plants in South African steppes have epigeous tubers so closely resembling the stones among which they grow that during the dry season, when they are leafless, it is almost impossible to distinguish them from the stones ; Wallace regards this as a case of mimicry.[7]

In South Africa one finds numbers of bulbous and tuberous plants, belonging to the Liliaceae, Amaryllidaceae, Oxalidaceae, and other families. There are also succulent plants, including Mesembryanthemum, Euphorbia, Aloë, and Pelargonium, which exhibit great diversity of form and are represented by numerous individuals (forming about thirty per cent. of the vegetation in certain parts of the Karroos). In addition there occur xerophytes that are poor in sap and belong to many different families, including Proteaceae, Restiaceae, and Mimosaceae (with Acacia).

EAST AFRICA

On mountainous country of German East Africa there is succulent-steppe which has been described by Volkens.[8] It is mainly formed of cactus-like species of Euphorbia, Stapelia, Sanseviera, and Kleinia. Growing between these succulent plants are shrubs, including such thorny

---

[1] See Massart, 1898. [2] Jönsson, 1902. [3] Schirmer, 1893.
[4] Massart, 1898. [5] Hochreutiner, 1904. [6] Flahault, 1906 *b*.
[7] See p. 124. [8] Volkens, 1897.

forms as Caralluma codonoides and Adenia globosa. Several kinds of shrubs have tuberous stems, from which the branches spring.

NORTH AMERICA

Many districts in the south-west of North America are true deserts or approximate to these (shrub-steppe).[1] By the establishment of the Carnegie Research Laboratory at Tucson in the desert of Arizona, and the co-operation of botanists (Coville, MacDougal, W. A. Cannon) working there, this desert will soon be the one of which most is known as regards the form, biology, and physiology of plants.

ADAPTATIONS IN DESERT

Rapidity of development. In all these deserts one notes the same astonishing rapidity of development of vegetation after the first showers of rain and the commencement of spring. Fresh, green shoots suddenly appear ; numerous and often beautiful flowers unfold on the parched shrubs or start forth from the hitherto arid soil.

Morphological features. In desert-plants one encounters a number of structural features indicative of extreme xerophily, such as diminution in size or arrest of leaves, production of aqueous tissue in stem or leaf, sunken stomata, thick-walled epidermis and strong development of cuticle, hairy coating, and other characters already described. Various protective devices are often combined in the same plant.[2]

Rolling plants in the desert. Not only in many deserts but also in the closely related steppes, into which deserts often pass insensibly, one finds certain species of plants that break loose from the soil and are buffeted hither and thither. These may be termed *rolling-plants*. Anastatica hierochuntica, commonly known as the ' rose of Jericho ', has long been placed in this category, but incorrectly so according to Volkens. In any case Odontospermum pygmaeum (Compositae) provides an example of the phenomenon, and, according to Michon and Schweinfurth, is the true ' rose of Jericho '. In South Africa there is the amaryllidaceous Brunsvigia whose infructescence, according to Bolus, is the sport of the wind, just as is the infructescence of Spinifex on East Indian dunes.[3] Finally, we may mention the crustaceous lichen, Parmelia esculenta, belonging to lithophilous vegetation in desert, which is torn from the rocks by storms, and is transported by the wind in large quantities as ' manna ', until finally deposited at some distant spot : this phenomenon is common on all the deserts extending from Central Asia to Algeria.

Hygrochasy is another character commonly exhibited by desert-plants. Stems, fruit-stalks, valves of fruits, and involucral bracts, are closely curled towards one another when dry, but open apart when moistened. By this means seeds are scattered only during the moist season. Such is the case with Anastatica hierochuntica, Lepidium spinosum, Odontospermum pygmaeum, and Ammi Visnaga. On the contrary in a moister climate many plants, including Daucus Carota, exhibit *xerochasy*.

---

[1] See Chapter LXXIII.
[2] The anatomy and morphology of desert-plants are dealt with by Volkens, 1897 ; Massart, 1898 ; Jönsson, 1902 ; Coville, 1893. The physiology of desert-plants has scarcely been touched ; but attention may be directed to papers by MacDougal, 1903, 1906, 1907 ; and Spalding, 1904.   [3] See p. 227.

CHAPTER LXXIII. SHRUB-STEPPE

MOST closely related to desert is shrub-steppe. Many shrubs and undershrubs do occur in desert. But where the external conditions are more favourable to existence, desert-shrubs congregate to form scrub ; additional species enter, and desert gives way to shrub-steppe.

According to the terminology of Schimper and some Russian authors, shrub-steppe is included within the term ' desert '. Shrub-steppe probably includes several formations, which our present limited knowledge does not permit us to diagnose and distinguish. Among such are those described in the succeeding paragraphs.

Vermuth-steppe.

In northern Turan and in the south-east of Russia there stretch boundless plains in which vermuth-shrubs (Artemisia maritima and A. frigida) are the dominant plants ; they form a belt round the deserts of Turan. Farther north and west, where the rainfall is more frequent, they are fringed by a belt of grass-steppe. Grass-steppe, shrub-steppe, and desert correspond to three stages of decreasing frequency of rainfall. Vermuth-steppe shows preference for a loam soil. This in arid regions is tolerably saline and consequently physiologically dry. In places transitional between vermuth-steppe and grass-steppe one sees the saline depressions occupied by vermuth-shrubs, while the slopes of the hills, from which the salt has been washed out, assume the garb of grass-steppe. In like manner vermuth-steppe forms a belt outside the true halophytes fringing salt-lakes. Where vermuth-steppe reigns cultivation of the soil is only possible by the aid of artificial irrigation ; yet the land is of utility to nomadic tribes in providing fodder for sheep and camels.

Grey, and to all appearance dead, vermuth-steppe stretches over boundless tracts. The artemisias display leaves coated with white hairs. Their roots, which descend to a depth of 4 metres, ensure the existence of these enduring plants, even when the scorching sun's rays exterminate nearly all other species present. Among the vermuth-shrubs grow such shrubs and perennial herbs as Alhagi camelorum, Xanthium spinosum, and Eryngium campestre. In spring-time many annual herbs and bulbous plants flower, and are already in fruit at the beginning of summer.[1]

North American Shrub-steppe.

As another example of vegetation in which scattered shrubs and under-shrubs, intermingled with herbs, play the chief part, we may mention the dry and barren land between the Rocky Mountains and Sierra Nevada. Here, according to Asa Gray, the dominant plants are species of Artemisia, woody Compositae with small capitula, and Chenopodiaceae. According to Watson there is not a spot devoid of vegetation, even in the driest season. Trees are wanting ; there is no carpet of grasses, but in its place are a few dominant species of fruticose and suffruticose plants, which apparently exterminate all other vegetation. Characteristic are the uniformly coloured, mainly grey or dark-olive tinted herbs. Most

[1] Radde, 1899 ; Nazarow, 1886.

abundant is the ' everlasting sage-bush ', Artemisia tridentata, an undershrub that extends over areas so vast that the eye can see nothing beyond :
yet it never grows so densely as to bar the way. Usually it does not
exceed 1 metre in height. Mingled with this species, on the frequently
salt soil, are Atriplex confertifolia, A. canescens, Artemisia spinescens,
Kochia prostrata, Eurotia lanata, Grayia polygaloides, and others.[1]

Composita-steppe in Cape Colony.

The summits of the table mountains of Cape Colony north of the
Karroos are clothed with steppe-vegetation in which undershrubs predominate and mainly belong to the Compositae. The most important
genera are Helichrysum, Senecio, Berkhaya, Euryops, Pentzia, and
Gazania. Common also are Leguminosae, Crassulaceae, and Scrophulariaceae. Bulbous and tuberous plants are represented by very numerous
species.[2] In Asiatic and North-American steppes Compositae are in like
manner dominant in certain places.

Succulent Steppe.

Within the cold-temperate zone succulent plants occur in considerable
numbers only on soil saturated with salts ; whereas in the subtropical
zone succulent shrubs and undershrubs, as well as succulent perennial
herbs, form an important part of the vegetation. In the Old World it is
species of Euphorbia that come to the fore, but in the New World,
Cactaceae.

North Africa. In *Morocco* there is bushland composed of species of
Euphorbia. The leading species, Euphorbia mauritanica, bears foliage
during winter, but is leafless in summer.

The lower-lying land of the *Canary Islands* is mountainous country
traversed by deep clefts whose slopes are occupied by a characteristic
steppe-vegetation. The largest shrubs belong to the cactus-like Euphorbia
canariensis, and to Kleinia neriifolia and Plocama pendula, all of which
are succulents. Among shorter shrubs, which are usually about 1 metre
in height, are several species of Euphorbia. Of these E. aphylla is
aphyllous, but the others are leafless only in summer. Intermingled with
the shrubs are xerophytic undershrubs, and above an altitude of 100
metres these are accompanied by great numbers of Crassulaceae, together
with grasses that possess rolled leaves, bulbous plants (including Urginea
and Scilla), and annual herbs.[3] The highland of *Cape Verde Isles* is
likewise Euphorbia-steppe.

North America. The rocky, dry plateaux of *Texas* and *Mexico* must
also be regarded as belonging to the same general type. But here are
found species absent from the Old World, including those of Agave,
Yucca, and, above all, Cactaceae. Cereus giganteus, the Mexican giant-
cactus, raises its mighty candelabra-like branches up to a height of
18 metres, and covers the hills to such an extent that at a distance these
seem to bristle with needles. Other Cactaceae produce short, richly
branched stems that are beset with white spines ; and yet others creep

[1] Pound and Clements, 1898–1900.   [2] Bolus, 1886.
[3] Christ, 1885 ; Vahl, 1904 *b*; Schröter, 1908.

over the ground forming entangled thickets. Many of them are supposed by the natives to be poisonous ; in any case it is an extremely laborious and painful task to extract the spines from one's skin when once they have penetrated it, especially as they are often armed with recurved hooks. Species of Opuntia, armed with red and yellow spines, grow along the roadsides and are always broken ; but each detached fragment, as it lies on the ground, strikes root and develops into a new plant. Nor are there wanting species of Agave and Yucca with their tall, dried-up inflorescences. Moreover, a feeble growth of deep-rooted grass maintains itself for months of the rainless season.

These succulent deserts in North America[1] also include many shrubs, such as the creosote bush (Covillea tridentata), and species of Ephedra, Acacia, and of Fouquieria.

Other Shrub-steppes.

North America. Bushland that probably should be placed here occurs in the north of *Mexico*, in *Texas*, and in *Arizona*. The *chaparral* occurring in these territories mainly consists of Mimoseae (including species of Acacia) and many other thorny shrubs ; in Texas it is largely formed by Prosopis juliflora, P. pubescens, Cercis and other Leguminosae, Prunus, Juglans nana, Morus, Rutaceae (with Xanthoxylum), Simarubaceae (with Castela), the zygophyllaceous Larrea mexicana, and others. According to Bray[2] two types of *chaparral* occur in western Texas, and owe their differentiation to climatic, geologic, and physiographical differences. According to the same botanist the *chaparral* country is drier than that of grass-steppe. In *chaparral* many bulbous and tuberous plants occur. In the shrub-steppe of southern *California* Parish[3] estimates that annual species form 36·5 per cent. of the flora. The shrubs are leafless here during summer.

South America. *Argentina* is also rich in dry, mostly thorny, bushland or in bush-forest. Under this heading we must include the vegetation that Grisebach[4] terms *chanar-steppe*, and Hieronymus terms 'espinar-forest'. In this vegetation the leaves are so small that the long, brown, long-thorned branches are more obvious to the eye than the leaves. The name *chanar* is derived from the leguminous chanar-shrub, Gourliea decorticans, which predominates in company with acacias and evergreen composites (including Baccharis, Tessaria, and others).

Of similar type is Lorentz's *monte-formation*,[5] in which species of Prosopis, Lippia, Acacia, and Cassia are intermingled with Cactaceae and Atriplex shrubs.

On inland Argentine dunes other types of bushland occur with species of Baccharis, Atriplex pamparum (half a metre in height), and the aphyllous, switch-stemmed shrub Heterothalamus spartioides and other Compositae.

Patagonian bushland on gravel soil is still more meagre and open, and is mainly formed by Leguminosae, Compositae, and Solanaceae.

Australia. *Scrub* occurs in central, western, and south-western parts

---

[1] Bray, 1901, 1906 ; Coville, 1893 ; MacDougal, 1903 ; Karsten und Stahl, 1903.
[2] Bray, 1901. [3] Parish, 1903. [4] Grisebach, 1872.
[5] 'Monte' signifies bushland or bush-forest.

of Australia, where drought prevails because the trade-wind has given up its moisture to the coastal mountains long before reaching these parts. These bushlands are about 3 or 4 metres in stature, and consist of entangled, often impenetrable shrubs, which are evergreen, though they are dirty green or brownish-green in tint. True thorny shrubs are not very common, though the leaves are often very narrow or divided with many stiff linear segments that terminate in spine-like points. Plants belonging to the ericoid or pinoid form are general, and especially so among Proteaceae ; other plants, namely, species of Acacia and Eucalyptus, have phyllodes or leaves with their surfaces vertical ; yet broad, stiff, rustling leaves also occur. Between the shrubs the ground is often bare, as grass and herbs are extremely scanty, or it may be clothed with a dense undergrowth of bushes. Many species, which vary with the district, compose this forbidding vegetation, for which no economic use has yet been found. The most notable families represented here are Proteaceae, Myrtaceae (including Eucalyptus, Melaleuca, Leptospermum, and others), Epacridaceae, Mimoseae (with Acacia), Myoporaceae, and others. There are special types (' associations ' or ' sub-formations ') of scrub, for instance :—

Mallee-scrub is largely composed of Eucalyptus-shrubs, whose stature is, on the average, nearly that of man. In dreary monotony this stretches like an unending sea of shrubs over the flat arid table-land, with the bare, yellow or rust-coloured soil showing between the confused maze of branches.

Mulga-scrub is mainly composed of thorny acacias, which form impenetrable associations where they grow close together.

Brigalow-scrub is largely formed by Acacia harpophylla.[1]

Scott-Elliot[2] remarks that Acacia shrub-lands ' surround the deserts in tropical and subtropical countries '. He describes how the vegetation changes when one steps from the ' ordinary tropical monsoon wood ' into the desert, how it passes over into ' isolated pioneer thorn shrubs or small trees dotted over the ground ' ; and how ' these isolated pioneers or scouts are almost invariably Acacias '. Such a ' transitional Acacia- and thorn-scrub region with a long dry season ' is encountered in many parts of Africa, Australia, and South and North America. This type of vegetation demonstrates the difficulty of distinguishing between tropical thorny bushland and subtropical shrub-steppe.

CHAPTER LXXIV. GRASS-STEPPE

TYPICAL steppes in the narrower sense of the term are *grass-steppes*, which occur in the form of extensive treeless plains clothed with grasses and other perennial herbs :

In the Old World—South Russia, Hungary, Central Asia—as **steppes**.

In North America—as **prairies.**

In South America—as **pampas.**

The vegetation is xerophilous in character, and does not form a dense,

---

[1] Michaelsen und Hartmeyer, 1907. Recently Diels (1906) has issued an excellent work dealing with South-West Australia.   [2] Scott-Elliot, 1905.

close carpet. Both these features serve to distinguish grass-steppe from the allied formation, meadow, whose close vegetation consists of fresh-green, soft-leaved, and broad-leaved grasses and other perennial herbs. On the other hand, vegetation clothing grass-steppe is closer and taller than that of desert.

There are usually two periods of rest, one caused by drought in summer, the other by cold in winter.

The soil varies widely in nature. But characteristic of steppe is the exceedingly common occurrence of loess. In places where grass-steppe bounds more open steppe, or even desert, the drifting dust is detained by the denser vegetation of grass-steppe, where it consequently comes to rest. It is in this manner, according to Richthofen, that the thick strata of loess have been deposited in northern China. But grass-steppe can itself give rise to loess, as in summer there are, between the grasses, many bare patches, from which the wind can raise the dust and deposit it elsewhere. A particular type of loess is the ' black earth ', which occurs in Russian and Turanian steppes, in the American prairies, in Morocco, and elsewhere. It abounds in humus.[1] Loess is very rich in nutritive salts, and ' black earth' is specially so.

Grass-steppe in the Old World.

South-eastern Russia.

The *steppes* of southern Russia and the *pusztas* of Hungary have vegetation that is identical in oecology and to some extent in flora.

The problem of the origin of steppes has given rise to a considerable literature ; von Baer, Ruprecht, Dokuchayev, and Tanfiljew are of opinion that they were always steppes, while Pallas and Palimpsestow believe that they developed from ruined forests.

There is also a discussion as to whether the distribution of forest and steppe is due to climate or to soil. Von Baer suggests that prolonged dryness is responsible for the lack of trees in steppe ; Middendorff's opinion was that hot, dry winds were responsible for this ; Beketow, Dokuchayev, the geologist, and Tanfiljew, the phyto-geographer, regard salinity of soil as the reason why forests have not advanced on steppe-soil. And Tanfiljew points out that forest actually does advance when the soil is gradually elutriated. On steppe-regions forest is met with especially on the ridges of watersheds, partly because these ridges are more elutriated than other spots.

But climatologists [2] all agree that the climate of steppe is in reality drier than that of forest. In particular, the frequency of rain is small. Rain falls in few but heavy downpours, so that the main mass of water runs away superficially without penetrating the soil. In regions where grass-steppe prevails tree-growth appears by no means to be absolutely excluded, but only rendered difficult by climatic causes. In Caucasia, Radde [3] noted that on the banks of rivers traversing the steppes trees in the riparian forests are stunted, and that many of their leaves dry up in summer. In these regions forest is confined to edaphically favourable localities. Here it occurs in river valleys, but likewise on the less saline

---

[1] Cp. Kostytscheff, 1890; Albert, 1907.
[2] Wöikof. Hann, and Köppen.   See Rikli, 1907 b, p. 107.   [3] Radde, 1899.

hills,[1] and rather on coarse-grained sand or gravel, into which water can more easily enter, than on heavy loam soil.[2]

Steppe has a continental climate. The summer is extremely hot and dry ; the winter very severe and long, with violent snow-storms. Spring commences late. The vegetative season lasts for only two or three months (from April to June in Europe) ; plants shoot rapidly forth from the soil, and, just like desert in the moist season, the steppe is fresh-green and shows a wealth of blossom. When summer arrives the vegetation assumes a greyish-green, faded tone, while the soil cracks and becomes dust. The moisture in autumn may again call forth verdure on the steppe ; and apart from certain species of Artemisia and other plants, autumn is specially the season of annual Chenopodiaceae and similar halophilous herbs. The snow during winter is an important source of water-supply to the vegetation.

The appearance presented by steppe also depends upon the relief of the surface, which is such as to allow the wind to play freely and violently over vast plains and thus to increase evaporation.

Growth-forms. The vegetation consists of a varied admixture of annual and perennial herbs, grasses, and undershrubs. It is easy to see that the external conditions must call into being xerophytic vegetation.

The *perennial* herbs ensure their lives by subterranean parts that are protected from complete desiccation under the soil.

Some of the plants are *spring-plants* with *bulbs or tubers ;* for instance, in the Orenburg District there is a gay show of Liliaceae (including Fritillaria, Allium, Scilla, Gagea, and Tulipa), Iris, Corydalis, and Adonis vernalis.

Many perennial herbs, which develop later, have deeper *tap-roots*, and often (particularly toward Asia) grey-haired shoots ; these include Labiatae, Cruciferae, species of Artemisia, Caryophyllaceae, Malvaceae, Papilionaceae, and many grasses which form the main mass of vegetation and impart its general tint. The last-named are *perennial tufted grasses ;* the tallest tufts consist of species of Stipa ; the leaves are narrow, stiff, and often setaceous, and often persist for months, though in a faded state.

Many *annual*, short-lived species occur here and there ; in this respect steppe contrasts not only with subglacial tracts, but also with the meso-phytic meadows and fields of north-temperate Europe.

Undershrubs occur, but shrubs and trees are wanting.

Investigation will probably show that some seeds are dispersed by wind, others by animals, as is the case in other types of vegetation.

The luxuriance and richness of this steppe vary widely with the geographical situation, and depend mainly upon the nature of the soil. On the most fertile South-Russian steppe, where soil is the ' black earth ' already mentioned, Festuca ovina and Koeleria cristata dominate, together with Medicago falcata, Thymus Serpyllum, and others. On less fertile steppe thyrsa-grass (Stipa pennata and S. capillata) is more abundant, while perennial herbs are less represented. The most sterile steppe is occupied almost solely by tall tufts of xerophytic thyrsa-grasses, and particularly by those of Stipa pennata. The degree to which the soil is

---

[1] Middendorf, 1867 ; Tanfiljew, 1905.　　　　　　　[2] Kostytscheff, 1890.

bare is shown by Cornies' interesting Plates,[1] on which are exhibited the carefully plotted and measured areas occupied by individual species.

Among phenomena belonging to this steppe may be mentioned ' rolling-plants ',[2] which are represented by Gypsophila paniculata and Rapistrum perenne. When these are dead they are dislodged by wind, become matted together into spherical, often gigantic masses, which the storm drives bounding over the plains in great leaps (hence the popular name ' steppe-witch ').

The Russian steppes[3] are continued eastwards into the grass-steppes of south-western Siberia.

In Hungary.

The Hungarian *pusztas*[4] are in general very similar to the steppes of southern Russia ; the two agree in oecology, succession of events, growth-forms present, and partly in species represented. Kerner distinguishes divers associations, for instance :—

The feather-grass (Stipa) association.

The golden-beard association, consisting of tall tufts of Andropogon Gryllus which grow *close together* and thus produce vegetation, deviating from that of typical steppe.

In Roumania and Servia.

The lowland of Roumania likewise shows a close affinity to Russian steppe.[5] On sandy soil in Servia there are steppes[6] whose vegetation is composed of xerophytic grasses, perennial herbs, bulbous and tuberous plants, shrubs, and annual herbs. The leaves are vertical or bent upwards. Many leaves are capable of becoming furled, or protect themselves from insolation by photometric movements. Other plants defend themselves by reduction of the leaf-surface. The roots are long.

In Iberia.

The dry character of Spain has brought into existence true steppes in several places. These have been described by Willkomm,[7] who has supplied the following statistics as to the share taken by different kinds of plants in the composition of the vegetation : herbs, $\frac{2}{5}$; undershrubs, $\frac{1}{4}$; grasses, $\frac{1}{9}$; shrubs, more than $\frac{1}{20}$; lichens and algae, $\frac{1}{27}$. Approximately one-sixth of the species have a vivid green colour, and the remainder show other tints.

Among interesting species in Iberian steppes we may mention the halfa-grass (esparto-grass, Stipa tenacissima) and Lygeum Spartum, which clothe wide areas of Spanish upland with their large stiff tufts and so represent the Stipa-grasses of Russia. They give rise to special associations on littoral steppe. Other Iberian steppe-grasses are Stipa parviflora and Avena filifolia.[8] According to Rikli the littoral steppes in the south-east of Spain belong to very xerophytic types, and are most

[1] Cornies, loc. cit.   [2] See p. 277.
[3] In regard to Russian steppes, consult Kusnezow's abstract of Russian literature published in Engler's *Bot. Jahrb.*, xxviii ; Tanfiljew, 1905.
[4] The name ' puszta ' signifies desert.   [5] Grecescŭ, 1898.
[6] Adamovicz, 1904.   [7] Willkomm, 1852, 1896.
[8] Rikli, 1907 *b*. In this paper figures of the leaf-anatomy are given.

closely allied to Drude's 'desert-steppes', in which the ground is very bare and shows itself on all sides.

In Asia.

On the Altai mountains there are herbaceous and grassy steppes which, with their waving thyrsa-grasses and species of Gypsophila, are similar to steppe on the 'black earth' of southern Russia.[1]

Grass-steppe (Prairie) in North America.

The North-American prairie is true steppe, and is evoked by the same physical factors: Continental climate; long severe winters with snow and minimum temperatures ranging from $-20°$ C. to $-50°$ C.; a summer, hot and dry, often rainless from the middle of July onwards, but with cold nights. The short vegetative season (May) is heralded by transitory falls of rain. Low atmospheric humidity and the occurrence of dry periods during the vegetative season cause the absence of forest and the presence of steppe. Here too, at least in certain districts, two periods of rest occur within the year ; and, here too, violent storms arise.

The prairies are vast plains on whose horizon the curvature of the earth is made manifest. The soil in the east appears to be like that in southern Russia, namely a black loess, intermingled with sand, which contains deep humus consisting of the remnants of countless preceding vegetations, and thus including a boundless store of wealth for future ages. Lesquereux suggested that the prairie is an ancient sea-bed which has been slowly dried up—the bed of the sea from which access of water has been cut off by the upheaval of the Andes in the Tertiary period. As regards supply of water, the prairies are more favourably situated than are the Asiatic steppes ; they are more thoroughly permeated by water, and are traversed by mighty rivers, with which arboreous vegetation is associated. With this exception prairies are treeless or show open tree-vegetation, which, according to Sargent, forms at most 10–20 per cent. of the vegetation clothing the ground.

Prairie is pure grassland, in which numerous Compositae (especially Heliantheae and Astereae), Leguminosae (especially Galegeae), and other perennial herbs, are intermingled with the grasses, which give to the landscape its special appearance ; 'buffalo-grass', according to Asa Gray, consists of Munroa squarrosa, Buchloë (Bulbilis) dactyloides and Bouteloua, also of many other genera, including Stipa, Andropogon, Aristida, Panicum, Hordeum, and Elymus. This 'buffalo-grass' land is clothed with a low, velvet-like, grassy carpet, which, if it be not exactly sward, is yet somewhat like it, and is green in early spring, but grey at other times. Even in winter it supplies fodder. Here is, or rather was, the home of the vast herds of bison and antelopes.[2]

In other places prairie is clothed with grasses, such as Spartina cynosuroides, Panicum capillare, and P. virgatum, which nearly attain man's height, as well as many Compositae, including Silphium and Helianthus.

---

[1] Krassnoff, 1888 ; Martyanov, 1882.   [2] Asa Gray, 1884.

Although recent investigations have enriched our information in regard to the prairies and have supplied floristic descriptions of diverse associations, very little has been done that bears on the oecology of the growth-forms present. Among the different prairies considerable differences reveal themselves. Certain eastern prairies should perhaps be properly regarded as mesophytic. Those west of Chicago, according to Cowles,[1] are rather allied to meadow, and differ entirely from steppe. Mayr [2] states that in places there is sufficient moisture for prairie to bear forest, and he is of opinion that the eastern parts were originally clad with forest which has been destroyed by prairie-fires ; and it is a fact that west winds prevail precisely at the season of prairie fires, namely in September and October.

According to Pound and Clements [3] the northern prairies are not essentially different from the southern. But in passing from west to east different belts reveal themselves, and correspond to the degree of humidity. The foot-hills at the base of the Rocky Mountains are clothed with Artemisia-steppe. To the east of these there succeeds a belt of sand-hills forming typical steppe, with ' bunch-grasses ' and some xerophytic shrubs and undershrubs. The dominant grasses are Andropogon scoparius, Aristida purpurea, A. basiramea, and Sporobolus cuspidatus. Still farther eastwards lies true ' prairie' formed by tufted grasses. Pound and Clements describe this vegetation as mesophilous, and only on the high hills as xerophytic. They mention several associations :—

Prairie-grass association, with Sporobolus asperifolius, Koeleria cristata, Eatonia obtusata, and Panicum Scribnerianum.

Buffalo-grass association, with Bulbilis on loam, and Bouteloa on sand.

Many associations are apparently to be distinguished. For example, Shantz [4] mentions associations (which he terms ' formations ') of Bouteloua oligostachya, of grama-grass, and of Muhlenbergia gracillima ; in these he recognizes smaller communities, and shows that they change in appearance with the season.[5] Tuberous plants are common on prairies, according to Thornber.[6]

In reference to the various types of western grasslands (' plains ') in Texas, reference should be made to Bray's works ; [7] but it may be mentioned that in these plains ' pure formations of prairie annuals ' occur, that an annual vegetation ' sweeps as a wave over the prairie in the early spring ', and that the annual individuals ' mass themselves in pure formations of incredible compactness and extent '.

Grass-steppe (Pampas) in South America.

The pampas represent another great grass-steppe tract. The name is a Quichua term, and signifies ' grass-clothed, completely treeless, level plains '.[8] The pampas occupy the vast, rockless, alluvial, South-American plains that stretch from the Atlantic coast to the Andes, and from Patagonia to the forests of Paraguay and Brazil. The boundless, level or

[1] Cowles, 1901 *b*. [2] Mayr, 1890. [3] Pounds and Clements, 1898.
[4] Shantz, 1906. [5] Concerning prairies in Kansas, see Hitchcock, 1898.
[6] Thornber, 1901. [7] Bray, 1901. [8] Brackebusch, 1893.

somewhat undulating, uniform treeless surface is clothed with perennial grasses and herbs, like ' a shoreless sea of grasses on whose horizon the eye finds no resting point, save where the sun rises and sinks '.[1] The genera represented are Melica, Stipa, Aristida, Andropogon, Pappophorum, Panicum and Paspalum. Between the grasses grow numbers of herbs belonging to many families : these include Verbena, Portulaca, Apocynaceae, Compositae, Eryngium, and others. Curiously enough, there are very numerous European species, which have succeeded in exterminating the inland vegetation for miles, and include not only such thistle-like Compositae as Cynara Cardunculus, Silybum Marianum, and Lappa, but also Lolium perenne, Hordeum murinum, H. secalinum, Medicago denticulata, and Foeniculum capillaceum. In the Flora of Buenos Ayres, according to Otto Kunze, at least three-quarters of the species are introduced, and largely Mediterranean in source. Floristic differences present themselves in the pampas, sometimes in the form of associations ; according to F. Kurtz,[2] different dominant species give rise to—

Verbena-pampa,
Junquillo-pampa (with Sporobolus arundinaceus),
Tupa-pampa (with Panicum patagonicum),
Zamba-pampa,
Chinata-pampa, and so forth.

West of the Parana, consequently in more continental districts, the likeness to Russian steppe is obviously greatest, as the grasses are taller and more stiff-leaved, and, as in the latter steppe, grow in tufts between which the soil is bare.

The soil is mostly a sand loess, which in many places is saline. The climate is like that of steppe and prairie. Rain may be withheld for long periods; whereupon the soil becomes a dry mass impervious to water, and downpours of rain are thus useless, as they flow away over it. Storms blow over the plains without hindrance. Yet the climate shows distinctive features ; there is neither a severe winter nor any long-lasting covering of snow ; moreover, dew is abundant. Consequently the grassy vegetation remains green for a long time, even throughout winter in certain districts.

Growth-forms. The oecology of the constituent species has not been thoroughly investigated. The number of *annual* species is very small.

Bulbous plants are represented in small numbers.

Tree-growth is not excluded, and in this respect an additional resemblance to grass-steppe reveals itself ; trees may be successfully planted even where there is no flowing water. C. Darwin[3] accordingly sought for some *geological cause* of the absence of trees. Köppen[4] was the first to solve the problem. He pointed out that here it is not the absolute amount, but the frequency, of rainfall which is the determinant factor. The rainfall in the pampas is not small (being 40–100 centimetres annually), but days with rain are very few in number, so that each month experiences periods of drought. The rainiest month seldom includes more than 5–10 days of rain on the average.

[1] See Grisebach, 1872.   [2] F. Kurtz, 1893.
[3] C. Darwin, 1845.   [4] Köppen, 1900.

Peruvian Loma.

An entirely peculiar type of grass-steppe is represented by the *loma* on the coasts of Peru.[1] The vegetative season is the short period in winter during which dense mists moisten the soil. The vegetation consists almost exclusively of bulbous and tuberous plants, in addition to annual herbs, and includes Amaryllidaceae, Liliaceae, species of Malva and Erodium, Convolvulaceae, species of Calceolaria, and Loasaceae. In many years the mist is withheld, and seeds do not germinate. In winter the *loma* is like a mountain-meadow ; in summer, like a desert.

CHAPTER LXXV. SIBLJAK

THE term *sibljak* is employed by Adamovicz[2] to designate bushland which is formed by photophilous, thermophilous shrubs, and is characteristic of the boundaries of steppe and forest. The *sibljak*-shrubs, namely, species of Cytisus, Prunus Chamaecerasus, Paliurus, Juniperus, and others, never grow in forest. Between them a number of steppe-plants grow as subordinate constituents. Just as *sibljak* is sharply delimited from forest, so in like manner it shows no transitional forms connecting it with the Mediterranean *maqui*. It is deciduous, whereas the latter is evergreen. *Sibljak* is encountered in the lower-lying tracts of the Balkan peninsula. Apparently it has often arisen on disforested soil, yet it must be regarded as a natural formation, which, although acquiring a wider distribution by the destruction of forest, does not owe its origin to this.

Similar bushlands occur in Roumania in the south of the Dobruja district.[3] And in Hungary this type of bushland occurs,[4] even as far west as the vicinity of Vienna,[5] also in Bohemia.[6]

In Caucasia, according to Radde,[7] bushland composed of Paliurus aculeatus attains a considerable distribution. In this bushland plants belonging to steppe and forest grow in confused array. Bushland composed of species of Glycyrrhiza occurring in Transcaucasia should probably be regarded as belonging to the same general type, as should also bushlands of deciduous shrubs in Aragon.[8]

Along river-banks woodland from adjoining forest-regions extends far into desert and steppe, as *fringing (riparian) forest*. In the Sahara and Kalahari, strips of forest composed of various thorny species of Acacia are formed. These forests are to be regarded as allied to thorn-forest in savannah.[9] In colder countries, for instance, in Turan and North America, deciduous trees form similar fringing forests ; they are, for the most part, composed of willows and poplars.

[1] Benrath, 1904.   [2] Adamovicz, 1902.   [3] Grecescu, 1898; Vahl, 1907.
[4] Pax, 1896.   [5] Beck von Mannagetta, 1890–3.   [6] Domin, 1905 a.
[7] Radde, 1899.   [8] Wilkomm, 1896.   [9] See p. 294.

SECTION XII

CLASS VIII. CHERSOPHYTES. FORMATIONS ON WASTE LAND[1]

CHAPTER LXXVI. WASTE HERBAGE

In many regions where the rainfall and atmospheric humidity are sufficient for the existence of forest or shrub-land, there may occur on particular *dry kinds of soil*, such as limestone rocks, stiff clay, and so forth, communities of xerophytic perennial herbs. These often have arisen as a consequence of the destruction of forest and shrub-land, yet they must often be regarded as natural formations. Despite many points of resemblance to steppe they must be distinguished from this. The more prolonged vegetative season and greater atmospheric humidity cause their vegetation to be composed of species and growth-forms foreign to true steppe. The communities in question are, as a rule, also much poorer in bulbous and tuberous plants than is steppe. As examples we may mention the following :—

Festuca valesiaca meadow on the Alps. Most steppe-like is that consisting of Festuca valesiaca, which occurs in dry Valais on sunny, arid, places where the soil is shallow. The dominant species, Festuca valesiaca, forms small cushions composed of numerous densely packed shoots, which display many capillary, folded, glaucous radical leaves and stiff haulms 10–30 centimetres in height. Between the grasses grow various perennial herbs, including bulbous plants (Gagea saxatilis and Muscari comosum). Among the grasses rolled leaves are common, and are possessed by Festuca valesiaca, Koeleria valesiaca, and Stipa pennata ; and some grasses, including Poa bulbosa and P. concinna, have tubers which serve as reservoirs for water or nutriment. The commonest devices adopted to guard against desiccation are the production of hairs (in Oxytropis Halleri and Artemisia valesiaca), succulence (in Sempervivum arachnoideum), and diminution of the assimilatory surface (in Onobrychis arenaria and Plantago serpentina).

Brome meadow on the Alps. Less xerophytic is brome meadow in which Bromus erectus dominates. It occurs particularly on dry sunny sites on calcareous soil. Among the companion-plants we may mention Galium mollugo, Festuca rubra, F. ovina and F. pratensis, Arrhenatherum elatius, Carex montana and C. verna, Prunella vulgaris, Salvia pratensis, and others.[2]

In Montenegro. Belonging to this category we must probably regard Montenegrin expanses of poor grassland occurring on stony ground, and thus providing a transition to fell-heath which is described in the succeeding chapter. Hassert[3] says of them that they clothe extensive

---

[1] See p. 136.   [2] Stebler und Schröter, 1889, 1892.   [3] Hassert, 1885.

tracts in the Banjani, and are readily distinguished from the lush meadows on the belt of schists, in that, owing to the lack of a stratum of earth, they form no continuous grassy sward but are traversed by weathering limestone ridges. True it is that at the time of spring-rains they show up in full green and are richly mottled with a gay show of blossom. Yet the sun's rays soon melt the last remnants of the winter-snow, the fissured limestone swallows up not only the summer's rain but also part of its own subterranean water; and in July one sees merely parched grassland which is trampled by cattle and yields only a very mediocre crop of pallid grass.

In **Eastern Germany, Hungary, Western Russia.** There is a community in these regions belonging to the same type, namely, that of the *sunny, Pontic (Pannonian) hills*, which, on account of its warm nature, is largely utilized for viticulture. In appearance it does not differ from many steppes in south-eastern Europe. In spring the slopes and undulating plains often display a wealth of blossom, and in many places Adonis vernalis plays a leading part. In addition species of Peucedanum, Dianthus Carthusianorum, Tunica prolifera, Scorzonera purpurea, and above all Stipa pennata and S. capillata, are characteristic plants.[1]

Alvar-vegetation of Sweden. In the interior of the island Öland is a plateau composed of Silurian limestone. And here is found characteristic vegetation composed of xerophytic perennial herbs and undershrubs. The first peculiarity that strikes the eye is the dwarfed growth of all the component plants. All the species are smaller here than they are elsewhere, and are represented by forms that are remarkable for the small size of their leaf-blades. The grasses, including Festuca ovina and F. oelandica, have rolled leaves, so likewise have some other common species, including Potentilla fruticosa. Many species are represented by very hairy varieties, such as Plantago lanceolata var. dubia and P. maritima var. gentilis, Medicago lupulina; Festuca ovina var. glauca is protected by a coating of wax. The leaves are, as a rule, directed sharply upwards and more or less isolateral; they also have a thick-walled epidermis and small intercellular spaces. But the branches trail over the ground in Artemisia campestris, A. rupestris, Herniaria glabra, Thymus Serpyllum, and several other species. Annual herbs are frequent. Bulbous plants are represented by only a solitary species, Allium Schoenoprasum. There is in fact a strong resemblance between this vegetation and that of steppe. But the moist air is responsible for the appearance, between the forms that are characteristic of steppe, of other growth-forms, such as mosses and the evergreen undershrub Calluna vulgaris, which are as different as possible from the forms prevailing on steppe.[2] Similar *alvar*-vegetation occurs also on Gotland and in Götland.[3]

In Spain. In Spain dry grassy wastes are stated by Willkomm[4] to occur. But it is not clear whether or no these belong to steppe.

In Madeira. On the highland of Madeira on dry shallow soil there occurs a peculiar type of waste herbage which is colonized almost exclu-

---

[1] See Gräbner, 1895, 1901; Domin, 1905; Vierhapper und Handel-Mazetti, 1905.
[2] J. Erikson, 1895; Hemmendorff, 1897; Grevillius, 1897; Witte, 1906.
[3] Sernander, 1908.   [4] Willkomm, 1896. See also Rikli, 1907.

sively by monocarpic herbs.   The dominant species are Airopsis praecox,
A. caryophyllea, Teesdalia nudicaulis, Lotus angustissimus, Erodium
Botrys, Galium parisiense, and others.   Of perennial herbs only one is
common—the bracken fern.   Not only this but also the monocarpic herbs
are dwarfed in growth.   The prevailing nanism and the domination of
monocarpic herbs are adaptations to the rainless summer and intense
atmospheric aridity.[1]

CHAPTER LXXVII.   BUSHLAND ON DRY SOIL

ALLIED to waste herbage is the formation represented by various types
of bushland that are more or less xerophytic, and occur in climatically
moist districts on dry soil, or have arisen as a consequence of the destruction of forest.   Of these, several kinds are denoted in the succeeding
paragraphs.

Hippophaetum is common, for example, in north-west Jutland,
grows mostly on sandy soil, and consists of low, spinose, tangled shrubs,
which are from half to a metre in height and possess matt leaves that
are silvery-grey with scale-like hairs.   The dense growth of the association
is due to the abundant production of root-suckers.[2]

Similar **thorn-bushland** occurs in southern Norway on Silurian soil,
in Sweden and Central Germany on sunny, stony spots.   It usually consists
of Prunus spinosa as well as Berberis, Crataegus, Rosa, and Rubus, or
mainly of Juniperus communis, or, for instance in Scotland, of Ulex
europaeus.   Bushland composed of Juniperus is encountered in many
other places, for instance, as J. excelsa, in the Transcaspian mountain-
flora.

Garide.   In the valley of the Rhone and among the Jura mountains there is a similar type of bushland to which the name *garide*
has been applied by Chodat.[3]   It is essentially composed of deciduous
shrubs, including Prunus Mahaleb, Berberis vulgaris, Ligustrum
vulgare, Amelanchier vulgaris, and Crataegus Oxyacantha.   On the
Jura mountains bushland composed of Buxus sempervirens occurs.[4]
Baumberger terms this type of bushland in the Jura mountains ' fell-
heath '.

Among other bushlands whose oecology has been so little investigated that it is impossible to decide whether they should be regarded as
belonging to the formation under discussion, or to steppe or sclerophyllous bushland, is :—

Palm-bushland.   In Mediterranean countries the dwarf-palm,
Chamaerops humilis, may form social communities extending over wide
areas and excluding nearly all other vegetation ; its tall rosettes of fan-
like leaves seem to spring directly from the ground.[5]   Mayr[6] describes
palm-bushland that occurs in America and is composed of Serenoa serru-
lata : this palm stretches over the ground and clothes sterile sand, on
which forest composed of Pinus australis and P. cubensis had formerly
grown until destroyed by fire or felling operations.   Serenoa has already
seized upon many square miles of country.   Even when fire attacks it

[1] Vahl, 1904 b.   [2] See p. 268; Warming, 1907–9.   [3] Chodat, 1902.
[4] Chodat, loc. cit.   [5] See Fig. in Börgesen, 1897.   [6] Mayr, 1890.

and the fan-like leaves are burnt or parched, the stems hidden underground send forth new leaves.

Fern-heath is likewise a kind of bushland, and is produced by the widespread bracken-fern, Pteris aquilina. In the south of England this fern may be seen as a dominant plant, often in company with Ulex europaeus ; and likewise on Madeira in the Erica arborea region. In Mediterranean countries and even in Brazil it is among the plants that take possession of ground on which forest has been destroyed. Bushland of bracken may be so dense as to be impenetrable without the aid of an axe, and as to exclude nearly all other kinds of plants. In Africa, for instance in Usambara, fern-heath likewise seems to appear on places devoid of forest.

Many other waste tracts probably belonging to this formation might be mentioned, notably those described by Adamovicz[1] as present in south-eastern Europe.

[1] Adamovicz, 1904.

SECTION XIII

CLASS X. PSILOPHYTES. SAVANNAH-FORMATIONS

CHAPTER LXXVIII. SAVANNAH-FORMATIONS

UNDER the heading of savannah-vegetation we may place a number of *tropical* and *subtropical* formations that are paralleled by steppe-formations. Moreover, as the external conditions become more favourable there is a successional series, quite analogous to that shown by steppe-formations, leading from desert and desert-like succulent steppe, through xerophytic bushland and grassland, to semi-mesophytic bushland, which provides a stepping-stone to mesophytic forest.

The formations belonging to the savannah-type are linked with those of steppe by transitions, and it is not easy to draw a sharp distinction between the two sets, yet the former differ essentially in the larger dimensions of the constituent plants, and above all, in the *presence of trees*.

We may regard as belonging to savannah-vegetation the following formations :—

1. **Thorny savannah-vegetation,** including : (*a*) **orchard-scrub,** (*b*) **thorn-bushland** and **thorn-forest.**

2. **True savannah:** tropical and subtropical savannah.

3. **Savannah-forest,** including bush-forest in Africa and ' campos serrados' in Brazil.

Appendix. **Evergreen tropical bushland:** West Indian Croton-bushland.

CHAPTER LXXIX. THORNY SAVANNAH

Orchard-scrub.

UNDER the name of ' orchard-steppe' Hans Meyer[1] describes a formation that covers many square miles in the Kilimanjaro region of East Africa. It is allied to desert-like steppe. The soil is bare and only at certain spots shows a slight growth of grass ; shrubs and herbs are lacking, at least in the dry season. Dotted over the surface at intervals of 3 or 4 metres are tolerably regular little trees which are from 2 to 4 metres in height. They are richly armed with thorns, and for the most part have ternate or pinnate leaves, but are leafless during the dry season.[2]

---

[1] Hans Meyer, 1892; Volkens, 1897.   [2] See Pfeil, 1888; Engler, 1895.

Orchard-scrub may sometimes abound in grasses and herbs during the rainy season.

Thorn-bushland and Thorn-forest.

Where the shrubs stand closer together orchard-scrub gives way to thorn-bushland. Thorn-bushland is widespread over the tropics where the rainfall is not considerable. According to Urban, in East Africa two types may be distinguished : deciduous thorn-bushland, and aphyllous bushland composed of succulent euphorbias. The two types frequently merge into one another. East African thorn-bushland is sometimes composed of many species, but at other times there are certain preponderant species, namely acacias. Despite prolonged drought epiphytes are not entirely wanting. Here and there a few herbs occur.

Characteristic of the shrubs in thorn-bushland are a thick cuticle or a coating of long persistent hairs, and the small development of the leaf-surface, in addition to the precocious change of leaves and stems into spines ; all of these being protective devices occurring in members of the most diverse families. Species of Acacia and other Leguminosae, on the contrary, guard against complete desiccation by the closing together of their glabrous leaflets. In Ugogo where thorn-bushland prevails there are no permanent streams, but only greenish, glittering, often soda-containing pools in the valleys, and carefully guarded water-holes in which rain-water must be stored often for eight months. In Ugogo springtime commences about the middle of November. The large bushes become green and conceal the red soil as well as their numerous thorns.

Thorn-bushland and thorn-forest are very widespread in the drier parts of Africa and Asia.[1]

Caa-Tinga.

In the more central and northern parts of Brazil, especially on calcareous soil, one encounters *Caa-Tingas*, forests which have been so well known ever since the travels of Martius. The majority of the trees protect themselves from the prolonged drought and heat by shedding their leaves, and for this reason *Caa-Tinga* forest is extremely hot during the dry season. Among the remarkable trees present the best known is the bombaceous Chorisia crispiflora, which has a barrel-like swollen trunk whose loose soft wood acts as a gigantic water-reservoir [2] : Spondias tuberosa apparently possesses in its swollen roots subterranean water-reservoirs. Smaller trees and bushes are evergreen, and in this case find in their coriaceous, thick, and stiff or white-haired leaves, protection against excessive transpiration. *Caa-Tinga* forest abounds in thorny and stinging (e.g. Jatropha) plants, in columnar Cactaceae, and in other succulent plants. It is forest green in the rainy season. Scarcely has the dry season given way to the first spring rains before the foliage hurriedly sprouts forth ; and in one or two days all is green. Similar behaviour characterizes the allied West Indian dry bushlands or bush-forests. The important phyto-geographical part played by water reveals itself in many ways ; if the permanent ground-water lies so near the

---

[1] Passarge, 1895 ; Warburg, 1893.   [2] Martius, 1840–7, Tab. 30 ; Ule, 1908.

surface as to be accessible to the roots, then *Caa-Tinga* forest may remain green throughout the dry season.

In the neighbourhood of Lagoa Santa, on calcareous soil, Warming [1] found bushland composed of Cactaceae and other thorny bushes.

On the eastern side of the Peruvian Andes, at a considerable altitude, the northern slopes are, according to Ule,[2] clothed with thorn-forest which is essentially composed of Cactaceae.

In various other parts of America the presence of cactaceous bushland is asserted by various writers. In America Cactaceae may replace the succulent euphorbias of the Old World.

Bamboo-bushland.

Bamboo-bushland is a form of vegetation represented, for instance, on dry uplands in the East Indies. Low thorny bamboos grow socially, interlacing with one another, clothing the ground with their fallen leaves, and sometimes excluding all other plants. Here and there in their company are Feronia and Aegle (two of the Aurantieae), rhamnaceous and mimosaceous shrubs, cactus-like species of Euphorbia, the asclepiadaceous Calotropis procera, and others. Also on the Andes and other South American mountains bamboo-bushland occurs, and sometimes consists of Chusquea aristata, which may extend almost up to the perpetual snow-line.

The above-mentioned Calotropis procera is a shrub possessing large, stiff, roundish, glaucous leaves, and abounding in latex. In Asia and in Africa, over a large extent of country surrounding Lake Chad, it produces the so-called ' *ochur*-vegetation '. It has been introduced into America, and is thus encountered in many spots in the West Indies and Venezuela, where it thrives wonderfully in the burning sun on the driest of soils.

CHAPTER LXXX. TRUE SAVANNAH

SAVANNAH is associated with moderately rainy tropical places ; most closely allied to grass-steppe, it owes its distinction from this solely to the tropical climate. The vegetation has only one resting period—the dry season—during which it shows itself yellowish-grey and parched, though by no means devoid of flowers. The plants are endowed with xerophytic epigeous organs, and withstand this dry season, during which the savannah is often devastated by fires. The rainy season coincides with summer ; at its commencement all the vegetation becomes fresh-green, and there is a great increase in the display of flowers ; especially those savannahs that have suffered from fires rapidly clothe themselves with fresh-green verdure and a wealth of blossom.

The majority of the plants are tall *grasses*, which have coarse and stiff leaves, and usually grow in tufts to a height varying from a third of a metre to a metre. When the vegetation is not too tall, between the tufts of grass one can see the soil, which is usually red clay. But in addition to the grasses many Cyperaceae occur in certain savannahs (some savannahs are flooded during a part of the year), for instance in Guiana ; also

---

[1] Warming, 1892.   [2] Ule, 1900.

numbers of perennial herbs and undershrubs occur ; and, *in contrast with true steppe, shrubs and trees* present themselves and are accompanied by a few lianes and epiphytes.

There is in reality a gradual transition from grass-steppe to savannah (which is termed *campo* by the Brazilians). Those *campos* that show the greatest abundance of trees are termed by the Brazilians *campos serrados*, which are low, open, sunny forests, composed of bent and tortuous trees together with a rich vegetation of grasses, perennial herbs and small scattered shrubs that cover the ground.

The vegetation is *xerophytic,*in many places because of the dry season that lasts for months and (in contrast with that of steppes in the Northern Hemisphere) coincides with *winter*, during which often no rain falls and dew appears to be the sole atmospheric source of water. But the xerophily is also due to the dry continental climate in general. The xerophytism is expressed in the features recounted in the following paragraphs, which more especially refer to the best-known South American savannahs— the Brazilian *campos*.[1]

A. TROPICAL SAVANNAH

Brazilian Campo.

With the exception of a small percentage the plants are *perennial*. This we must attribute to the suppression of annual plants in the struggle with tall, dense, perennial plants, also possibly to savannah-fires, and to other causes.

Bulbous, tuberous, and true *succulent* plants are much more uncommon than in steppe, at least in American savannahs ; this is certainly due to the circumstances that there is not such a short, suddenly commencing vegetative season, nor such a prolonged and extreme dry season as are met with in desert.

The *grasses* forming the main mass of the vegetation are *caespitose*, but only very rarely produce runners ; their leaves are usually narrow, stiff, rough, hairy, and sometimes encrusted with wax ; and some are tunic-grasses.[2]

The perennial herbs, as well as many of the undershrubs and shrubs, exhibit a peculiar type of growth, for they produce under the soil tuberous, irregular, woody bodies (*xylopodia*),[3] which apparently consist of stem and root, but are mainly of a stem nature,[4] and send up numerous, usually unbranched or feebly branched, shoots.

The caespitose mode of growth is also very common among woody species ; individual shrubs may extend over areas several square metres in surface.

Runners and epigeous travelling shoots are wanting or, in any case, very rare among the herbs.

The *trees* throughout are *low in stature*, the tallest in the densest *campos* being only about the height of our orchard-trees, and resembling these in their tortuous trunks and branches : their cortex generally includes very thick, light cork, which exhibits long cracks and is often blackened by fires.[5]

[1] Warming, 1892. [2] See p. 116. [3] Lindman, 1900. [4] See p. 124.
[5] Figured by Martius (1840–7) and by Warming, 1892.

Lichens, mosses, and *algae* are entirely absent from the soil, and, at the most, are scantily represented on stones and trees.

The *leaves* of the dicotylous plants show xerophily, particularly in their stiffness (for they are often so rigid and dry as to rattle in the wind), in their lie, often in small size (though many are broad, oblong, or obovate, or compound ; ericoid and pinoid types being almost unrepresented), and in their very hairy nature or, if glabrous, in their waxy or lacquered coatings.

Ethereal oils are present in a whole series of the plants, and, in South America, especially in Verbenaceae, Labiatae, and Myrtaceae.

Many trees in tropical savannah are leafless during the dry season, or shed their foliage during this period. Yet this is the time at which many species are in flower. In this respect the Brazilian *campos* are not typical, but rather form a transition towards subtropical savannah, as the leafless season is very short.[1] This arises from the circumstance that winter on the Brazilian upland is not so hot as in tropical lowlands, for which reason there is less transpiration ; in addition the dry season in the *campos* is not without a scanty supply of rain.

The physiognomy of savannah is subject to great variety, which depends partly upon the height of the vegetation and partly upon the share taken on the one hand by grasses and perennial herbs, and on the other by shrubs and trees.

In Brazil and Venezuela there are some savannahs—*campos*—in which trees, rising above ground that is clothed with vegetation varying from half a metre to a metre in height, are so close together as to produce a kind of forest, which is open, sunny, shadeless, and hot, and allows the pedestrian to walk freely and the horseman to ride in all directions : such are the Brazilian *campos serrados*. There are other savannahs in which trees are extremely scanty and low, or are entirely wanting, and the carpet of grass and perennial herbs is very short and almost sward-like. In Matto Grosso, in the interior of Brazil, according to Pilger[2] the dry season is not entirely rainless. Most of the trees are deciduous and blossom after defoliation. Along the streamlets there extends forest composed of Mauritia vinifera and evergreen trees and shrubs, with perennial herbs, sedges, and grasses carpeting the ground.

Llanos.

Belonging to a special type of savannah are the *llanos*, which form boundless plains to Venezuela and have been so admirably described by Humboldt.[3] In the *llanos* trees are very few in number ; indeed they may be entirely wanting save in the moistest places, where palms, including Mauritia flexuosa and a species of Corypha, together with other plants, give rise to forest, which does not itself really belong to savannah ; in other places isolated trees of the proteaceous Rhopala or other species occur. Grasses form a vegetable covering, often as tall as man, and with them are mingled Compositae, Leguminosae, Labiatae, and so forth. Large portions of the *llanos* are under water during the rainy season,

---

[1] Warming, 1892.   [2] Pilger, 1902.
[3] In addition to Humboldt's works, that of C. Sachs should be consulted.

thanks to floods caused by the Orinoco ; yet the prolonged dry season gives to the vegetation a xerophytic stamp that is obvious, though its details have not been investigated.

Patanas.

The *patanas* of Ceylon, according to Pearson,[1] are xerophytic grassy slopes and plains of considerable extent, covered more especially with various species of grasses which, to some extent, belong to the same genera (Panicum, Paspalum, Sporobolus, Aristida, and others) as those on savannahs and *pampas*. Trees are almost entirely represented by comparatively few individuals belonging to two species (one myrtaceous and one euphorbiaceous). There are both dry and moist *patanas*. There seems to be no doubt that the dry *patanas*, which occupy altitudes less than 4,500 feet, are closely allied to American savannahs. And, just as in the case of these, various theories have been promulgated to account for the existence of extensive, comparatively barren, patana-areas in the midst of the luxuriant subtropical growth of the montane region. Grass-fires appear to play an important part in the matter ; nevertheless, the peculiar climate must also be blamed for such conversion of savannah-forest into savannah. Above the altitude of 4,500 feet there appear moister *patanas* which have a sour humus soil, and may be compared with *moor-formations* in temperate climes.

Lalang-vegetation.

In eastern Asia there are certain scrub-like associations of lalang-grass, Imperata arundinacea.[2] In Java there is no weed more resistant and pernicious than this grass, which is 1 or 2 metres in height and seizes upon places where forest has been destroyed. It is doubtful whether or no lalang-vegetation, which can be found interspersed with trees, is most closely allied to savannah.

African Savannah.

African savannahs in many places appear very similar to those of South America. Pechuel-Loesche[3] describes such savannahs in the Congo, and terms them *campine*. In East Africa Engler[4] distinguishes several associations (or possibly formations) belonging to savannah :—

(i) Treeless grass-savannah, which is in turn divisible into—

(a) Low grass-savannah formed by low grasses.

(b) Tall grass-savannah formed by tall grasses. According to Passarge[5] grass-savannah, without trees, occurs in the back country of Kamerun only on plateaux.

(ii) Bush-savannah occupied by grasses, shrubs, and little trees.

(iii) Tree-savannah with tall trees, among which Adansonia digitata is a veritable giant that is widespread over Africa as a whole. The trees of African savannahs are deciduous.

In regard to ' tree-steppe ' (the ' veld ') in Rhodesia, readers should consult the work of L. S. Gibbs.[6]

[1] Pearson, 1899 ; Holtermann, 1906.   [2] Junghuhn, 1852-4.
[3] Pechuel-Loesche, 1882.   [4] Engler, 1895 ; see also 1906.
[5] Passarge, 1895.   [6] L. S. Gibbs, 1906.

Other Tropical Savannahs.

In the savannahs of northern Australia and New Caledonia species of Eucalyptus play a prominent part.

B.   SUBTROPICAL SAVANNAH

Savannah in Cape Colony.

In Kaffraria, according to Thode,[1] savannahs very similar to those in South America occur in districts where the rain falls in summer and there is a marked dry season ; but they are probably evergreen and thus differ from true tropical savannah.   In such places, and particularly on craggy, stony country, some of the remarkable South African succulent plants occur, and include Euphorbia tetragona, which attains a height of several metres, as well as species of Aloë and Senecio ; in addition, bulbous plants are present.   Grasses, belonging to the genera Danthonia, Panicum, and Eragrostis, form the main mass of the vegetation, and are available as fodder for cattle throughout the year.   Between the grasses grow numbers of perennial herbs and undershrubs.   ' This varicoloured carpet of flowers, in which yellow and white tints preponderate, gives to the physiognomy a gay appearance which recalls the North American prairie, and is lacking for only a few weeks of the dry season.'[2]   At spring-time, as in steppe and desert, bulbous plants and orchids prevail ; during summer, Asclepiadaceae, Scrophulariaceae, and Gnaphalieae ; later on Malvaceae and Oxalidaceae appear ; while at all times Leguminosae and Compositae are to be found.   Scattered over this carpet, as in South America, either singly or in groups, are woody plants, which are mainly species of Acacia ; among these Acacia horrida, the Karroo-thorn, is especially noticeable ; ' Whichever way the traveller may turn, his eye encounters the finely divided pinnate foliage of acacias.'[2]

Other Subtropical Savannahs.

Not only in the southern Kalahari, but also in subtropical Australia and in the north of Argentina (in the Gran Chaco) does true savannah occur with tall grasses, which give rise to tufts but not to a continuous covering.

CHAPTER LXXXI.   SAVANNAH-FOREST

In favoured spots savannah-trees congregate more closely, other trees become added to them, and savannah passes over into forest, as we have already noted in the case of the Brazilian *campos serrados*.   Such forest, the ' savannah-forest ' of Schimper, is also known as bush-forest. It is as a rule of medium height, thin, and casting but little shade.   The ground is richly clothed with grass and perennial herbs.   The trees are leafless in the dry season.   Vast stretches of country in Africa are covered with savannah-forest of this kind.[3]   The jungles of India also seem to

---

[1] Thode, 1890.   [2] Thode, loc. cit. ; Weiss, 1906.
[3] Engler, 1895 ; Schweinfurth und Diels, op. cit; Pechuel-Loesche, 1882.

belong to this formation, and Kurz,[1] has described several forms of savannah-forest occurring in Burma.

Eucalyptus-forest.

Allied to tropical savannah-forest is the Eucalyptus-forest in subtropical parts of eastern Australia where the rain falls in summer. In this open, well-lighted, evergreen forest the trees stand so far apart that their crowns do not meet. The ground is clothed with a continuous carpet of grass, including an admixture of perennial herbs, which sprout forth at the beginning of the rainy season and form a fresh, lush sward. In the dry season many of the plants vanish, the longest to remain being the grasses and Compositae, as in Brazilian *campo*. The country, viewed from a distance, appears to be densely clad with forest ; yet one can ride through the forest, and travel freely in all directions. There is a close likeness to Brazilian *campos serrados*, but the trees are taller and more slender, while the number of species present seems to be small.

Aphyllous forest.

Apparently very closely allied to savannah-forest is the peculiar *tjemoro-forest*, which is formed by species of Casuarina, and occurs on dry bare soil on the mountains of eastern Java and the Sunda Isles. Atmospheric precipitations are small in amount, and are not retained by the porous soil : the vegetation consequently must be xerophytic. Casuárina derives its protection against excessive transpiration from the characteristic construction of its shoot : the shoots are Equisetum-like, almost aphyllous, terete, and have a dark, matt, green surface, while their stomata often lie within deep furrows on the twigs. Schimper[2] has supplied a description of a forest of this type occurring on the Javanese mountain Ardjuno at an altitude of 2,500–3,000 metres. The ground is covered with the brown, dead, acicular twigs of Casuarina, just as it is in a European pine-forest with pine-needles. On this covering grow some herbs, including the narrow-leaved Festuca nubigena and Euphorbia javanica, which occur in great numbers. Cushions of a small unscented violet (Viola serpens), Plantago asiatica, small, white-flowered Umbelliferae (species of Pimpinella), small species of Gnaphalium, and, above all, Pteris aquilina, give a European cast to the flora. On less sloping spots the vegetation becomes more vigorous, and several kinds of shrubs, including species of Antennaria and Rubus pruinosus, are added to it. Among the herbs, Sonchus javanicus recalls the European S. arvensis, while Valeriana javanica is very like the European V. officinalis ; in addition, other European genera are represented by Ranunculus prolifer, Galium javanicum, Alchemilla villosa, Cynoglossum javanicum, Thalictrum javanicum, and Agrimonia javanica. Mosses are very scanty.

Trees in relation to Savannah and Steppe.

All savannahs, prairies, and possibly grass-steppes, suggest the question : ' Why should trees be absent, or be few and scattered ? ' The

[1] S. Kurz, 1875.　　　　[2] Schimper, 1893.

present condition of our knowledge does not permit of our giving any final answer to the query. But it may be regarded as certain that the phenomenon is intimately associated with edaphic and climatic conditions. Both in steppe and in savannah we have seen that very xerophytic bushland is the most accommodating, and that, as the external conditions are gradually ameliorated, there succeed, in order, grass-formations, semi-mesophytic bushland (*sibljak*, and savannah-forest), and finally mesophytic forest. And side by side with secular changes in soil and climate corresponding modifications have affected the vegetation. In certain cases other factors, for example fires, have co-operated in producing the result. The Brazilian upland presumably was originally clothed with forest, but as the land increased in extent the central and oldest parts gradually acquired a continental and drier climate, and forest was changed into the *campos*.[1] In this case the peculiar forms assumed by the trees and many other plants are not caused by climate alone, but also by fires on the *campos*. The savannah in Java and Sumatra has arisen, according to Junghuhn,[2] through destruction of forests.[3] The *llanos* are clothed with a relatively recent vegetation, which has immigrated from the mountains of Guiana and Venezuela.[4] Between the antiquity of a vegetation and the number of its species a certain proportion exists. The *llanos*, *pampas*, and *prairies* are, in accordance with the foregoing remarks, obviously more recent, and at the same time much poorer in species, than are the ancient highlands of Brazil and Guiana. Moreover, the poverty in species of the forests of northern Europe is certainly to be ascribed to the relative recency of this vegetation since the Glacial Epoch.[5]

APPENDIX

Evergreen Tropical Bushland.

In some tropical spots, where the rainfall is small but there is no marked dry season, xerophytic evergreen bushland and bush-forest come into existence. Such is the case in the Madras Province, in southern Java,[6] and in the West Indies.

Dry tropical bushland in the West Indies is oecologically closely related to the Mediterranean *maqui*,[7] though differing widely from this as regards flora. The Danish and some other isles of the Antilles that have but a slight rainfall are largely clothed with a grey, desolate, useless, and scorching bushland, between whose thorny, tangled shrubs and low trees one cannot penetrate without the aid of an axe. A *grey felt of hairs* coats many species, including those of Croton, which in places dominate to such an extent as to form extensive, almost pure associations (*crotoneta*), for instance, in the eastern parts of St. Croix[8]; a similar coating clothes the aromatic verbenaceous Lantana, species of Cordia, and Melochia tomentosa. Other plants, on the contrary, display fresh-green glossy foliage; one usually sees these species standing out from the grey bushy vegetation as isolated dark-green patches, which provide a striking contrast that is

---

[1] P. V. Lund, 1835; Warming, 1892, 1899.   [2] Junghuhn, 1852–4.
[3] Concerning the *patanas* of Ceylon, see p. 298.   [4] Ernst, 1886, p. 313.
[5] Warming, 1899.   [6] Junghuhn, loc. cit.; Schimper, 1893.
[7] See Chapter LXXXIV.   [8] Eggers, 1876.

particularly noticeable when the observer can examine the whole tract from above. Here are found many thorny shrubs, including Acacia Farnesiana and A. tortuosa, Parkinsonia aculeata, and Randia aculeata ; as well as Cactaceae (including Cereus, Opuntia, and Melocactus) and species of Agave. There are not a few laticiferous plants, including Plumeria, Rauwolfia, and Calotropis ; also many shrubs with upwardly directed leaves, or, as in the case specially of acacias, with leaves that execute movements according to the intensity of the light, or, finally, showing other methods of protection against excessive transpiration. In these tropical bushlands there are some lianes, also epiphytic Bromeliaceae and Orchidaceae, although atmospheric aridity checks any luxuriance or abundance of growth. After more prolonged periods of drought, according to Börgesen,[1] the leaves hang flaccid and more or less wilted ; yet there is no regular shedding of the foliage.

Allied to this type of bushland is subalpine bushland on tropical mountains. This is composed of gnarled little trees with strongly xerophytic foliage. Epiphytic Vascular plants are lacking, though lichens often abound on the trees.[2]

[1] Börgesen and Paulsen, 1900.
[2] Junghuhn, 1852–4 ; Schimper, 1898 ; Engler, 1891 ; Volkens, 1897 ; O. Stapf, 1894.

SECTION XIV

CLASS XI. SCLEROPHYLLOUS FORMATIONS. BUSH AND FOREST

CHAPTER LXXXII. SCLEROPHYLLOUS VEGETATION AND FORMATIONS

THE term 'sclerophyllous' is employed by Schimper[1] in connexion with *xerophytic bushland and bush-forest* in *subtropical regions where the rain falls in winter*. It refers to the small, thick, coriaceous, entire leaves which are so extremely common in these regions. Such regions with winter-rain are the Mediterranean countries, California, the south-western part of Cape Colony, the coastal parts of West and South Australia, and Chile between 30° and 38° S. In these places the winter is mild, even if slight frosts occur. The rain falls in few but violent showers, but the winter enjoys much sunshine. The summer is dry, and only infrequently experiences a light shower of rain. In the interior of the larger masses of land the rain in winter is often very scanty, for instance, in the Sacramento Valley in California, also in the interior of Chile, Spain, and Asia Minor. In such places steppe prevails. But where the winter-rain is more abundant the country becomes clothed with low bushland; well-grown forest, on the contrary, is rare.

The prolonged summer-drought is hostile to vegetation. Hence the rarity of larger trees. The trees are small, with gnarled trunks and boughs; and most of them may occur in the guise of dwarf-trees and shrubs. The leaves of the trees and shrubs are, as a rule, evergreen and protected from desiccation in various ways, yet their structure is not so extreme as that of desert-plants. The most obvious feature is the reduction in size of leaf, for the leaves are small; leaves of medium size occur only rarely, and then only on shrubs in specially favoured localities. Among the various types of leaf may be mentioned: the ericoid leaf, possessed by Erica, Elytropappus Rhinocerotis, and Cliffortia falcata and others; the pinoid leaf, occurring in many families in Cape Colony; the linear rolled leaf of various Labiatae; ternate and pinnate leaves, exhibited by Leguminosae, Pistacia, and Cunonia capensis. Most of the leaves are stiff, thick, strongly cutinized, and rich in sclerenchyma, and also have their intercellular spaces reduced. According to Guttenberg,[2] there is in nearly all Mediterranean evergreen sclerophyllous leaves a characteristic mechanism, in the form of supporting cells and cell-walls that act as struts to prevent collapse of the assimilatory tissue. The lower face of the leaf is often very hairy, rarely are both faces tomentose. Bud-scales are uncommon.

Winter and spring form the true vegetative season of sclerophyllous

---

[1] Schimper, 1898.　　　　　　　　[2] Guttenberg, 1907.

vegetation, even though brief cold periods sometimes cause a lull. From January onwards, in Mediterranean countries, perennial spring-herbs commence to flower, and in January shrubs begin to send forth their new leaves.[1] In summer only a few extremely xerophytic perennial herbs blossom ; the numerous bulbous and tuberous plants lie hidden in the soil, and annual herbs are bearing ripe fruit. According to Bergen[2] and Guttenberg[3] many evergreen plants do actually enter into a condition of rest in summer.

Sclerophyllous woodland is divisible into at least four formations, according to dryness of climate and soil :

1. **Garigue.** In dry rocky localities this forms an open kind of vegetation, which is transitional between fell-field and woodland. It is composed of perennial herbs and undershrubs, and is widely distributed over Mediterranean countries.

2. **Tomillares.** In regions where the air is very dry and the rainfall very small this suffrutescent vegetation prevails.

3. **Maqui.** In rainier regions where the vegetation is bushland composed of taller shrubs.

4. **Sclerophyllous forest** is associated with the most favoured localities.

CHAPTER LXXXIII. GARIGUE. TOMILLARES

Garigue.

On dry hilly and mountainous parts of southern France and of other Mediterranean countries, on the southern Alps, and on rocks in Greece and Syria, one encounters a widespread type of vegetation which is known in France as *la garigue*. This has been repeatedly investigated by Flahault.[4] 'Garigue is a vegetation belonging to forest-soil, but lacks trees.' The soil contains no humus, and the rocks, possibly largely calcareous, often lie exposed ; but small shrubs ($\frac{1}{2}$ to $1\frac{1}{2}$ metres in height), undershrubs, and herbs seize upon the soil and clefts of the rocks, and, despite their seeming scantiness, deck these in motley array. Never do they approach so close together as to form continuous bushland. The colour of the landscape is determined rather by the soil than by vegetation. Here it is that *true Mediterranean flora* develops. Winter scarcely checks its development ; certain species, Ruscus aculeatus, for example, grow throughout the year, and in the middle of winter one finds many plants in flower. In spring, including April and May, vegetation is at its height. Summer, with its lack of rain and its heat, causes a resting period, during which, if plants are to endure, they must guard against excessive transpiration. This they accomplish by varied devices, for instance, by reduction of the transpiring surface, coatings of woolly hairs, the excretion of ethereal oils, and the production of subterranean bulbs and tubers. Many are the low shrubs and undershrubs, which include the spiny Genista Scorpius, aromatic Labiatae, such as Lavandula Spica, Thymus vulgaris, and Rosmarinus officinalis; furthermore, glandular-haired, fragrant, large-flowered species of Cistus, Pistacia Terebinthus and P. Lentiscus, Phillyrea

[1] Vaupell, 1858.   [2] Bergen, 1903.   [3] Guttenberg, 1907.
[4] Flahault, 1888, 1893.

angustifolia, Euphorbia dendroides; the woody umbelliferous Bupleurum fruticosum, Plantago Cynops, the boraginaceous Lithospermum fruticosum—'The hotter and drier Nature is, the more numerous are the woody species.' Bulbous and tuberous plants are also present in great numbers; species of Narcissus, Iris, Asphodelus, Muscari, Tulipa, and orchids deck the rocks in springtime. Annual plants are relatively numerous, as the climate is hot and there is sufficient open ground available to them. The herbs mainly belong to the Gramineae, Compositae, Papilionaceae, and Labiatae; they are so numerous as to determine the physiognomy of the vegetation. Among grasses worthy of special mention is the social, setaceous-leaved, Brachypodium ramosum. Aromatic plants are extremely abundant; everywhere one notes the strong scent of Labiatae, species of Cistus, Terebinthinae, Ruta, the leguminous Psoralea bituminosa, and of Compositae.

The *garigue* is a complex of associations evoked by differences in the soil.[1] Species of Asphodelus and Acanthus seem to impress a particular stamp upon the *garigue* of Attica. It is a type transitional between *maqui* and fell-field.

The same formation occurring in other Mediterranean districts has been described under the names of 'formation of rock-plants',[2] 'fell-heath',[3] and 'stone-heath'.[4] Apparently it very often owes its origin to the devastation of *maquis* or forests; in particular, grazing goats often check the regeneration of devastated *maquis*.[5]

Tomillares.

These are communities of undershrubs, especially Labiatae, that are common in Mediterranean countries, and especially occur on dry Spanish plateaux.[6] The name is derived from the Spanish word for Thymus, 'tomillo.' Various associations can be distinguished according to the dominant plants, namely, Thymus-*tomillares*, Lavandula-*tomillares*, Salvia-*tomillares*. Despite the wide distribution of this formation almost nothing is known of its oecology.[7]

CHAPTER LXXXIV. MAQUI: SCLEROPHYLLOUS SCRUB.

Mediterranean Maqui.[8]

THE Mediterranean *maqui* is so styled in French literature, which acquires the term from Corsica; in Spain the name is 'monte bajo', in Italy 'macchia' (plural 'macchie'), and in Greece 'xerovuni'

---

[1] L. Blanc, 1905. [2] Willkomm, 1852, 1896. [3] Rikli, 1903.
[4] Beck von Mannagetta, 1901.
[5] As regards recent literature bearing on *garigue*, special reference should be made to the works of Flahault (1888, 1893), L. Blanc (1905), and Rikli (1903, 1907).
[6] Willkomm, 1896; Chodat, 1905; Rikli, 1907.
[7] Regarding *tomillares* on the Balkan Peninsula, see Adamovicz, 1907.
[8] In regard to Mediterranean *maqui*, the following works should be consulted: Kerner, 1886; Willkomm, 1896; Flahault, 1901; Beck von Mannagetta, 1901; Bergen, 1903; Rikli, 1903, 1907; Vahl, 1904, 1906; Adamovicz, 1905; Guttenberg, 1907; Ginzberger und Maly, 1905.

Maqui is composed of shrubs, the majority of which are *evergreen*, though a few are deciduous. In the evergreen species the leaves are coriaceous, glossy, or grey-haired, usually elliptical or ovoid, and entire : to this type belong those of Myrtus communis, Buxus, Nerium, Olea europaea, Laurus, Quercus Ilex, Phillyrea latifolia, Arbutus Unedo, Pistacia Lentiscus, and Ilex Aquifolium. To the ericoid type belong Erica arborea and E. corsica. To the switch-type belong many species, including Spartium junceum, whose large yellow flowers stand out from the bushland towards the end of spring ; also species of Genista, for example, in Corsica, the hard-thorned G. corsica. Among cladode-bearing forms are Ruscus and Asparagus, the latter of which to some extent occurs as a liane, as does Smilax aspera. Common, too, are species of Cistus, of which C. ladaniferus in places covers square miles of country in Spain. It is one of the *aromatic* plants which play so striking a part in dry districts of the western Mediterranean countries, and include other *undershrubs*, such as Thymus vulgaris, species of Lavandula, Calamintha, Rosmarinus, Stachys, Teucrium, and other Labiatae, also Myrtus communis, and the terebinthinous Pistacia. Leaves that are hairy, or involute, or narrow, or show other structural features already mentioned, indicate dryness of habitat. Of spiny plants there are not a few, including wild olive-trees, Ilex Aquifolium, and Prunus spinosa.

The climate is not favourable to deciduous shrubs that are leafless during the true vegetative season. These shrubs are consequently common only in the northern fringe of the Mediterranean region. Elsewhere they are confined to the banks of streams or other moist localities. It should be noted that some shrubs which are deciduous in Central Europe become evergreen in the Mediterranean region.[1]

Finally, it may be added that the numbers of bulbous plants, including Crocus, Romulea, and Hyacinthus, occur and blossom in the early spring.

The *maqui* attains a height of one, two, or sometimes even three metres, and gives rise to an ‘ almost impenetrable tangled mass ’. The air is hot, and the *maqui* rich in flowers and scent, at least in spring, from February to March. Its impenetrable nature is partly due to the abundance of twining and climbing plants, among which are species of Rubus, Smilax aspera, Rosa sempervirens, Rubia peregrina, and several species of Clematis.

Maqui is widespread over Mediterranean countries, from Spain to Palestine, and particularly clothes wide tracts of warm limestone rocks. It forms a parched, sterile type of vegetation for which no use has been found. Its flora is much the same everywhere within this area.

Some *maquis* include many species, but others show a limited number of social species occurring in vast numbers together. Among the associations into which Mediterranean *maquis* may be subdivided according to the species dominant, we may note—

Aphyllous *Retama-bushland* in the south of Spain is perhaps to be regarded as allied to shrub-steppe.

Dwarf-palm maqui also occurs in the south of Spain.[2]

---

[1] Schimper, 1898 ; Beck von Mannagetta, 1901 ; Bergen, 1903 ; Rikli, 1903 ; Vahl, 1904, 1906. [2] See p. 291 ; also Börgesen, 1897.

Cistus-maqui is the most widely distributed over the whole Spanish peninsula ; such *maquis* are known as ' jarales '.

Erica-maqui occurs in the north of Spain where the air is more humid ; in it species of Erica preponderate, but species of Ulex and Sarothammus, as well as such deciduous shrubs as Crataegus monogyna, also play a prominent part. The species of Erica—E. scoparia, E. ciliaris, E. vagans, and others—are shrubs which vary from two to six feet in height. In the south of France these Erica-*maquis* are known as *Les Landes*. From heaths they differ in the size of the shrubs and in the absence of raw humus. *Maquis* and *garigues*, according to Rikli,[1] also characterize the underwood of forests of Quercus Ilex and Pinus halepensis.

Pseudo-maqui. Closely allied to *maqui* is Adamovicz's [2] *pseudo-maqui*, a xerophytic evergreen bush-formation, which occupies montane and submontane regions in Mediterranean countries, and can endure a more severe winter. The tallest constituents are species of Juniperus and Quercus, Buxus sempervirens, and Prunus Laurocerasus.

Makaronese Maqui.

On the Azores, Madeira, and the Canary Isles winter is so mild that even the higher parts of the mountains lie within the subtropical climatic zone. Summer is practically rainless, but diurnal winds cause at a certain altitude a cloud-belt, which moistens the vegetation, increases the atmospheric humidity, and decreases insolation by acting as a screen. In this moist belt *maqui* is formed by shrubs which are from three to five metres in height. Among these, some are Mediterranean species, such as Erica arborea and E. scoparia, but the majority are special species differing from the Mediterranean ones in the larger size of their leaf-blades. The leaves are mostly of medium size, but otherwise belong to the sclerophyllous type. The commonest species are Laurus canariensis, Ilex canariensis, Heberdenia excelsa, and Myrica Faya. Lianes are scarcely represented, while bulbous and tuberous plants are almost entirely absent. On the southern slopes of the Canary Isles, also on Madeira and the Azores above the cloud-belt, the *maqui* is formed by small-leaved shrubs.[3]

Maqui in Cape Colony.

The vegetation of this is very similar to that of Mediterranean *maqui*. It, likewise, is composed of low evergreen shrubs with small, often ericoid or pinoid, stiff, not uncommonly brownish-green or grey leaves. In winter, that is from May to October, the soil is often saturated with rain, and the shrubs sometimes are dripping with water : this is the season of their growth. Then succeeds the long dry season which they must endure. Very many species have assumed a habit so similar—namely, the ericoid habit—that it is very difficult to distinguish between them in their flowerless condition, although they belong to families so utterly diverse as the Ericaceae (with about 400 species of Erica), Proteaceae, Rhamnaceae, Santalaceae, Polygalaceae, and Rutaceae (tribe Diosmeae). Sedges and grasses play only a subordinate part. But there is a wealth of bulbous

[1] Rikli, 1903.   [2] Adamovicz, 1905.   [3] Schacht, 1859 ; Vahl, 1904 *b*.

and tuberous plants belonging to the Iridaceae, Liliaceae, and to Oxalis ; and associated with them are species of Pelargonium, Crassulaceae, and others.[1]   This vegetation represents a transition from the low dwarf-shrub heath of the north to xerophytic bushland of the tropics.

Maqui in Chile.

In Chile there is *espinal* or ' espinar-forest ', in which an important part is played by the rhamnaceous Colletia, whose evergreen, opposite branches are converted into spines ; nor are Cactaceae and Bromeliaceae lacking.   Meigen [2] describes dry bushland occurring near Santiago as being composed of evergreen, small-leaved shrubs, among which Quillaja Saponaria is the commonest.   Here and there a tree rises above the shrubs.   Climbing plants are frequent, as are bulbous and tuberous plants belonging to the Liliaceae, Amaryllidaceae, Iridaceae, and Oxalidaceae.   While, thanks to the difference in their appearance, Cereus guisco and the bromeliaceous Puya chilensis attract notice.[3]

Maqui in other Countries.

In California *maqui* is known under the name of *chaparral*.[4]   *Maqui* is also very widespread in West and South Australia.[5]

CHAPTER LXXXV. SCLEROPHYLLOUS FOREST

MOST of the trees whose home lies in regions where the rain falls in winter are capable of assuming the form of shrubs.   In depressions of the land, where water remains for a longer time, the shrubs become taller, and in such spots *maqui* often gives way to forest.   In specially favoured spots—for instance, in deep gullies, rainy mountain declivities, and such like—one finds true forest, in which there occur not only trees belonging to *maquis*, but also some trees characteristic of forest ; and together with them live ground-plants adapted to grow in shade.

Mediterranean Oak-forest.

In Mediterranean countries one finds low forest of evergreen species, consisting, for example, of oaks and of Quercus Ilex in particular.   This species has lanceolate, spinose, tomentose leaves, and is a true xerophyte growing in dry, stony soil, or even in rock.   Associated with it are numbers of arboreous, fruticose, and suffruticose plants, and perennial herbs, all of which are xerophytic in structure, and some of which may reappear in the sunny *garigues* or *maquis*.   Among these is Quercus coccifera, a low shrubby oak, which by means of its root-suckers seizes upon wide stretches of *garigue*, and gives rise to low, useless bushland : others worthy of mention are Juniperus Oxycedrus, species of Cistus, Arbutus Unedo, Viburnum Tinus, Paliurus australis, and Ilex Aquifolium.   One also encounters small lianes, such as Lonicera implexa, Smilax aspera, and Rosa sempervirens.

[1] L. S. Gibbs, 1906.   [2] Meigen, 1893.   [3] See also Reiche, 1907.
                    [4] See p. 280.   Diels, 1906.

At greater altitudes, on wet, cold, clay soil, evergreen species are replaced by Quercus sessiliflora var. pubescens, whose stiff, hairy leaves betray its xerophytic nature.[1]

In Algeria, according to Trabut,[2] where the annual rainfall exceeds 60 centimetres, Quercus Suber gives rise to forest. In the forest-district agriculture is practicable without artificial irrigation. As the forest includes, in addition to evergreen trees, such deciduous ones as Castanea, Populus tremula, Alnus glutinosa, Fraxinus, and Ulmus, its flora recalls Central Europe rather than the *maquis*. On the Atlas Mountains forest is formed by Quercus Ballota.

Other Mediterranean Forests.

In Mediterranean countries one finds *olive-forests* and olive-plantations, composed of the marked xerophyte Olea europaea.[3]

On the Austrian coast Laurus nobilis produces forest. This species, which has leaves rather large for a Mediterranean plant, grows for the most part as underwood in oak-forests.[4]

Laurineous Forest on the Canary Isles.

These laurel-forests, which have been described by Christ,[5] develop in the cloud-belt, and specially in valleys and gullies, where even in summer heavy mist arises daily or almost daily. The ground is covered with a dense green carpet of ferns and mosses. The forest consists of laurineous trees, including Persea indica, Laurus canariensis, Ocotea foetens, and Phoebe barbusana, with which are copiously mingled Ilex canariensis, Erica arborea, Myrica Faya, and others. The underwood is composed of Rhamnus glandulosa, Viburnum rigidum, and others; lianes are represented by species of Smilax. The leaves belong to the ' laurel-form ', that is to say, they are undivided, entire, and coriaceous; but other purely xerophytic types are also to be seen. A peculiar, deep-green shade prevails in the forest under the dark canopy of laurineous trees. The refreshing humidity in these forests contrasts sharply with the scorching heat of the open slopes, and is accentuated by the scent of violets, moss, and earth that comes from forest-soil. This last is almost exclusively carpeted with countless ferns, and thus recalls forests in New Guinea and other Pacific islands; apart from these, herbs are scanty.

The same type of forest is encountered on Madeira.[6]

This laurineous forest approaches mesophytic forest most closely.

Other Sclerophyllous Forests.

Sclerophyllous forest occurs in various other parts of the world where the rain falls in winter : in Australia, Chile, and California.[7] In Australia it is largely composed of Eucalyptus, Acacia, and Proteaceae. In California it is formed by species of Quercus (Q. macrocarpa) and Sequoia sempervirens.

---

[1] Flahault, 1893.   [2] Trabut, 1888.   [3] See p. 128; also
Beck von Mannagetta, 1901.   [4] Kerner, 1886.   [5] Christ, 1885.
   [6] Vahl, 1904 b.
[7] Schomburgh, 1875 ; Diels, 1906; Reiche, 1907 ; Mayr, 1890.

SECTION XV

CLASS XII. CONIFEROUS FORMATIONS. FOREST

CHAPTER LXXXVI. EVERGREEN CONIFERAE

THE scale-like leaves of the Cupressaceae and the needles of the Abietaceae are representative of coniferous foliage, which is characterized by smallness of surface, strong cuticularization of the epidermis, frequent sunken position of the stomata, and other features calculated to depress transpiration. Evergreen, coniferous trees exhale much less water-vapour than dicotylous trees, but the amount of transpiration varies with the species.[1] In accordance with their xerophytic nature conifers have few or insignificant root-hairs.

Evergreen Coniferae are pronounced xerophytes when judged by their morphology, anatomy, and physiological characters. The reason why they are not grouped together with sclerophyllous vegetation, but are dealt with here in a separate section, is that sclerophyllous formations constitute a natural group adapted to definite climatic conditions, namely, to a *subtropical* climate where the rain falls *in winter*. As characteristic peculiarities of sclerophyllous Dicotyledones we may regard, in addition to the small size of the leaf, the rarity of arboreous growth and the normal absence of bud-scales.

Coniferous forest occurs in the most diverse climates, which, however, do not include the torrid dry climate. It is most extensively distributed in the cold-temperate zone of the Northern Hemisphere, and becomes less prominent in warmer climates.

The soil upon which coniferous forest occurs varies widely, yet, so far as reliable information is available, it is always *physically or physiologically dry*—a fact that harmonizes with the xerophytic structure of Coniferae. In warmer regions coniferous forest has, as a rule, seized upon permeable, warm, sandy soil ; but in colder places it grows upon nearly all kinds of soil, varying from dry rocks and sand-fields to wet, marshy ground. This accords with the evergreen nature of most Coniferae. The cold winter is a physiologically dry season, against which trees can protect themselves by defoliation or by xerophytic structure. The larger plants that have to endure a severe winter, and are too tall to derive protection from a covering of snow, need protective devices capable of saving them from death due to lack of water in winter.[2] Added to this is the circumstance that in cold regions most Coniferae produce raw humus and therefore an acid, physiologically dry soil.

The xerophytic structure of Coniferae is, as M. C. Stopes[3] has pointed out, a phyletic character. Her suggestion is that the xerophytism of the Coniferae is a result of the imperfect nature of the water-conducting

[1] See M. C. Stopes, 1907.　　[2] Schimper, 1898.　　[3] M. C. Stopes, loc. cit.

tissue : 'The gymnospermic "xerophily" is not the result of even inherited adaptations to dry conditions, is not in fact an ecological adaptation in the usual sense, but is a result of their histological structure (which is incapable of allowing a rapid flow of water through the wood), their wood consisting entirely of tracheids, which are usually pierced by "bordered pits"; the diameter of the tracheids is less than that of the vessels of the flowering plants, and the whole structure of the wood is simpler and more uniform.' The Gymnosperms are a very ancient and primitive group 'in which the woody conducting-system has not reached the state of specialization and efficiency which was afterwards attained by Angiosperms.' Miss Stopes comes to the following conclusion : ' It appears then that the xerophytic characters of the Coniferales in very many cases are not adaptations to xerophytic conditions in their own lives, nor are they "inherited" from the remote past as vestigial characters no longer in touch with present-day necessities, but are the result of physiological limitations of the type of wood in this ancient and incompletely evolved group. In other words, their 'xerophytism' is not oecological, but phylogenetic.'

It may seem to be beyond doubt that the xerophytism of Coniferae, as well as the almost complete prevalence of the evergreen habit, are phylogenetic characters derived from the archetype, and that the xerophytic structure of the foliage is correlated with the primitive structure of the conducting-tissue. But Miss Stopes is incorrect in assuming that Coniferae nowadays do not for the most part grow under dry conditions of life. As already indicated, all evergreen trees compelled to experience a cold winter must necessarily be capable of withstanding a season that is rigidly dry in a physiological sense ; and when they grow upon raw humus or peat the soil is relatively physiologically dry, even though it drip with water.

The primitive structure of the wood of the Coniferae may accentuate their xerophytic external form, and may be the phylogenetic cause of this ; yet it cannot be regarded as wholly answerable for the xerophytism which now prevails, since coniferous forest at present mainly occurs on physically or physiologically dry soil.

Inasmuch as so many of the coniferous trees, Picea excelsa, for instance, are shade-enduring trees, the vegetation clothing the ground is very scanty because the forest is very shady, in virtue of the numerous foliaged shoots, whose leaves allow the passage of no light and remain attached throughout the year. The deep shade therefore endures all the year. Consequently, under various conifers, Picea excelsa and Abies pectinata for instance, raw humus readily arises.

The humble *plants clothing the soil in northern coniferous forest* are all *perennials*, which vary in construction of shoot and other characters.

Dwarf-shrubs and *undershrubs* are numerous, and include species of Vaccinium, Ledum, Calluna, Empetrum, and Juniperus ; and to these we may append species of Pirola. The majority of these shrubs, like many of the herbs, are evergreen.

Creeping rhizomes, or roots that emit buds, are possessed by not a few species, including those of Pyrola, Monotropa, Maianthemum, Goodyera repens, Oxalis Acetosella, Trientalis europaea, Vaccinium Vitis-idaea and V. Myrtillus, Pteris aquilina, and Aspidium Dryopteris : this may possibly be ascribed to the loose texture of the soil with its thick carpet of moss and fallen needles.

Plants that *travel above ground* are represented by Linnaea, Lycopodium clavatum, L. annotinum, Veronica officinalis, and others. But the majority are strictly confined to their points of settlement, and possess a multicipital primary root or a vertical root-stock bearing several stems.

Grasses are very scanty in some forests, but numerous in others. Cryptogamia are common.

Most of the herbs show no xerophytic structure ; they are mesophytes fitted to live in the shade and moist air of forest. But among the dwarf-shrubs the evergreen ones are xerophytic.

One peculiarity, in which northern coniferous forest contrasts with dicotylous forest, is the abundance of dwarf-shrubs with *fleshy fruits*, e.g. species of Vaccinium, Arctostaphylos Uva-ursi, Empetrum, and Juniperus communis. This must probably be associated with the residence within coniferous forest of birds that carry from place to place seeds and fruits which they have swallowed or which adhere to their bodies ; in this way Linnaea, species of Pyrola, and Goodyera have been introduced into Danish pine-plantations only about a century old.[1]

Among the many kinds of coniferous forest those of Europe have been the most completely investigated. Some of these are treated in the succeeding paragraphs.

Pine-forest (Pinetum).

The Scots pine, Pinus sylvestris, can grow on very diverse soils, which vary from dry warm sand, or rock with a thin layer of loose earth, to moist, or even wet, soft bog-soil. It is an extremely *accommodating* tree, capable of growing on very sterile soil, and in this respect resembles ling. It is a *light-demanding* tree, whose inner branches therefore perish early, so that the trunk gives way to a bare bole ; the leaves remain attached for only three or four years, and then only at the twig-ends and on the crown. The ground is consequently often densely clothed with plants, composed sometimes of one kind and at another time of another kind of vegetation,[2] which varies according to the dryness and other qualities of the soil, but is always in essence xerophytic and oecologically closely allied to *lichen-heath* or *dwarf-shrub heath*. Sometimes Cladonia rangiferina and other fruticose lichens, such as other species of Cladonia and Cetraria islandica, thrive here better than on windy spots, and extend over the ground as a whitish-grey carpet, in which are interspersed low, contorted ling-plants and others, including Linnaea, Arctostaphylos Uva-ursi, species of Pyrola, Lycopodium annotinum, L. clavatum, Potentilla sylvestris, Viola canina, and Maianthemum Bifolium. In other cases mosses are commoner, or Juniperus communis, Vaccinium Myrtillus, V. uliginosum, V. Vitis-idaea, Calluna, Populus tremula, and Empetrum, are more frequent and taller ; in addition, Picea excelsa may occur as a shrub forming underwood. A list of plants characterizing pine-forest in North Germany has been compiled by Höck.[3] There are northern pine-forests whose soil has an extremely dry covering, which consists of Arctostaphylos Uva-ursi, Juniperus, Calluna, Betula nana, Anten-

---

[1] Warming, 1904.   [2] Varieties of association, see p. 146.   [3] Höck, 1893.

naria dioica, and others, also of masses of lichens (Cladonia) and mosses (Grimmia). In other places different mosses (Hylocomium and Dicranum), Luzula pilosa, and grasses—particularly two narrow-leaved xerophytic species, Aira flexuosa and Festuca ovina—and some of the plants already enumerated, combine to form a much closer and softer covering. These different varieties of pinetum are, according to the dominant species, termed *pinetum cladinosum, pinetum hylocomiosum, pinetum herbidum*, and so forth.

The ground-vegetation of pine-forest thus *consists of xerophytes ;* for the soil is usually poor and dry, while light and wind can usually penetrate with ease and tend to dry up the vegetation. Yet even in this forest one or another kind of mesophyte can gain a footing. The pine-forests of southern Russia obviously differ not a little from Scandinavian pine-forest, since many tall perennial herbs rise from their soil.

Birches are sometimes interspersed in pine-forest, as they, as well as the Scots pine, are light-demanding trees.[1]

Spruce-forest.

Picea excelsa, the common spruce or Norway spruce, thrives upon various kinds of soil, as is the case with Scots pine, but is more exacting in its demands, as it does not endure dryness so well ; beneath it no such xerophytic vegetation grows. It is a shade-enduring tree ; the branches and needles accordingly remain attached for a much longer time (the needles living for eight to thirteen years) than in the case of Scots pine, while the crown acquires the familiar dense conical form. The vegetation clothing the ground harmonizes therewith. Underwood is wanting, and the ground in the darkest spruce-forest is often quite bare, as in its close carpet of needles, often several centimetres in thickness, there grow only a few stunted mosses, though crowds of pileate fungi develop in autumn. Where the light is stronger the mosses become more vigorous ; and in good forests the ground-vegetation may combine to form a continuous, close, uniform, green, *soft carpet of mosses.* These are mainly species of Hylocomium, whose cushions lie loose above the soil and conceal humus that is occupied by earthworms ; in addition there occur, among others, Polytrichum and Dicranum, both of which genera have the power of producing raw moss-humus. In the mossy carpet and on the loose soil there are often numerous scattered Spermophyta, many of them possessing creeping rhizomes, e. g. Oxalis Acetosella, Trientalis europaea, Circaea, Vaccinium Myrtillus, V. Vitis-idaea, species of Anemone, Viola sylvatica, Linnaea, and species of Pyrola (in addition to ferns and lycopods). Some of these plants are pronounced sciophytes, and some, including Monotropa and Goodyera, are at the same time saprophytes. For lichens spruce-forest is mostly too dark, so that neither soil nor stems are densely clothed with them ; yet an exception in this respect is provided by spruce-forest on sterile soil and at higher altitudes, where Usnea festoons the branches with its wisps and gives to the forest a characteristic appearance.

[1] A considerable literature dealing with European pine-forest exists ; the most recent papers are those by Domin, 1905 *b* ; G. Andersson and Hesselman, 1907 ; A. Nilsson, 1896, 1897 *a*, 1897 *b*, 1902 ; Hesselman, 1906 ; S. Birger, 1904.

In northern parts of Europe the relations are often somewhat different. The soil is occupied by xerophytic dwarf-shrubs belonging to pine-forest ; an underwood of Salix, Betula, Alnus, Sambucus nigra, and others may develop ; while lichens are present, though scanty.

There are, therefore, different varieties of spruce-forest, which differ in the nature of the ground-vegetation, according as mosses, grasses, perennial herbs, and so forth are abundant.

Spruce-forest retains humidity more efficiently than pine-forest does. Raw humus occurs not infrequently in spruce-forest ; the carpet of spruce-needles covering the ground may be interlaced by fine spruce-roots and produce peat, beneath which hard pan occurs, just as in ling-heath or beech-forest. Raw humus produced from spruce is lighter in colour and less firm than that derived from Calluna or beech.[1]

Common spruce produces prostrate branches, which often trail over considerable distances, and may produce adventitious roots as well as new leading shoots. In this way the spruce can acquire several stems and give rise to scrub.[2] In this respect it goes farther than the Scots pine ; for the latter retains the tree-form until external conditions prevent its growth, whereas the spruce assumes deformed and prostrate shapes, not only when it passes beyond the limit of forest in Lapland,[3] but also on the shores of Norway.

Silver Fir Forest.

Abies pectinata, the common silver fir, is a conifer producing vast and stately forests in the central and southern mountainous districts of Europe. In the mountainous parts of Saxony ('Saxon Switzerland') it competes with Picea excelsa ; on the poorer sandstone, which weathers with difficulty, Picea has the upper hand, but on the basalt, with its thicker weathered strata, Abies dominates. Yet both species may occur on the two kinds of soil, and the final issue is the result of their struggle.[4]

Mountain-Pine Woodlands.

Pinus montana gives rise to stately forests in the Pyrenees and French Alps, but farther east it dwindles to scrub (elfin wood = 'Krummholz'), and, expelled from more favoured localities by other species, has to content itself with the poorest habitats.[5] It is a shade-enduring tree, though not to the same extent as the Norway spruce,[6] so that in its forests the ground is poor in plants.

Pinus montana grows not only on the driest and most sterile mountain-slopes, but also in wet moor, and in both these situations it gives rise to scrub or bush-forest.[7] Beneath it on moors grow such shrubs as Ledum palustre, Andromeda polifolia, Calluna, Vaccinium uliginosum, V. Myrtillus, Vaccinium Vitis-idaea, V. Oxycoccos, also such low herbs as Eriophorum and Carex, mosses such as Hylocomium, Dicranum, and Sphagnum, and, finally, lichens. In fact we are dealing with a *Calluna-moor*,[8]

---

[1] P. E. Müller, 1887 a ; Ramann, 1905.   [2] J. M. Norman, 1894–1901.
[3] Kihlman, 1890 ; and see p. 215.
[4] P. E. Müller, 1871 ; Schröter, 1904–8.   [5] See p. 18.
[6] P. E. Müller, 1887 b.   [7] See p. 216.   [8] See Chap. LII.

on which trees are growing. Many of these plants are xerophytic in construction.[1] Pinus sylvestris in like manner passes on to moors. It is only the most accommodating and hardy of plants, both arboreous and fruticose, that can grow on such extreme kinds of soil.

Mixed Coniferous Forest.

In many coniferous forests there is an admixture of species, for example, in Sweden, of Pinus sylvestris, Picea excelsa, birch-trees, Populus tremula, and others. This admixture of species is apparently more marked the farther one travels to the east in Europe, possibly because the land here has been occupied by plants for a longer time than it has in the north-western parts, and because the migration of species has, for the most part, been from east to west. In the Russian Government of Perm spruce-forest formed by Picea excelsa and P. obovata includes an admixture of Larix sibirica, Pinus Cembra, Abies sibirica, and others, in addition to dicotylous trees. The type of ground-vegetation present depends, as elsewhere, upon the conditions of illumination ; one finds the same carpet of mosses with scattered Spermophyta, belonging even to the same species, as in Denmark, or a rich vegetation of grasses, ferns, perennial herbs, and so forth.

Other Evergreen Coniferous Forests.

Various other associations of evergreen coniferous forest exist, some of which are :—

The *Siberian* primeval forest ('Taiga ') on the upper Lena, according to Cajander,[2] consists of pines, or of larches (Larix sibirica), or of spruces (Picea obovata) ; pine-forest reigns on sunny, dry slopes ; larch-forest on fresher slopes and in the lowlands ; spruce-forest in the moistest valleys. The forest in plains at the mouth of the Lena is very moist, and abounds in epiphytes ; it may be described as *larch-moor* (Larix dahurica) that has very thin peat.

The *Mediterranean* pineta of Pinus Pinea, with their interesting flora, partly xerophilous and partly marshy halophilous.

The *South-European* extensive forests[3] of Pinus halepensis, which suppresses the holm-oak (Quercus Ilex) in places where the rocks are tolerably weathered.

The *Mount Lebanon* cedar-forest.

The *North American* Abies-forests and Pinus-forests, of which the northernmost grow on icy soil and to some extent differ from European coniferous forest in physiognomy.[4]

The *Canary Isles* have their *pinares*, which are forests consisting of Pinus canariensis occurring at an altitude of 1,600 to 2,000 metres, especially on the drier, windy, slopes (the laurel-forest selects moist gullies within the cloud-belt). The *pinar*, Pinus canariensis, has a conical stem, which bears boughs right down to the ground, and thin needles, 15 centimetres long, which hang down in graceful curves. In these *pinares* one hears

---

[1] See Chap. XLVI.   [2] Cajander, 1903.
[3] Flahault, 1893 ; Beck von Mannagetta, 1901.
[4] Mayr, 1890 ; Kearney, 1901 ; Whitford, 1905.

only the whispering wind ; no bird-song trills through the air. The ground-vegetation differs as much as the high-forest from that in northern Europe ; it consists largely of species of Cistus and Genista, xerophytic genera that play so prominent a part in Mediterranean *maquis*; thus the ground-vegetation in these forests is a reflection of the *maquis* and *garigues*, just as that of northern forests is essentially a replica of lichen-heath, and dwarf-shrub heath or of fell-field. In addition to the shrubs named Daphne Gnidium, Asphodelus ramosus, the fern Notochlaena Marantae, and two species of the leguminosous Adenocarpus are common. At greater altitudes the ground-vegetation is composed of Adenocarpus viscosus and annual grasses.

In *Brazil*, extending approximately from the tropic of Capricorn and southwards, are nearly pure *pinheiros*, which are forests composed of Araucaria brasiliensis. This broad-needled tree has a dark-green, pine-like crown.[1] The vicinity of the tropics is suggested by the presence of epiphytic Spermophyta.

Deciduous Coniferous Forest (Larch-forest).

Larches are the hardiest of all conifers in relation to frost, since the acicular form of the foliage is combined with its deciduous nature. Larch-trees give rise to forests (of Larix sibirica) round the Polar region of Siberia, they endure a greater degree of dryness than the Norway spruce, and can utilize a very short vegetative season, possibly because their vigorously transpiring leaves can assimilate much more rapidly than can those of evergreen species. Larches are therefore less dependent upon the cold of winter than upon the heat of summer. In addition, to a marked degree they are *light-demanding* trees ; consequently their forests are well-lighted, and the ground is decked with numerous spermophytic herbs, as well as ferns and mosses. For instance, in larch-forest (composed of Larix decidua) in the Alps, one finds Arnica montana, Solidago alpestris, Campanula barbata, many orchids, and so forth. On the Altai Mountains larch-forest seems, indeed, to be exterminated by this mesophytic vegetation of herbs and grass. Here, according to Krassnoff,[2] gigantic centenarian larch-trees stand singly or in groups, far from one another in the forest ; and on the old humus-soil produced by the fine needles there has arisen herbage so tall and luxuriant that one can easily hide within it. This herbaceous vegetation consists of species of Aconitum, Delphinium, Paeonia, Clematis, and others. Each year millions of larch-seeds fall into this sea of herbage, yet only a few find places where they can germinate : the forest is apparently doomed to extinction.

[1] See Martius, 1840–7; Schenck, 1903 *a*.   [2] Krassnoff, 1888.

SECTION XVI

CLASS XIII. MESOPHYTES

CHAPTER LXXXVII. MESOPHYTIC VEGETATION AND FORMATIONS

MESOPHYTES[1] are plants that show a preference for soil and air of moderate humidity, and avoid soil with standing water or containing a great abundance of salts. No single factor is of paramount import to mesophytes. They select places where the atmospheric precipitations are distributed over the seasons of the year more equably than in the homes of xerophytes. In the habitats of mesophytes the soil seems always to be rich in alkaline humus.

The morphology and anatomy of mesophytes have already been discussed on p. 135. In temperate regions one often finds a difficulty in interpreting the adaptations shown in these plants. Many species are very plastic in epharmosis; such is the case with the beech[2] and many common European plants. In their power of adjusting themselves to differences in the surroundings, mesophytes perhaps exceed all other plants; but we know too little in regard to this matter. The diversity of leaf-form is as a whole greater than in other formations. Divided and compound leaves, also leaves with the margin indented in various ways, are more common than among xerophytes.

Compared with certain communities of xerophytes and halophytes, no mesophytic community is so open or poor in plant-life; this must be ascribed to the more favourable conditions of life. In the humblest and simplest communities grasses and other herbs play the weightiest part; among such communities are meadows and pastures. Richer than these is vegetation composed of tall perennial herbs and mesophytic bushes, in which several storeys of plants occur. And richest of all is tropical rain-forest.

Mesophytic communities find their homes largely within temperate regions, and particularly within the northern forest-belt, where the rain mainly falls in summer and autumn; they therefore occur specially in the zones of evergreen coniferous forest and of deciduous dicotylous forest, but they also occur in Polar countries and within the tropics. Especially in temperate regions they are often associated with *cultivated* land; the soil and climate that they require admirably suit them for cultivation by man. As a result of cultivation of the soil the natural communities, which beyond doubt were originally very few in number, have been destroyed and broken up into a number of new communities, which are largely cultivated, semi-cultivated, or ruderal in character.

[1] See Chapter XXX, pp. 131, 135. [2] Stahl, 1880, 1883.

These new communities are constantly struggling with one another, and are equally difficult to diagnose and designate. *Communities of cultivated plants* consist mainly of annual and biennial species, and are likewise mesophytic communities ; but they are not treated in this work.[1]

Shade-plants belonging to the lower storeys of forest and bushland are usually mesophytic in structure.

Natural mesophytic communities may be classified thus :—

A. COMMUNITIES OF GRASSES[2] AND HERBS

1. **Arctic and Alpine mat-grassland and mat-herbage.**
2. **Meadow.**
3. **Pasture on cultivated soil.**

B. COMMUNITIES OF WOODY PLANTS

1. **Mesophytic bushland.**
2. **Deciduous dicotylous forest:** including monsoon-forest.
3. **Evergreen dicotylous forest:** including antarctic forest, subtropical rain-forest, tropical rain-forest, palm-forest.

CHAPTER LXXXVIII. ARCTIC AND ALPINE MAT-GRASSLAND AND MAT-HERBAGE

IN Polar countries and above the tree-limit on many mountains there occur on the slopes green tracts of monocotylous and dicotylous herbs. This vegetation, though it may be allied as regards flora to the adjoining fell-fields, nevertheless always includes a number of other species, because the external conditions are more favourable to plant-life. Dwarf-shrubs and undershrubs are absent or rare ; grasses are much more numerous.

This vegetation reveals itself as a *fresh-green, dense* covering, which, if it be typical, is also *low* and *soft*, and for this reason may be described as *mat-vegetation*. Roots and rhizomes usually are closely interlaced, so that there arises either a soil tending towards raw humus or one approximating to that of European littoral meadows, with whose vegetation this shows the greatest physiognomic agreement. Among Dicotyledones shoots with broad rosulate leaves are common, presumably in accordance with the low height of the vegetation and the rich supply of light ; this is also true of subglacial communities with which mat-vegetation has other features in common, to wit, deep, pure, floral tints and certain xerophytic features. The majority of species are perennial. Grasses may preponderate. Amongst them undershrubs are interspersed. Mosses are intermingled in greater or smaller numbers ; but lichens are wanting, or rare and scanty.

---

[1] A comprehensive survey of formations influenced by man and by grazing animals was published by Bern-á-tzky, 1904. The ' waste-formation ' and its ' associations ' in Nebraska, are discussed by Pounds and Clements, 1898–1900. Negri, 1905, distinguishes in *Stazione culturale* four ' associations ', and in *Stazione ruderale*, three.

[2] The term ' grass ' is here used in a wide, oecological sense, and includes Gramineae, Cyperaceae, Juncaceae, Eriocaulonaceae, Xyridaceae, and similar, mainly tropical, Monocotyledones.

The mat-vegetation of Polar countries and that of Central European and other mountains display so close an oecological agreement that the two cannot be treated apart. Yet mat-vegetation should perhaps be subdivided into—

Mat-grassland—mainly of Gramineae.

Mat-herbage—mainly of dicotylous perennial herbs.

As a first step towards the formation of mat-vegetation we may regard that of the *snow-patches*.[1] These are, according to C. Schröter,[2] gently inclined, flat or concave, spots that occur on mountains and in Polar countries and are saturated with water from melted snow. If they be depressions within which the snow remains lying for a long time, then there is gradually deposited a thick, black, humus, which owes its origin to the snow. For this carries down from the air a quantity of organic dust; in addition, particles are deposited by the wind ; thus the snow is copiously ' manured '. This habitat is sharply characterized by lowness of temperature, abundance of humus, and permanent saturation of the soil ; and the community of plants settling upon it in Switzerland is extremely constant and is composed of only a few species, and, above all, of Salix herbacea, with which are found Alchemilla pentaphylla, Gnaphalium supinum, Ligusticum Mutellina, Plantago alpina, species of Soldanella, Sibbaldia procumbens, and others.

Arctic Mat-grassland.

In mat-vegetation of arctic countries grasses often preponderate over other Monocotyledones or over Dicotyledones. Brotherus mentions luxuriant mat-grassland in Kola consisting of Poa pratensis and Festuca rubra, in addition to which there are many perennial herbs, including species of Trollius, Ranunculus, Cochlearia, Geranium, Melandryum, and Cerastium, Rubus Chamaemorus and R. arcticus, Cornus suecica, also species of Archangelica, Matricaria, Solidago, and Rhinanthus. Similar mat-grassland is stated to occur in Nova Zembla, and recurs in Greenland, near Eskimo dwellings, also in Iceland. On this island cultivation often has played a part, inasmuch as the application of manure is a factor of great moment. As Thoroddsen says, ' The welfare of the country depends on its grass.' Here the commonest species include Anthoxanthum odoratum, Alopecurus geniculatus, Aira caespitosa, Poa trivialis, P. pratensis, and Agrostis alba ; there is, of course, an admixture of monocotylous and dicotylous perennial herbs.

But travellers do not sharply distinguish between those fields that are mainly clothed with grasses and those that are largely covered with dicotylous herbs. By the term ' pasture ' is obviously meant a field which has a fresh-green, close, and low vegetation suitable for grazing purposes.

Mat-herbage.

In arctic mat-grassland there is probably always a greater or less admixture of monocotylous and dicotylous perennial herbs. Where the dicotylous herbs gain the mastery over the grasses there arises a different type of vegetation which one may designate mat-herbage, herb-field, or,

[1] See p. 257.   [2] C. Schröter, 1904–8.

because it mostly occurs on slopes, herb-slope.[1] It is more widely distributed than typical grassland in Polar countries and on high mountains ; indeed one may encounter communities in which grasses scarcely develop at all. Such richly-flowered, fresh-green herbage occurs commonly in Greenland in sheltered spots, where the soil remains uniformly moist, and not only in the lowlands but sometimes also at considerable altitudes. It is low, dense, and soft, while its perennial herbs are largely rosette plants. In addition to perennial herbs, such dwarf-shrubs as Salix herbacea, S. polaris, and Cassiope hypnoides, are mingled with grasses. Fresh-green mosses, including Hypnum and Aulacomnium, also play a part.[2] The same type of community reappears in Iceland, in Scandinavia, and in Finland.

Oases in Tundra. Oases in tundra obviously represent examples of richly-flowered taller mat-herbage. They have been described by Middendorff[3] as present in Siberia, for instance on the slopes bounding the river Taimyr, where they are sheltered from raw winds and where the soil is a black humus. Caltha palustris, Geum glaciale, species of Potentilla, Ranunculus, Polemonium, Eritrichium, Oxytropis, Pedicularis, Saxifraga and Delphinium, and Papaver nudicaule, together with many other tall perennial herbs, with their numerous flowers and gay tints, cause these oases to stand out as bright spots relieving the general desolation. Similar vegetation is described by von Baer and Heuglin as occurring in Nova Zembla. Nathorst's slopes[4] on Spitzbergen, Kjellman's flower-field[5] in Siberia, and Pohle's flower-mats[6] in Kanin, are oecologically allied types of vegetation, and perhaps are luxuriant fell-fields. The 'oases' and the 'flower-carpet' described respectively by Middendorff and von Baer appear to differ from the mat-herbage of Greenland and the other places mentioned by their taller and less close vegetation, between which one can easily see the dark soil. How rich such herbaceous communities may be is proved by the fact mentioned by Heuglin that in Nova Zembla there were about fifty species of plants growing on a tract only a few square yards in extent. And Stefánsson mentions mat-vegetation in the Vatn valley in the north of Iceland, where twenty-four species occurred on a square ell.

In mat-herbage the leaves of the herbs may be large and luxuriant, as in Alchemilla vulgaris and species of Ranunculus and Potentilla ; this is due to the great atmospheric humidity, the long duration of the weak light, and to the fact that the soil is sheltered, well-lighted, and rich in humus. With the exception of certain species of Gentiana, the species are perennial, and green only during the vegetative season. Concerning growth-forms it may be noted that the caespitose habit with a persistent primary root or with a vertical multicipital root-stock seems to preponderate, but that travelling shoots also occur ; we have, however, only little information on these matters. Rosette-shoots are very general.[7]

[1] Warming, 1887; Rosenvinge, 1889–90.   [2] Warming, loc. cit., p. 38.
[3] Middendorff, 1867.   [4] 'Sluttningar.'
[5] 'Blomstermark.'   [6] 'Blumenmatten.'
[7] In regard to arctic mat-vegetation, the following works should be consulted : Middendorff, 1867; von Baer, 1838; Nathorst, 1883 a; Kjellman, 1882, 1884; Warming, 1887; Rosenvinge, 1889; Pohle, 1903; Cajander, 1903, 1905 a; A. Cleve, 1901; C. Schröter, 1904-8; Brockmann-Jerosch, 1907; Heuglin, 1874; H. Jonsson, 1905; Stefánsson, 1894.

Mat-vegetation of the Alps.

The distinction between mat-vegetation and meadow is not great, but mat-vegetation is shorter, so that it is essentially suited for grazing purposes and not for mowing. Mat-vegetation passes over into certain subglacial communities, as is natural, since it often occurs among these and forms their continuation lower down mountains, where, in fact, the conditions are more favourable to growth. As a typical example one may mention Kerner's[1] Carex ferruginea-'formation', which includes Solda-nella alpina, Gentiana acaulis, alpine auriculas, alpine anemones, Nigri-tella, Globularia nudicaulis, Phaca frigida, Lotus corniculatus, and other herbs, with Sesleria coerulea, Festuca violacea, F. pulchella, and other grasses ; in it one can also find stray dwarf-shrubs or prostrate under-shrubs, such as Erica carnea, Salix reticulata, S. retusa, Dryas, or others.

In the same category may be reckoned Stebler and Schröter's[2] Leon-todon mat-vegetation, which is composed of Leontodon hispidus, L. autumnalis, L. pyrenaicus, Crepis aurea, Homogyne alpina, Ligusticum Mutellina, species of Potentilla, Geum, Sibbaldia, and Plantago, also Soldanella alpina, Veronica alpina, and Polygonum viviparum, as well as grasses.

As mat-vegetation many botanists regard a number of communities which are, to some extent, of a different oecological stamp, and probably should be regarded as belonging to a different type of community. Stebler and Schröter discuss, inter alia, the following communities :—

1. Nardus stricta-association (nardetum), which grows on poor dry soil, and often alternates with rhododendrons or with dwarf-shrub heath. Scattered about this vegetation are Potentilla aurea, P. sylvestris, Calluna vulgaris, Leontodon pyrenaicus, Trifolium alpinum, Geum montanum, Arnica montana, Homogyne alpina, and Lycopodium alpinum ; such grasses as Aira, Anthoxanthum odoratum, and Festuca rubra ; Luzula albida and L. spicata ; masses of lichens belonging to the genera Cladonia and Cetraria ; and species of Vaccinium. This association, in many respects, is allied to heath or to a community of chersophytes.

2. Carex firma-association, in dry spots on limestone mountains, at altitudes of 2,000 to 2,900 metres, assumes the form of the topmost continuous carpet of dense, low, caespitose plants with short, stiff leaves. Accompanying Carex firma are Elyna spicata, Festuca pumila—which forms fine-leaved tufts—Carex nigra, and other grass-like plants ; also, ' scattered like pearls in the emerald sward,' a number of species of Saxifraga and Gentiana, Alsine verna, Campanula Scheuchzeri, Primula integrifolia, and others.

These two communities obviously betray some degree of xerophytism, and possibly it would be most correct to reckon the former as xerophytic and as constituting a special type of subglacial community.

According to Brockmann-Jerosch,[3] in Switzerland there are many types of meadows which correspond to various combinations of factors. Ex-posure plays a great part, because of the strong insolation and because

---

[1] Kerner, 1863.   [2] Stebler und Schröter, 1892.
[3] Brockmann-Jerosch, 1907.

on high mountains the difference between direct and indirect sunlight is greater than in plains. He distinguishes dry meadows with and without shade, on level and on sloping ground ; he recognizes fresh or lush meadows, which are or are not manured. There result many different associations that are characterized by different species.

Under the heading of alpine mat-vegetation possibly may be included Flahault's *prairies pseudo-alpines*, which develop on old forest-soil.[1]

In Denmark littoral meadow seems to be the community most closely allied to arctic and alpine mat-grassland and mat-herbage. Littoral meadow has dense, low, often soft vegetation with roots and shoots densely tangled, and thus resembles many, but not all, the above-mentioned types of mat-vegetation.[2] Certain alpine mat-grassland is similar to littoral meadow in showing many xerophytic characters—such as narrow, approximately terete leaves that betray a tendency towards succulence —which reappear on saline soil.

All high mountains display such mat-grassland and mat-herbage at the limit of forest. On the Andes, according to Brackebusch,[3] there are ' alpine meadows '—mainly pastures—on which abundant rainfall induces luxuriant growth of grass on the fertile soil, which is often interrupted by masses of rock. The flora varies with the latitude and altitude. In addition to many species of grasses there is an abundance of perennial and annual (?) herbs and small shrubs, which display a glorious show of blossom and belong to the Ranunculaceae, Malvaceae, Cruciferae, Polygalaceae, Geraniaceae, Caryophyllaceae, Rosaceae, Passifloraceae, and others. But there are also many humble Cactaceae, ferns, mosses, and lichens, interspersed in the vegetation, which therefore does not completely correspond to typical alpine mat-vegetation or meadow. Oecologically it may be most closely allied with certain xerophytic alpine mat-vegetation. R. Fries[4] mentions as widely distributed on the Argentine Andes evergreen Hypsela-meadows, in which the vegetative organs, pedicels, and peduncles, are very short, and the rosette-type of shoot prevalent.

CHAPTER LXXXIX. MEADOW

THE mesophytic communities previously described as occurring in arctic countries and on mountains must be regarded as natural, in so far as they have not been modified by man or have had their character only affected to an extremely slight degree by the grazing of cattle, sheep, and goats.

In all countries of moderate temperature and humidity mesophytic communities of grasses and herbs occur ; even in the north of Russia, according to Pohle,[5] there are natural alluvial meadows showing a wealth

[1] Flahault, 1901. In regard to alpine mat-vegetation and meadow, the following should be consulted : Kerner (1863), Stebler und Schröter (1892), Beck (1890), Radde (1899), Pax (1898), C. Schröter (1904–8), Engler (1901), Brockmann-Jerosch (1907), Hayek (1907), Szabó (1907).　　　　[2] See p. 230.
[3] Brackebusch, 1893.　　　　[4] R. Fries, 1905.　　　　[5] Pohle, 1903.

of true grasses, including Poa, Alopecurus, and the like, and of tall perennial herbs. In all countries in which atmospheric precipitations and moisture are equably distributed throughout the year, and in which man has practised cultivation for so long as to make his influence felt, there are artificial communities of grasses and herbs, to wit, *meadows* and *pastures*, that essentially owe their origin and composition to man. Many of these communities in cultivated countries grow on soil that was formerly clothed with forest, which has been destroyed by human agency; they are thus ' secondary ' productions. If we were to leave these to themselves, in time, they would certainly give way to forest. By manuring, cropping, and grazing, meadows have been more or less modified, especially as regards their constituent species. There are, on the contrary, meadows that are not the result of cultivation ; such, for instance, occur on high mountains, or near streams, where floods or ice exclude tree-growth.

In this and the next chapter we shall deal mainly with North-European meadow and pasture, which are two types of communities associated respectively with moderately moist and moderately dry soil. Natural meadow is more frequent than natural pasture, which is rare.

As providing types of *meadow* the lowlands of northern Europe may be cited.

Meadow stands as a link between mesophytic and hydrophytic communities ; some types of it are more closely allied with the latter, others distinctly belong to the former. The soil has a definite degree of humidity, which is 60 to 80 per cent. of full saturation. Its ground-water lies deeper and varies more in level than in swamp, and also flows more freely, so that the soil is periodically aerated. The soil is often rich deep humus, but may be sandy, particularly in the case of new meadows.

Meadow is a community of *tall, long-stemmed, perennial herbs*, and especially of *true grasses*. The covering of vegetation is *closely continuous* and compact, has a dense tangle of roots and rhizomes, and its plants are usually tall, being a foot or more in height, so that the soil is invisible. The vegetation owes its density in no slight degree to mowing and grazing. Mowing very materially affects the conditions of life of meadow, in that it prevents the maturation of seeds, promotes branching, and modifies the composition of the flora. The same effect as is produced by mowing is naturally brought about by regular summer-floods in the neighbourhood of many rivers. In the middle of the vegetative season all the plants are thus robbed of their vegetative epigeous organs.

The vegetation in summer is of a fresh-green tint, and, when estimated according to individuals, and often according to species, is mainly composed of Gramineae that have flat, fresh-green leaves which do not roll up in dry weather. The genera of grasses present include Aira, Avena, Dactylis, Festuca, Poa, Holcus, Anthoxanthum, Alopecurus, Phleum, Briza, Agrostis, and others. Often from twenty to thirty species of these are uniformly distributed throughout the same meadow. In addition to grasses the vegetation is formed by many monocotylous and dicotylous perennial herbs, belonging to the Ranunculaceae, Papilionaceae, Compositae, and so forth. But trees and shrubs are almost excluded (the latter may be represented by Salix repens), as are annual species, excepting on mole-heaps, where, for instance, Saxifraga tridactylites occurs. Meadow is distinguished by its wealth of blossom and consequent abundance of insects, as well as by its fresh-green tint; in the former respect it contrasts with the likewise green and very similar meadow-moor, which is poor in flowers. Between the herbs, especially when these are short, one often finds many *mosses*, including species of Hypnum, Aulacomnium, Mnium, and Bryum. Lichens, on the contrary, are wanting.

The *resting period* of the vegetation is occasioned only by frost; and, although in winter it is yellowish-grey and faded, meadow oecologically approaches very near to evergreen vegetation, for under the old leaves fresh green ones lurk, and many yellow leaves in a mild winter recover their green colour. In mild winters meadow may be green without interruption. Grass commences to grow at a temperature of 11° to 12·5° C.

ADAPTATIONS

The vegetation shows the following adaptive characters :—

1. The vast majority of the constituent species are *perennial*. For monocarpic plants (hemi-parasitic Rhinantheae excepted) there is neither sufficient light nor sufficient space; yet a few annuals and biennials, such as Linum catharticum and Cirsium palustre, occur.

2. Some species have *creeping rhizomes*, and are thus excellently fitted for producing a carpet of vegetation. Among grasses thus equipped are Poa pratensis, Festuca rubra, Agrostis vulgaris, and A. alba; among sedges are several species of Carex; among other perennial herbs are Lathyrus pratensis, Valeriana dioica, Epilobium palustre, Mentha, Lycopus, and Equisetum palustre.

3. Yet the majority of the grasses present are *caespitose;* such is the case with Aira caespitosa, Avena pubescens, Dactylis glomerata, Alopecurus pratensis, Anthoxanthum, Festuca elatior, Poa trivialis, Briza media, and Holcus lanatus. Indeed little or no power of vegetative locomotion characterizes most of the perennial herbs, such as Myosotis palustris, Rumex acetosa, Succisa pratensis, Geranium pratense, Polygonum Bistorta, Lychnis Flos-cuculi, Parnassia, and species of Ranunculus, Caltha, Trollius, and Primula. The reason for this is probably the resistance opposed to species with travelling shoots by the numerous, tough, and tangled roots and rhizomes of the grasses. Bulbous and tuberous plants are more uncommon, but are represented by Orchis and Colchicum autumnale.

4. The *leaves* are thin, flat, broad, flexible, and glabrous; they possess neither thick-walled epidermis nor any other special means of protecting themselves against excessive transpiration. The leaves of the grasses show stomata on both faces and are incapable of rolling up. Mechanical tissue is developed to little or no extent.

FLORA

The flora of different meadows varies widely, in harmony with differences in moisture of soil, geographical situation, and agricultural practice (grazing, mowing, ditching, irrigation, and so forth). Weber,[1] for instance, recognizes several 'sub-formations' of natural meadow. Among these we must regard as mesophytic meadow that which occurs

---

[1] Weber, 1892.

on soil higher and drier than marsh and exhibits the following ' sub-formations ' (associations) :—

1. Poa pratensis, which is 2 to 3 metres above the mean water-table.
2. Poa trivialis, 1 to 1·5 metres above the same.
3. Aira caespitosa, which in June and July is ·4 to ·7 metre above the level of water in pools.

The ' sub-formations ' of Carex panicea, of C. gracilis, and of Molinia coerulea, on the other hand, must be regarded rather as belonging to moor.

The ' sub-formation ' of Festuca elatior belongs to the transition between marsh and drier ground.

In grassland derived from true marsh that has been diked the ' sub-formations ' of Agrostis alba, of Poa pratensis, and of Lolium perenne are to be placed in the category of mesophytic meadow or pasture (see p. 231).

DISTRIBUTION

In mountainous country true meadow occurs in many lands, for instance in Norway and Switzerland. Günther Beck's [1] ' valley-meadows ' are among such ; they are for the most part mown twice in the year, and contain twelve species of grasses together with many other herbs. In Switzerland alone there are a number of different meadow-associations.[2] Schröter [3] classifies these into :—' dry meadow', ' wet meadow ', ' fresh meadow '.

Belonging to the same type are Adamovicz's [4] Servian types of 'forest-meadow ', ' mountain-meadow ', ' valley-meadow '.

In reference to ' orchid-meadow ' and other types of Bohemian meadows readers should consult Domin's papers.[5] Other communities of tall perennial herbs are mentioned by Hayek [6] and C. Schröter.[7] The latter refers to the ' richly foliaged stems whose broad, horizontally extended, shady leaves allow nothing to spring up from the ground '.

Certain prairies seem to approximate to meadow. For instance, dealing with prairies in Nebraska, Pound and Clements [8] write : ' They are to be regarded in general as mesophytic ; two sorts may be distinguished, high prairies and low prairies ; the latter have much in common with meadows and pastures ; the former bear no small resemblance to the sand hills in certain respects. The principal grasses of the former are sod-formers, of the latter, bunch-grasses.'

In *eastern Asia* there occur meadows in which the grasses are taller, and the dicotylous herbs so much so that in places they may be several feet in height. The characteristic appearance of meadow is thus lost, and there arise *communities of tall perennial herbs* of which Asia provides examples in several regions. Kittlitz has pictured mixed societies of stately, tall, perennial herbs, including huge species of Heracleum, which rise above the luxuriant meadow soil : in addition we may cite the ' park-lands ' of eastern ·Asia, where grassland has become occupied by trees and shrubs and thus acquired a likeness to savannah.[9] The same formation reappears in Sweden, in a form that may be termed ' wood-meadow ',[10] also in the United States.[11] Park-land, in which meadow alternates with

[1] G. Beck, 1890.   [2] Stebler und Schröter, 1892 ; cp. Stebler und Volkart, 1904.
[3] Schröter, 1904 ; cp. Vierhapper und Handel-Mazetti, 1905.
[4] Adamovicz, 1898.   [5] Domin, 1904, 1905 a.   [6] Hayek, 1907.
[7] C. Schröter, loc. cit.   [8] Pound and Clements, 1898.
[9] Grisebach, 1872.   [10] Hesselman, 1904.   [11] Cowles, 1901 b.

dicotylous woodland, seems in general to be the form of vegetation natural to river-plains in the temperate zone. As our information in regard to these communities is very incomplete, it is impossible to assign them to their correct position.

The same is true of the ' grassland of the creeks ' in Usambara, of which Engler [1] writes : ' A little above the sea-level there extend inland, often for miles, great sand-fields or stony tracts, which are for the most part under water during the rainy season.' Here grow Cyperaceae, Eriocaulonaceae, Ipomoea Pes-caprae, and other plants.

CHAPTER XC. PASTURE ON CULTIVATED SOIL

FROM meadow to pasture is no great step. The difference depends particularly upon moistness of soil. Pasture is higher and drier ; it is exposed to no greater amount of moisture than that brought by ordinary atmospheric precipitations. The vegetation of pasture is shorter and more open than that of meadow ; often it cannot be mown, but only grazed over. The driest pasture, in which deep-rooted herbs preponderate over grasses, may be termed ' waste herbage ' (German ' Trift '), and merges with the chersophytic vegetation described in Section XII.

Pastures in the plains of northern *Europe*, and other regions that were formerly clothed with forest, are almost without exception artificial products : were the human race to die out they would once more be seized by forest, just as their soil was originally stolen from forest. Exceptions to this rule are provided only by small patches of meadow in old forests, that have been regularly grazed over and manured by wild animals. Spiranthes spiralis is described as being characteristic of such stations in certain parts of northern Germany. Pasture usually consists mainly of grasses, which over a large part of Europe belong to the same species, namely Festuca rubra, Lolium perenne, Anthoxanthum, Poa pratensis, Agrostis vulgaris, species of Bromus, Triticum repens, Holcus mollis ; even in the Italian *pascoli* one finds many of these species. But an essential part is also played in pasture by dicotylous herbs such as Taraxacum, Leontodon, Bellis, Chrysanthemum Leucanthemum, Achillea Millefolium, Campanula rotundifolia, species of Plantago, Ranunculus, Cerastium, Trifolium, Daucus, Pimpinella, and Carum. Many mosses, including species of Hypnum, may be interspersed.

The composition of the flora is of slight interest, because pasture has been so metamorphosed and modified by cultivation, in accordance with the use to which the farmer puts it. A number of associations will certainly in the future be distinguished. For instance, R. Smith [2] recognized in Scotland pasture of the basalt hills, pasture of the Silurian hills, and pasture of the Pentlands. Furthermore, experience has taught us that water plays a very important part in regard to pastures, and that the constituent plants are easily affected by it. On p. 45 it has already been noted that Feilberg showed how the vegetation on the plains near Skagen in Jutland changes with the level of the water-table. According

---

[1] Engler, 1894.   [2] R. Smith, 1900 a.

to this admirable observer there is, between the grass of Jutland and of Seeland respectively, a difference which is to be attributed to the circumstance that in spring rain falls more frequently in the former place. In addition, Weber's observations cited on p. 324 show that the vegetation is dependent upon the level of the water-table.[1]

The *Icelandic* pastures, modified by cultivation to a relatively less extent, have been investigated by Feilberg and Stefánsson.[2] The most important grasses present are Festuca rubra, Poa alpina, P. pratensis, and Aira caespitosa ; many others appear on manured spots and near springs.

In regard to grasslands on the *Faroe Isles* Ostenfeld's[3] work should be consulted.

Pasture also occurs in the *tropics*, where it is always artificial. In *Brazil*, on old forest soil, one very often encounters pasture which is an extremely dense association formed by the glutinous Melinis minutiflora (*capim gordura*). A few other plants, including shrubs, may be interspersed, but it is this grass that dominates and at flowering time lends to the landscape a reddish-brown colour.[4]

In the *West Indies* there is pasture which is composed partly of indigenous species and partly of introduced species of Panicum and Paspalum, also of Avena domingensis, Pennisetum setosum, Sporobolus, and others. Mingled with the grasses are certain sedges, including species of Kyllinga and Fimbristylis. There also are species of Cassia, Sida, Cipura, and other herbs and small shrubs. The shrubs would soon suppress the herbaceous vegetation were they not regularly clipped. The pasture is found on old forest-soil, and did not originally occur on the islands.

The *Sandwich Isles* possess unusually extensive grasslands, composed, according to Hillebrand,[5] of Paspalum, Panicum, and particularly of Cynodon Dactylon, though the last-named was introduced within the last few decades. These grasslands have at least been materially modified by man, and probably owe their origin entirely to cultivation. They are described as dense mat-vegetation.

In regard to the ' cogonales ' on *Luzon* Whitford's[6] work should be consulted.

In *Australia* intact virgin grassland seems to occur ; it is composed partly of such grasses as Poa, Glyceria, Briza, Festuca, and Panicum, but partly of Liliaceae and others. Especially common is kangaroo-grass (Anthistiria ciliata and A. imberbis), whose leaf-structure recalls that of European meadow-grasses. But these grasslands to some extent show steppe-like characters.

[1] See Gräbner, 1898 a, 1898 b.   [2] Stefánsson, 1894.
[3] Ostenfeld, 1908.   [4] Warming, 1892.
[5] Hillebrand, 1888.   [6] Whitford, 1906.

CHAPTER XCI. MESOPHYTIC BUSHLAND

In discussing certain bushlands occurring far north and on high mountains, it was pointed out [1] that perhaps these would most correctly be regarded as mesophytic, although they display strongly developed means of checking transpiration.

In *Greenland* and other *arctic places* such bushland presents itself in the form of willow-bushland—salicetum.[2] One finds it at the bottoms of valleys and in ravines, in sheltered sunny spots, particularly where flowing *water* or water trickling from rocks provides a uniform supply of moisture, and where *humus* accumulates and earthworms burrow. In the south of Greenland it is Salix glauca that gives rise to extensive, frequently almost impenetrable bushland, which is some metres in height, but farther north attains scarcely one metre and produces more or less prostrate branches. Beneath the willows there flourish large and broad-leaved, fresh-green, perennial herbs, such as Archangelica officinalis, Oxyria, Taraxacum officinale, Alchemilla vulgaris, species of Potentilla, Epilobium angustifolium, and Arabis alpina, also Poa alpina and other broad-leaved grass-like plants, ferns, and large lax mosses, including Hylocomium, Hypnum, and Dicranum. Here and there leaf-peat is formed.[3]

On *Norwegian mountains* there appears a *willow-region*, which differs from that in Greenland in that the bushland is formed by many different species of Salix, including S. Lapponum, S. lanata, S. Arbuscula, S. glauca, S. phylicifolia, and S. nigricans, also in that the ground entertains a still richer herbaceous flora. This bushland provides a transition to oxylophytic bushland. Bonnier and Flahault [4] apply to it the term of ' willow-prairie ', and point to these extensive willow-bushlands as providing a distinction from the Alps, where most of the same species of Salix occur but do not dominate to the same extent. Bushland of the same type occurs in Lapland and Siberia. Willow-bushland is very general throughout temperate Europe on the *banks of streams* outside the marshy belt, and is represented, for instance, in Servia by Adamovicz's 'willow-meadow' [5]; it even occurs on flat islets in the Amazon.[6] Other bushlands above the limit of forest are composed of birches or birches and willows, which are accompanied by alders, other shrubs, and such tall perennial herbs as Aconitum, Ranunculus, Digitalis, Geranium sylvaticum, Vicia, and Lathyrus, or, in the interior of Lapland, by Veratrum, Senecio nemorensis, and others. Birch-bushland here and there merges with birch-forest.

Among mesophytic bushlands that of *green alder* may be mentioned. Alnus viridis, in the Alps at altitudes of 1,200 to 2,000 metres, on naturally irrigated spots, gives rise to dense bushland with tall perennial herbs clothing the ground.[7]

The lowlands of temperate regions abound in bushlands similar to those formed by willows. Evergreen *holly-bushland* appears on the south-west coasts of Norway.

Mesophytic and xerophytic bushlands merge with one another. As

[1] p. 215. [2] Warming, 1887 ; Pohle, 1903. [3] Pohle, loc. cit.
[4] Bonnier et Flahault, 1879. [5] Adamovicz, 1898.
[6] Grisebach, 1880, p. 388.
[7] C. Schröter, 1904–8 ; Brockmann-Jerosch, 1907, p. 275.

an intermediate type one may safely regard bushland described by Günther Beck [1] as composed of Prunus spinosa, Crataegus, Rosa, Cornus, Berberis, blackberry, raspberry, and other plants, which deck themselves with a snow-white show of blossom in spring and bear glossy berries or drupes in autumn. Countless perennial herbs clothe the ground, and the plants that demand light in high-forest congregate in such bushland. In many places these species that give rise to bushland produce underwood beneath such light-demanding trees as Fraxinus, Populus tremula, and Prunus Padus.

Allied to the preceding are certain bushlands or thinly wooded forests described by Blytt as occurring on *rubble-heaps* in the south of Norway. These are composed of Corylus, Ulmus, Tilia, Fraxinus, Acer, Sorbus, Quercus, Rosa, Crataegus, and others, under the shelter of which develops a rich flora of more southern forms, including aromatic Labiatae, Geranium, Hypericum, Dentaria bulbifera, Lathyrus sylvestris, L. vernus, L. niger, and various grasses. By the term 'rubble-heap'[2] is meant a soil composed of loose stones that have been cast down to form an accumulation (talus). When such a heap abounds in species and vigorous individuals, this is due to several causes : First, wind causes an accumulation of inorganic and organic particles among the stones. Secondly, water accumulates beneath the stones and can evaporate only with difficulty. Thirdly, this stony soil readily becomes heated. Fourthly, such rubble-heaps nearly always occur on sloping spots at the foot of walls of rock, where they are easily heated, if the aspect of the slope be not too unfavourable.

Mesophytic bushlands owe their existence to divers causes. Those mentioned as occurring in Polar lands and high up mountains occur in spots where the conditions for growth (temperature) are unfavourable to forest, but are too favoured for mere mat-vegetation. Other bushlands are the result of cultivation, since they represent the remnants of forests which have been felled by human agency and are still kept under by hostile circumstances that result directly or indirectly from human interference : belonging to this type are oak-bushland in Jutland,[3] bushland in the Balkan peninsula composed of oaks, Rhus Cotinus and others,[4] also the hornbeam-scrub mentioned by Focke [5] as occurring above the level of marsh on German coasts bounding the North Sea. 'Bush-pasture' in the Alps is likewise a 'zoogenetic' admixture of more than one type of vegetation, according to Brockmann-Jerosch.[6]

CHAPTER XCII. DECIDUOUS DICOTYLOUS FOREST

DECIDUOUS dicotylous forest is that type of forest in which the trees are leafless for a longer or shorter season of the year, and are foliaged during some (five to eight) months.[7] This behaviour is closely correlated

---

[1] G. Beck, 1890–3.   [2] See p. 246.
[3] Vaupell, 1863; Warming, 1907.
[4] Grisebach, 1872; Adamovicz, 1898; Vahl, 1907.   [5] Focke, 1893, p. 261.
[6] Brockmann-Jerosch, 1907.
[7] The ash-tree in Denmark may bear foliage for only four months. The beech-tree in Madeira apparently is in leaf for eight months.

with climate, and is exhibited most frequently in temperate and cold regions where there is a winter, but also within the tropics where the dry season is protracted. Tropical forests have already been discussed in connexion with xerophytes; in them the old leaves are often stiff or hairy. In mesophytic deciduous forests, on the contrary, the leaves are thin, flexible, and relatively translucent, possessed of a thin-walled epidermis, dorsiventral in structure, and are often plastic (as in Fagus) in relation to external conditions. They usually arrange their blades perpendicular to the strongest diffuse light. Their shapes are manifold. There are undivided, divided, and compound leaves ; but their division is not so complete nor into so many leaflets as in the case of species belonging to tropical rainforest.

There is therefore a season of foliation and one of defoliation. At the former time one sees only the young, usually fresh-green, shoots ; but in the tropics, reddish tints due to anthocyan also show themselves, and may even occur in Central Europe, for instance, on species of Quercus and Acer. The foliage gradually assumes a darker tint of green as summer progresses ; before leaf-fall, yellow and red colours appear, partly because the chlorophyll is decolorized, as in yellow leaves, and partly because anthocyan is produced, as in red leaves (which are displayed in their full glory by North American trees).

Leaf-fall is usually associated with the commencement of the *cold* season ; one and the same species may lengthen or abbreviate its vegetative season according to the local climate. The more proximate cause of the phenomenon must probably be sought in the desiccation that is threatened by the cold soil ; the causes of leaf-fall are certainly the same when it is evoked by cold or by direct drought.

During the resting season protection against intense transpiration is provided by *bud-scales* clothing the youngest parts of the shoots, and by cork investing older parts.

Mesophytic deciduous trees often have a rich system of branches with many small twigs ; nearly all the buds, excepting those near the base of the year's shoot, develop into branches, but conditions of illumination may affect the result. In this way there arises a canopy of leaves more continuous than is wont in a tropical tree.

Compared with evergreen trees deciduous ones do not live under conditions so favourable to existence, since a large part of their lives is passed in inactivity : they rarely attain the gigantic dimensions reached by evergreen trees in tropical rain-forest.

In *mesophytic forests of temperate countries* a leading part is played by the Amentiferae, also by Fraxinus, Acer, Tilia, Populus, and Ulmus ; while in warmer lands many additional trees are gradually added to these. In the forests of North America and Eastern Asia numerous other genera occur.[1]

In the *dicotylous forests of northern Europe* the trees are mostly windpollinated, and bear flowers very early in the year, before or during the act of foliation ; the flowers hibernate inside buds or naked. Some southern forms, lime-trees for instance, do not blossom before summer, and are insect-pollinated.

In dicotylous forest there are at least two, and commonly several, *storeys*

[1] See p. 335.

of vegetation. The number and nature of the plants associated with definite trees varies with the amount of shade that these cast : but this is more fully discussed in the sequel.

Herbs on the floor of the forest are mostly tall and possessed of elongated internodes : they are not rosette-plants. The leaves of plants forming the underwood and vegetation on the ground are similar to those of the trees forming the high-forest, but are thinner and still less xerophytic ; some are definitely sciophylls, whose structure approaches that of a hydrophyll. This is mainly due to shade and a moist atmosphere, but possibly also to the damp humus-soil. The leaves are therefore generally large, broad, flat, thin, unpolished, and glabrous—for instance, in Oxalis Acetosella, Anemone nemorosa, Impatiens Noli-me-tangere, Lactuca muralis, species of Corydalis, Circaea, Paris, Adoxa, Mercurialis, and Convallaria. Grasses of the forest have broad, flexible, mostly arcuate leaves, which are devoid of any power of rolling up, and bear unprotected stomata on both faces or largely on the upper face ; as examples may be mentioned Milium effusum, Poa nemoralis, Melica uniflora, M. nutans, Dactylis glomerata, Festuca gigantea, Bromus erectus, and Brachypodium sylvaticum.

Many plants in moist shady forest are, according to Wiesner,[1] ombrophobous, and it is impossible to wet the surfaces of their leaves ; but others, including Sanicula europaea, are ombrophilous.

Epiphytes are represented mainly by mosses and lichens, never by Spermophyta definitely adapted to an epiphytic mode of life. Of *lianes* there are but few—Lonicera Periclymenum, Hedera, Humulus, and Clematis.

The soil of the forest entertains many *saprophytes*, including fungi that show themselves especially in the autumn of moist years. Among Spermophyta there are only a few holosaprophytes—Monotropa, Neottia, Epipogum, and Corallorrhiza—but probably many hemisaprophytes, including Orchidaceae and species of Pyrola. Mycorhiza occurs in connexion with many trees and with saprophytes.

ASSOCIATIONS

As examples of associations belonging to dicotylous forest in temperate countries, we may mention beech-forest, oak-forest, and birch-forest.

Beech-Forest.

Beech-forest is most luxuriantly developed in Denmark and in the west of Germany on humus-soil. Fagus sylvatica, the beech, is very definitely a shade-enduring tree ; its tall, slender, smooth, light-grey trunk bears a crown that is rendered dense and shady by the distichous phyllotaxis, the numerous dwarf-shoots, the mosaic of leaves, and by the power possessed by the leaves of assimilating in weak light. The light reaching the ground is very subdued, for which reason there is no underwood, and the ground in many beech-forests is extremely poor in plants— a result that is also partly due to the thick carpet of fallen foliage.

[1] See p. 32.

The soil varies widely in nature, and the vegetation on the ground varies accordingly; the most important varieties are: (a) soil with mild humus; (b) soil with sour humus.[1] As a whole, the beech prefers a good, deep, marly soil.

(a) The *mild humus-soil* of beech-forest is friable, porous, and, being excavated by earthworms as well as other small animals, it is well ventilated. The volume of its pores in the superficial layers is 50–60 per cent.; its particles are freely mobile. In the height of summer the ground is often almost covered with brown, faded beech-leaves, which, together with fallen twigs, cupules, and so forth, form a thick carpet that is sharply cut off from the disintegrated substratum. Only here and there, where light penetrates, does one find Spermophyta, including Asperula odorata, Oxalis Acetosella, Anemone nemorosa, A. ranunculoides, Hepatica, Viola sylvatica, Mercurialis perennis, Melica uniflora, Milium effusum, Stellaria nemorum, species of Corydalis, and Hedera Helix. Mosses are wellnigh unrepresented where the thick layer of dead leaves lying on the floor of the forest checks them; those that do sometimes occur, for example Bryum argenteum, form a very thin coating over the soil.

The vegetation on the ground of beech-forest has the characteristic that it is a *spring-vegetation* with a very brief vegetative season. It must utilize the light before the high-forest is foliaged, or while this shows only young leaves. The acts of flowering, assimilation, and fruiting, follow in rapid succession, so that at midsummer a number of species are only represented above ground by slight traces. Such is the behaviour of the plants that are particularly characteristic of beech-forest in northern Europe—namely, of such plants as species of Anemone, Corydalis, Gagea, and some species of Primula.

But there are other plants that remain green for a longer period: among such are Mercurialis perennis, Oxalis Acetosella, Stellaria Holostea, S. nemorum, Pulmonaria officinalis, Luzula pilosa, Carex digitata, C. remota, and the grasses Milium effusum, Melica uniflora, Dactylis glomerata, Poa nemoralis, and others.

In some of the early-flowering species the embryo has reached only a very early stage of its development when the seed falls, as in Eranthis hyemalis, and it may even be unicellular, as in Ficaria and Corydalis cava. This may possibly be correlated with the fact that these spring-plants enjoy only a short vegetative season; the seed acquires nutriment from the parent plant, but the subsequent development that is normally continued on the parent plant does not take place until the detached seed experiences a secondary process of ripening.

In accordance with the brevity of the vegetative season and with the early date of flowering, nearly all the species are perennial herbs; yet such annuals as Impatiens Noli-me-tangere and Cardamine impatiens are not wanting.

The loose texture of the soil favours the development of horizontal *subterranean travelling* shoots. These are possessed by numerous species, including Aspidium Dryopteris, Anemone nemorosa, A. ranunculoides, Asperula odorata, Mercurialis perennis, Dentaria bulbifera, Stellaria nemorum, S. Holostea, Oxalis Acetosella, Adoxa Moschatellina, Stachys

---

[1] Consult P. E. Müller, 1878, 1884, 1887 a, 1894; Höck, 1895.

sylvatica, species of Circaea, Paris quadrifolia, Convallaria majalis, species of Polygonatum, Cephalanthera, Epipactis, Listera ovata, Melica uniflora, also the saprophytes Neottia, Corallorrhiza, Epipogum, Limodorum, and Monotropa, the last of which has roots that emit shoots.

Shoots that *travel above ground* are possessed by Glechoma hederacea, Lysimachia nemorum, Lamium Galeobdolon, and Lycopodium annotinum.

Tubers occur on species of Corydalis, Arum maculatum, Cyclamen (for example, in beech-forest on the Alps), Phyteuma spicatum, species of Orchis, and Ophrydeae.

Bulbs present themselves on Gagea, Allium ursinum, Lilium Martagon, Galanthus, and Scilla bifolia.

Plants that are *confined to their point of fixation* are : Campanula Trachelium, Epilobium montanum, Sanicula europaea, Hieracium murorum, Pulmonaria officinalis, species of Primula, Actaea spicata, Brachypodium spicatum, Festuca gigantea, Luzula pilosa, Aspidium Filix-mas, A. spinulosum, and Athyrium Filix-femina.

Lichens do not occur.

(b) *Beech-forest upon a soil with sour humus* has an entirely different vegetation (P. E. Müller's trientale-vegetation[1]) on the ground. Fortunately such beech-forest is of restricted occurrence, excepting where it is exposed to conversion into heath. The firm soil is permeated with roots and fungal hyphae and the volume of its pores is relatively small ; not being excavated nor consequently aerated by earthworms it produces humous acids ;[2] moreover, it is dried up by the sun's rays, and its leaf-covering is often soon blown away. On it usually flourishes a dense vegetation of Aira flexuosa, a setose-leaved, xerophytic grass producing compact soft tufts ; in addition there are Trientalis europaea, Maianthemum Bifolium, the hemiparasitic Melampyrum pratense (the two last-named also occur on mild humus), and a rich vegetation of mosses. The close, soft mossy carpet consists of Polytrichum formosum, Hypnum Schreberi, H. cupressiforme, H. purum, Hylocomium triquetrum, H. splendens, Dicranum scoparium, Leucobryum glaucum, species of Mnium, and other species ; Sphagna may also, but only rarely, be found on the frequently wet, somewhat marshy soil. Calluna and Vaccinium Myrtillus are often present, and thus the soil approaches that of ling-heath.

With the way thus paved, natural regeneration of the beech is no longer possible, the beech-forest ultimately disappears in many places, and is replaced by ling-heath.[3]

The northern limit of beech-forest continues from the south of Norway, through eastern Prussia to the Caucasus ; the companion-plants of the beech, of course, vary with the locality.

Oak-forest.

The common oak (Quercus pedunculata and Q. sessiliflora) is a tree making moderate demands in regard to light, has quincuncial phyllotaxis, and shows somewhat irregular branching. Its tortuous branches give rise to a crown that is neither so dense nor so shady as that of the beech. In Denmark the beech suppresses the oak, because, among other reasons,

[1] P. E. Müller, 1878, 1884, 1887 a, 1894.   [2] See pp. 62, 78.
[3] P. E. Müller, loc. cit. ; Gräbner, 1895, 1901.

the former is a shade-enduring tree whose foliage shoots out earlier in spring than that of the oak. Only in humid places—for instance, on the low-lying clay soil of Laaland and on the more sterile soil of western Jutland—can the oak successfully compete with the beech.

The high-forest is of very mixed nature, because the demands for light made by the oak are moderate. In German and Austrian oak-forests, Tilia, Acer, Populus tremula, Ulmus, Fraxinus, and Carpinus are interspersed; in France, Fagus and Castanea are often subordinate members of oak-forest.

In opposition to beech, the oak has beneath it an abundant *underwood*, which is often formed by a dense bushy growth of Corylus, Crataegus, Acer campestre, Prunus spinosa, Carpinus, Rhamnus Frangula, Euonymus europaea, Salix, Viburnum Opulus, Rubus Idaeus, Lonicera Xylosteum, and others; of these shrubs the particular species present vary with the nature of the habitat. In certain cases, Juniperus, Pteris, and even Calluna, may occur, especially in oak-forest on poor sandy soil. In Austrian oak-forest there also occur Viburnum Lantana, Ligustrum vulgare, Staphylea pinnata, Daphne Mezereum, and others.

The soil of oak-forest may be a good, black or greyish-brown, friable, mild humus, inhabited by earthworms. Beneath and between the shrubs forming the bushy underwood flourish numbers of grasses and herbs, which, however, do not produce a continuous carpet of vegetation. Among them are species of Anemone and Viola, Vicia Cracca, Lathyrus macrorrhizus, Hypericum perforatum, H. quadrangulum, Potentilla sylvestris, Campanula rotundifolia, and Achillea Millefolium. In addition, Pteris aquilina plays a prominent part. The majority of plants on the ground of the forest blossom in spring. The soil may also partake of the nature of sour humus; but the sour humus of oak-forest is very different from that of beech-forest.[1] Now and again oak-forest grows on marshy, badly ventilated soil, or upon fine-grained, dense, sandy soil, or in the flood-zone bounding rivers in sheltered spots.

Birch-forest.

The common birch (Betula pubescens and B. verrucosa) is a tree making strong demands for light, as is indicated by its loose, open crown. It can grow on very diverse kinds of soil : in clefts of rocks, on dry gravel or sand, on moist humus, even on wet, boggy soil. The flora growing on the ground in birch-forest varies widely with the nature of the soil, and is often very rich because plenty of light reaches it. In some cases grass forms a continuous covering over the soil. In other cases the vegetation forms a kind of heath with a dense growth of Cladonia rangiferina, Polytrichum juniperum and other mosses, Molinia coerulea, Salix repens, Calluna, species of Carex, and others (Gräbner's ' birch-heath ').

The birch is often accompanied by Pinus sylvestris, also often by Populus tremula and Salix. The first case illustrates the difficulty of grouping communities into those of xerophytes, mesophytes, and so forth ; for the evergreen xerophyte grows side by side with the mesophyte.

---

[1] P. E. Müller, 1878, 1884, 1887 a, 1894.

Ash-forest.

On the east coast of *Jutland*, in lower *Austria*, and elsewhere, Fraxinus excelsior, the ash-tree, on a loose moist soil produces true forests, with a dense ground-vegetation composed of tall herbs that, apart from this, usually occur in open, moist fields or meadows.

Alder-forest.

Alnus incana in the north of Sweden gives rise to forest with a ground-vegetation of Spiraea Ulmaria, Geranium sylvaticum, Geum rivale, Aira caespitosa, Milium effusum, Urtica dioica, mosses, and others.[1]

Other Deciduous Forests.

In a similar manner several other indigenous European trees give rise to pure or mixed associations, whose character differs with the humidity of the soil and the conditions of illumination in the forest.

In *Sweden* there is a mixed forest, *wood-meadow*, consisting of such deciduous trees as oak, elm, maple, lime, and hornbeam, beneath which there is an undergrowth of shrubs and tall herbs that is very luxuriant and rich in species ; it is open, well lighted, and park-like, has a rich humus-soil, and possibly represents forest that has been more or less changed by cultivation.   Hesselman[2] has studied it in detail, particularly as regards the demands for light, the assimilatory activity and transpiration of the perennial herbs, and in general the mode in which these perform their vital functions under various external conditions.

In the *region of the Danube*, about half-way from the source of this, the forests are remarkably mixed in nature, and include a profuse mixture of Fagus, Carpinus, Quercus sessiliflora, Acer, Betula, Prunus Cerasus, Pyrus communis, Populus, Tilia, and Coniferae.   The underwood consists of Berberis, Cornus sanguinea, C. Mas, Euonymus europaea, E. verrucosa, species of Prunus, Juniperus communis, and others.   In addition, dwarf-shrubs belonging to the Ericaceae, and Polygala Chamaebuxus, occur.[3] This complexity indicates closer proximity to the tropics, and probably also has a geological cause, namely, that the land has been free from ice for a longer time than, say, Scandinavia, so that the immigration of species has been more effective here than in the case of Scandinavia.

In *Mediterranean countries* on mountains there are other forests, formed by Castanea sativa, Platanus orientalis, and so forth.   Populus nigra and P. alba give rise to ' poplar-meadow ' in Servia.[4]

North America has its belt of deciduous dicotylous forest corresponding to that of Europe.   A strong admixture of species is characteristic of North American forests.   In addition, the underwood is denser and taller. Lianes are more abundant.   Yet the physiognomy of the forest is much the same as in Europe.   Excepting in the south, epiphytes are confined to mosses and lichens.   The yellow and red autumn tints are unusually deep, particularly in species of Quercus, Crataegus, and others.   The flora is different.   Many genera are indigenous to the temperate zone,

[1] Grevillius, 1894.   [2] Hesselman, 1904.
[3] G. Beck, 1890-3; Kerner, 1863; Vierhapper und Handel-Mazetti, 1905.
[4] Adamovicz, 1898.

and are represented here by numerous species : such is the case with Quercus, Juglans, Carya, Betula, Alnus, Ulmus, Celtis, Fagus (F. ferruginea), Castanea, Carpinus, Ostrya, Populus, Salix, Acer, and Fraxinus. But, in addition, many genera that are subtropical or reminiscent of the tropics, and include deciduous species, extend into this forest, especially in the southern and eastern parts, because the connexion of the land with more southern regions has permitted facile migration since the Glacial Epoch. Among such genera foreign to the north of Europe are Magnolia, Liriodendron, Robinia, Gleditschia, Gymnocladus, Catalpa, Morus, Liquidambar, Sassafras, Platanus, and Aesculus.

In North America dicotylous forest is represented by many different associations, according as different species dominate. In Michigan, according to Livingston,[1] *beech-maple* forest represents the highest stage, and grows on the best soil; then succeed *maple-elm* association, *oak-hickory* association, *oak-hazel* association, and *oak-pine* association, all of which stages denote deterioration in quality of soil.[2]

Antarctic deciduous beech-forest. In Patagonia there are deciduous forests composed of Nothofagus antarctica, N. obliqua, N. procera, N. Montagnii, and N. Pumilio. They occur especially in the higher parts of mountains in a moderately moist zone, and on the drier eastern slopes of the Andes, where they rise like islets out of the eastern Patagonian steppe-region, whereas evergreen forests composed essentially of Nothofagus betuloides occur in the rainy belt.[3]

Japanese forest, like North American, is also very rich in species, and thus contrasts with ordinary European forest ; a luxuriant mountain-forest shows us in flower a hundred species of trees and shrubs, belonging to at least seventy-six genera. In this case the cause of the multiplicity is certainly geological. The forest region of Fujiyama, according to Rein,[4] includes mainly dicotylous forest, but here and there conifers give rise to woods. The dicotylous forests consist largely of deciduous oaks, beeches, and maples, to which are added species of Zelkova, Juglans, Pterocarya, Betula, Tilia, Fraxinus, Magnolia, Cercidiphyllum, Acanthopanax, and Aesculus. The flora shows a strong affinity to that of eastern North America. There are numbers of lianes belonging to the genera Actinidia, Celastrus, Vitis, Rhus, Wistaria (W. chinensis), Akebia, and Clematis. The underwood is very rich. Japanese forests are remarkable for the beauty of their autumn tints. Somewhat similar forests occur in Caucasia, Talish, and in the north of Asia Minor.[5] Japanese forest differs from European in being so rich in species and in the number of lianes present.

APPENDIX. MONSOON-FOREST

The term monsoon-forest ' is due to Schimper, who thus designates tropical, deciduous high-forest. It occurs within *the tropics* in districts where the annual rainfall exceeds 180 centimetres and there is a prolonged dry season. Such a combination is somewhat infrequent, and appears to be encountered only on the open slopes of mountains exposed to the

[1] Livingston, 1903.
[2] For further details consult also Mayr, 1890; Cowles, 1901 ; Whitford, 1901, 1905.
[3] Consult Neger, 1897, 1901 ; Dusén, 1905; Reiche, 1907.
[4] Rein, 1879, 1881 ; Yokoyama, 1887.   [5] Radde, 1899.

summer-monsoon in southern Asia. According to Brandis,[1] on the west
coast of cis-Gangetic India up to a certain altitude forests are deciduous;
at a greater altitude, where atmospheric precipitations in winter are more
abundant, monsoon-forest is replaced by rain-forest.[2]

The trees in monsoon-forest are tall. In addition to the important
teak-trees, there are species of Sterculia and Erythrina, Grewia elastica,
Milletia Brandisiana, and others. Underwood is scantily represented, but
large bamboos play a considerable part. Epiphytes are not numerous.

Monsoon-forest is represented on the Misantla plateau in Mexico, and
has been described by Karsten.[3] This forest consists mainly of deciduous
plane-trees. Cecropia and Croton also occur. Underwood is formed by
Anonaceae and Urticaceae; while clumps of Heliconia are scattered here
and there. Lianes and epiphytes are scanty in comparison with their
abundance in rain-forest. The lianes are largely species of Vitis and
Menispermaceae. Among epiphytes species of Philodendron are common.

The forests that Schweinfurth mentions as occurring on the western
slopes in Abyssinia may be monsoon-forests.

CHAPTER XCIII. EVERGREEN DICOTYLOUS FOREST

EVERGREEN dicotylous forests occur not only in rainy tropical and
subtropical regions, but also in the cold-temperate zone of the Southern
Hemisphere. Though it is true that in many of these forests species occur
which are entirely leafless for a shorter or longer period, yet most of the
trees retain their foliage until after a new crop of leaves has appeared,
or for more than twelve months.

In most regions dry periods can step in at any season of the year,
and in the region of tropical rain-forest, for instance in Java, the day
itself may show dry periods (in the morning until rain falls at 2 or 3 o'clock
in the afternoon) during which the air is relatively dry, and transpiration may become a source of danger.[4] Consequently nearly all leaves of
trees composing the high-forest are protected in some way from excessive
transpiration.

The leaf, not only for this reason but also because it lives for more
than a year, by no means displays such uniformity of structure as in
deciduous dicotylous forests of temperate countries.

In evergreen, tropical dicotylous forest defoliation and foliation are
neither so general, nor so synchronous in the different species, as in
temperate regions; so that there is no such seasonal change in leaf tint.
As the foliage ages it is gradually shed; yet this process is especially
active in certain months, for instance, in July to September in central
Brazil.[5] Throughout the year the tint of the forest is of a darker green
than that of our European forests; although certain species show striking
colours when they are leafing (the young leaves often being red), their
effect is lost in the general mass of foliage of the other species. Bud-scales
are usually wanting.

As the green leaves presumably are active throughout the year (indeed
some species produce new foliage all through the year) it is easy to see

[1] Brandis, 1898.   [2] See S. Kurz, 1875; Büsgen, Jensen, und Busse, 1905.
[3] Karsten, 1903 b.   [4] Haberlandt, 1892.   [5] Warming, 1892.

that the plant is capable of producing much more food than European dicotylous trees can ; hence the rapidity of growth and the huge dimensions of many tropical trees.

Mesophytic evergreen forest is represented by :—

Antarctic forest.
Subtropical rain-forest.
Tropical rain-forest.

There are also several special kinds of forest produced by certain tropical plant-types, such as palm-forest and fern-forest.

ANTARCTIC FOREST

Antarctic forest is known to us through the descriptions supplied by Darwin, J. D. Hooker, and Dusén.[1] It extends from a latitude of 36° S. in southern Chile to Tierra del Fuego, where it clothes the country from the sea up to the altitude of 1,700 to 2,000 metres on the western side of the mountain-chain. The climate shows a low annual mean temperature, which is 5 to 7° C., and only a difference of about 9° C. between the mean temperatures of winter and summer respectively ; but the rainfall is very great, and is distributed over nearly all months of the year. Under these conditions there is developed extremely luxuriant forest. In the northern districts this forest includes a great abundance of lianes and epiphytes, as well as underwood in which tree-ferns and bamboos play a part, and thus passes over into sub-tropical rain-forest. Towards the south this character is lost, yet the forest is dark-green throughout the year. Beech is the common forest-tree, and includes the evergreen species Nothofagus betuloides, N. Dombeyi, and N. nitida, in addition to species of the same genus that are leafless in winter. The leaves of these beeches are small and myrtle-like, but numerous ; consequently their general appearance is utterly different from that of Fagus sylvatica. In addition to the beeches, Coniferae (Libocedrus tetragona), Drimys Winteri, Myrtaceae, and Proteaceae, supply forest-trees.

The buds are protected by scales. The forest is deeply shady and has but little underwood. As epiphytes Hymenophyllaceae and other ferns occur, but lichens are sparse. The ground of the forest is clothed with a dense, uninterrupted carpet of water-soaked mosses and liverworts, among which Hymenophyllaceae grow. This type of forest oecologically stands nearest to spruce-forest. The evergreen character of the foliage must be ascribed to brevity of the warm season.[2]

Antarctic beech-forest occurs on the mountains of New Zealand as well as in Patagonia.

From a floristic standpoint it is noteworthy that side by side with Nothofagus (which is closely related to Fagus) there grow social trees belonging to Proteaceae, Myrtaceae, Podocarpus, Libocedrus, Fitzroya patagonica (a huge coniferous tree), as well as other tropical and austral types ; parasitic on the beech-trees are species of Myzodendron. Forests in New Zealand show a strong floristic likeness to Patagonian forest.[3]

[1] C. Darwin, 1845; J. D. Hooker, 1847 ; Dusén, 1898–1908.
[2] Consult Dusén, loc. cit. ; Reiche, 1907 ; Neger, 1897 a and b, 1901 ; Aboff, 1902.   [3] Cockayne, 1908 a, 1908 b, and other cited papers ; Schenck, 1905.

SUBTROPICAL RAIN-FOREST

It has already been pointed out that the farther north one goes in Patagonian forest the more numerous do species become. Lianes, elaborate epiphytes, and bamboos increase in numbers, and antarctic forest gives way to subtropical rain-forest. In Chilian rain-forest species of Nothofagus still occur, but they are supplemented by numerous other species, including Laurelia sempervirens, Drimys chilensis, Persea Lingue, and Podocarpus nubigena. Underwood is abundant. Among epiphytes, beside mosses and ferns, two Gesneraceae, Sarmienta repens and Mitraria coccinea, also two species of Luzuriaga are common. In contrast with tropical rain-forest the leaves of most of the trees are upwardly directed and coriaceous, while ' drip-tips ' are rare.[1] In rain-forest, on Juan Fernandez, according to Johow,[2] the trees have coriaceous or membranous leaves devoid of ' drip-tips ' ; lianes are rare ; epiphytic ferns abound, and in the underwood many tree-ferns grow.

Fern-forest. Tree-ferns are dependent upon a humid atmosphere ; they are indicative of an atmosphere continuously saturated with aqueous vapour, and of a uniform climate. Forests in New Zealand, Australia, and Tasmania, are rich in tree-ferns. In fact, these together with other ferns and thin-leaved herbs may form the main mass of the vegetation.[3] On several of the more raised West Indian islands, for instance on Jamaica, which is extraordinarily rich in ferns, one finds on the mountains at a certain altitude, namely, in the cloud-belt, a vegetation that may be termed fern-forest and includes such forms as Cyathea and Alsophila ; this possibly gives us a blurred picture of one of the most ancient types of vegetation.

Tree-ferns are often very abundant in subtropical rain-forest that occurs at a certain altitude on tropical mountains, where it replaces tropical rain-forest.

Subtropical rain-forest not only occurs in the subtropical places already mentioned where rain falls throughout the year, but it is also met with both on mountains within the tropical zone, where at a certain altitude it gives way to tropical rain-forest, and also in regions where the winter is more or less rainless although the whole rainfall is large in amount. These latter regions are found on the eastern sides of all continents, namely in the United States, south Brazil, eastern South Africa, east Australia to Tasmania, south China, and south Japan. In these regions, as Schimper[4] has pointed out, subtropical rain-forest prevails where the annual rainfall is great ; whereas savannah-forest or savannah occupies drier country. In such places rain-forest may approach tropical forest in luxuriance.[5]

TROPICAL RAIN-FOREST

Encircling the Earth in equatorial countries there is a belt of forest of which one thinks upon hearing the expression ' primeval forest '. A *primeval forest* is virgin forest that has preserved its original character

---

[1] Philippi, 1858 ; Neger, 1897 *a* and *b*, 1901.   [2] Johow, 1896.
[3] Consult Hochstetter, 1863 ; Tenison-Woods, 1874 ; Cockayne, 1908 ; Diels, 1896, 1905.   [4] Schimper, 1898.   [5] Rein, 1881 ; Mayr, 1890.

because man has left it almost or entirely undisturbed. Its trees remain standing until they die a natural death or succumb in the struggle with their neighbours, and thereafter their corpses sink to the ground, moulder away, and leave a bare space where other species recommence to battle. There are still primeval forests, not only on the warm humid plains bounding the Amazon, but also on the storm-beaten rocky soil of Lapland and Scandinavia,[1] as well as in Germany and Bohemia.

Tropical rain-forest is confined to regions where great heat prevails, and a flood of light streams down from the sun that stands high overhead, and where there are almost daily violent discharges of rain that are derived from the huge volumes of air saturated with moisture which rise vertically and are cooled in the upper atmospheric strata. Between the crowns of the trees warm mists often arise, while, at least at certain seasons of the year, water drips from the leaves during most of the day, and the air may be almost saturated with moisture ; for instance, at Buitenzorg in Java the relative humidity of the air is about 95 per cent. from 2 to 3 p.m. until the next morning.

The soil of rain-forest is always a rich humus, black and porous, filled with mouldering remains of trunks, twigs, leaves, flowers, and fruits, and presumably excavated by subterranean animals. Yet the layer of humus is not so thick as is often assumed ; layers several metres in thickness are not usual.[2] Some writers regard the soil as being thoroughly wet at all times, but others more correctly state that the rain percolates rapidly downward owing to the porous nature of the soil.

In such circumstances the plant-world develops with a luxuriant diversity unrivalled elsewhere. Tropical rain-forest constitutes the climax in the development of vegetation for the whole world. Its characteristics are discussed in the succeeding paragraphs.

Utilization of space. One usually finds so many storeys of plants that the whole nearly forms a single complex of vegetation. As Humboldt expressively remarks : 'forest is piled upon forest.' The trees forming the highest storey have tall thick trunks which are unbranched up to a height of 40 to 50 metres or more. Beneath them are trees of moderate stature with branches not reaching those of the higher tier. Beneath these, in turn, succeed slender, thin-stemmed, low palms, tree-ferns, and shrubby Urticaceae, Piperaceae, Myrsinaceae, Rubiaceae, and others. Scattered about are huge herbs which reach 4 or 5 metres in height and belong to the scitamineous and araceous types. If there still remain space available on the ground that is reached by the light, it is occupied by dark-green ferns, Selaginellae, mosses, and similar sciophytes. But often the light is too feeble to permit of more than a very small number of plants developing on the ground, which then may be almost bare of vegetation,[3] with its black humus covered only by fallen, decaying, wet leaves, twigs, and remnants of fruits, between which only bizarre *saprophytes* (Burman-niaceae, Gentianaceae, and others[4]) or *root-parasites* (Rafflesiaceae and Balanophoraceae) find places. Large pileate fungi are seldom met with; but there are hordes of *epiphytes*[5] clothing trunks and branches, and belonging to the Orchidaceae, Araceae, Piperaceae, Bromeliaceae (in

[1] Andersson and Hesselman, 1907.   [2] Reinhardt ; Whitford, 1906.
[3] Martius, 1840-7.   [4] See p. 89.   [5] See p. 87.

America), the cactaceous Rhipsalis (in America and Africa), and other spermophytic families, as well as ferns, mosses, and so forth. Trees of the forests situate in the cloud-belt in Java and the Moluccas are enveloped in a soaking mossy felt, which may be thicker than the trunks themselves and imparts to them a peculiar, dark appearance. Here, too, lies the proper home of the Hymenophyllaceae, ferns whose anatomical structure reveals them as true ' mist-plants '. Even the leaves of evergreen species may be densely clothed with epiphyllous algae, liverworts,' and small lichens. Among the plants requiring the heaviest rainfall, according to Schimper,[1] we may regard woody epiphytes, of which many develop in rainy primeval forests. The fiery Rhododendron javanicum decks the tree-crowns in mountain-forests of Java, and together with it one sees species of Ficus, of the melastomaceous Medinilla, of the loganiaceous Fagraea, and of the araliaceous Sciadophyllum. In Javanese mountain-forests one commonly encounters the huge epiphytic ferns Asplenium Nidus and Platycerium alcicorne ; large specimens of Lycopodium Phlegmaria and other species of Lycopodium, also of Psilotum flaccidum, which hang loosely down from the trees in wisps, like so many horsetails. Finally, there is a wealth of *lianes*, whose flowers and fruits one can rarely see, and whose long, often curiously shaped stems span the distance between soil and tree-crowns, or hang down from the latter or partly trail along the ground. Many plants in addition to trees provide innumerable points of support, enabling lianes to reach the tree-crowns. The amount of light prevailing in the forest accounts for the luxuriant vegetation ; light can pass through the open crowns of the topmost storey on to the lower crowns, and downwards through these. The twilight prevailing is much less dark than in European beech-forest. All the species, as Junghuhn[2] expresses it, seem to ' abhor a vacuum ' and to combine in an endeavour to utilize all the space available.

Number of species. The number of species in tropical rain-forest is extraordinarily large. The absence of any social method of growth on the part of species has often been mentioned by writers, and provides a sharp contrast to the uniformity of forest in northern Europe. This is well illustrated by the fact that in Brazil, on three geographical square miles round Lagoa Santa, there are about 400 species of trees in the forest[3] ; also by the fact mentioned by Whitford[4] that in the Philippines, on an area of 1,200 square metres, there grew 896 trees exceeding 3 metres in height and belonging to 120 species. This multiplicity of species is partly due to a geological cause, namely, the antiquity and uninterrupted development of tropical life[5] : but it is also due to a physical cause, namely, favourable life-conditions ; for there are examples showing that a moist, rich soil entertains a larger number of species than does adjoining dry soil (forest and *campos* of Brazil).[6] It seems to be justifiable to assume that the rate of production of new species is, in general, dependent upon favourable conditions of life, and that it is therefore greater in tropical than in cold or temperate lands.

[1] Schimper, 1898.   [2] Junghuhn, 1853–4.   [3] Warming, 1892.
[4] Whitford, 1906.   [5] Wallace, 1891 ; Warming, 1899.   [6] Warming, 1892.

ADAPTATIONS

Tree-form. The majority of trees show nothing remarkable in their shape, but some forms encountered are worthy of note. Haberlandt [1] has mentioned and figured some of these, including the umbrella-form, the candelabra-form, the tiered-form ; but several others might be mentioned, including the palm-form. The mode of branching is far more diversified, and apparently more irregular than in the trees of northern Europe ; very commonly the twigs bear clusters of leaves only near their ends, and each stem emits only few branches.[2]

Roots. Plank-buttresses are formed by the roots of many species. These buttresses are much taller than they are thick, and are continued up from the base of the trunk, sometimes to a height of two or three metres, as large, often bent, plank-like growths. The cross-section of the trunk immediately above the soil thus acquires a stellate shape, and the space round the base of the trunk is divided into stalls. These roots provide firm and broad foundation for trees that possess huge trunks and large crowns. Such plank-buttresses occur in certain species of Bombaceae, in Ficus, Myristica, Carallia, Sterculia, Canarium, and others. According to Schimper they are particularly characteristic of humid forest, and are wanting where the rainfall is small.[3]

Prop-roots similar in design to those of Rhizophora [4] are possessed by Iriartea and some other palms, and by Pandanus. They are terete flying buttresses springing from a certain height up the trunk, descending at acute angles to the ground, and showing a radiating arrangement similar to that in Rhizophora. The number of such props possessed by a tree is sometimes considerable, for example, more than twenty. In another guise prop-roots occur on Ficus religiosa and other trees, as they spring from the boughs and enable a single tree to spread over a vast area, and produce a forest which has an extremely thick canopy of leaves and casts deep shade. It is this shade that may possibly be the cause of the luxuriant growth of the roots in question.

Bark. The cortex is thin, but otherwise varies greatly. In this respect a sharp contrast is provided by trees growing on the Brazilian *campos,* for these have very thick coats of cork and cortex, though they may be growing only a few metres away from the forest-trees.[5]

Thorny stems. Thorny stems are not uncommon, and occur in Hura, Erythrina, and others, but are commonest in palms. In addition one encounters trees, such as Xanthoxylum, with peculiar laminated cones of cork occurring on their stems.

Buds. The buds do not possess dry bud-scales such as occur in the trees of northern Europe, or, at most, bud-scales are infrequent and mainly encountered in drier forests.[6] The buds are protected by stipules, leaf-sheaths, and outgrowths of the petiole ; in addition water, resin, or mucilage is often excreted between the bud and its envelope.[7]

Flowers. Remarkably few flowers are to be seen, although tropical forest always abounds in flowers, which are as a rule high overhead in the

[1] Haberlandt, 1893.   [2] See the figures of J. Schmidt, 1903.
[3] See also Whitford, 1906.   [4] See p. 236.   [5] Warming, 1892.
[6] Warming, loc. cit., figs. on pp. 409–11.
[7] Percy Groom, 1892 ; Schimper, 1898 ; Raunkiär, 1905, 1907.

tree-crowns. But on looking down upon forest from some eminence, one sees dotted over it large yellow, white, violet, or red spots, which indicate blossoming trees and lianes. In many cases, for instance among Lauraceae and the majority of Papilionaceae, the flowers are small but rendered easily visible to insects by their great abundance. In many species it is remarkable that the flowers spring from thick trunks or branches, or even from the base of the bole, because year after year they develop from the same dormant buds. The most widely known example of such *cauliflorous* species is Theobroma Cacao, the chocolate tree ; other examples are supplied by Myrtaceae, Sapotaceae, Leguminosae, Ficus Roxburghii, Crescentia Cujete, and species of Swartzia.[1] Wallace suggests that the flowers of these cauliflorous species may be adapted for pollination by Lepidoptera, which flutter within the calm forest. Whether or no this is the case has not been determined. But judging by floral construction it does not seem to be true of Theobroma, whose flowers are probably pollinated by other kinds of insects or are self-pollinated.

Periodicity. In tropical rain-forest there is generally neither summer nor winter, neither spring nor autumn : the periodic habit of development so distinct in other plant-communities is lacking or very feebly represented. Some species acquire new foliage throughout the year : and even if certain species do exhibit a distinct resting period, or are entirely leafless for a short time, they are lost among the crowds of others that show either no period of rest, or one at a different season of the year. Probably nearly all the species have their own definite time of flowering ; but this is not synchronous for the different species. The forest (like South American savannah) is rich in flowers all through the year. Thus rain-forest shows no periodicity in its life as a whole.[2]

Leaves. Foliage-leaves in tropical rain-forest nearly always remain attached to the tree for longer than one year, generally for thirteen or fourteen months.[2] They probably often remain active for many months, perhaps for more than a year—a fact that is of profound oecological importance to the plants, and explains the huge dimensions of these as well as the large production of organic matter. The old leaves bend, sometimes, according to Haberlandt,[3] by active movements, in order to provide space for the younger foliage. The play of colour of the foliage has already been discussed on page 337.

Leaf-shape. In tropical rain-forest leaves assume an extraordinary number of shapes. On trees we find not only the forms familiar to us in Europe—namely, ovate, elliptical, and so forth, simple or simply compound—but also a number of new forms ; for example, the pinnate or palmate foliage of palms ; the large, undivided, characteristically veined leaves of Scitamineae; the pinnately compound leaves of Leguminosae, and particularly the bipinnate or tripinnate mimosaceous leaves, whose countless leaflets execute phototactic movements ; the digitate leaves of Bombaceae and of the araliaceous Panax ; the palmately divided peltate leaf of Cecropia ; the long-stalked, large, cordate or ovate-cordate leaves of Araceae ; and the bamboo-leaves, which dispose themselves at the tip of the branch in a digitate fashion ; and so forth. Yet the commonest

[1] Wallace, 1891 ; Haberlandt, 1893 ; Whitford, 1906.
[2] See Warming, 1892; Holtermann, 1902; Volkens, 1903.   [3] Haberlandt, 1893.

type of leaf is perhaps of the ' laurel-form ', that is to say, it is a large, glabrous, glossy, elliptical, or lanceolate leaf, of which an example is provided by Ficus elastica. Glossy coriaceous leaves are in general a strikingly characteristic feature of tropical forest ; whereas foliage in the forests of northern Europe is matt and more translucent. Haberlandt[1] estimates that entire leaves are more frequent than in northern Europe. The leaves are often of huge dimensions, for instance in the humid coastal forests of Brazil and in Amazonian forests. Moreover, they are of a darker tint than in temperate climes, because they, and particularly the palisade parenchyma, are thicker than in European trees. Other leaves, on the contrary, especially in lower storeys of the forest, are very thin because the light is weak and the air humid.

Regulation of the amount of water in the plant. According to the investigations of Haberlandt[2] and others, plants in Javanese rain-forest, and possibly in the higher storeys of tropical rain-forest in general, are exposed to conditions far more extreme than are experienced by European vegetation. From about 6 to 7 a.m. until about 1 p.m. the temperature rises, while the atmosphere gradually becomes drier under the direct insolation, and the relative atmospheric humidity finally sinks to 70 per cent. The second diurnal period commences at about 2 to 3 p.m. with storms and violent discharges of rain, and during this period the air is so charged with moisture (93 to 95 per cent.) that transpiration is almost arrested. Thus during two-thirds of the day the air approaches saturation-point. During the course of the day the plants are therefore threatened from two entirely different sides, and against the double danger, which particularly concerns the process of assimilation, they guard themselves in various ways.

The significance of hydathodes possessed by leaves has already been discussed on p. 101.

But the opposite danger arises from great atmospheric aridity and from the consequent intense transpiration during the morning. It is true that the transpiration as a whole is small (according to Haberlandt,[3] two or three times less than in plants in Central Europe, but this estimate is regarded by Stahl[4] as being too low,[5] yet in the morning it is intense and brings with it the risk of fading, or, at least, of so material a diminution of turgidity as to inhibit photosynthesis. This explains the remarkable fact that many plants in tropical rain-forests exhibit protective devices against excessive transpiration similar to those displayed by xerophytes. A thick-walled, strongly cuticularized epidermis, sunken stomata, mucilage-cells, storage-tracheids, aqueous tissue, and so forth, present themselves. The aqueous tissue of Ficus elastica is well known. The leaves of a number of palms, and the large thin leaves of Scitamineae, display aqueous tissue towards the upper face or sometimes towards both faces ; and this tissue may be quite as voluminous as the chlorenchyma.[6] In Javanese rain-forest, according to Haberlandt, several species, including Gonocaryum pyriforme and Anamirta Cocculus, contain in their chlorenchyma mechanical cells, precisely as do certain xerophytes mentioned on p. 128. And these cells have the same significance in both cases—they guard the chlorenchyma against shrinkage due to desiccation.

¹ Haberlandt, 1893.   ² Haberlandt, 1892, 1897.   ³ Haberlandt, loc. cit.
⁴ Stahl, 1894.   ⁵ See also Giltay, 1897, 1898 ; Holtermann, 1902, 1907 ;
Burgerstein, 1904.   ⁶ Figures are given by O. G. Petersen, 1893.

Haberlandt [1] found that some hydathodes possess the power of absorbing dyes, and he therefore concluded that they also serve to absorb water and hand it on to the plant. This is presumably possible only at a definite time of day, namely, when the first showers of rain fall, some hours after midday. When the plant has transpired too vigorously the hydathodes would thus aid the plant in rapidly recovering its turgidity. Hydathodes may accordingly act as *regulators* of the plant's water-supply, by expelling any excess of water, and by absorbing water when pressing necessity so demands.

What has been said so far refers to plants forming the upper storeys, so that their leaves are at the roof of the forest and are directly insolated. But different relations may be expected to subsist in regard to humbler plants standing beneath others inside the forest. And here we do find plants that are eminently adapted to live in shade and in moist air ; among such are the Hymenophyllaceae, whose filmy fronds show only one or few layers of cells, no proper epidermis, and no intercellular spaces, and whose stems bear root-hairs.[2]

Adaptation to rain. Other structural features seem to relate to the discharges of rain, and concern—

a. The *violence* of these, which in this respect are unparalleled in the European climate.

b. The *frequency* of the rainfall.

Adaptation to the mechanical action of rain.[3] The great distance at which one can hear the noise of rain falling upon a tropical forest gives us some idea as to the violence of the impact ; but the trees are adapted to withstand this. Many simple leaves are firm and coriaceous, and the epidermis may be so impregnated with silica that the lamina is rigid and brittle, as in Medinilla magnifica, and the blade may resemble a ' green lacquered sheet of metal '. The leaves of mimosas, acacias, and other Leguminosae, also of palms, are divided into many leaflets, or segments, so that they oppose less resistance to rain. In addition very often they can execute movements causing their leaflets to close together, so that they expose a smaller surface or even merely an edge to the falling rain-drops. In other plants the leaf shows folds or is trough-like, and thus acquires additional mechanical power of resistance ; this is most distinct in palm-leaves whose pinnately or palmately arranged segments are folded, inasmuch as their lateral parts are inclined upwards or downwards. The leaf-stalk is often directed *upwards*, presumably for a reason different from that in connexion with xerophytic communities, namely, in order to oppose greater resistance to the force of rain. In many other cases the same object is accomplished by the pendent arrangement of the young leaf-blades and young twigs ; in Araceae many large leaves retain this lie, but others subsequently erect themselves. The huge leaves of palms and Scitamineae have large sheaths which embrace the stem, and thus add strength to stem and leaves and enable these to resist the violence of wind and rain.

[1] Haberlandt, 1894, 1895, 1904.
[2] For further information concerning transpiration in moist tropical climates readers should consult the works of Haberlandt, 1892, 1897 ; Stahl, 1894 ; Burgerstein, 1904 ; Giltay, 1897, 1898 ; and Holtermann, 1902, 1907.
[3] Wiesner, 1895 a, 1897.

Adaptation to the frequency of rain. Frequent falls of rain may have an injurious effect upon plants by causing the leaf-blades to become too wet and heavy.[1] But transpiration and photosynthesis are also hindered by epiphyllous algae, lichens, fungi, liverworts, and, according to Haberlandt, even bacteria, which settle upon the leaves. The older leaves of many evergreen trees in humid tropical forest are covered with a mass of epiphyllous species.[2] Therefore it would seem to be of advantage to plants in rain-forest that their leaves should *dry rapidly.* Jungner[3] working in Kamerun, and Stahl[4] in Java, concluded that this aim is accomplished by the following devices :—

1. A smooth *cuticle* which cannot be rendered wet ; this device is very widespread.

2. *Drip-tips,* as Stahl terms the long, often suddenly narrowed leaf-apices. These are familiar in Ficus religiosa, and occur among various ferns, Monocotyledones, and Dicotyledones, not only in simple but also in compound leaves. The drip-tips, which occur only on leaves capable of being rendered wet on the surface, enable rain to flow away rapidly. They are downwardly directed. The longer and sharper the tip the more rapidly does the leaf dry ; the sabre-like tip is the most sufficient abductor of water, which sometimes drips away in a continuous jet. Drip-tips occur neither on leaves whose surface cannot become wet nor among xerophytes.

3. *Channelled nerves* often occur, and serve to convey water to the leaf-tip. The arcuate course of the nerves along melastomaceous leaves is of use in this connexion.

4. *Velvety leaves* are encountered in the shadiest and moistest parts of rain-forest, in connexion with herbs forming the ground-vegetation and species belonging to the lower storeys. The epidermal cells project as countless papillae which give to the leaf a velvety appearance. Between the papillae the water rapidly spreads by capillarity as a thin film extending over the whole leaf-blade, so that the water can evaporate more quickly than if it were not thus spread out. The opinion has also been expressed most recently that these papillae serve to increase the amount of light supplied to the leaf.[5]

FLORA

The flora of tropical rain-forest is so diversified as to be beyond detailed discussion in this work. The predominant trees belong to the Leguminosae, Lauraceae, Myrtaceae, Moraceae, and other families.

DISTRIBUTION

Rain-forest is not exacting as regards soil, and extends over vast areas where the rainfall is very large (above 180 centimetres annually) and the dry season does not exceed a few months. Where the rainfall is less, and prolonged dry seasons prevail, rain-forest is confined to humid low-lying land bounding rivers, where it forms fringing forest ; while the elevations are occupied by savannah or savannah-forest.[6]

---

[1] Concerning ombrophilous and ombrophobous species see p. 32 ; and Wiesner, 1893 *a*.   [2] See p. 83.   [3] Jungner, 1891.   [4] Stahl, 1893; also see p. 116.
[5] Haberlandt, 1905, 1908; Guttenberg, 1905.
[6] Schimper, 1898 ; Warming, 1892 ; Passarge, 1895 ; Pechuel-Loesche, 1882.

ASSOCIATIONS

Associations formed by single dominant species are exceedingly rare in tropical forest. Owing to the profuse admixture of species tropical rain-forest over the whole world seems to constitute only a single comprehensive community. In different regions rain-forest is represented by different types, which may perhaps be regarded as sub-formations. For instance, there are rain-forests with but little underwood, and others with abundant underwood. Sometimes lianes are scanty, at other times abundant, and the same is true of epiphytes. Huber,[1] dealing with Amazonian forests, has shown that detailed examination results in the recognition of distinct associations and varieties of association.[2]

APPENDIX. PALM-FOREST

In tropical South America one encounters certain forests that are composed mainly of palms and occur on the river-banks or on still moister soil. Thus in Brazil there are the 'buritysales', that is to say, forests formed by the burity-palms, Mauritia vinifera and M. flexuosa. Lund[3] describes these forests in the following words : ' The valleys are clothed with a fresh, luxuriant carpet of grass, and at the bottom, where a stream always flows, they are adorned with groups of the incomparably beautiful burity[-palm] ' ; Martius in his ' Tabulae ' figures forests composed of both species. Gran Chaco, in north-western Argentina, includes vast plains clothed with palm-forest composed of Copernicia cerifera. This palm is a light-demanding tree that can form only open and shadeless forests, which are therefore well lighted and presumably have a rich vegetation on the ground. These palm-forests perhaps belong rather to savannah-forest than to rain-forest.

[1] J. Huber, 1906.
[2] In regard to tropical rain-forest in general, readers should consult the works of Haberlandt, 1893 ; Schimper, 1898 ; Whitford, 1906 ; and Koorders, 1907.
[3] Lund, see Warming, 1892.

SECTION XVII

STRUGGLE BETWEEN PLANT-COMMUNITIES

CHAPTER XCIV. CONDITIONS OF THE STRUGGLE

HITHERTO we have treated plant-communities as if they were static entities, in a condition of equilibrium and with their evolution concluded, and were living side by side at peace with one another. Yet such is by no means the condition of affairs. *Everywhere and unceasingly* a struggle is taking place not only within the several plant-communities but also between them, so that each of these is continually striving to invade the territory of the others. Moreover, *each slight change in the environment* upsets the condition of equilibrium hitherto existing, and at once occasions *a disturbance and change in the reciprocal relations subsisting. Extremely slight* changes in the environment often evoke remarkably great changes in the vegetation, by favouring certain species and suppressing others. 'Rise and fall of the *water-table* should be considered in inches, not in feet,' writes the experienced practical man Feilberg.[1] The zonal distribution of vegetation round small lakes and pools that one observes in West Jutland,[2] the distribution of Weber's 'sub-formations' of meadow,[3] and that of the several 'types and sub-types' of heath, all tell the same tale. Moreover, P. E. Müller[4] shows that minute *climatic* changes suffice to cause forest to give way to another kind of vegetation. From Gräbner[5] we learn that relatively small distinctions in the climate of different parts of the North-German plain cause the local floras to be sharply delimited. Attacks by insects or fungi, dry or rainy years, and so forth, may bring about changes. The struggles in question have been the subject of extremely little investigation, so that a wide and attractive field of research lies open.

The struggle between communities is of course dependent upon that between species, to which allusion has already been made.[6] This struggle is caused by endeavour on the part of species to extend their area of distribution by the aid of such means of migration as they possess. 'Situation wanted' is the cry in all communities, whether these be human or vegetable. Millions upon millions of seeds, spores, and similar reproductive bodies are annually scattered abroad in order that species may settle in new stations ; yet millions upon millions perish because they are sown in places where physical conditions or nature of the soil check their development or where other species are stronger.

Not until recent times was attention drawn to the ceaseless struggle among species. Darwin it was who directed our notice to this struggle,

---

[1] Feilberg, 1890.   [2] Raunkiär, 1889 ; Warming, 1890, 1906, 1907.
[3] Weber, 1892.   [4] P. E. Müller, 1887 b.   [5] Gräbner, 1895, 1901.
[6] See pp. 70, and 93.

which forms the basis of one part of his hypothesis concerning the origin of species.[1] Yet other writers had previously noted this struggle in nature ; for instance, A. P. de Candolle wrote : ' Toutes les plantes d'un pays, toutes celles d'un lieu donné, sont dans un état de guerre les unes relativement aux autres.'[2]

The struggle and competition among plants are brought into greater prominence by *changes* that continue without interruption in soil, climate, or other conditions affecting plant-life, including changes in the animal-world.

Without such changes the results of the struggle would be neither so distinct nor so rapid. The changes in question are—

i. Production of new soil.

ii. Changes in old soil, or in the vegetation covering it, and in the factors discussed in the first Section of this work, but particularly those caused by man, who is thus responsible for ' semi-cultivated ' formations. Man intervenes directly when he utilizes the soil for his own purposes, by converting forests into arable ground, or by draining moors, but he also intervenes indirectly when permitting cattle to graze, or when he mows, manures the soil, and so forth.

In regard to the question now under discussion reference should be made to Clement's interesting work, ' The Development and Structure of Vegetation ' (1904). In discussing the migrations and invasions of plants Clements distinguishes between migration and ' ecesis '. ' Migration merely carries the spore, seed, or propagule into the area to be invaded ; ecesis is the adjustment of a plant to a new habitat, it is the decisive factor in invasion, inasmuch as migration is entirely ineffective without it.' In discussing invasion he treats of barriers, endemism, polygenesis, also manner and kinds of invasion.[3]

The struggles between communities and the development of these are elucidated by means of examples in the succeeding chapter.

CHAPTER XCV. THE PEOPLING OF NEW SOIL

WHEN new soil arises anywhere it is soon invaded by plants. And it is of deep interest to follow the successive phases in the development of the vegetation. In this way one will acquire evidence of a long series of struggles among the successive immigrants ; these struggles sometimes do not end in any decisive result before the lapse of many decades.

New soil arises in the following places : on coasts, where the sea deposits fresh material ; at the mouths of rivers ; even in river-beds, where masses that have been washed down are deposited. New soil also arises by the following agencies : action of glaciers, talus, volcanic eruptions, fire that devastates the original vegetation, and human action, but

[1] See also J. D. Hooker. [2] A. P. de Candolle, 1820.
[3] See also MacMillan, 1893, 1896, 1897 ; Shantz, 1906, 1907 ; Cowles, 1899, 1901 ; Clements, 1905 ; Adams, 1905 ; A. Nilsson, 1899 ; Whitford, 1901. The means of migration of plants are dealt with in papers by Guppy, 1903, 1906 ; Hemsley, 1885 ; Jouan, 1865 ; Schimper, 1891 ; Hildebrand, 1873 ; Sernander, 1901 ; Vogler, 1901 *b* ; J. Holmboe, 1898 ; Ravn, 1894 ; and many others.

particularly where cultivated land is left to itself. In the last cases the soil is not new to the same extent as in the first cases ; it is not barren, but includes a greater or smaller number of seeds and the like. The following examples will serve to illustrate the development of various types of vegetation.

Vegetation on Sand.

Psammophytic vegetation on the coasts of northern Europe [1] first arises on the flat foreshore, which is sometimes several hundred feet in width and receives from the sea deposits of sand : this assemblage of psammophilous halophytes constitutes the vegetation of shore-sand. Thereafter the wind raises up in such places dunes (shifting dunes), which are colonized by true dune-plants, such as sea-marram. These plants, if they emerge successfully from their struggles with the wind, prepare the place for a fresh kind of vegetation ; between them and under their shelter new species can now flourish. When these latter grow up and constantly produce a denser vegetation, the spaces become too confined for dune-plants, which gradually perish and give way to the vegetation of grey (fixed) dunes, or to sand-fields, or, in many cases, to dwarf-shrub heath.[2]

G. Beck[3] has described the kinds of vegetation that succeed one another on sand-banks cast up by high water on the Danube. First, on the bare moist sand are found some herbs, including species of Polygonum and Chenopodium, among which seeds of Salix, Populus, Alnus, and Myricaria germanica then germinate. The next colonists are a number of other herbs, particularly belonging to species with travelling rhizomes ; some settle upon moister spots, others upon drier, and produce a ' shifting-sand vegetation '. The willows, poplars, alders, and other trees in the meanwhile grow up and produce bush-forest—' willow-meadow '—which suppresses the herbs by means of its shade. But where humus can form and is not carried away by high water the willows and alders are van-quished, and there arises another type of forest—' poplar-meadow '—com-posed of Populus and Ulmus. Similar phenomena are witnessed in the willow-flora skirting the Vistula.

All the world over one sees on like habitats like struggles. And in this connexion we may allude to Stéfansson's [4] account of the development of the vegetation in Vatn Valley in Iceland, where islets of mud and sand are formed in the river, and become gradually colonized by Eriophorum, Carex, and grasses. These plants vanquish each other in a definite order.

The origin of heath-moor upon sand has been described by Gräbner.[5] The first plants to appear are Cyanophyceae, whose threads permeate the sand to a depth of three millimetres ; then Polytrichum juniperum, Radiola millegrana, Juncus capitatus, and other annual and perennial herbs occur ; finally, Sphagnum, Ledum, Calluna, and others appear.

---

[1] See Sect. X, also Sect. VII.
[2] Warming, 1891 ; Gräbner, 1895, 1901 ; Gerhardt, 1900 ; Cowles, 1899 ; Adamovicz, 1904.
[3] G. Beck, 1890.　　　　[4] Stéfansson, 1894.　　　　[5] Gräbner, 1901.

Production of Marsh.

On the shores of the North Sea, of Kattegat, and of the Baltic Sea, where there is a tide and yet sufficient protection against the violence of the waves, there is deposited at high tide extremely fine mud composed of particles of clay, sand, and humus. In this production of land an important part is played by vegetation, inasmuch as the mud deposited and the fixed Cyanophyceae (especially Microcoleus Chthonoplastes) find shelter and resting-places between the shoots of, first, associations of Zostera marina [1] in the deeper water of sand-flats, and thereafter of Salicornia herbacea [2] in less-deep water. Slowly the soil is raised, until eventually the daily high tides can no longer wash completely over it. Then the zone of Salicornia is seized upon by other plants, and gradually, as the land becomes higher and drier, it is occupied by associations of Festuca, of Juncus Gerardi, and others belonging to littoral meadow.[3] In littoral meadow no earthworms live ; but if it be diked and washed out by rain, the raw humus is replaced by mild humus and earthworms occur.[4] In the course of time the soil of littoral meadow will invariably be washed out, and its vegetation will undergo a corresponding change.

The development of the vegetation clothing reclaimed land at the mouth of the Rhone has been described by Flahault and Combres.[5] On the low, moist, saline, alluvial land of Camargue Arthrocnemium macrostachyum establishes itself in the first place. Round it small quantities of sand and organic dust accumulate, and thus raise the soil to a slight extent. Soon the tufts of Arthrocnemium are supplemented by Salicornia fruticosa, Atriplex portulacoides, and Aeluropus litoralis. Fresh material blown by the wind becomes lodged among the prostrate stems of these plants, and gives rise to low hummocks which are 2 or 3 metres in diameter and 10 centimetres in height. Some humus now arises ; rain-water washes out the hummocks ; and other plants, including annuals, establish themselves. The vegetation may thus become entirely changed in nature, and may include such conifers as Juniperus phoenicea and Pinus Pinea.

Lowering of Water-level.

New soil also arises where the level of lake-water sinks so as to lay bare rocks or other substrata that were previously submerged. Mälar Lake in Sweden has provided a case of this kind, which has been investigated by Callmé, Grevillius, and Birger [6] ; in about twenty years this new soil has given birth to forest.

Volcanic Eruptions.

Volcanic eruptions may bring into existence plantless tracts. The lava-fields of Iceland were at first devoid of vegetation, and some of them are still extremely poor in plants. Grönlund [7] states that at Myvatn, in the north-east of Iceland, on extensive lava-fields which arose in 1724–9, often there are only crustaceous lichens, including species of Gyrophora and

[1] See pp. 175, 177, 230.   [2] See pp. 225, 230.   [3] See p. 230; also Warming, 1890, 1906.   [4] P. E. Müller, 1878.   [5] Flahault et Combres, 1894.   [6] Callmé, 1887 ; Grevillius, 1893 ; Birger, 1906 a.   [7] Grönlund, 1884–90.

Stereocaulon ; and among the very few mosses that occur Rhacomitrium lanuginosum deserves special mention. Not until lava is weathered does it become a soil suitable for plants. Comes [1] has described the colonization and decomposition of the lava of Vesuvius by algae and lichens, the later immigration of mosses and ferns, and the final appearance of xerophytic Phanerogamia, including perennial herbs, shrubs, and trees, which eventually give rise to forest.

The devastation of Krakatoa in 1883 provides another example. The immigration of plants was investigated by Treub,[2] who came to the conclusion that the ashes and pumice first became clothed with a thin film of Cyanophyceae, and particularly of Lyngbya Verbeekiana and L. minutissima, which prepared the way for the germination of the abundant fern-spores. ' Three years after the eruption the new flora of Krakatoa consisted almost entirely of ferns. Spermophyta occurred only singly, dotted here and there near the shore or on the mountain.' These were largely brought by water and birds. The later development of this flora was studied by Penzig and Ernst.[3] Beccari found that Tamboro volcano on Sambava, which in 1815 had been denuded of vegetation, in 1874 was clothed from base to summit with virgin forest.

Landslips.

Rocky soil is laid bare by landslips or by human activity. In the Alps and many other mountainous countries one sees huge masses of stones surrounding the base of a mountain at a definite angle of inclination : these are masses of rubble or *talus*.[4] The developmental succession of the plants thereon is as follows : First there are lithophytes—lichens, algae, and mosses [5]—whose rhizoids penetrate the stone more or less deeply according to its hardness and porosity, and cause it to crumble. Between and on these plants masses of dust, brought by rain and wind, together with the mouldering remains of the plants themselves, provide a scanty stock of humus, in which larger plants can gain footing.[6]   The richness of the vegetation depends upon the steepness and the liability to weathering of the substratum. In steep places the vegetation remains open and low—essentially composed of Thallophyta and mosses (rock-vegetation [7]) ; on less-steep ground, where the stones soon become covered with plants and humus, forest often arises eventually.[8] Near Eisenach torrents of rain have produced deep chines and stony terraces. On these, according to Senft,[9] the vegetation shows the following developmental succession. First, the bare stony heaps were clothed with mosses, including Hypnum sericeum and Barbula muralis ; then followed some xerophytic grasses, such as Festuca ovina and Koeleria cristata, also shallow-rooted perennial herbs (a type of vegetation belonging to dry places). Thereafter succeeded other xerophytic herbs, such as

[1] Comes, 1887 ; see Engler, 1899, p. 179.   [2] Treub, 1888.
[3] Penzig, 1902 ; A. Ernst, 1907.   [4] See p. 246.   [5] See p. 241.
[6] The majority of rubble-heaps cannot be described as entirely new soil, because their descent is slow and is accompanied by humus, seeds, and the like, but when a mountain-slope simultaneously gives way to landslip, this becomes gradually clothed with common species derived from the neighbouring plant-communities (Blytt).
[7] See p. 241.   [8] See p. 329.   [9] Senft, 1888.

Helianthemum annuum, Ononis spinosa, O. repens, Origanum vulgare, and Anthyllis Vulneraria, also such shrubs as Crataegus, Juniperus, and Viburnum Lantana. Juniperus itself gives rise to a dense bushy growth. When the vegetation had reached this stage several other kinds of shrubs with fleshy fruits established themselves, and in twelve years gave rise to an impenetrable bushland. Finally, Sorbus, Fagus, and other trees appeared, and forest arose. The soil was constantly changed and improved by the death of the previous occupants ; each kind of vegetation suppressed another, until finally forest vanquished bushland, which could maintain itself solely at the margin of the forest as a mere fringe.

The slopes of marly diluvial hills laid bare by water usually clothe themselves at first with a community consisting mainly of members of the segetal and ruderal flora, and especially with annuals ; and not until later is there found the flora that is characteristic of these sunny slopes and is formed by longer-lived plants.

Fires in Forest and Grassland.

New soil is not always entirely free from propagative bodies. Its mode of origin determines this question. For instance, soil whose vegetation has been devastated by fire is seldom entirely sterilized thereby ; it preserves seeds, living roots, and rhizomes in great numbers, and new plants can sprout forth from these. Yet the vegetation may be interfered with to such an extent that a new kind of vegetation can make its entrance.[1]

Tropical and subtropical grasslands, including steppes and savannahs, in various parts of the world are designedly fired by the inhabitants : in some places for hunting pursuits ; in others for grazing purposes, because the burning of the old, dry carpet of grass and perennial herbage causes a rapid upgrowth of new grass. Several of these kinds of communities, and savannahs in particular, entertain scattered trees.[2] It naturally suggests itself that where one tree can grow many trees could flourish and give rise to forest. And when, as a matter of fact, no forest exists this lack has been ascribed to fires. M. Christie, Mayr, and Redway[3] suggest that the prairies may be treeless because fires prevent trees from growing up ; fires may also be the cause of the absence of earthworms and snails. Asa Gray expresses the opinion that, between soil receiving rain sufficient to bring forth forest, and soil receiving rain inadequate for this, there is a debatable tract where relatively slight causes decide whether forest or prairie shall prevail.

The Brazilian campos are regarded by P. V. Lund[4] as derived from forest that has been metamorphosed into savannahs (campos) by fire. Reinhardt[5] and Warming[6] take another view, although neither of them denies the important modifying influence of fires ; moreover Volkens[7] adopts this opinion.

---

[1] There is a rich literature dealing with fires in prairies, savannahs, and forests. See Warming, 1892 ; L. S. Gibbs, 1906 ; Pearson, 1899 ; also see p. 298.
[2] See pp. 296 and 299.
[3] M. Christie, 1892 ; Mayr, 1890 ; Redway, 1894.
[4] P. V. Lund, 1835.   [5] Reinhardt, 1856.
[6] Warming, 1892.   [7] Volkens, 1897.

Fire is one of the means by which man upsets natural conditions, and is of direct service in the cause of agriculture in the tropics, where man usually prepares land for cultivation by the felling and firing of forest.[1] So long as the soil is cultivated—and this is often only for a few years—man has constantly to contend with wild plants, for instance with the stool-shoots and suckers of old forest trees, if he is to protect cultivated plants. Hardly has the soil been left to itself before it is invaded by wild plants. The first to settle are numerous annual and other herbs, also shrubs, which form a commonplace vegetation of weeds, whose seeds and fruits blow hither from all sides or are brought by birds. Thus arises a community which is gradually converted into weed-bushland (a 'secondary formation'). But soon the forest-plants start to grow afresh ; shoots sprout forth from stumps and roots, and perhaps from seeds that lay hidden in the soil : and after a number of years forest once more occupies the site.

But there are cases in which a secondary formation does not revert to the primary one. According to Pearson,[2] the *patanas* in Ceylon have been derived from savannah-woodland, but are now to be regarded as permanent grasslands that cannot ever be changed back into forest because the soil has been modified.

According to Kihlman,[3] forest fires prevent the spread of the common spruce in certain parts of the northern forest-zone. The Scots pine has expelled it from regions where it was formerly abundant. The farther north a station lies the more serious is the effect of forest fires, because the maturation of seeds becomes increasingly difficult. Between Kola and Lake Imandra Kihlman discovered an elevation, three kilometres in length, which had been devastated by fire several years earlier : here the spruces, which were formerly dominant, had all perished, but there were isolated pine-trees that had survived the ordeal. With this exception the soil was occupied by a young, tolerably dense association of birches, among which one sought in vain for conifers. In this case it appears that aided by fire the birch will suppress the spruce because its seeds ripen more readily. Hult[4] showed the very grave extent to which forest-fires in Blekinge affect the competition among the different kinds of vegetation.

Krassnoff,[5] travelling in the inner valleys of the Altai mountains, observed forests that had been burned over a distance of ten or eleven kilometres. Although a long time had elapsed since the fire raged, no new forest had arisen, but in its place there was a waving sea of herbs, 3 feet in height, which, however, did not form a continuous covering over the soil ; the herbs included Helleborus, Aconitum, Thalictrum, Ligularia, Paeonia, and Pedicularis. Forest in this case seems to have been suppressed by a community of an entirely different class.

Fires on heaths provide another example of the production of new soil. Often the first vegetation to appear differs from that occurring subsequently, but finally Calluna recovers the ground ; yet often Calluna from the outset gives rise to an association, and the plants burnt emit shoots from the ground, while countless seedlings sprout forth. Fires on

---

[1] Warming, 1892. [2] H. H. W. Pearson, 1899; see p. 298.
[3] Kihlman, 1890. [4] Hult, 1885. [5] Krassnoff, 1888.

moors provide another opportunity for watching the struggles among different kinds of vegetation ; in some places they are succeeded immediately by the appearance of Senecio sylvaticus and Epilobium angustifolium.　The names applied to the latter in Denmark and America respectively, ' Ildmärke ' and ' fireweed ', denote that in both countries it is among the first plants to settle on burnt sites.

Other Sources of New Soil.

New soil is produced by such a removal of sods as lays the ground bare. In North Europe after heath-shrubs have been removed, together with the upper layers of soil, for use in place of straw or for absorbing manure, the bare ground first clothes itself with mosses, including Polytrichum, and small annual herbs, including Radiola, Centunculus, and Cicendia, between which ling-seedlings, and often young birches, pines, and other trees raise themselves.[1]

A peculiar cause of the genesis of new soil is the natural death of old vegetation.　This occurs in the case of ling-heath in Jutland and North Germany,[2] because Calluna lives for only between ten and twenty years, and often dies of old age.　When Calluna-plants die simultaneously over a large area, because they have attained the same age, the soil is laid bare and the heath is regenerated by means of seedlings.

In like manner on all other sites where old vegetation covering the ground is decimated, there arises a community which differs from the original one, but is, as a rule, eventually suppressed by the latter.　When wind makes a breach in an old, long-established, fixed dune, another type of vegetation arises ; in fact, by this means, way is made for the re-entrance of sea-marram.[3]　Where water here and there collects to form puddles in littoral meadow, the submerged mat-herbage is destroyed, and there appears a different type of vegetation, which consists essentially of annual halophytes, such as Salicornia and Suaeda maritima.[4] Where an avalanche passing through a forest has left behind it a treeless track, this is mostly clothed with plants differing from those elsewhere in the forest.

Where cultivated land is left to itself, not only in the cases already discussed, but also in general, there appears new soil that is rapidly colonized by a horde of plants, which are essentially weeds.　For example, in Jutland, fields that are neglected because their poor soil has yielded only a scanty crop of corn gradually become converted into heath. Again, in Blekinge (in Sweden), according to Hult's admirable investigations,[5] the new soil first becomes covered with weeds and plants whose seeds are easily transported by wind.　After some years the weeds have vanished and the field has become a grassy plain with a tolerably rich flora that includes forty to sixty species of Spermophyta.　Subsequently trees and shrubs are found, and forest arises.　On poorer soil ling seizes the field, but where better soil lies at a slight depth below and no hard pan occurs, the ling may be vanquished by forest.

Everywhere one sees the same struggle going on : only one more example of this is given here.　In Corsica, when cultivated soil formerly covered with maqui is left to itself, the first plants to appear are herbs,

[1] Gräbner, 1895.　　　　　[2] Gräbner, loc. cit.　　　　[3] Warming, 1907.
[4] Warming, 1906,　　　　　[5] Hult, 1885.

A a 2

including Papaver hybridum, Helianthemum guttatum, Trifolium agrarium, Galactites tomentosa, and Jasione montana.[1] After some years this herbage is suppressed by Cistus monspeliensis, and the rest of the *maqui*-vegetation gradually appears ; first, Daphne Gnidium establishes itself, and is succeeded by the other species : while eventually Cistus monspeliensis is consigned to the position that it normally takes in *maqui*.

Summary of Results.

It is difficult to make general statements in regard to vegetation appearing on new soil, because very few detailed investigations bearing on the subject have been conducted. Published work [2] seems to justify the following conclusions :—

i. In many cases, possibly always, the first colonists are algae and lichens, as well as mosses (for example, arenicolous algae on the shore, algae and lichens on lava-fields, rocks, and so forth) ; these prepare the way for Vascular plants. The early vegetation is open. Some time elapses before a coherent covering of vegetation is produced. So far as Vascular plants are concerned, the individuals are at first very scattered, but gradually increase in numbers.

ii. The number of species present is small at first ; it increases, until after the lapse of a certain length of time it is greater than ultimately. For, at first, many species find suitable spots, but are subsequently suppressed when the vegetation forms a continuous covering and more tyrannous species have entered. Various parts of the recently colonized ground are often clothed with plants in a very dissimilar manner. Gradually the vegetation becomes more uniform and poorer in species.

iii. Very frequently annual and biennial species are more numerous at first than later on, because on open ground they find the conditions more favourable to them than on overgrown ground ; many of them belong to the local weed-flora. Afterwards perennial herbs or woody plants preponderate.

iv. The first species to enter are those which occur in the vicinity and possess the best means of dispersal by wind or by birds. Rubble-heaps in the Alps are first colonized by species with wind-dispersed seeds.[3] In Norway, when a coniferous forest is destroyed, the first immigrants are the birch and poplar (with fruits and seeds respectively that are easily conveyed by wind) and Sorbus (which has berry-like fruits).[4]

v. So far as the immigration of trees is concerned, light-demanding trees precede shade-enduring ones ; the reverse is impossible. Shrubs are suppressed by trees that enter subsequently.

vi. The differentiation of sharply defined communities proceeds gradually. The earliest commingled individuals in reality belong to different natural communities, which only little by little distribute themselves in the most suitable stations. One can therefore speak of *initial, transitional,* and *final communities.*

To the preceding statements there are naturally exceptions, as is shown in the succeeding paragraphs.

[1] See Fliche, 1888.
[2] See Hult, 1885 ; Grevillius, 1893 ; A. Nilsson, 1902 ; Cowles, 1899, 1901 ; Clements, 1904, 1905 ; Ernst, 1907. See also : Moss, 1906.
[3] Kerner, 1863.   [4] Blytt, 1882 ; Hult, 1885.

Annual species may find soil more favourable to them at a later stage than at the beginning. Fliche[1] has given a suggestive account of the change that took place in the course of time in young forest plantations at Champfêtu. At first the young forest was so well lighted that a vigorous dense vegetation of perennial social species, together with mosses, was able to appear. Little by little the number of woody species increased ; Quercus, Carpinus, and Fagus outgrew the rest, and enfeebled or suppressed the ground-vegetation. As the soil changed in various ways, step by step with the increasing vegetable detritus, annual species found an increasingly favourable habitat within this mixed forest.

The capacity of spreading possessed by species depends not only upon their means of dispersal, but also upon other factors. One is usually apt to over-estimate the speed at which migrations take place. The able French forest-botanist, Fliche, in his investigation of a particular station, came to the following conclusions as regards the speed at which certain species travel : The greatest distance in metres to which the seeds are conveyed is, in Fagus sylvatica, 500 to 600 ; in Castanea sativa, 500 to 550 ; in Pinus sylvestris, 115 ; in Sorbus Aucuparia, 1,400 to 2,100. These distances are short : the fleshy fruits of Sorbus show the greatest range, the winged seeds of Pinus the least, although the latter present the appearance of being the best equipped for long journeys. Taking into consideration the age at which the trees named bear fruits, Fliche estimated the time that it would take for them to travel over the 280 kilometres separating Nancy from Paris at 18,640, 12,925, 48,680, and 1,330 to 2,000 years respectively. Too great reliance must not be placed on these figures ; yet they show migration, excepting through the agency of birds, is astonishingly slow ; and they are of significance because only few observations have been made on this question.

Agricultural experience coincides with these results. On land that has been drained by diking the soil does not bear a continuous covering of vegetation before the lapse of many years, unless man aids the process by sowing grass-seeds. Certain species easily transported by the wind are the first to settle. According to Mayr[2] the prairie tract of North America is only about 500 kilometres wide, yet there is not a single species of tree common to the Atlantic and Pacific flora, with the exception of those northern species that flank the prairies on the north. This shows the difficulty with which wind and birds convey seeds over long distances, at any rate over land. Hult arrived at the same conclusion from his study of mosses in Finland : migrations are very slow, and are regulated by secular climatic and geological changes.

In harmony with these results stands A. de Candolle's proof that certain parts of the Alps are far richer in plants than others are, because the former places were either not covered with ice during the Glacial Epoch, or were freed from it at an earlier date. In like manner extremely old parts of South America, namely the uplands of Brazil and Guiana, seem to be far richer in species than are the younger parts—the pampas and savannahs. Within the former region itself forest is much richer in species than is savannah ; but it is not known whether this is due to the greater antiquity of the forest-tract or to the circumstance that in

---

[1] Fliche, 1883.   [2] Mayr, 1890.

this tract the more favourable conditions promote the production of new species to a greater extent than in savannah.[1]

Birds all tend to carry seeds over long distances to a greater extent over sea than over land, because in the former case they find no resting-places on which they can alight and discharge the seeds. It is certain that ocean-currents can carry seeds very far.[2] Sernander[3] has classified Scandinavian plants according to their means of dispersal ; and of special interest are his investigations concerning drift cast up by sea and fresh-water, and concerning wind-dispersal ; in his work he deals with the transport not only of seeds and fruits, but also of vegetative shoots. He proves that various plants are adapted for dispersal at definite seasons, and that the different species can travel to greater or smaller distances.[4]

CHAPTER XCVI.   CHANGES IN VEGETATION INDUCED BY SLOW CHANGES IN SOIL FULLY OCCUPIED BY PLANTS

THE struggles discussed in the preceding chapter are to some extent between different communities, of which one community prepares the soil for the other, and, so to speak, persistently works out its own destruction : as illustrations of this we may cite the production of marsh, also the change from shifting to stationary dune. When slow changes in the nature of a station take place the consequent struggles are due, in the vast majority of cases, to changes in the nature of the soil, and especially in the water-content of this. Such struggles are exemplified in the following cases :—

Many plants besides Salicornia and Zostera act as capturers of mud. Aquatic mosses, algae, and other fresh-water hydrophytes, in rivers and lakes detain mud among themselves. In expanses of fresh water in Europe there is a developmental succession of which the following is an outline : The plants are ranged in zones, which are determined partly by depth of water and partly by nature of soil. In deeper water, besides *plankton*, *limnaea*-communities dominate ; under water, Myriophyllum, Characeae, and others spread themselves ; while on the surface rest the swimming-leaves of Potamogeton, Nuphar, and Ranunculus. Nearer to the bank, in shallower water, commences marsh-vegetation ; farther outwards reigns *reed-vegetation*, which is formed by the tallest and most vigorous species, namely, by Scirpus lacustris, Phragmites communis, and others.[5] In the course of time the remains of all these kinds of vegetation, together with inorganic particles that are brought hither by currents of water and wind, accumulate at the bottom of the water so that its bed rises little by little. By this means the site is prepared for other marsh-plants that can grow only in more shallow water : among such are Sium latifolium, species of Carex, Ranunculus Lingua, Menyanthes, Lythrum,

---

[1] Warming, 1892.

[2] See Warming, 1887, 1903, p. 674 ; Hemsley, 1885 ; Schimper, 1891 ; Guppy, loc. cit.

[3] Sernander, 1901.

[4] In regard to the contents of this chapter also, the papers by Cowles (1899, 1901) and Clements (1904) should be consulted.

[5] See pp. 154, 183, 186.

Oenanthe Phellandrium, Iris, Butomus, Acorus, and Equisetum limosum.[1] Gradually reed-swamp gives way to *low-moor*.[2] The bed of the water continues to rise, thanks to species of Carex and other plants belonging to low-moor. When it has been raised so high that the water is filled up to or even above its level with plants and their remains, then in the peat-like soil of the marsh one finds several grasses, also monocotylous and dicotylous herbs. Thus arises *meadow*, which, however, in most cases becomes *bushland* (with willow-bushes and alder-bushes) and *forest*, if it be not disturbed by the hand of man.[3]

The developmental succession does not necessarily proceed along the exact lines just indicated. Low-moor may give way to *Sphagnum-moor*, as it is colonized by Sphagna which continue the development[4]; the Sphagnum-moor builds itself up on top of the marsh-moor, constantly rising in height far above the water-table. Development does not necessarily stop at this stage. The drier soil becomes fitted for other plants, and particularly for woody plants ; Sphagnum-moor that is dried from any cause leads the way to *ling-heath*, as Calluna, species of Vaccinium, and other heather-plants invade the drier surface.[5] A moor thus converted into heath, 100 square kilometres in extent, occurs in the north of Jutland. Finally, such ling-heath may give way to *forest*, as Betula and Pinus sylvestris make their entrance. If the soil be rendered dry, possibly by artificial means, these trees may be replaced by others, such as Picea excelsa and Quercus.[6]

Another type of development is witnessed when the water-level suddenly and considerably falls. Feilberg has given an example of this. When Lake Söborg in Seeland was artificially emptied its original marsh-vegetation, including Menyanthes, Phragmites, Equisetum limosum, and others, was first displaced by Carex acutiformis, Agrostis vulgaris, and Poa trivialis. As the moisture diminished Poa pratensis occupied large tracts, but in turn was vanquished by Festuca rubra. If the soil be brought under cultivation, as the subsoil is loosened and the soil covered with a thin layer of loam, fodder-grasses, including Dactylis glomerata, Festuca elatior, and Poa trivialis, as well as Trifolium repens, make their entrance.

On moraine-soil in northern Europe many small moors have been produced in small lakes and pools that arose at the Glacial Epoch. Beneath the moor is a thin layer of clay which arose from the mud washed from the neighbouring elevations. In this are deposited numerous remains of the subglacial *tundra-vegetation*, which arose on land immediately after the Glacial Epoch, and was the *Dryas-vegetation*, consisting of Dryas octopetala, Salix reticulata, S. polaris, Betula nana, Oxyria digyna, Arctostaphylos alpina, Polygonum viviparum, and others. These fossil remains were discovered in 1870 by Nathorst in Schonen, later in Denmark and other countries. In these water-filled depressions the following was the course of development. Plankton and limnaea-vegetation first developed, and at the margin reed-swamp or moor (with Sphagnum and

---

[1] In regard to the British Isles see Scott-Elliot, 1900.   [2] See Chap. XLVII.
[3] See p. 217.   [4] See Chap. XLIX.   [5] See Chap. LII.
[6] In regard to developmental succession of this kind see Klinge, 1890 ; Steenstrup, 1841 ; Kerner, 1863 ; Hult, 1881 ; Magnin, 1893, 1894 ; Stebler and Schröter, 1892 ; Weber, 1894; Früh and Schröter, 1904; Gräbner, 1909 ; and C. MacMillan, 1893, 1896.

Hypnum) began to extend into the water. Gradually development advanced from the margin of the depression to the centre, in the form of a supernatant Sphagnum-moor on which Eriophorum, species of Carex, and other plants grew. As the climate became milder the higher surrounding ground acquired vegetations of trees in the following order : Populus tremula and Betula, Pinus sylvestris and Quercus.[1] Trunks of these trees were overturned by wind and buried in the moor together with their leaves, fruits, and so forth : thus arose the forest-moors which were rich in trees and are so common in the north of Seeland. Above them often lies low-moor or Sphagnum-vegetation ; but many of them are covered with meadow or, thanks to cultivation, with pastures and cornfields.

There are many other means by which water-filled depressions are occupied by upward growth : some of these have not been sufficiently investigated, others cannot be discussed here. In peat-hollows one may sometimes see rhizomes or even prostrate horizontal assimilatory shoots of Equisetum limosum growing from the margins or sides towards the centre and gradually preparing the way for other plants.

As a whole the development of vegetation in Denmark and in many other countries for centuries or even millennia past has proceeded in the direction of gradual drying, and is still so moving. Aquatic vegetation is being suppressed, lakes and ditches are vanishing, streams are being narrowed : in favour of these statements there is much historical, archaeo-logical, and geological evidence. The choking up of lakes in Denmark and on the Baltic coast is generally influenced by the direction of the wind, as Forchhammer[2] pointed out some decades ago. Klinge[3] has noted that the same is true in Russian provinces bounding the Baltic Sea. The western banks of the lakes are mostly shallow, flat, and marshy, while the eastern banks have steep, stony sides. The reason for this is that the western banks are more protected and sheltered from the prevailing south-west and west winds than are the eastern banks, where the violence of the waves prevents the process of filling up by plants. On the western banks the marsh-vegetation can consequently advance and the banks recede, whereas the eastern banks rather move in a landward direction.

Another interesting example of the production of land by the activity of vegetation and of the coincident displacement of one type of vegetation by another is provided by the behaviour of *mangrove*-vegetation.[4] The zone farthest out to sea is formed by species of Rhizophora. Thousands of their aerial roots weaken the force of the waves ; organic and other transported particles collect here and are deposited. In this way the rhizo-phoras prepare the soil for other mangrove-plants that are not capable of advancing so far out to sea. On the landward side the final stage is the development of xerophytic littoral forest, such as Barringtonia-forest, on dry soil.[5] Thus in favoured spots mangrove-vegetation constantly advances farther in the seaward direction.

A peculiar developmental succession evoked by increasing dryness is exhibited in Lapland,[6] where *Sphagnum-moor* undergoes the following changes. Bog-mosses gradually disappear, as their tufts become over-grown by other mosses that are content with less moisture, and by lichens.

[1] Steenstrup, 1841. See p. 203.
[2] Forchhammer, in lectures at the University of Copenhagen, 1860.
[3] Klinge, 1890.   [4] See p. 234.   [5] See p. 228.   [6] Kihlman, 1890.

Fruticose lichens first appear, together with some dwarf-shrubs, so that lichen-heath arises. At a later stage both the former and the latter become sickly and perish ; at the same time greyish-white patches of Lecanora tartarea show themselves and gradually cover everything with their brittle fissured crusts, out of which project puny shoots of Empetrum, Vaccinium Myrtillus, Ledum, and others. In various parts of Lapland the highest parts of the undulating covering of mosses are shrouded in these crusts. Development is not always concluded at this stage ; for the buried plants gradually decompose and become earth-like, and the incrustation of Lecanora loses its firm attachment. Fissures produced by frost or drought provide points of attack for the wind, and the incrustations are disintegrated. The black peat then lies open to colonization by any plants ; but the cohesion of its particles is too slight to permit of vegetation making permanent settlement. Storms rage without ceasing among the loose masses, excavate large hollows, just as in sand-dunes, and drifting dust results.[1] At the bottom and sides of the hollows, which often reach down to the original moraine-soil, a new type of vegetation may then establish itself.[2]

Several examples have been described in the foregoing paragraphs showing that a leading part is played by the depth of the water-table, or the level to which the water present can rise. Too great stress cannot be laid on the fact that the amount of water in soil is of the most profound importance, and that extremely slight, almost imperceptible, differences in that amount often exercise a decisive influence.[3]

The examples given have demonstrated transitions from hydrophytic to mesophytic or xerophytic communities. The reverse course of development may be seen when for any reason the amount of water in soil increases —for instance, by the damming of a stream or brook by dunes, by blocking an outlet, and so forth. According to Blytt's theory,[4] dry and humid periods of great length alternate, and to these correspond the alternating strata of tree-trunks and moss—the trees growing on the moor in the dry periods, and the mosses dominating and suppressing the forest during the moist periods. Facts militating against Blytt's theory have already been mentioned.[5]

The great moors of northern Germany are regarded as having arisen after the conversion of the large tracts of forest into marsh. In Sweden the conversion of forest into swamp and its devastation by Sphagnum are still frequent.[6] In North America the habitations constructed by beavers are allowed to occasion floods : this provides an example of the intervention of animals.

The developmental succession on rocks is well known. First, bare rock clothes itself with algae and crustaceous lichens ; these prepare a substratum suitable for fruticose lichens (cladinetum, and so forth) or for communities of mosses ; in the more or less thick cushion of the last-named some Phanerogamia occur ; and then a callunetum may develop, and perhaps finally a coniferous forest.[7]

All other changes taking place in the conditions prevailing in any habitat will have like results, that is to say, will induce changes in the vegetation by putting certain species in a position to suppress other older

[1] See pp. 61, 208.   [2] See Cajander, 1904 b, 1905 b.   [3] See p. 44.
[4] Blytt, 1882.   [5] See p. 203.   [6] See A. Nilsson, 1897.   [7] See p. 242.

settlers. These changes may be of very divers kinds, and so slow as to be almost imperceptible. What factors play the most important part in the development of vegetation in such cases it is often extremely difficult to decide ; it is usually not a single factor, but a whole series of factors which interpose and co-operate.

Change in depth of water-table and in the amount of water in soil represent one factor, change in the chemical nature of soil is another. It has already been mentioned [1] that in Russia steppe and forest compete with each other; if Tanfiljew [2] be correct, it is the slow but constant elutriation of the soil that gives victory to forest.

In Central Europe steppes existed at one time, namely, after the Tundra-period that succeeded the Glacial Epoch. [3] These steppes subsequently became forests, and probably differed from modern steppe-like communi-ties in the amount of salt present in the soil. The causes for this change of vegetation are not known, but must be sought in climatic or physico-geographical changes or influences. In more recent times the forests over wide areas have had to give way to arable land.

In *forest* itself a *change in the constituent species* has taken place, and is still proceeding. Steenstrup's [4] investigations of moors taught us that in Denmark one kind of vegetation succeeded another. [5] Nathorst [6] amplified these investigations by discovering arctic tundra-vegetation underlying the moors. While Vaupell [7] has elucidated the latest phases of the struggle, namely, that between oak and beech. Here, too, we may allude to P. E. Müller's [8] researches dealing with the struggle between forest and dwarf-shrub heath.

What causes are responsible for the successive changes of vegetation in the course of thousands of years it is difficult to decide. [9] An important part has perhaps been played by changes of climate, which on the whole has become milder. It is not probable that any vast, secular rotation of crops takes place, or that one species renders the soil more suited to a successor and less so for itself, as do certain low organisms. According to Whitford, [10] in Michigan soil is improved by coniferous forest. When a considerable amount has accumulated dicotylous forest establishes itself. Conversely, by the removal of leaf-mould the soil may be im-poverished with a consequence that dicotylous forest is vanquished by coniferous forest. Yet, under certain conditions, as there is a progressive improvement of the soil by an accumulation of humus, more exacting species will be favoured at the expense of the more accommodating species which were the first to occur : this holds true only if the soil be not robbed of the plant-remains (in the shape of crops of any kind, hay or wood) which contain its chief nutritive ingredients. Among forest-trees, oak and beech are exacting, whereas birch and Scots pine are accommo-dating. There can be no doubt that an important part has also been played by the varying relations between trees and light. [11] In the compe-tition between oak and beech in Denmark, human action (felling of trees, and drainage) has affected the result in favour of the beech, so that the oak has been able to maintain itself only on moister spots and poorer parts of

[1] See p. 282.   [2] Tanfiljew, 1894, 1905.   [3] Nehring, 1890.
[4] Steenstrup, 1841.   [5] See p. 359.   [6] Nathorst, 1870.
[7] Vaupell, 1857, 1863.   [8] P. E. Müller, 1878, 1884 (1887 a).
[9] See also Adams, 1905.   [10] Whitford, 1901.   [11] See p. 18.

Jutland. In these places the beech does not thrive, attaining comparatively low stature and maturing its seeds imperfectly ; thus it is that the oak gains the mastery. Moreover, on sandy soil the beech readily produces raw humus, and is incapable of natural regeneration.

For some centuries ling-heath has spread at the expense of forest in Denmark and northern Germany. Jutland was formerly clad with oak-forests, or less probably with one continuous forest ; now, patches of oak-scrub on heaths are the solitary relics of such forest. The forest has been devastated by careless and ignorant felling, by utilization of the wood in the production of iron from bog-ore—an industry largely pursued in Jutland during the Middle Ages—and by the west wind. As soon as the soil dries, a covering of raw humus arises and the vegetation changes, as was shown by P. E. Müller.[1] The earthworms disappear and the soil consolidates. In the layer of raw humus humous acids are produced ; and in the subsoil, as rain-water washes out the mud, layers of hard pan arise.[2] At the same time the vegetation on the ground of the forest is utterly changed. In the beech-forest, with its wealth of humus, there grows a vegetation consisting of Anemone, Corydalis, Asperula odorata, and others.[3] When the soil acquires raw humus it is invaded by Aira caespitosa, Trientalis, Maianthemum, and others, and becomes well fitted for the growth of Calluna vulgaris. This enters and gradually converts the vegetation into heath, since the beech is incapable of natural regeneration. This metamorphosis takes place especially on windy hills and on the western sides of forests.

Both oak-forest and beech-forest succumb in the struggle with heath, when felling of the trees on the washed-out soil allows the wind free play. Continental parts of Europe are unfavourable to the production of raw humus ; here bad forestry and wind do not lead to the production of heath, but, when afforestation is checked by strong winds or by drought, there arise many allied communities familiar in the south-east of Europe—namely bushland, steppe, scrub, or communities of perennial herbs belonging to the sunny (Pontic) hills in Germany.[4]

A development in the reverse direction takes place when ling-heath is irrigated with water rich in nutriment. Even one year after the commencement of irrigation ling vanishes, and after a lapse of three years the heath may be replaced by a carpet of grass and the soil may be inhabited by earthworms.[5]

---

[1] P. E. Müller, 1887 a, 1899.   [2] See p. 62.   [3] See p. 332.
[4] See Sect. XII, p. 290.
[5] The evolution of the European flora has been dealt with by many botanists, including Ad. Engler (1879–82), G. Anderson (1896), Clement Reid (1899), and Marie Jerosch (1903). Concerning change of species in North American forests, E. Bruncken's work (1904) should be consulted, as should those of Clements (1904 a, 1905, 1907), C. MacMillan (1895), and Transeau (1905 a, 1906) on 'Zonation'.

CHAPTER XCVII.   CHANGE OF VEGETATION WITHOUT
CHANGE OF CLIMATE OR OF SOIL

NUMEROUS facts have proved that there are many species which are still migrating and have not attained the distribution that soil, climate, their means of travelling, and other relations, would permit. Such species are able to emerge triumphant from struggles in many communities, without requiring the aid of any change in the inanimate surroundings. Senecio vernalis in northern Germany has spread towards the west as a pestilent weed, within a period scarcely more than twice a man's life.[1] We have already mentioned[2] the hordes of European plants that have entered Argentina and have destroyed the indigenous vegetation here and there. On the other hand, American plants introduced into Europe have locally suppressed native species : among North American immigrants are Elodea canadensis in fresh water of Central Europe, Opuntia and Agave americana in Mediterranean countries, and a number of weeds, including Oenothera biennis and Erigeron canadensis. In the same way several hundred foreign species have reached New Zealand, where some of them defeat the native vegetation.[3] It is essential that climate and soil shall suit immigrant plants ; otherwise they fail to gain an entrance, even when protected by man, as is proved by unsuccessful attempts to introduce certain trees. One species that is perhaps still moving westward is Picea excelsa. Travelling from the east, it has entered the northern part of the Scandinavian peninsula and advanced towards the south, but has not yet reached the south of Sweden nor Denmark. In several places in Norway it has swept through passes and vanquished the Scots pine, but has not been able to establish itself everywhere, so that there are remarkable gaps in its distribution. That the common spruce can defeat the Scots pine largely depends upon its hardier nature and upon its faculty of enduring shade.[4]

In every region, usually without any change in the physical environment, the vegetation indubitably undergoes slow changes, which may be appreciable only after the lapse of ages and are the result of struggles among species. This conclusion is forced upon us when we see a tract of soil that has been laid bare successfully occupied by a long series of different kinds of vegetation. In this connexion we may mention Hult's[5] description of the district of Blekinge in the south of Sweden : here most of the ' formations of vegetation are merely transitional stages towards a few final communities, whose distribution within the area is ultimately determined by soil '. Nevertheless we must assume that the struggles in question are rare in very ancient countries, whose vegetation is not appreciably disturbed by man or animals, and which has been exposed for long ages past to immigration from adjoining countries : in this case a certain condition of equilibrium must have been attained. Yet nearly all changes in vegetation that we see taking place—for instance, in forests in various parts of the earth—seem to be due to recent physical changes, and particularly to such as were caused by the destruction of forest through

[1] Ascherson, 1863.   [2] See p. 287.   [3] See Cheeseman, 1882.
[4] For details consult G. Andersson, 1896.   [5] Hult, 1885.

human agency. Some changes in the nature of forests may have been simply due to the entry of new species. Possibly it is thus that we must explain the changes which take place in Russian forests, where oak [1] is sooner or later suppressed by beech, and this in turn is vanquished by spruce ; similar changes occur in the north of Germany.[2]

If the vegetation occupying an extensive area be left entirely to itself certain formations will suppress all others after the lapse of sufficient time, and will represent the *final stage* of development.

We have already discussed the victory of ericaceous heath over forest. Borggreve and E. H. L. Krause [3] regard heath as a ' semi-cultivated ' community that owes its origin exclusively to cultivation of the soil : this view is certainly incorrect. Ericaceous heath in certain spots in the north of Europe is beyond doubt a natural vegetation in its final stage, not only on the mountain slopes of Blekinge, but even on the poor sandy soil of western Jutland. Heath is quite as independent and natural in origin as oak-forest, and has existed in Denmark, in Jutland, since the Stone Age.[4] This truth is not affected by the fact that heath has materially spread at the expense of forest, as a result of deforestation. It is, however, worthy of note that Hult [5] has shown that ericaceous heath is exterminated by forest at certain spots in Blekinge.

As other final stages in the development of vegetation in Blekinge, Hult mentions the following :—

i. Pine-forest on dry sand, on moraine-soil with detritus, and on peat-moor.

ii. Spruce-forest on less extensive moors near shores.

iii. Birch-forest with Betula pubescens on deeper moors and on low-moors.

iv. ' Grove-dell-formation ' on rivers and streams.

v. Thorn-bushland on the warmest dry places.

vi. Beech-forest on all other soil.

All other formations change gradually into these : not only grassland, but also Menyanthes-' formation ', marsh, and low-moor change thus, ' even on the rocks there develops a long series of transitional types of vegetation ' before the climax of forest is attained.[6]

With the exception of thorn-bushland all these final stages are types of forest, to wit, associations whose distribution over the ground is determined by nature of soil. Forest is in all places the last natural stage of evolution of vegetation, excepting where the development of trees is checked by a substratum of rock, lack of nutriment, water, cold or drought (either due to lack of water or to wind). In such places the final stages are represented by fell-field, dwarf-shrub heath, tundra, meadow, steppe, desert, scrub, or other communities.

[1] Korzchinsky, 1891.   [2] Grisebach, 1872.
[3] Borggreve, 1872 ; E. H. L. Krause, 1892.
[4] Sarauw, 1898.   [5] Hult, 1885.
[6] See Whitford, 1901 ; Cowles, 1901 b ; Kerner, 1863.

CHAPTER XCVIII.   THE WEAPONS OF SPECIES

THERE is scarcely any biological task more attractive than that of determining the nature of the weapons by which plants oust each other from habitats. Yet we are far from having exhaustively solved the problem even with regard to a single species ; for instance, we do not completely comprehend the struggles between the beech and oak, or between other economically important forest-trees. Obviously the matter is not settled by asserting that lack of available space is decisive, or that in the plant-world, as in all other communities (including the human race), everything turns on the question of nutrition. Such statements scientifically analysed resolve themselves into a series of the most difficult questions which science could propound, and which could be answered only after many-sided investigations. For instance, there arise such questions as : ' Is it lack of one or another nutritive body or of water in the soil ? Or the excess of another substance ? Is it want of heat or of light or of an appropriate combination of these ? Or can roots and rhizomes grow so close together as to bar the way to other plants in a purely mechanical manner, or so as to rob them of water and nutriment ? '[1] And so forth.

We see perennial herbs extinguishing annuals that have settled on ground which was bare but a short time before ; but with what weapons the former conquer we cannot say with any certainty. We see silicicolous vegetation of sand (Ornithopus perpusillus, Teesdalia, Spergula, Rumex Acetosella, Pteris aquilina, and others) disappear when the sterile field is supplied with lime (either by special addition of lime as a food-material, or by a change in the lime already present so that this becomes more easily available to plants) ; and we see this vegetation gradually return as the carbonic acid in the water dissolves and carries away the lime ; but we can give no deeper explanation of these phenomena.

Living beings forming a community have their lives linked and interwoven into one common existence in so manifold, intricate, and complex a manner that change at one point may bring in its wake far-reaching changes at other points. In this direction a wide field lies open for investigators.

To Stebler and Volkart[2] we owe an interesting piece of work on the influence of shade on the distribution of plants.

Not only do the manifold relationships shown by species to the oecological factors, light, water, heat, and so forth,[3] play a great part in these changes, but so likewise do the different *biological* characters of growthforms ;[4] and of these characters we cannot say that they are the direct result of the factors above mentioned. If forest becomes the final stage of the vegetation in a whole series of habitats, this result is due, *inter alia,* to the longevity and large size of forest-trees, which can raise themselves above, and cast their shade over, herbs and shrubs, and can produce numerous seeds year after year. In this way forest-trees easily gain the mastery over many other growth-forms, even if only a solitary individual originally succeeded in gaining entrance to the community. In such

[1] Concerning the struggles between roots of trees, see Fricke, 1904.
[2] Stebler und Volkart, 1904.   [3] See Section I.   [4] See Chap. II.

struggles a very great part is played by such conditions as, whether one species demands more light or withstands more shade, or better endures moist soil or moister air or drier wind than other species ; but likewise of great import are the conditions as to whether a species grows more or less rapidly than its rivals, or whether in this respect it behaves differently in its youth and old age. Other crucial conditions are not only whether the nutritive contents of the soil are more suited to one species or to another, but also whether the one bears more seeds, becomes reproductive at an earlier age, possibly multiplies more freely in a vegetative manner by suckers or propagative buds [1] ; whether its seeds preserve their power of germination for a long or short time; whether or no the seeds germinate more easily, what arrangement and pose are assumed by the branches, and what is the general type of architecture ; whether or no the roots and rhizomes are strongly branched and densely matted ; and so forth. Thus the biological and other characters of growth-forms, in addition to many of the factors discussed in Section I, are of great significance in the struggles between species ; sometimes one species gains the lead over another by means of some almost imperceptible advantage.

But in addition to the vital characters peculiar to different species other conditions are of importance in these struggles, for instance : attacks by parasitic fungi, and by insects or other animals (mice in forests, and the like), presence or absence of animals burrowing in the soil (for instance, earthworms [2]), and, in short, the conditions prevailing in regard to all organisms beneficial or harmful to plants.

The general statement can be made that a species has the greater probability of emerging victorious from its struggle the greater the extent to which it finds itself in its *optimal area*, or in other words, the more numerous are the oecological conditions best suited to it. Consequently a species has always to engage in its hardest and most exhausting struggles at the boundaries of its distributional area, if it has here reached the utmost limit of its wanderings as determined by climate. The more suited is the climate to a species, the less exacting is this as regards soil and other conditions, and the more capable it is of competing with rivals. As a pertinent example may be cited the fates of pine and spruce, as already described on p. 354. If a species of tree be burned down or felled on a station lying within its optimal area, it will as a rule reoccupy the denuded spot if this be not artificially interfered with ; but if it meets with this fate outside the area of its best growth, then it will not reappear, but its place will be taken by a species of tree in whose optimal area the station is situate.[3]

It may be mentioned that some species when producing formations are not in their optimal sites. Alders attain their most luxuriant development on well-drained soil. But they are usually expelled from this by other competing trees. Only in swamps, where they do not thrive so well, are they dominant. In like manner Calluna vulgaris flourishes upon rich soil better than on poor soil, but it is excluded from the former by competing species. On Madeira Vaccinium maderense shows its strongest growth in the region of laurel-*maqui*, where it is rare, and does not produce

---

[1] See the remarks on social species, pp. 92, 139. [2] See pp. 78, 363.
[3] Mayr, 1890.

pure associations until above the limit of Laurus canariensis, where it is common.

One circumstance of significance in determining the distribution of species, is—*which species is the first to arrive.* If the conditions be equally favourable to several species, the result of any struggle between them will depend upon which species succeeds in seizing the ground ; in such a case the first settlers may be able to retain possession (as *beati possidentes*).

Possibly in this way we can explain the distribution of phragmiteta, scirpeta, and other associations in European reed-swamps, or the distribution of various dwarf-shrubs on dwarf-shrub heaths.

Sometimes species can avoid competition by developing at a different time or by having their subterranean organs at different depths. Woodhead [1] names these ' complementary associations ', and describes an English wood, near Huddersfield, where Holcus mollis, Pteris aquilina, and Scilla festalis amicably give rise to a ' complementary society '. Holcus lies nearest the surface, and is succeeded by Pteris, while Scilla lies deepest in the soil ; Scilla appears above ground first and then Pteris. ' Their soil-requirements, their mode of life, their periods of active vegetative growth, their times of flowering and fruiting, are for the most part different.'

The results of struggles among plants are therefore :—

i. The arrangement of species in natural communities.

ii. Unceasing change in the composition of vegetation all the world over.

iii. Occurrence of rare species.

iv. Possibly, the origin of new species.

CHAPTER XCIX.   RARE SPECIES

THE struggle among species finds expression in the occurrence of rare species, which are so interesting to many botanical collectors.

A species may be rare in an area for various reasons, including :—

i. Because a suitable habitat is lacking, for instance, rocky soil in plains.

ii. Because it is an immigrant that has just reached the district in question, though it may be becoming more abundant here every year, as is the case with Elodea canadensis.

iii. Because it is a ' relic-plant ', that is to say, a relic of a former but now suppressed vegetation.

The extensive migration of plants that took place after the Glacial Epoch,[2] has perhaps left its traces in many supposed relic-plants, which have maintained themselves here and there, but now occur only sporadically in small numbers and are gradually dying out. The localities where they have survived are those agreeing most closely with the conditions that prevailed in the Tundra Epoch, namely, cold, wet low-moors and Sphagnum-moors. Among such survivals in Denmark and northern Germany various botanists include Cornus suecicus, Rubus Chamae-

---

[1] Woodhead, 1906.   [2] See Chap. XCVI.

morus, Polygonum viviparum, Saxifraga Hirculus, Scheuchzeria palustris, Primula farinosa, and Carex chordorrhiza. These species will beyond doubt gradually become rarer or disappear entirely from the floras, as has already been the fate of some relic-plants.

It is very difficult to prove of a species that it is a relic-plant, or even to show this with reasonable probability, and in many cases this character has been ascribed on insufficient grounds to species that are perhaps recent immigrants.[1]

CHAPTER C. ORIGIN OF SPECIES

THROUGH the whole of the preceding text has run the conception that the structure and whole course of development of species *stand in perfect harmony* (*epharmony* [2]) *with the environment, and that species are adapted to the surrounding conditions.* To Darwin we owe an everlasting debt in that he drew our attention to the epharmony prevailing in every niche and cranny of Nature. The question naturally arises : ' How has this epharmony arisen ? ' Various replies have been given to this query. Old-fashioned teleologists would answer : ' Through the direct creative action and wisdom of God.' Modern research, relying upon the facts of evolution, whose foundation was laid by Darwin, gives other explanations. Among these are the following three :—

(i) According to *Darwin* there arise an infinite number of indefinite, diversified, and small variations in individuals ; and, inasmuch as the world does not provide sufficient space for all the individuals born, Nature must exert some selective action, and by means of the ' struggle for existence ' will lead to the preservation of the most useful or fitting characters. By means of continuous selection, in the course of time more considerable distinctions will be evoked. And in this way adaptation has come into being. This explanation has recently been assailed on many sides, and does not now find so many supporters as it had when first promulgated by Darwin.

(ii) Recently *Korschinsky* [3] and *H. de Vries* [4] have put forward a somewhat different explanation of the origin of species. According to them new forms arise by sudden great changes (in a ' heterogenetic ' manner, as Korschinsky terms it, or by ' mutations ', as de Vries has it) and instantly become constant. As the new forms are very diversified, and may possess characters useful, indifferent, or even harmful to the possessor, epharmony can only come into force by naturally eliminating the forms with disadvantageous characters, and by leaving those with useful or indifferent ones.

That new forms can arise by mutation is a fact ; but we do not yet know the extent to which they can differ from the parent-form, nor how far they are able to acquit themselves in their struggles with other forms. Possibly in this way there have arisen many forms, also many characters that are ' indifferent ' in effect on the plant's existence and might be termed ' useless ' (' systematic ' or ' morphological ' characters).

---

[1] Warming, 1904.   [2] Vesque, 1882 *a*.   [3] Korschinsky, 1899.
                            [4] H. de Vries, 1901, 1903.

Here we must pass over such factors in the origin of species as the crossing of different species, as correlation among different parts of the organism, as Vesque's[1] *variabilité phylétique* (an inherited variability dependent on the ancestry of species and responsible for the progressive evolution of plants and animals).

(iii) In 1809 Lamarck published, in his famous *Philosophie Zoologique*, speculations on the origin of species. Like Darwin and other believers of the theory of evolution he assumed that forms showed variation ; and, like Darwin, he devoted consideration to domesticated plants and animals. According to Lamarck, evolution in Nature, or the natural modification of forms, goes on ceaselessly. Time and nature of environment are the two weightiest factors responsible for the natural production of all the various forms. The environment acts on the organism, so that when it changes, or when the organism migrates and becomes exposed to different surroundings, the animal feels the necessity (*besoin*) to adapt itself to the new conditions, naturally makes different use of its members or ceases to use them, and thus causes them to undergo change. Geoffroy St. Hilaire laid greater stress upon the direct action of the surrounding medium on the organism, but Lamarck also assumed this to take place especially in the case of plants. In neo-Lamarckism the former type of *active* adaptation is scarcely discussed, but *passive* adaptation or direct self-regulation (self-adaptation, epharmosis[2]) receives greater consideration. Lamarck assumed without discussion that the new forms arising in this manner transmitted the acquired characters to their offspring, and were thus preserved by *generation ;* but this is the weak point of the hypothesis, and at present is a much-discussed question which is still open.

Warming assumes that plants possess a peculiar inherent force or faculty by the exercise of which they *directly adapt themselves* to new conditions, that is to say, they change in such a manner as to become fitted for *existence in accordance* with their new surroundings. He thus assumes that between external influences and the utility of variation there is a definite connexion, which is of obscure nature (*self-regulation* or *direct adaptation, epharmosis*).

It is certainly a fact that the plant is extremely plastic, and that external factors can evoke numerous changes in it. This is proved by numbers of facts and experiments recorded during the last few decades, by Costantin, Volkens, Lothélier, Stahl, Vöchting, Schenck, Lesage, G. Karsten, Frank, Dufour, Vesque, Bonnier, Askenasy, Klebs, Massart, Göbel, Lewakoffsky, Gräbner, and others, who have investigated the morphological and anatomical plasticity of individuals[3]. The result of these investigations has been to show that by change in external conditions there is set up a course of development *tending to adapt the plant to its environment in precisely the same manner as plants or plant-communities growing under natural conditions normal to them are adapted.*

The preceding remarks are elucidated by examples discussed in the succeeding paragraphs.

The characteristics of sun-leaves and shade-leaves have already been dealt with in Chapter VI. Change in the illumination can evoke tor-

---

[1] Vesque, 1882 *a*, 1889–92.   [2] Vesque, loc. cit. ; see Diels, 1906.
[3] Göbel, 1908.

sions, and alterations in the form or position of chloroplasts, and can cause leaf-blades to assume another lie ; in like manner it can give lead to the development of a type of morphological and anatomical structure which is characteristic of the plants concerned,and which must be regarded as being of use to them. Even the peculiar shapes of leaf-like Cactaceae are mainly induced by light, as has been proved by Vöchting[1] and Göbel.[2] The etiolation of photophilous plants in darkness is presumably to be regarded as a beneficial adaptation. The modelling action of light is further illustrated by the differentiation of the vegetative organs of Marchantia, and by the production of archegonia on the shaded face of fern-prothallia.

There are definite and constant differences between shoots in the soil and shoots exposed to light, both in general and in a single species, or between the structure of roots and shoots exposed to light. Costantin[3] cultivated the same root or shoot in soil and in air ; and he proved that the manifold external conditions left their different anatomical impresses on the organs, and that the differences observed were identical with those characteristic of plant-members normally living under corresponding conditions.

The same truth is revealed by Lesage's experiments showing how plants adapt themselves to a saline soil.[4]

Experiments have also been conducted upon the action of heat on plant-members. Prillieux and Vesque[5] show that when the temperature of soil is raised the osmotic force of roots is increased, so that plants become succulent, thus acquiring the water-reservoirs and considerable volume, associated with a small transpiring surface, which help them to exist on warm, dry, rocky soil, and the like. Thermal conditions may be responsible for the increased production of wax that P. Nielsen and Raunkiär[6] have observed in hot summers on the haulms of Hordeum, Triticum (section Secale), and other grasses : this increase presumably depresses the transpiration in a manner beneficial to the plants exposed to the changed conditions.

The anatomical and morphological characters of hydrophytes and xerophytes have already been discussed in this work. Costantin,[7] Schenck,[8] Askenasy,[9] Lothélier,[10] Dufour,[11] Volkens,[12] Glück,[13] and Gerschon,[14] have proved that different organs—roots, stems, leaves, and hairs—in a single species undergo morphological and anatomical change, according as these are developed in air or in water, or in dry or in moist air. They also proved that the structural features thus induced were those generally characteristic of terrestrial and aquatic plants, or of xerophytes and hydrophytes respectively, or at least that development tended in those directions. It is clearly a case of self-regulation when the intercellular spaces become smaller as the factors inducing transpiration become stronger, or the reverse. Certain species are very plastic ; for instance, within a few weeks it is possible to change the land-form of Polygonum amphibium into the aquatic form.[15]

Change in the food-supply causes a modification in the general form

[1] Vöchting, 1894.   [2] Göbel, 1908.   [3] Costantin, 1883, see also 1898.
[4] See p. 219.   [5] See p. 124.   [6] Verbally communicated.
[7] Costantin, 1883–5 ; see also 1908.   [8] Schenck, 1884.   [9] Askenasy, 1870.
[10] Lothélier 1890.   [11] Dufour, 1887.   [12] Volkens, 1884.
[13] Glück, 1905, 1906.   [14] Gerschon, 1905.   [15] Massart, 1902.

of plants, as is well known to farmers. It seems also capable of inducing distinctions in floral structure, as an increased supply of food leads to the production of a larger receptacle, larger flowers, and more numerous floral leaves (for instance, more numerous carpels in Papaver).

Not only are external structural features affected, but also likewise are internal ones : not only the length of roots and internodes, the size and thickness or length of leaves, the greater or smaller production of hairs, and so forth, but also the relative thickness of cortex and stele and pith of axes, of palisade and spongy parenchyma of leaves, the depth of epidermis, the thickness of cuticle, the number and volume of vascular bundles, the amount of lignification, the volume of xylem, of tracheae, of tracheids, and of mechanical tissue, the size of intercellular spaces, the amount of chlorophyll, the development of stomata, of endodermis, and so forth.

The plant thus has a demonstrable faculty of reacting to external influences in a varied manner. Sometimes one of its parts may be directly affected, without the others undergoing any change. Even one and the same leaf adapts itself differently in harmony with different surroundings ; for example, the upper parts of the leaves of Stratiotes project above the water and, *inter alia*, acquire stomata, and become less transparent as well as darker green than the submerged parts.[1]

There is not only plasticity of form but also plasticity of *biological character*.[2] Gardeners know from experience that weakling plants are more easily killed by frost than are other individuals of the same species. Annual or biennial species may be induced by external influences to become perennial. The times of resting, of foliation, of defoliation, and of flowering may be changed. Cleistogamous flowers may be called into being by cold and dull weather. A number of facts bearing on this matter have been collected by Henslow.[3] The *metabolism* of the plant is, as a whole, subject to the laws of adaptation or self-regulation. Saccharomyces adjusts itself to the presence or absence of oxygen ; and the turgor of the root is similarly adjusted to the resistance encountered.

Von Wettstein[4] in his valuable work upon *seasonal dimorphism* expressed the opinion that regular mowing (therefore also regularly recurring natural influences) caused the production of reputedly definite forms which showed their own times of flowering and belonged to characteristic types. The work of the same botanist on the assumption of an annual habit by cultivated plants, including Phaseolus, should also be consulted.[5]

Plants are not all equally plastic. Differences among species in this respect depend sometimes on disposition that is due to their affinity, sometimes on the stage of evolution at which the species or genus is (certain genera, including Hieracium and Rubus, seem to be in a condition of active evolution), and sometimes on the degree to which acquired characters are fixed by heredity. Some will change more in one direction, others more in another. Even individuals belonging to one species are not all equally variable.

[1] Costantin, 1883-5.
[2] See Vöchting, 1893 ; Gräbner, 1893, p. 148 ; Göbel, 1908.
[3] Henslow, 1894, 1895, 1908.   [4] Von Wettstein, 1900.
[5] Von Wettstein, 1897-8.

Direct adaptation or self-regulation appears to find its field of operation largely in vegetative organs and in the sphere of metabolism. The flowering shoot in its development follows laws that, in some ways, differ entirely from those concerning vegetative organs ; at least, so far as is known, its reactions to climate and soil are far slighter. This is essentially due to the facts that the flowering shoot is short-lived and that metabolic processes play a less important part than in connexion with vegetative shoots.

It seems to be beyond doubt that characters peculiar to growth-forms have arisen through direct adaptation to the environment or natural self-regulation operating through countless generations, and that at the same time the acquired characters have been fixed to a greater or less extent by heredity (which is antagonistic to new adaptations). In this matter Lamarck had a keener eye for the truth than many modern investigators appear to possess. Nevertheless this suggested course of evolution must be regarded as a *hypothesis*, as the inheritance of ' acquired characters ' has not been completely proved. Yet when one keeps in mind not only the facts and experiments here recounted, but also epharmonic convergence, vicarious species or forms at different altitudes or in different soils, physiological races, and additional facts supplied by Von Wettstein [1] and others, then our doubts seem resolved. From day to day new experiments go to prove that the direct action of external factors can modify the plant's nature and can evoke *hereditary* distinctions in form : an admirable example of this is provided by MacDougal's [2] recent experiments on the effect of injecting chemical bodies into the ovary. Direct adaptation is beyond doubt one of the most potent evolutionary factors in the organic world, and appears to play the leading rôle in the adaptation of growth-forms and formations. And its study will dispel some of the darkness shrouding the mystery of life, though we cannot hope to reach the heart of that mystery.[3]

[1] Von Wettstein, 1898. [2] MacDougal, 1906 a, p. 129.
[3] Special literature: Lamarck, 1809; Darwin, 1859; Göbel, 1907 ; Klebs, 1905 ; Von Wettstein, loc. cit.; Diels, 1906; Warming, 1908 b.

374

LITERATURE

ABOFF, N. 1902. Essai de Flore raisonnée de la Terre de Feu. Ann. Mus. La Plata, i.
ABROMEIT, 1900. Dünenflora. See Gerhardt.
ADAMOVICZ, L. 1898. Die Vegetationsformationen Ostserbiens. Engler's Jahrb., xxvi.
—— 1902. Die Sibljak-Formation. Engler's Jahrb., xxxi.
—— 1904. Die Sandsteppen Serbiens. Ibid. xxxiii.
—— 1905. Über eine bisher nicht unterschiedene Vegetationsformation der Balkanhalbinsel, die Pseudomacchie. Verh. Zool.-Bot. Ges. Wien, lvi.
ADAMS, C. C. 1905. An ecological survey in Northern Michigan. Rep. Univ. Mus. Lansing, Michigan.
AGASSIZ, L. 1850. Observations on the Vegetation in ' Lake Superior ', Boston.
AITCHISON, J. E. T. 1887. The Botany of the Afghan Delimitation Commission. Trans. Linn. Soc., ser. 2, iii.
ALBERT, F. 1900. Las dunas del centro de Chile. Act. Soc. Sci. Chili, x.
—— 1907. Bezeichnung der Humusformen des Waldbodens. Zeitschr. für Forst- und Jagdwesen.
ALTENKIRCH, G. 1894. Studien über die Verdunstungsschutzeinrichtungen in der trockenen Geröllflora Sachsens. Engler's Jahrb., xviii.
ANDERSSON, G. 1896. Svenska växtvärldens Historia. Stockholm.
—— 1900. Om växtlifvet i de arktiska trakterna. Nord. Tidsskr.
—— 1902. Zur Pflanzengeographie der Arktis. Geogr. Zeitschr.
—— 1903. Das nacheiszeitliche Klima von Schweden. Ber. Zürcherischen bot. Ges., viii.
—— and HESSELMAN. 1900. Bidrag till Kännedomen om Spetsbergens och Beeren Eilands Kärlväxtflora. Bihang Sv. Vet. Akad. Handl., xxvi, 3.
—— 1907. Vegetation och flora i Hamra kronopark. Skogsvaardsföreningens tidsk., 2. Stockholm.
ANDRUSSOW. 1893. Tiefseeuntersuchungen im Schwarzen Meere. Mittheil. Geogr. Ges. Wien, xxxiii.
APSTEIN, C. 1896. Das Süsswasserplankton. Kiel and Leipzig. See Zacharias.
ARESCHOUG, F. 1878. Jemförande undersökningar öfver bladets anatomi. Lunds Universitets Aarsskrift.
—— 1882. Der Einfluss des Klimas auf die Organisation der Pflanze, insbesondere auf die anatomische Structur des Blattes. Engler's Jahrb., ii.
—— 1895. Beiträge zur Biologie der geophilen Pflanzen. Lunds Univ. Årsskr., xxxi.
—— 1902. Untersuchungen über den Blattbau der Mangrovepflanzen. Biblioth. Botan., lvi.
—— 1906. Über die Bedeutung des Palisadenparenchyms für die Transpiration der Blätter. Flora, xcvi.
ASCHERSON, P. 1859. Die Salzstellen der Mark Brandenburg. Zeitschr. Deut. geolog. Ges., xi.
—— 1863. Eine Exkursion unter dem 39. Breitegrad. Abhand. Bot. Ver. Prov. Brandenburg, v.
—— 1871. Geographische Verbreitung der Seegräser. Petermann's Mittheil.
—— 1883. Botan. Exkursion in die Reisfelder Oberitaliens. Ibid. xxv.
—— und P. Graebner. 1898-99. Flora des nordostdeutschen Flachlandes. Berlin.
ASKENASY, E. 1870. Über den Einfluss des Wachstumsmediums auf die Gestalt der Pflanzen. Bot. Zeitg., xxviii.

AUBERT. 1892. Recherches sur la respiration et l'assimilation des plantes grasses. Rev. Gén. de Bot., iv.

BACHMANN, E. 1904. Die Beziehungen der Kieselflechten zu ihrem Substrat. Ber. Deut. Bot. Ges., xxii.

BAER, K. E. VON. 1838. Expédition à Novaia Zemlia et en Laponie. Bull. Acad. Imp. Sc. St. Pétersbourg, iii.

BALFOUR, I. B. 1878. On the genus Halophila. Trans. Bot. Soc. Edin., xiii.

———— 1888. Botany of Socotra. Trans. Roy. Soc. Edinb., xxxi.

BARY, A. DE. 1877. Comparative Anatomy of the vegetative organs of the Phanerogams and Ferns. Engl. ed. Oxford, 1884.

BASINER. 1848. In Baer and Helmersen, Beiträge zur Kenntniss des Russischen Reiches, xv.

BATALIN, A. 1884. Wirkung des Chlornatriums auf die Entwickelung von Salicornia herbacea L. Bull. Congrès Internat. St.-Pétersbourg.

BATTANDIER, A. 1883. Sur quelques cas d'hétéromorphisme. Bull. Soc. Bot. France, xxx.

———— 1887. Quelques mots sur les causes de la localisation des espèces. Ibid. xxxiv.

BECK VON MANNAGETTA, G. 1884. Flora von Hernstein in Niederösterreich und der weiteren Umgebung. In M. A. Becker, Hernstein in Niederösterreich. Wien.

———— 1890-3. Flora von Niederösterreich. Wien.

———— 1901. Die Vegetationsverhältnisse der illyrischen Länder. Engler u. Drude, Vegetation d. Erde, iv.

———— 1902. Über die Umgrenzung der Pflanzenformationen. Oesterr. Bot. Zeitschr., lii.

BEIJERINCK. 1895. Über Spirillum desulfuricans als Ursache von Sulfatreduction. Centralbl. f. Bakteriol. Parasitenkunde, i.

BÉKÉTOFF, A. De l'influence du climat sur la croissance de quelques arbres résineux. Mém. Soc. Imp. Nat. Cherb., xv.

BELT. 1874. The Naturalist in Nicaragua. London.

BENECKE, W. 1892. Die Nebenzellen der Spaltöffnungen. Bot. Zeitg., l.

———— 1901. Über die Diels'sche Lehre von der Entchlorung der Halophyten. Pringsh. Jahrb., xxxvi.

BENEDEN, P. J. VAN. 1869-70. Le commensalisme dans le règne animal. Bull. Acad. Roy. Belg., 2e sér., xxviii, xxix.

BENRATH. 1904. Eine Reise durch die Cordilleren Mittelperus. Geogr. Zeitschr.

BERGEN, J. Y. 1903. The Macchie of the Neapolitan coast region. Botan. Gazette, xxxv.

———— 1903. Transpiration of Spartium junceum and other xerophytic shrubs. Ibid., xxxvi.

———— 1904 a. Transpiration of sun leaves and shade leaves of Olea europea and other broad-leaved evergreens. Ibid. xxxviii.

———— 1904 b. Relative transpiration of old and new leaves of the Myrtus type. Ibid.

———— 1905. Tolerance of drought by Neapolitan cliff flora. Ibid. xl.

BERGGREN, E. 1887. Om Rotbildningen hos australa Coniferer. Bot. Notiser.

BERGON, P. 1907. Biologie des Diatomées. Bull. Soc. Bot. France, liv.

BERNÁTSKY, J. 1899. Beiträge zur Kenntnis der endotrophen Mycorrhizen. Termész. Füz., xxii.

———— 1901. Pflanzenökologische Beobachtungen auf Süd-Lussin. Ibid. xxiv.

———— 1904. Anordnung der Formationen nach ihrer Beeinflussung seitens der menschlichen Kultur und der Weidetiere. Engler's Jahrb., xxxiv.

———— 1905. Über die Halophytenvegetation des Sodabodens im ungarischen Tieflande. Ann. Mus. Nation. Hungarici, iii.

———— 1907. The geographical plant relations of the neighbourhood of Lake Balaton. (In Hungarian.) German translation by V. Borbas in Resultate d. wissensch. Erforsch. des Balatonsees, ii. Budapest.

BESSEY, E. A. 1905. Vegetationsbilder aus Russisch-Turkestan. Karsten u. Schenck, Vegetationsbilder, iii. 2.

———— C. E. 1900. Some agricultural possibilities of Western Nebraska. Ann. Rep. Nebraska St. Bd. Agric.

BIRGER, S. 1904. Vegetationen och Floran i Pajala Socken. Arkiv för Botanik.
—— 1906 a. Die Vegetation einiger 1882–6 entstandenen schwedischen Inseln. Engler's Jahrb., xxxviii.
—— 1906 b. Die Vegetation bei Port Stanley auf den Falklandsinseln. Ibid. xxxix.
BITTER, G. 1899. Über das Verhalten der Krustenflechten beim Zusammentreffen ihrer Ränder. Pringsh. Jahrb., xxxiii.
BLACKMAN, F. F., and A. G. TANSLEY. 1905. Ecology in its physiological and phytotopographical aspects. New Phytologist, iv.
BLANC, L. 1905. La végétation aux environs de Montpellier. Bull. Soc. Bot. France, lii.
BLANKINSHIP, J. W. 1903. The Plant-formations of Eastern Massachusetts. Rhodora, v.
BLYTT, A. 1882. Die Theorie der wechselnden kontinentalen und insularen Klimate. Nebst Nachtrag. Engler's Jahrb., ii.
—— 1893. Zur Geschichte der nordeuropäischen, besonders der norwegischen Flora. Ibid. xvii.
BOBISUT, O. 1904. Zur Anatomie einiger Palmblätter. Sitzungsber. Wiener Akad., cxiii.
BÖHM, J. 1863. Über die Ursache des Saftsteigens in den Pflanzen. Sitzungsber. Wiener Akad., xlviii.
BÖRGESEN, F. 1894. Bidrag til Kundskaben om arktiske Planters Bladbygning. Bot. Tidsskr., xix. (Sur l'anatomie des feuilles des plantes arctiques. Journ. de Bot., ix, 1895.)
—— 1897. Beretning om et Par Exkursioner i Sydspanien. Bot. Tidsskr., xxi.
—— 1900. A contribution to the knowledge of the marine Alga vegetation on the coasts of the Danish West-Indian Islands. Ibid. xxiii.
—— 1905. The Algae-vegetation of the Faeröese coasts. Botany of the Faeröes. iii. Copenhagen.
—— 1906. Algenvegetationsbilder von den Küsten der Färöer. Karsten u. Schenck, Vegetationsbilder, iv. 6.
—— 1907. An ecological and systematic account of the Caulerpas of the Danish West Indies. Kongl. Danske Ved. Selsk. Skr., T. R., iv.
—— and C. JENSEN. 1904. Utoft Hedeplantage. Bot. Tidsskr., xxvi.
—— et O. PAULSEN. 1900. La végétation des Antilles danoises. Rev. Gén. de Bot., xii (1898, in Bot. Tidsskr., xxii).
BOLUS, H. 1886. Sketch of the Flora of South Africa. Offic. Handb. Cape of Good Hope.
BONNIER, G. 1879. Quelques observations sur les relations entre la distribution des Phanérogames et la nature chimique du sol. Bull. Soc. Bot. France, xxvi.
—— 1884. Sur quelques plantes annuelles ou bisannuelles qui peuvent devenir vivaces aux hautes altitudes. Ibid. xxxi.
—— 1890 a. Influence des hautes altitudes sur les fonctions. Comptes Rendus, Paris, cxi.
—— 1890 b. Cultures expérimentales dans les Alpes et les Pyrénées. Rev. Gén. de Bot., ii.
—— 1894 a. Les plantes arctiques comparées aux mêmes espèces des Alpes et des Pyrénées. Ibid. vi.
—— 1894 b. Adaptation des plantes au climat alpin. Ann. Sc. Nat., 7e sér., xx.
—— 1894 c. Remarques sur les différences que présente l'Ononis natrix cultivé sur un sol calcaire ou sur un sol sans calcaire. Bull. Soc. Bot. France, xli.
—— 1895. Recherches expérimentales sur l'adaptation des plantes au climat alpin. Ann. Sc. Nat., 7e sér., xx.
—— et C. FLAHAULT. 1879. Observations sur les modifications des végétaux suivant les conditions physiques du milieu. Ann. Sc. Nat., 6e sér., vii.
BOODLE, L. A. 1904. Succulent leaves in the Wallflower (Cheiranthus Cheiri). New Phytologist, iii.
BORBAS, V. See BERNÁTSKY, 1907.
BORGGREVE, 1872. Über die Einwirkung des Sturmes auf die Baumvegetation. Abh. Naturw. Ver. Bremen, iii.
BOTANY OF THE FÄRÖES. By various botanists. Copenhagen, 1901–1908.

BOYSEN-JENSEN, P. 1909. Über Steinkorrosion an den Ufern von Fureso. Internat. Revue der gesammten Hydrobiologie, ii.

BRACKEBUSCH. 1893. Über die Bodenverhältnisse des nordwestlichen Teiles der Argentinischen Republik mit Bezugnahme auf die Vegetation. Petermann's Mitteil., xxxix.

BRAND, F. 1896. Über die Vegetationsverhältnisse des Würmsees und seine Grundalgen. Bot. Centralbl., lxv.

—— 1905. Über die sogenannten Gasvakuolen und die differenten Spitzenzellen der Cyanophyceen, sowie über Schnellfärbung. Hedwigia, xlv.

BRANDIS, D. 1872. On the Distribution of the Forests in India.

—— 1887. Die Beziehungen zwischen Regen und Wald in Indien. Meteorol. Zeitschr.

—— 1889. Address : Specifische Individualität in dem Eintritt und in der Dauer der Blüthezeit bei Phanerogamen. Sitzungsber. d. niederrhein. Ges. Bonn.

—— 1890. Der Wald in den Vereinigten Staaten von Nordamerika. Verhandl. Naturh. Vereins d. preuss. Rheinlande und Westfalen.

BRANDT, K. 1904. Über die Bedeutung der Stickstoffverbindungen für die Production im Meere. Beihefte, Bot. Centralbl., xvi.

BRAY, W. L. 1901. The ecological relations of the vegetation of western Texas. Botan. Gazette, xxxii.

—— 1906. Distribution and adaptation of the vegetation of Texas. Bull. Univ. Texas, lxxxii.

BREWER, W. H. 1864. Notice of plants found growing in hot springs in California. Proc. Calif. Acad., iii.

BRICK, C. 1888. Beiträge zur Biologie und vergleichenden Anatomie der baltischen Strandpflanzen. Schrift. Naturf. Ges. Danzig, vii.

BRIGGS, L. J. 1900-1. Investigations on the physical properties of soils. U.S. Dept. Agric. Field Oper., Div. Soils.

BROCKMANN-JEROSCH, H. 1907. Die Flora des Puschlav (Kanton Graubünden) und ihre Pflanzengesellschaften. Leipzig.

BROTHERUS. 1886. Botanische Wanderungen auf der Halbinsel Kola. Botan. Centralbl., xxvi.

BROUN, A. F. 1905. Some notes on the ' Sudd ' Formation of the Upper Nile. Journ. Linn. Soc. Lond., xxxvii.

BROWN, F. B. H. 1905. A botanical survey of the Huron River Valley, iii. The plant societies of the Bayou at Ypsilanti, Michigan. Botan. Gazette, xl.

BRUNCKEN, ERNEST. 1902 a. Studies in plant distribution. Bull. Wisconsin Nat. Hist. Soc., ii.

—— 1902 b. On the succession of forest-types in the vicinity of Wisconsin. Ibid. ii.

BUCHENAU, F. 1886. Vergleichung der nordfriesischen Inseln mit den ostfriesischen in floristischer Beziehung. Abh. Naturw. Ver. Bremen, ix.

—— 1889 a. Über die Vegetationsverhältnisse des Helms (Psamma arenaria, Röm. et Schult.) und der verwandten Dünengräser. Ibid. x.

—— 1889 b. Die Pflanzenwelt der ostfriesischen Inseln. Ibid. xi.

—— 1903. Der Wind und die Flora der ostfriesischen Inseln. Ibid. xvii.

BUHSE, F. A. 1850. Nachrichten über drei pharmacologische wichtige Pflanzen und über die grosse Salzwüste in Persien. Bull. Soc. Nat. Imp., Moscou, xxiii.

BUNGE, AL. 1880. Pflanzengeographische Betrachtungen über die Familie der Chenopodiaceen. Mém. Acad. d. St. Pétersbourg, 7e sér., xxvii, no. 8.

BURGERSTEIN, A. 1904. Die Transpiration der Pflanzen. Jena.

BUSCALIONI, L., e POLLACCI. 1903. Le antocianine ed il loro significato biologico nelle piante. Atti Inst. Bot. Univ. Pavia, 2a ser., viii.

BÜSGEN, M., JENSEN, Hj., u. BUSSE, W. 1905. Vegetationsbilder aus Mittel- und Ost-Java. Karsten u. Schenck, Vegetationsbilder, iii. 3.

BUSSE, W. 1905. Über das Auftreten epiphyllischer Kryptogamen im Regenwaldgebiet von Kamerun. Ber. Deut. Bot. Ges., xxiii.

—— 1906 a. Das südliche Togo. Karsten u. Schenck, Vegetationsbilder, iv. 2.

—— 1906 b. Westafrikanische Nutzpflanzen. Ibid. iv. 5.

—— 1907. Deutsch-Ostafrika. Ibid. v. 7.

CAJANDER, A. K. 1903. Beiträge zur Kenntniss der Vegetation der Alluvionen des nördlichen Eurasiens. Die Alluvionen des unteren Lena-Thales. Act. Soc. Sc. Fenn., xxxii.
—— 1904 a. Studien über die Vegetation des Urwaldes am Lena-Fluss. Fennia.
—— 1904 b. Ein Beitrag zur Entwickelungsgeschichte der nordfinnischen Moore. Fennia, xx.
—— 1905 a. Die Alluvionen des Onega-Thales. Act. Soc. Sc. Fenn., xxxiii.
—— 1905 b. Beiträge zur Kenntniss der Entwickelung der europäischen Moore. Fennia, xxii.
CALLMÉ. 1887. Om de nybildade Hjelmaröarnes vegetation. Bihang Sv. Vet. Akad. Handl., xii.
CANNON, W. A. 1905. On the water-conducting systems of some desert plants. Botan. Gazette, xxxix.
CAVARA. 1901. La vegetazione della Sardegna meridionale. Nuov. Giorn. Bot. Ital., viii.
CHATIN, A. 1856. Anatomie comparée des végétaux. Paris.
CHEESEMAN, F. F. 1882. Trans. Auckl. Inst.
CHODAT, R. 1896. Flore des neiges du Col des Écaudies. Bull. Herb. Boiss., iv.
—— 1897–8. Études de biologie lacustre. Ibid., v, vi.
—— 1898. Algues vertes de la Suisse. Études de biologie lacustre. Ibid., vi.
—— 1902. Les dunes lacustres de Sciez et les Garides. Bull. Soc. Bot. Suisse, xii.
—— 1905. Une excursion botanique à Majorque. Bull. Trav. Soc. Bot. Genève, xi.
—— 1906. Observations sur le Macroplancton des étangs du Paraguay. Bull. Herb. Boiss., vi.
CHRIST, H. 1879. Das Pflanzenleben der Schweiz. Basel.
—— 1885. Vegetation und Flora der Canarischen Inseln. Engler's Jahrb., vi.
—— 1897. Die Farnkräuter der Erde.
CHRYSLER, M. A. 1904. Anatomical notes on certain strand plants. Botan. Gazette, xxxvii.
—— 1905. Reforestation at Wood's Hole, Ma. A study in succession. Rhodora, vii.
CHUN, C. 1900. Aus den Tiefen des Weltmeeres.
CIESLAR, A. 1904. Die Rolle des Lichtes im Walde. Mitt. Forstl. Versuchsw. Österreichs, xxx.
CLEGHORN, H. 1856. Note on the sandbinding plants of the Madras beach. Hooker's Lond. Jour. Bot., viii.
CLEMENTS, E. S. 1905. The relation of leaf structure to physical factors. Trans. Amer. Micros. Soc.
CLEMENTS, F. E. 1897. See POUND.
—— 1902 a. A system of nomenclature for phytogeography. Engler's Jahrb., xxxi, Beibl. lxx.
—— 1902 b. Greek and Latin in biological nomenclature. University Studies, Univ. Nebraska, iii.
—— 1904 a. The development and structure of vegetation. Bot. Surv. of Nebraska. Studies in the Vegetation of the State, iii.
—— 1904 b. Formation and Succession Herbaria. University Studies, iv, Lincoln.
—— 1905. Research methods in ecology. Lincoln, Nebr.
—— 1907. Plant physiology and ecology. London.
CLEVE, A. 1901. Zum Pflanzenleben in nordschwedischen Hochgebirgen. Bihang Sv. Vet. Akad. Handl., xxvi.
CLEVE, P. 1894. De svenske hydrograf. undersökningar, aren 1893–4. Ibid. xx.
—— 1897. A treatise on the Phytoplankton. Upsala.
—— 1901. The seasonal distribution of Atlanic Plankton organisms. Göteborg.
CLOS, D. 1889. Du nanisme dans le règne végétal. Mém. Acad. Sci. Toulouse, xi.
COCKAYNE, L. 1901. A short account of the plant-covering of Chatham Island. Trans. New Zealand Inst., xxxiv.
—— 1904. A botanical excursion during midwinter to the Southern Islands of New Zealand. Ibid. xxxvi.

COCKAYNE, L. 1905 a. Notes on the vegetation of the Open Bay Islands. Ibid. xxxvii.

—— 1905 b. On the significance of spines in Discaria Toumatou, Raoul. New Phytologist, iv.

—— 1907. Some observations on the coastal vegetation of the South Island of New Zealand. Trans. New Zealand Inst., xxxix.

—— 1908 a. Report on a botanical survey of the Tongariro National Park. Dept. of Lands, C. 11. Wellington.

—— 1908 b. Report on a botanical survey of the Waipoua Kauri forest. Ibid. C. 14.

COHN, F. 1892. Über Entstehung von Kalk- und Kieselgestein durch Vermittelung von Algen. Jahresber. Schles. Ges.

—— 1893. Über Erosion von Kalkgestein durch Algen. Ibid.

COMES. 1887. Le lave, il terreno Vesuviano e la loro vegetazione. Translated in Sammlung gemeinverständlicher wissenschaftlicher Vorträge, iv. 80, 1889.

CONTEJEAN, C. 1881. Géographie botanique. Paris.

CORNIES. 1844. In Baer and Helmerson, Beiträge zur Kenntnis d. russischen Reiches, xi.

CORNISH, V. 1897. The formation of sand dunes. Geogr. Jour.

COSTANTIN, J. 1883. Études comparées des tiges aériennes et souterraines des Dicotylédones. Ann. Sc. Nat., 6e sér., xvi.

—— 1884. Recherches sur la structure de la tige des plantes aquatiques. Ibid., 6e sér., xix.

—— 1885 a. Recherches sur l'influence qu'exerce le milieu sur la structure des racines. Ibid., 7e sér., i.

—— 1885 b. Observations critiques sur l'épiderme des feuilles des végétaux aquatiques. Bull. Soc. Bot. France, xxxii.

—— 1885 c. Recherches sur la Sagittaire. Ibid.

—— 1885 d. Influence du milieu aquatique sur les stomates. Ibid.

—— 1886. Études sur les feuilles des plantes aquatiques. Ann. Sc. Nat., 7e sér., iii.

—— 1887. Observations sur la flore du littoral. Jour. de Bot., i.

—— 1897. Accommodation des plantes aux climats froid et chaud. Bull. Sci. France et Belgique, xxxi.

—— 1898. Les végétaux et les milieux cosmiques. Paris.

COULTER, S. M. 1903. An ecological comparison of some typical swamp areas. Rep. Missouri Bot. Gard., xv.

COVILLE, F. V. 1893. Botany of the Death Valley Expedition. U.S. Dept. Agric. Contrib. U.S. Nat. Herb., iv.

—— and D. T. MACDOUGAL. 1903. Desert Botanical Laboratory of the Carnegie Institution, Washington, DC.

COWLES, H. C. 1899. The ecological relations of the vegetation on the sand dunes of Lake Michigan. Botan. Gazette, xxvii.

—— 1901 a. The influence of underlying rocks on the character of vegetation. Bull. Am. Bur. Geogr., ii.

—— 1901 b. The physiographic ecology of Chicago and vicinity; a study of the origin, development and classification of plant societies. Botan. Gazette, xxxi.

CROSSLAND, C. 1903. Note on the dispersal of mangrove seedlings. Ann. of Bot., xvii.

CZECH, C. 1869. Über die Funktionen der Stomata. Bot. Ztg., xxvii.

DANGEARD. 1888. Note sur la gaîne foliaire des Salicornieae, Benth. et Hook. Bull. Soc. Bot. France, xxxv.

DARWIN, C. 1845. Journ. of Researches . . . during the Voyage of H.M.S. Beagle. London.

—— 1859. Origin of Species. London.

—— 1875. The movements and habits of climbing plants. London.

—— 1880. The Power of Movement in Plants. London.

—— 1881. The formation of vegetable mould through the action of worms. London.

DARWIN, F., and D. F. M. PERTZ. 1896. On the effect of water-currents on the assimilation of water-plants. Proc. Cambr. Phil. Soc., ix.

De Candolle, A. P. 1820. Essai élémentaire de géographie botanique. Dict. Sc. Nat., xviii.

―――― 1832. Physiologie végétale.

De Candolle, Alphonse. 1856. Géographie botanique raisonnée. Paris.

―――― 1874. Constitution dans le règne végétal de groupes physiologiques applicables à la géographie ancienne et moderne. Bibl. Univ., l.

Dehérain, P. 1892. Traité de chimie agricole.

Delbrouck, C. 1875. Die Pflanzen-Stacheln. Hansteins Bot. Abhandl., ii.

Delden, A. van. 1903. Beitrag zur Kenntniss der Sulfatreduction. Centralbl. f. Bakteriol. u. Parasitenkunde, xi.

Delpino, F. 1867. Pensieri sulla biologia vegetale.

Detmer, W. 1877. Beiträge zur Theorie des Wurzeldruckes. Preyer's Samml. physiol. Abhandl., T. 8.

―――― 1897. Landschaftsformen des nordwestlichen Deutschlands. Berlin.

Detto, C. 1903. Über die Bedeutung der ätherischen Öle bei Xerophyten. Flora, xcii.

De Vries, H. 1901. Die Mutationstheorie. Leipzig.

Diels, L. 1896. Vegetationsbiologie von Neu-Seeland. Engler's Jahrb., xxii.

―――― 1898 a. Die Epharmose der Vegetationsorgane bei Rhus L. Gerontogeae, Ibid. xxiv.

―――― 1898 b. Stoffwechsel und Struktur der Halophyten. Pringsh. Jahrb., xxxii.

―――― 1905. Über die Vegetationsverhältnisse Neu-Seelands. Engler's Jahrb., xxxiv.

―――― 1906. Die Pflanzenwelt von West-Australien südlich des Wendekreises. Engler u. Drude, Die Vegetation der Erde, vii.

Domin, K. 1904. Die Vegetationsverhältnisse des tertiären Beckens von Veseli, Wittingau und Gratzen in Böhmen. Beiheft Bot. Centralbl., xvi.

―――― 1905 a. Das böhmische Mittelgebirge. Engler's Jahrb., xxxvii.

―――― 1905 b. Das böhmische Erzgebirge und sein Vorland. Arch. f. d. Naturwiss. Landesdurchf. v. Böhmen, xii.

Drude, O. 1876. Über ein gemischtes Auftreten von Haiden- und Wiesen-Vegetation. Flora, lix.

―――― 1884. Die Florenreiche der Erde. Petermann's Mitteil., lxxiv. Ergänzungsheft.

―――― 1886-7. Atlas der Pflanzenverbreitung. Berghaus, Physikal. Atlas. Neue Ausg., 5. Abteil.

―――― 1887. Entwurf einer biologischen Eintheilung der Gewächse. A. Schenk, Handbuch der Botanik, iii, p. 487.

―――― 1889. Über die Prinzipien in der Unterscheidung von Vegetationsformationen, erläutert an der centraleuropäischen Flora. Engler's Jahrb., xi.

―――― 1890. Handbuch der Pflanzengeographie.

―――― 1896. Deutschlands Pflanzengeographie. Bd. I.

―――― 1902. Der Hercynische Florenbezirk. Engler u. Drude, Die Vegetation der Erde, vi.

―――― 1904. Die Beziehungen der Ökologie zu ihren Nachbargebieten. Address in S. Louis. Isis ' in Dresden, 1905.

―――― 1905. Pflanzengeographie. Neumayr's Anleitung zu wissenschaftlichen Beobachtungen auf Reisen.

―――― 1908. Die kartographische Darstellung mitteldeutscher Vegetationsformationen. Bericht d. V. Zusammenkunft d. freien Vereinigung d. systemat. Botaniker u. Pflanzengeographen. Leipzig.

Dufour, L. 1886. Note sur les relations qui existent entre l'orientation des feuilles et leur structure anatomique. Bull. Soc. Bot. France, xxxiii.

―――― 1887. Influence de la lumière sur la forme et la structure des feuilles. Ann. Sc. Nat., 7e sér., v.

Dusén, P. 1898. Über die Vegetation der feuerländischen Inselgruppe. Engler's Jahrb., xxiv.

―――― 1903. The vegetation of Western Patagonia. Rep. Princeton Univ. Exped. Patagonia, viii.

―――― 1905. Die Pflanzenvereine der Magellanswälder. Wissensch. Ergeb. d. Schwed. Exped. nach d. Magellansländern, 1895-7, iii.

DUVAL-JOUVE. 1868. Des Salicornia de l'Hérault. Bull. Soc. Bot. France, xv.
—— 1875. Histotaxie des feuilles des Graminées. Ann. Sc. Nat., 6ᵉ sér., i.
EBERDT, O. 1887. Beitrag zu der Untersuchung über die Entstehungsweise des Palissadenparenchyms. Inaug.-Diss., Freiburg i. B.
—— 1888. Über das Palissadenparenchym. Ber. Deut. Bot. Ges., vi.
EGGERS, H. 1876. St. Croix's Flora. Vidensk. Meddel. Naturh. Forening. Kjöbenhavn.
ELLIS, D. 1907. A contribution to our knowledge of the thread bacteria. I. Centralbl. f. Bakt. u. Parasitenkunde, xix.
ENGELMANN. 1887. Die Farben bunter Laubblätter und ihre Bedeutung für die Zerlegung der Kohlensäure im Lichte. Bot. Zeitg., 1887.
ENGLER, AD. 1879–82. Versuch einer Entwicklungsgeschichte der Pflanzenwelt. Leipzig.
—— 1891. Über d. Hochgebirgsflora d. tropischen Afrika. Abh. Berliner Akad.
—— 1894. Über die Gliederung der Vegetation von Usambara. Abh. Berliner Akad. See also Engler's Jahrb., xvii (1893).
—— 1895. Die Pflanzenwelt Ostafrikas und der Nachbargebiete. Deutsch Ost-Afrika. Bd. V. Berlin.
—— 1899. Entwicklung der Pflanzengeographie in den letzten hundert Jahren. Wiss. Beitr. zum Gedächtn. d. 100 jähr. Wiederkehr des Antritts von A. v. Humboldts Reise nach Amerika am 5. Juni 1799. Berlin.
—— 1901. Die Pflanzenformationen und die pflanzengeographische Gliederung der Alpenkette erläutert an der Alpenanlage des neuen königlichen botanischen Gartens zu Dahlem-Steglitz bei Berlin, mit 2 Orientierungskarten. Notizbl. Kgl. Bot. Gart. Berlin, iii, App. vii.
—— 1903. Die Frühlingsflora des Tafelberges bei Kapstadt. Ibid., App. xi.
—— 1904. Über die Vegetationsverhältnisse des Somalilandes. Sitzungsber. Berliner Akad.
—— 1906 a. Beiträge zur Kenntniss der Pflanzenformationen von Transvaal und Rhodesia. Sitzungsber. Berliner Akad.
—— 1906 b. Vegetationsverhältnisse von Harar und des Gallahochlandes. Ibid.
—— 1908 a. Pflanzengeographische Gliederung von Afrika. Ibid.
—— 1908 b. Die Pflanzenwelt Afrikas. Engler u. Drude, Die Vegetation der Erde, ix.
—— 1908 c. Die Vegetationsformationen tropischer und subtropischer Länder. Engler's Jahrb., xli.
ERIKSON, J. 1895. Alfvarfloran paa Öland. Bot. Notiser, Lund.
—— 1896. Studier öfver Sandfloran i östra Skåne. Bihang Sv. Vet. Akad. Handl., xxii.
ERNST, A. 1873. Travels and Adventures in South and Central America. Hartford.
—— 1886. Vegetation d. Savanen v. Caracas. Gartenflora.
—— 1907. The new Flora of the Volcanic Island of Krakatau. Engl. Ed., Cambridge, 1908.
FEDTSCHENKO, B., und A. FLEROFF. 1907. Russlands Vegetationsbilder. St. Petersburg.
FEILBERG, P. 1890. Om Gräskultur paa Klitsletterne ved Gammel Skagen. Søborg. Autographie.
—— 1891. Om Enge og vedvarende Gräsmarker. Tidsskr. Landökon. Kjøbenhavn.
FEIST. 1884. Schutzeinrichtungen dicotyler Laubknospen. Leipzig.
FINK, B. 1903. Some common types of Lichen formations. Torrey Bulletin, xxx.
—— 1896–1903. Contributions to a knowledge of the Lichens of Minnesota. Minnesota Botan. Stud., i, ii, iii.
FISCHER, A. 1891. Beiträge zur Physiologie der Holzgewächse. Pringsh. Jahrb., xxii.
FLAHAULT, CH. Nouvelles observations sur les modifications des végétaux suivant les conditions physiques du milieu. Ann. Sc. Nat., 6ᵉ sér., ix.
—— 1888. Les herborisations aux environs de Montpellier : II. Les Garigues. Jour. de Bot., ii.
—— 1893. Distribution des végétaux dans un coin de Languedoc. Montpellier.

FLAHAULT, CH. 1894. Projet de carte botanique forestière et agricole de la France. Bull. Soc. Bot. France, xli.

—— 1896. Catalogue raisonné de la Flore des Pyrénées orientales. Perpignan.

—— 1897. Essai d'une carte botanique et forestière de la France. Ann. de Géogr., vi.

—— 1899. La naturalisation et les plantes naturalisées en France. Bull. Soc. Bot. France, xlvi.

—— 1900. Projet de nomenclature phytogéographique. Comptes rendus du Congrès Internat. de Botanique à l'Exposition Universelle de 1900, Paris.

—— 1901 a. La flore et la végétation de la France, avec une carte de la distribution des végétaux en France. H. Coste, Flore de la France.

—— 1901 b. Premier essai de nomenclature phytogéographique. Bull. Soc. Languedoc. de Géogr.

—— 1901 c. Les limites supérieures de la végétation forestière et les prairies pseudo-alpines en France. Rev. des Eaux et des Forêts.

—— 1904. Rapport . . . au sujet des jardins botaniques de l'Aigoual. Montpellier.

—— 1906 a. Les progrès de la géographie botanique depuis 1884. Progr. Rei Bot., i.

—— 1906 b. Rapport sur les herborisations de la Soc. bot. de France pendant la Session d'Oran. Bull. Soc. Bot. France, liii.

—— et BONNIER. 1879. See BONNIER.

—— et P. COMBRES. 1894. Sur la flore de la Camargue et des alluvions du Rhône. Bull. Soc. Bot. France, xli.

FLEROFF, A. T. 1907. Wasser- und Bruchvegetation aus Mittelrussland. Karsten u. Schenck, Vegetationsbilder, iv, 8.

FLICHE. 1888. Un reboisement. Ann. Sci. Agron., i.

—— et GRANDEAU. Recherches chimiques sur la bruyère commune. Ibid.

FOCKE, W. O. 1871. Über die Vegetation des nordwestdeutschen Tieflandes. Abh. Naturw. Ver. Bremen, ii.

—— 1890. Die Herkunft der Vertreter der nordischen Flora im niedersächsischen Tieflande. Ibid. xi.

—— 1892. Beiträge zum Verständnis des heimischen Pflanzenlebens. Ibid. xii.

—— 1893. Die Heide. Abh. Naturw. Ver. Bremen, xiii.

FOREL, F. A. 1866. Note sur les galets sculptés des lacs. Bull. Soc. Vaud. des Sc. Nat.

—— 1878. Faunistische Studien in den Süsswasserseen der Schweiz.

—— 1891. Allgemeine Biologie eines Süsswassersees. In Zacharias, Die Tier- und Pflanzenwelt des Süsswassers. Leipzig.

—— 1892–1902. Le Léman. Vol. i–iii.

FRICKE. 1904. ' Licht- und Schattenholzarten,' ein wissenschaftlich nicht begründetes Dogma. Zentralbl. f. d. ges. Forstwesen, xix.

FRIES, R. E. 1905. Zur Kenntniss der alpinen Flora im nördlichen Argentinien. Nov. Acta Reg. Soc. Sci. Upsal., Ser. 4, i.

FRITSCH, F. E. 1906. Problems in aquatic biology, with special reference to the study of Algal periodicity. New Phytologist, v.

—— 1907. A general consideration of the subaërial and freshwater Algal Flora of Ceylon. Pt. I. Proc. Roy. Soc., B., lxxix.

FRÜH, J. 1901. Die Abbildung der vorherrschenden Winde durch die Pflanzenwelt. Geogr.-Ethnogr. Ges. Zürich.

—— u. C. SCHRÖTER. 1904. Die Moore der Schweiz, mit Berücksichtigung der gesammten Moorfrage. Bern.

FUCHS, T. 1882. Untersuchungen über den Einfluss des Lichtes auf die bathymetrische Vertheilung der Meeresorganismen. Verh. Zool.-Botan. Ges. Wien, xxxii.

GAIDUKOW, N. 1903 a. Die Farbenänderung bei den Prozessen der komplementären chromatischen Adaptation [and other papers]. Ber. Deut. Bot. Ges., xxi.

—— 1903 b. Über den Einfluss farbigen Lichts auf die Färbung der Oscillarien. Script. Bot. Hort. Univ. Petropol., xxii.

—— 1904. Die Farbe der Algen und des Wassers. Hedwigia, xliii.

GAIN, E. 1893. Contribution à l'étude de l'influence du milieu sur les végétaux. Bull. Soc. Bot. France, xl.

—— 1895 a. Recherches sur le rôle physiologique de l'eau dans la végétation. Ann. Sc. Nat., 7ᵉ sér., xx.

—— 1895 b. Action de l'eau du sol sur la végétation. Rev. gén. de Bot., vii.

GANONG, W. F. 1894. On the absorption of water by the green parts of plants. Botan. Gazette, xix.

—— 1897. Upon raised peat bogs in the province of New Brunswick. Trans. Roy. Soc. Canada, Ser. 2, iii.

—— 1899. Preliminary outline for a plan for a study of the precise factors determining the features of New Brunswick vegetation. Bull. Nat. Hist. Soc. New Brunswick, xvii.

—— 1902. A preliminary synopsis of the grouping of the vegetation (phytogeography) of New Brunswick. Ibid. xxi.

—— 1903. The vegetation of the Bay of Fundy salt and diked marshes. Botan. Gazette, xxxvi.

—— 1904. The cardinal principles of ecology. Science, xix.

—— 1906. The nascent forest of the Miscou beach plain. Botan. Gazette, xlii.

GAUCHERY, P. 1899. Recherches sur le nanisme végétal. Ann. Sc. Nat., 8ᵉ sér., ix.

GEIKIE, J. 1898. The Tundras and steppes of prehistoric Europe. Scot. Geogr. Mag.

GERHARDT, J. 1900. Handbuch des deutschen Dünenbaues. Berlin.

GERSCHON, S. 1905. Variationen von Jussieua repens. Abh. K. Leop.-Kar. Akad. Naturf., lxxxiv.

GIBBS, L. S. 1906. A contribution to the botany of Southern Rhodesia. Jour. Linn. Soc. Lond., xxxvii.

GILG, E. 1891. Beiträge zur vergleichenden Anatomie der xerophilen Familie der Restiaceae. Engler's Jahrb., xiii.

GILLOT, F. 1894. Influence de la composition minéralogique des roches sur la végétation ; colonies végétales hétérotropiques. Bull. Soc. Bot. France, xli.

GILTAY, E. 1886. Anatomische Eigentümlichkeiten in Beziehung auf klimatische Umstände. Nederl. Kruidk. Arch., iv.

—— 1897. Vergleichende Studien über die Stärke der Transpiration in den Tropen und im mitteleuropäischen Klima. Pringsh. Jahrb., xxx.

—— 1898 a. Über die vegetabilische Stoffbildung in den Tropen und in Mitteleuropa. Ann. Jard. Bot. Buitenzorg, xv.

—— 1898 b. Die Transpiration in den Tropen und in Mittel-Europa, II. Pringsh. Jahrb., xxxii.

—— 1900. Die Transpiration in den Tropen und in Mittel-Europa, III. Ibid. xxxiv.

GINZBERGER, A., und K. MALY. 1905. Exkursion in die illyrischen Länder. Internat. Botan. Kongress, Wien.

GLÜCK, H. 1905, 1906. Biologische und morphologische Untersuchungen über Wasser- und Sumpfgewächse. Jena.

GÖBEL, K. 1886. Über die Luftwurzeln von Sonneratia. Ber. Deut. Bot. Ges., iv.

—— 1889–91. Pflanzenbiologische Schilderungen, I, II.

—— 1898–1901. Organography of Plants. Engl. Ed., Oxford, 1900–5.

—— 1908. Einleitung in die experimentelle Morphologie der Pflanzen. Leipzig u. Berlin.

GOLA, G. 1905. Studi sui rapporti tra la distribuzione delle piante e la costituzione fisico-chimica del suolo. Ann. di Botan. del Prof. R. Pirotta, iii.

GRÄBNER, P. 1893. Über gelegentliche Kleistogamie. Abhandl. Bot. Ver. Brandenburg, xxxv.

—— 1895 a. Studien über die norddeutsche Heide. Engler's Jahrb., xx.

—— 1895 b. Zur Flora der Kreise Putzig, Neustadt Wpr. und Lauenburg in Pommern. Schr. Naturf. Ges. Danzig, ix.

—— 1898 a. Gliederung der westpreussischen Vegetationsformationen. Ibid., N. F., ix.

—— 1898 b. Über die Bildung natürlicher Vegetationsformationen im norddeutschen Flachlande. Arch. d. Brandenburgia, iv. Naturw. Wochenschr., xiii.

GRÄBNER, P. 1901. Die Heide Norddeutschlands. Leipzig.
—— 1909. Die Pflanzenwelt Deutschlands. Leipzig.
GRAN, H. H. 1900. Hydrogr. biol. Studies of the North Atlantic Ocean and the coast of Nordland. Rep. Norweg. Fish. and Mar. Invest., i, 5.
—— 1901. Studien über Meeresbakterien. I. Bergens Museums Aarb., x.
—— 1902. Das Plankton des norwegischen Nordmeeres. Rep. on Norwegian Fish and Marine Investigations, ii. Bergen.
—— 1905. Diatomeen. Nordisches Plankton, hrsg. von K. Brandt, xix.
GRAY, A. 1884. Characteristics of the North American Flora. Amer. Jour. Sci. and Arts, ser. 3, xxviii.
GRECESCU. 1898. Conspectue florei Romaniei. Bucuresti.
GREVILLIUS, A. Y. 1893. Om vegetationens utveckling på de nybildade Hjelmaröarne. Bihang Sv. Vet. Akad. Handl., xviii.
—— 1894. Biologisch-physiognomische Untersuchungen einiger schwedischen Hainthälchen. Bot. Zeitg., lii.
—— 1897. Morphologisch-anatomische Studien über die xerophile Phanerogamenvegetation der Insel Öland. Engler's Jahrb., xxiii.
GRISEBACH, A. R. H. 1838. Linnaea, xii.
—— 1872. Die Vegetation der Erde. Leipzig.
—— 1875. Pflanzengeographie. In Neumayer, Anleitung. 1. Aufl.
—— 1880. Gesammelte Abhandlungen. Leipzig.
GRÖNLUND, C. 1884–90. Planteväxten paa Island. Festskr. Naturh. For. Kjöbenhavn.
GROOM, P. 1892. On bud-protection in Dicotyledons. Trans. Linn. Soc. Lond., Ser. 2, iii.
—— 1893. The influence of external conditions on the form of leaves. Ann. Bot., vii.
—— 1895 a. Contribution to the knowledge of monocotyledonous saprophytes. Jour. Linn. Soc. Lond., xxxi.
—— 1895 b. On Thismia Aseroë, Beccari, and its mycorhiza. Annals of Botany, vol. ix.
GRUBER. 1882. Anatomie und Entwickelung des Blattes von Empetrum. Diss., Königsberg.
GRÜSS, J. 1892. Beiträge zur Biologie der Knospe. Pringsh. Jahrb., xxiii.
GUBLER, A. 1851. Observations sur quelques plantes naines. Comptes Rendus Soc. Biol. Paris, iii.
GUPPY, H. B. 1893. The river Thames as an agent in plant dispersal. Jour. Linn. Soc. Lond., xxix.
—— 1906. Observations of a naturalist in the Pacific. II. Plant-Dispersal. London.
GUTTENBERG, H. v. 1905. Die Lichtsinnesorgane der Laubblätter von Adoxa Moschatellina, L. und Cynocrambe prostrata, Gaertn. Ber. Deut. Bot. Ges., xxiii.
—— 1907. Anatomisch-physiologische Untersuchungen über das immergrüne Laubblatt der Mediterranflora. Engler's Jahrb., xxxviii.
HABERLANDT, G. 1881. Vergleichende Anatomie des assimilatorischen Gewebesystemes der Pflanzen. Pringsh. Jahrb., xiii.
—— 1892. Über die Transpiration einiger Tropenpflanzen. Sitzungsber. Wiener Akad., ci.
—— 1893. Eine botanische Tropenreise. Leipzig.
—— 1894. Anatomisch-physiologische Untersuchungen über das tropische Laubblatt. Sitzungsber. Wiener Akad.
—— 1895. Über die Ernährung der Keimlinge und die Bedeutung des Endosperms bei viviparen Mangrovepflanzen. Ann. Jard. Bot. Buitenzorg, xii.
—— 1894–5. Über wassersecernierende und -absorbierende Organe. Sitzungsber. Wiener Akad., ciii. civ.
—— 1897. Über die Grösse der Transpiration im feuchten Tropenklima. Pringsh. Jahrb., xxxi.
—— 1904. Physiologische Pflanzenanatomie. Leipzig.
—— 1905. Die Lichtsinnesorgane der Laubblätter. Leipzig.
—— 1908. Über die Verbreitung der Lichtsinnesorgane der Laubblätter. Sitzb. Wiener Akad., cxvii.

HACKEL, E. 1890. Über einige Eigentümlichkeiten der Gräser trockener Klimate. Verh. Zool.-Bot. Ges. Wien, xl.

HÄCKEL, E. 1866. Generelle Morphologie. Berlin.

―――― 1890. Planktonstudien. Jena.

―――― 1891. Planktonstudien. Jena. Zeitschr. f. Naturw., xxv.

HANN, J. 1897. Handbook of Climatology. Engl. Ed., New York, 1903.

―――― 1901. Lehrbuch der Meteorologie. Leipzig.

HANSEN, E. 1881. Sur le Saccharomyces apiculatus et sa circulation dans la nature. Compt. Rend. Lab. Carlsberg, i, 3.

HANSTEEN, B. 1892. Algeregioner og Algeformationer. Nyt. Mag. f. Naturvidskaben, xxxii.

HANSTEIN, R. 1868. Über die Organe der Harz- und Schleimabsonderung in den Laubknospen. Bot. Zeitg., xxvi.

HARSHBERGER, J. W. 1897. The vegetation of the Yellowstone hot springs. Amer. Jour. Pharm., lxix.

―――― 1900. An ecological study of the New Jersey strand flora. Proc. Acad. Nat. Sci. Philad., lii.

―――― 1901. An ecological sketch of the flora of Santo Domingo. Ibid. liii.

―――― 1902. Additional observations on the strand flora of New Jersey. Ibid. liv.

―――― 1903. An ecological study of the flora of mountainous North Carolina. Botan. Gazette, xxxvi.

―――― 1904. A phyto-geographic sketch of extreme south-eastern Pennsylvania. Torrey Bull., xxxi.

―――― 1905. The plant formations of the Bermuda Islands. Proc. Acad. Nat. Sci. Philad., lvii.

―――― 1905. The plant formations of the Catskills. Plant World, viii.

HARTZ, N. 1895. Östgrönlands Vegetationsforhold. Meddel. om Grönland, xviii.

HARVEY, Le R. H. 1903. A study of the physiographic ecology of Mt. Ktaadn, Maine. Univ. of Maine Stud., v.

―――― 1908. Floral succession in the prairie-grass formation of South-Eastern South Dakota. Botan. Gazette.

HASSERT. 1895. Petermanns Mitth., Ergänzungsh. cxv.

HAYEK, A. v. 1905. Exkursion auf den Wiener Schneeberg, II. Internat. Botan. Kongress, Wien.

―――― 1907. Die Sanntaler Alpen. Steiner Alpen. Abh. Zool.-Bot. Ges. Wien, iv.

HEDGCOCK, GEORGE G. 1902. The relation of the water content of the soil to certain plants, principally mesophytes. Bot. Surv. of Nebraska, vi. Lincoln, Nebr.

HEGI, G. 1905. Beiträge zur Pflanzengeographie der bayerischen Alpenflora. Ber. Bayer. Bot. Ges., x.

HEINRICHER, E. 1884. Über isolateralen Blattbau. Pringsh. Jahrb., xv.

―――― 1885. Über einige im Laube dikotyler Pflanzen trockenen Standortes auftretende Einrichtungen, welche muthmaasslich eine ausreichende Wasserversorgung des Blattmesophylls bezwecken. Bot. Centralbl., xxiii.

―――― 1897-1904. Die grünen Halbschmarotzer. Pringsh. Jahrb., xxxi, xxxii, xxxvi, xxxvii, xxxviii.

HEMBERG, E. 1904. Tallens degenerationszoner i södra och västra Sverige. Skogsvårdsför. Tidskr.

HEMMENDORFF, E. 1897. Om Ölands Vegetation. Upsala.

HEMSLEY, W. B. 1885. On the dispersal of plants by oceanic currents and birds. Challenger Rep., Botany, i.

HENSEN, V. 1887. Über die Bestimmungen des Planktons. Ber. Kommis. Wiss. Unters. Deut. Meere, v, vi.

―――― 1890. Einige Ergebnisse der Plankton-Expedition der Humboldt-Stiftung. Sitzungsber. Berliner Akad.

HENSLOW, G. 1894. The origin of plant-structures by self-adaptation to the environment, exemplified by desert or xerophilous plants. Jour. Linn. Soc. Lond., xxx.

―――― 1895. The origin of plant-structures. London.

―――― 1908. The heredity of acquired characters in plants. London.

HERDER, F. v. Die neueren Beiträge zur pflanzengeographischen Kenntnis Russlands. Engler's Jahrb., viii, ix.

HESSELMAN, H. 1879. Några iaktagelser öfver växternas spridning. Bot. Notiser.

— — 1904. Zur Kenntnis des Pflanzenlebens schwedischer Laubwiesen. Beiheft z. Bot. Centralbl., xvii.

— — 1906. Om svenskaskogar och skogssamhällen. Stockholm.

HEUGLIN, M. TH. v. 1874. Reisen nach dem Nordpolarmeer in den Jahren 1870–71. iii.

HILDEBRAND, F. 1873. Über die Verbreitungsmittel der Pflanzen. Leipzig.

— — 1882. Die Lebensdauer und Vegetationsweise der Pflanzen, ihre Ursachen und ihre Entwickelung. Engler's Jahrb., ii.

— — 1884. Die Lebensverhältnisse der Oxalis-Arten. Jena.

— — 1902. Über Ähnlichkeiten im Pflanzenreich. Leipzig.

HILGARD. A report on the relations of soil to climate. S.S. Dept. Agric., Weather Bur., iii, Washington, 1892.

HILL, E. J. 1900. Flora of the White Lake region, Michigan. Botan. Gazette, xxix.

HILLEBRAND, 1888. Die Vegetationsformationen der Sandwich-Inseln. Engler's Jahrb., ix.

HITCHCOCK, A. S. 1898. Ecological plant geography of Kansas. Trans. Acad. Sci. St. Louis, viii.

— — 1899. A brief outline of ecology. Trans. Kansas Acad. Sci., xxvii.

— — 1904. Methods used for controlling and reclaiming sand dunes. U.S. Dept. Agric., Bull. lvii. See also : Nation. Geogr. Mag., 1904.

HOCHREUTINER, B. P. G. 1904. Le Sud-Oranais. Ann. du Conserv. et du Jard. Bot. Genève, vii–viii.

HOCHSTETTER, F. VON. 1863. New Zealand. Engl. Ed. Stuttgart, 1867.

HÖCK, F. 1892. Begleitpflanzen der Buche. Bot. Centralbl., lii.

— — 1893 a. Nadelwaldflora Norddeutschlands. Forschungen zur Deutschen Landes- und Volkskunde, hrsg. von Kirchhoff, vii.

— — 1893 b. Begleitpflanzen der Kiefer in Norddeutschland. Ber. Deut. Bot. Ges., xi.

— — 1895. Brandenburger Buchenbegleiter. Verh. Bot. Ver. Prov. Brandenburg, xxxvi.

— — 1896. Laubwaldflora Norddeutschlands. Forschungen zur Deutschen Landes- und Volkskunde, hrsg. von Kirchhoff.

— — 1898. Eine Genossenschaft feuchtigkeitsmeidender Pflanzen Norddeutschlands. Allg. Bot. Zeitschr.

HOLM, THEO. 1885. Recherches anatomiques et morphol. sur deux monocotylédones submergées. Bihang Sv. Vet. Akad. Handl., ix.

— — 1887. Novaia Zemlias Vegetation. Dijmphna-Togtets Zool.-Bot. Udbytte. Kjöbenhavn.

— — 1891. On the vitality of some annual plants. Amer. Jour. Sci., xlii.

HOLMBOE, J. 1898. Nogle iakttagelser over fröspredning paa ferskvandis. Bot. Notiser.

HOLTERMANN, C. 1902. Anatomisch-physiologische Untersuchungen in den Tropen. Sitzungsber. Berliner Akad.

— — 1907. Der Einfluss des Klimas auf den Bau der Pflanzengewebe. Leipzig.

HOMÉN. 1897. Der tägliche Wärmeumsatz im Boden. Helsingfors.

HOOKER, J. D. 1847 a. Botany of the antarctic voyage of H.M. Discovery Ships *Erebus* and *Terror*.

— — 1847 b. On the vegetation of the Galapagos Archipelago as compared with that of some other tropical islands of the continent of America. Linn. Trans., xx.

— — 1867. On the struggle for existence amongst plants. Pop. Sci. Rev., vi.

— — 1896. Lecture on insular floras. London.

HUBER, G. 1905. Monographische Studien im Gebiete der Montigglerseen (Südtirol), mit besonderer Berücksichtigung ihrer Biologie. Arch. f. Hydrobiol. u. Planktonk., i.

HUBER, J. 1906. La végétation de la vallée de Rio Purus (Amazon). Bull. Herb. Boissier, sér. ii, vi.

HUITFELD-KAAS, K. 1906. Planktonundersögelser i Norske Vande. Christiania.

HULT, R. 1881. Försök till analytisk behandling af växtformationerna. Meddel. Soc. Faun. Flor. Fenn., viii.

—— 1885. Blekinges vegetation. Ibid. xii.

—— 1886. Mossfloran i trakterna mellan Aavasaksa och Pallastunturit. Acta Soc. Faun. Flor. Fenn., iii.

—— 1887. Die alpinen Pflanzenformationen des nördlichsten Finlands. Ibid. xiv.

HULTBERG. Anatomiska undersökningar öfver Salicornia. Lunds Universitets Aarsskrift, xviii.

HUMBOLDT, A. 1805. Aspects of Nature. Engl. Ed., Lond., 1849.

—— 1805 (1807). Essai sur la géographie des plantes. Paris.

—— 1806. The Physiognomy of Plants. Engl. Ed., Lond., 1849.

HUNGER, W. 1899. Über die Funktion der oberflächlichen Schleimbildungen im Pflanzenreiche. Leiden.

HUTH, E. 1887. Die Klett-Pflanzen, mit besonderer Berücksichtigung ihrer Verbreitung durch Tiere. Biblioth. Botan., ix.

—— 1889. Die Verbreitung der Pflanzen durch die Exkremente der Tiere. Samml. naturw. Vortr., iii, Berlin.

ISTVANFFI, G. VON. 1898. Die Kryptogamenflora des Balatonsees. Result. d. wiss. Erforsch. d. Balatonsees. 2. Bd., ii.

—— 1905. Flore microscopique des thermes de l'île Margitsziget. Budapest.

JACCARD, P. 1907. La distribution de la flore dans la zone alpine. Revue gén. d. sciences pures et appliquées. 18ᵐᵉ année (Ref. in Bot. Centralbl., 107).

JAHN, E. 1886. Über Schwimmblätter. Beit. z. wiss. Bot., x.

JEROSCH, M. 1903. Geschichte und Herkunft der schweizerischen Alpenflora. Leipzig.

JOHOW, F. 1884 a. Über die Beziehungen einiger Eigenschaften der Laubblätter zu den Standortverhältnissen. Pringsh. Jahrb., xv.

—— 1884 b. Die Mangrovensümpfe. Kosmos.

—— 1885. Die chlorophyllfreien Humusbewohner Westindiens, biologisch-morphologisch dargestellt. Ibid. xvi.

—— 1889. Die chlorophyllfreien Humuspflanzen nach ihren biologischen und anatomisch-entwicklungsgeschichtlichen Verhältnissen. Ibid. xx.

—— 1896. Estudios sobre la flora de las Islas de Juan Fernandez. Santiago.

JÖNSSON, B. 1878–9. Bidrag till kännedomen om bladets anatomiska byggnad hos Proteaceerna. Lunds Univ. Aarsskr., xv.

—— 1902. Zur Kenntnis des anatomischen Baues der Wüstenpflanzen. Ibid. xxxviii.

—— 1903. Assimilationsversuche bei verschiedenen Meertiefen. Nyt Mag. f. Naturw., xli.

JONSSON, H. 1895. Optegnelser fra Vaar- og Vinterexkursioner i Öst-Island. Bot. Tidsskr., xix.

—— 1905. Vegetationen i Syd-Island. Ibid. xxvii.

JOST. Lectures on Plant Physiology. Engl. Ed., Oxford, 1908.

JOUAN, H. 1865. Recherches sur l'origine et la provenance de certains végétaux phanérogames observés dans les îles du Grand-Océan. Mém. Soc. Sci. Nat. Cherbourg, xi.

JUNGHUHN. 1852–4. Java. German Ed. by Hasskarl. 2 Bde., Leipzig.

JUNGNER. 1891. Anpassungen der Pflanzen an das Klima in den Gegenden der regenreichen Kamerungebirge. Bot. Centralbl., xlvii.

KARSTEN, G. 1891. Über die Mangrove-Vegetation im Malayischen Archipel. Biblioth. Bot., xxii.

—— 1894. Morphologische und biologische Untersuchungen über einige Epi-phytenformen der Molukken. Ann. Jard. Bot. Buitenzorg, xii.

—— Das Phytoplankton des Antarktischen Meeres nach dem Material der Deutschen Tiefsee-Expedition, 1898–9. Wiss. Ergeb. Deut. Tiefsee-Exped., ii.

—— 1905–6. Das Phytoplankton des Atlantischen Oceans. Ibid.

—— 1907. Das indische Phytoplankton. Ibid.

—— und SCHENCK, H. 1903–8. Vegetationsbilder. Jena.

—— 1903 a. Vegetationsbilder aus dem Malayischen Archipel. Karsten u. Schenck, Vegetationsbilder, i, 2.

KARSTEN, G., und SCHENK, H. 1903 b. Mexikanischer Wald der Tropen und Subtropen. Ibid. i, 4.
—— 1903 c. Monokotylenbäume. Ibid. i, 6.
—— 1904. Die Mangrovevegetation. Ibid. ii, 2.
—— und E. STAHL. [1903. Mexikanische Cacteen-, Agaven- u. Bromeliaceen-Vegetation. Ibid. i, 8.
KEARNEY, T. H. 1900. The plant-covering of Ocracoke Island. Contrib. U. S. Nat. Herb., v.
—— 1901. Report on a botanical survey of the Dismal Swamp region. Ibid. v.
—— 1904. Are plants of sea beaches and dunes true halophytes? Botan. Gazette, xxxvii.
—— and F. C. CAMERON. 1902. Some mutual relations between alkali soils and vegetation. Rep. U. S. Dept. Agric., lxxi.
KELLER, C. 1887. Humusbildung und Bodenkultur unter dem Einfluss tierischer Thätigkeit.
KELLER, ROB. 1903. Vegetationsbilder aus dem Val Blenio. Mitteil. d. Naturw. Ges. Winterthur.
—— 1904. Vegetationsskizzen aus den Grajischen Alpen. Ibid.
KERNER VON MARILAUN, A. 1858. Über die Zsombek-Moore Ungarns. Abhandl. Zool.-Bot. Ges. Wien, viii.
—— 1863 a. Das Pflanzenleben der Donauländer. Innsbruck.
—— 1863 b. Über das sporadische Vorkommen sogenannter Schieferpflanzen im Hochgebirge. Verhandl. Zool.-Bot. Ges. Wien, xiii.
—— 1869. Die Abhängigkeit der Pflanzengestalt von Klima und Boden.
—— 1886. Österreich-Ungarns Pflanzenwelt. Die österr.-ungar. Monarchie in Wort und Bild. 2. Band, 1. Abt., Wien.
—— 1887-91. The Natural History of Plants. Engl. Ed.
KIHLMAN, A. O. 1890. Pflanzenbiologische Studien aus Russisch Lappland. Act. Soc. Faun. Flor. Fenn., vi.
KIRCHNER, O. 1892. See SCHRÖTER u. KIRCHNER.
KIRCHNER, LOEW, und SCHRÖTER. 1904-. Lebensgeschichte der Blütenpflanzen Mitteleuropas.
KISSLING, P. B. 1895. Beiträge zur Kenntniss des Einflusses der chemischen Lichtintensität auf die Vegetation. Halle a. S.
KITTLITZ, F. H. VON. 1850-2. Twenty-four Views of the Vegetation of the Coasts and Islands of the Pacific. Engl. Ed., Lond., 1861.
KJELLMAN, F. R. 1875. Végétation hivernale des Algues à Mosselbay (Spitzberg). Compt. Rend., lxxi.
—— 1878. Über Algenregionen und Algenformationen im östlichen Skager Rak. Bihang Sv. Vet. Akad. Handl., v.
—— 1882. Om växtligheten på Sibiriens Nordkust. Vega-Expeditionens vetenskapl. iakttagelser, i.
—— 1883. Norra Ishafvets Algflora. Ibid. iii.
—— 1884. Ur polarväxternas lif. Nordenskiöld, Studier och Forskningar.
—— 1906. Om främmande alger ilanddrifna vid Sveriges västkust. Archiv f. Botanik, v.
KLEBAHN, H. 1895. Gasvacuolen, ein Bestandtheil der Zellen der wasserblüthebildenden Phycochromaceen. Flora, lxxx.
KLEIN, J. 1880. Zur Kenntniss der Wurzeln von Aesculus Hippocastanum. Flora, lxiii.
KLEIN, L. 1899. Die Physiognomie der mitteleuropäischen Waldbäume. Karlsruhe.
—— 1904. Charakterbilder mitteleuropäischer Waldbäume, I. Karsten u. Schenck, Vegetationsbilder, ii, 5, 6. 7.
KLINGE, J. 1890. Über den Einfluss der mittleren Windrichtung auf das Verwachsen der Gewässer. Engler's Jahrb., xi.
KNOBLAUCH, E. 1896. Ökologische Anatomie der Holzpflanzen der südafrikanischen immergrünen Buschregion. Habilitationschr.
KNUTH, P. Handbook of Flower Pollination. Engl. Ed., Oxford, 1906-8.
KNY, L. 1878. Methoden zur Messung der Tiefe, bis zu welchen Lichtstrahlen in das Meerwasser eindringen. Bot. Zeitg., xxxvi.
—— 1895. Über die Aufnahme tropfbar flüssigen Wassers durch winterlich entlaubte Zweige von Holzgewächsen. Ber. Deut. Bot. Ges., xiii.

KNY, L. V. ZIMMERMAN, C. 1885. Die Bedeutung der Spiralzellen von Nepenthes. Ber. Deut. Bot. Ges., iii.

KÖPPEN, V. 1884. Die Wärmezonen der Erde nach der Dauer der heissen, gemässigten und kalten Zeit und nach der Wirkung der Wärme auf die organische Welt betrachtet. Meteorol. Zeitschr.

—— 1900. Versuch einer Klassification der Klimate, vorzugsweise nach ihren Beziehungen zur Ozeanenwelt. Geograph. Zeitschr.

KÖRNICKE, M., und F. ROTH. 1907. Eifel und Venn. Karsten u. Schenck, Vegetationsbilder, v. 1, 2.

KOFOID, C. 1903. The Plankton of the Illinois river. I.: Bull. of the Ill. Lab., vi. II.: Ibid. 1908, viii.

KOHL. 1886. Die Transpiration der Pflanzen und ihre Einwirkung auf die Ausbildung pflanzlicher Gewebe.

KOLKWITZ u. MARSSON. 1902. Grundsätze für die biologische Beurtheilung des Wassers. Berlin.

KOORDERS, S. H. 1895. Beobachtungen über spontane Neubewaldung auf Java. Forstl. Naturw. Zeitschr., iv.

—— 1907. Ein von der Holländisch-Indischen Sumatra-Expedition entdecktes Tropen-Moor. Note by Potonié in Naturw. Wochenschr., xlii.

KORSCHINSKY, S. 1891. Über die Entstehung und das Schicksal der Eichenwälder im mittleren Russland. Engler's Jahrb., xiii.

—— 1899. Heterogenesis und Evolution. Flora, lxxxix, 1901.

KOSTYTSCHEFF. 1890. Der Zusammenhang zwischen den Bodenarten und einigen Pflanzenformationen. Scripta Bot. Hort. Univ. Petropolitanae, iii.

KRASAN, F. 1882. Die Erdwärme als pflanzengeographischer Factor. Engler's Jahrb., ii.

—— 1883. Die Berghaide der südöstlichen Kalkalpen. Ibid. iv.

—— 1884. Über die geothermischen Verhältnisse des Bodens und deren Einfluss auf die geographische Verbreitung der Pflanzen. Verh. Zool.-Bot. Ges. Wien, xxxiii.

KRASSNOFF, A. 1886. Geobotanical Researches in the Kalmuk Steppe. (In Russian.) Bull. Russ. Geogr. Soc., xxii. Extended reference by F. von Herder in Engler's Jahrb., x (1889), Litteraturbericht.

—— 1888. Bemerkungen über die Vegetation des Altai. Engler's Jahrb., ix.

KRAUS, G. 1905. Anemometrisches von Krainberg bei Gambach. Verh. Phys.-Med. Ges. Würzburg, N.F., xxxvii.

—— 1906 a. Über den Nanismus unserer Wellenkalkpflanzen. Ibid. xxxviii.

—— 1906 b. Die Sesleria-Halde. Ibid.

—— 1908. Erfahrungen über Boden u. Klima auf dem Wellenkalk. Ibid. xl.

KRAUSE, E. H. L. 1891. Die Eintheilung der Pflanzen nach ihrer Dauer. Ber. Deut. Bot. Ges., ix.

—— 1892 a. Die Heide. Engler's Jahrb., xiv.

—— 1892 b. Beitrag zur Geschichte der Wiesenflora in Norddeutschland. Ibid. xv.

—— Die natürliche Pflanzendecke Norddeutschlands. Globus, lxi.

—— Die Existenzbedingungen der nordwestdeutschen Heidefelder. Ibid. lxx.

KRONFELD. 1890. Über die biologischen Verhältnisse der Aconitumblüthe. Engler's Jahrb., xi.

KRÜGER, P. 1883. Die oberirdischen Vegetationsorgane der Orchideen in ihren Beziehungen zu Clima und Standort. Flora, lxvi.

KURTZ, F. 1893. Dos viages botánicos al Rio Salado superior. Boletín Acad. nac. de Cordoba.

KURZ, S. 1875. Preliminary report on the forests and other vegetation of Pegu. Calcutta.

KUSNEZOW, N. J. 1898. Übersicht der in den Jahren 1891–4 über Russland erschienenen phyto-geographischen Arbeiten. Engler's Jahrb., xxvi.

—— 1901. Die Vegetation und die Gewässer des europäischen Russlands. Ibid. xxviii.

LAGERHEIM, G. DE. 1892. Die Schneeflora des Pichincha. Ber. Deut. Bot. Ges., x.

LAMARCK. 1809. Philosophie zoologique. Paris.

LAMSON-SCRIBNER, F. 1892. Mt. Kataadn and its Flora. Botan. Gazette, xvii.

LAUTERBACH, C. 1889. Untersuchungen über Bau und Entwicklung der Sekretbehälter bei den Cacteen. Bot. Centralb., xxxvii.

LAZNIEWSKI, W. v. 1896. Beiträge zur Biologie der Alpenpflanzen. Flora, lxxxii.

LEIST, K. 1889. Einfluss des alpinen Standortes auf die Ausbildung der Laubblätter. Bern.

LEIVISKÄ, J. 1908. Die Vegetation an der Küste des Bottnischen Meerbusens. Fennia, xxvii.

LESAGE, P. 1890. Recherches expérimentales sur les modifications des feuilles chez les plantes maritimes. Rev. Gén. de Bot., ii.

―――― 1894. Sur les rapports des palissades dans les feuilles avec la transpiration. Comptes Rendus, Paris, cxviii.

LEWIS, F. J. 1904. Geographical distribution of Vegetation on the Basins of the Rivers Eden, Tees, Wear and Tyne. Pts. I, II. Journ. Roy. Geogr. Soc., xxiii.

―――― 1905-07. Plant remains in the Scottish peat mosses. Pts. I-III. The Scottish Southern Uplands. Trans. Roy. Soc. Edin., xli, xlv, xlvi.

LIDFORSS, B. 1903. Über den Geotropismus einiger Frühjahrspflanzen. Pringsh. Jahrb., xxxviii.

―――― 1907. Die wintergrüne Flora. Lunds Univ. Aarsskr., N. F., Afd. 2, ii.

―――― 1908. Weitere Beiträge zur Kenntniss der Psykroklinie. Lunds Univ. Aarsskr., N. F., Afd. 2, iv.

LINDMAN, C. 1883. Om drifved och andra af hafsströmmar uppkastade naturföremaal vid Norges kuster. Göteborg.

―――― 1899. Zur Morphologie und Biologie einiger Blätter und belaubter Sprosse. Bihang Sv. Vet. Akad. Handl., xxv.

―――― 1900. Vegetationen i Rio grande do Sul, Sydbrasilien. Stockholm.

LIVINGSTON, B. E. 1901. The distribution of the plant societies of Kent County, Mich. Ann. Rep. Mich. St. Bd. Geol. Surv.

―――― 1903. The distribution of the upland plant societies of Kent County, Mich. Botan. Gazette, xxxv.

―――― 1904. Physiological properties of bog water. Ibid. xxxix.

―――― 1905. The relation of soils to natural vegetation in Roscommon and Crawford Counties, Michigan. Ibid.

―――― 1906. The relation of desert plants to soil moisture and to evaporation. Carnegie Inst. Washington, l.

―――― and G. H. JENSEN. 1904. An experiment on the relation of soil physics to plant growth. Botan. Gazette, xxxviii.

LJUNGSTROOM, E. 1883. Bladets byggnad inom familjen Ericineae. Lunds Univ. Aarsskr., tom. xix.

LOESKE, L. 1900. Die Moosvereine im Gebiete der Flora von Berlin. Abhand. Bot. Ver. Prov. Brandenburg, xlii.

LOHMANN. 1902. Die Coccolithoporidae. Archiv f. Protistenkunde, i.

LORENZ, T. R. 1863. Physikalische Verhältnisse und Vertheilung der Organismen im Quarnerischen Golfe. Wien.

LOTHELIER, A. 1890. Influence de l'état hygrométrique de l'air sur la production des piquants. Bull. Soc. Bot. France, xxxvii.

―――― 1891. Influence de l'éclairement sur la production des piquants des plantes. Comptes Rendus, Paris, cxii.

―――― 1893. Recherches sur les plantes à piquants. Rev. Gén. de Bot., v.

LOVÉN, H. 1891. Naagra rön om Algernas anding. Bihang Sv. Vet. Akad. Handl., xvii.

LUBBOCK, J. 1885. Ants, Bees, and Wasps. Ed. 7, London.

―――― 1899. On Buds and Stipules. London.

LUDWIG, F. 1895. Lehrbuch der Biologie der Pflanzen. Stuttgart.

LUND, P. W. 1835. Bemærkninger over Vegetationen paa de indre Höjsletter af Brasilien. K. Danske Vid. Selsk. Skr., vi.

LUNDSTRÖM, A. 1884. Die Anpassungen der Pflanzen an Regen und Tau. Act. Soc. Reg. Upsal., Ser. iii, xii.

―――― 1887. Anpassungen der Pflanzen an Thiere. Ibid. xiii.

LYNGBYE, H. C. Rariora Codana. Vid. Meddel. Naturh. For. Kjöbenhavn, 1879-80.

MACDOUGAL, D. T. 1903. See F. V. COVILLE and D. T. M.
—— 1906 a. The vegetation of the Salton Basin. Year-Book Carnegie Inst. Washington, v.
—— 1906 b. The Delta of the Rio Colorado, &c. Contrib. New York Bot. Gard., lxxvii.
—— 1907. The Desert Basins of the Colorado Delta. Bull. Amer. Geogr. Soc.
MACMILLAN, C. 1893. On the occurrence of sphagnum atolls in central Minnesota. Minnesota Botan. Stud., Bull. ix.
—— 1896. On the formation of circular Muskeg in Tamarack swamps. Torrey Bull., xxiii.
—— 1897. Observations on the distribution of plants along shore at Lake of the Woods. Minnesota Botan. Stud., i.
—— 1899. Minnesota Plant Life. St. Paul, Minn.
MAGNIN, A. 1893 a. Recherches sur la végétation des lacs du Jura. Rev. Gén. de Bot., v.
—— 1893 b. La végétation des Monts Jura. Jour. de Bot., viii.
—— 1894. Contributions à la connaissance de la flore des lacs du Jura suisse. Bull. Soc. Bot. France, xli.
—— 1904. Les lacs du Jura. Paris.
MAGNUS, W. Studien an der endotrophen Mykorrhiza von Neottia Nidus-Avis, L. Morph. Jahrb., xxxv.
MARKTANNER-TURNERETSCHER, G. 1885. Zur Kenntniss des anatomischen Baues unserer Loranthaceen. Sitzungsb. Wiener Akad., xci.
MARLOTH. 1887 a. Zur Bedeutung der salzabscheidenden Drüsen der Tamarisconeen. Ber. Deutsch. Bot. Ges., v.
—— 1887 b. Das südöstliche Kalahari-Gebiet. Engler's Jahrb., viii.
—— 1888. Die Naras. Acanthosicyos horrida Welw., var. Namaquana. Ibid. ix.
MARSSON. 1907, 1908. In works issued by the Kaiserl. Gesundheitsamt, vols. xxv, xxviii. Biologische Untersuchungen des Rheines.
MARTIN, K., und K. REICHE, 1899. Sümpfe und Nadis in Chile. Verh. Deut. Wiss. Ver. Santiago, iv.
MARTINS, C. 1857. Expériences sur la persistance de la vitalité des graines flottant à la surface de la mer. Bull. Soc. Bot. France.
MARTIUS. 1840-7. Tabulae physiognomicae. Flora Brasil., i-ix.
MARTJANOW, N. 1882. Materials towards the Flora of Minussinsk. (In Russian.) Summary by Herder in Engler's Jahrb., ix.
MASCLEF. 1888. Études sur la géographie botanique du Nord de la France. Jour. de Bot., ii.
MASSART, J. 1893. La biologie de la végétation sur le littoral belge. Bull. Soc. Roy. Bot. Belgique, xxxii.
—— 1898 a. Un voyage botanique au Sahara. Ibid. xxxvii.
—— 1898 b. Les végétaux épiphylles. Ann. Jard. Bot. Buitenzorg, Suppl. ii.
—— 1902. L'accommodation individuelle chez Polygonum amphibium. Bull. Jard. Bot. de l'État à Bruxelles, i.
—— 1904. Les conditions d'existence des arbres dans les dunes littorales. Bull. Soc. Cent. For. Belgique.
—— 1906 a. See WERY, J.
—— 1906 b. Les lianes, leurs mœurs, leur structure. Bull. Soc. Cent. For. Belgique.
—— 1908. Essai de géographie botanique des districts littorals et alluviaux de la Belgique. Rec. Inst. Bot. Léo Errera, vii.
MAYR, H. 1890. Die Waldungen von Nordamerika.
MAZÉ, P. 1899. Les microbes des nodosités des Légumineuses. Ann. Inst. Pasteur, xiii.
MEIGEN, F. 1893. Skizze der Vegetationsverhältnisse von Santiago in Chili. Engler's Jahrb., xvii.
—— 1894. Biologische Beobachtungen aus der Flora Santiagos in Chile. Trockenschutzeinrichtungen. Ibid. xviii.
—— 1900. Beobachtungen über Formationsfolge im Kaiserstuhl. Deut. Bot. Monatschr., xviii.

MENTZ, A. 1900 a. Botaniske Iagttagelser fra Ringköbing Fjord. Rambusch, Ringköbing Fjord. Köbenhavn.
—— 1900 b. Studier over Likenvegetationen paa Heder. Bot. Tidsskr., xxiii.
—— 1902. Træk af Mosvegetationen paa jydske Heder. Bot. Tidsskr., xxiv.
MESCHAJEFF, V. 1883. Über die Anpassungen zum Aufrechthalten der Pflanzen und die Wasserversorgung bei der Transpiration. Bull. Soc. Nat. Imp. Moscou.
MEYEN. 1836. Outlines of the geography of plants. Engl. Ed. Roy. Soc. Lond., 1846.
MEYER, H. 1892. Der Kilimandscharo. Leipzig.
MEZ, C. 1904. Physiologische Bromeliaceen-Studien, I. Die Wasser-Oekonomie der extrem-atmosphärischen Tillandsien. Pringsh. Jahrb., xl.
—— 1905 a. Neue Untersuchungen über das Erfrieren eisbeständiger Pflanzen. Flora, xciv.
—— 1905 b. Einige pflanzengeographische Folgerungen aus einer neuen Theorie über das Erfrieren eisbeständiger Pflanzen. Engler's Jahrb., xxxiv, Beibl.
MIALL, L. C. 1898. A Yorkshire Moor. Roy. Inst. Gt. Brit., Feb. 1898.
MIDDENDORFF, A. T. v. 1867. Reise in dem äussersten Norden und Osten Sibiriens. St. Petersburg.
MINDEN, VON. 1899. Beiträge zur anatomischen und physiologischen Kenntnis Wasser-secernierender Organe. Biblioth. Bot., Heft xlvi.
MIRA, F. 1906. Las Dunas de Guardamar. Mem. Real Soc. Españ. Hist. Nat., iv.
MIYAKE, K. 1902. On the starch of evergreen leaves and its relation to photosynthesis during the winter. Botan. Gazette, xxxiii.
MIYOSHI, M. 1902. Über das massenhafte Vorkommen von Eisenbakterien in den Thermen von Ikao. Bot. Centralbl., lxxi.
MOHL, H. VON. 1848. Über das Erfrieren der Zweigspitzen mancher Holzgewächse. Bot. Zeitg.
MOHR, C. 1901. Plant life of Alabama. Contrib. U.S. Nat. Herb., vi.
MOLISCH, H. 1903. Die sogenannten Gasvacuolen und das Schweben gewisser Phycochromaceen. Bot. Zeitg., lxi.
MONTAGNE, C. 1844. Mémoire sur le phénomène de la coloration des eaux de la Mer Rouge. Ann. Sci. Nat., 3e sér., ii.
MOORE, S. LE M. 1895. The phanerogamic botany of the Matto Grosso Expedition, 1891–2. Trans. Linn. Soc. Lond., 2nd ser., iv.
MOSELEY, W. N. 1875. Notes on the Fresh-water Algae obtained at the boiling springs of Furnas, St. Michaels, Azores, and their neighbourhood. Journ. Linn. Soc. Lond., xiv.
MOSS, C. E. 1906. Succession of Plant Formations in Britain. Rep. Brit. Ass.
—— 1907. Geographical distribution of vegetation in Somerset: Bath and Bridgewater District. Jour. Roy. Geogr. Soc.
MÜLLER, P. E. 1871. Om Ædelgranen i nogle franske Skove. Tidsskr. f. Pop. Fremst. Naturvid. Kjöbenhavn.
—— 1878, 1884. Studier over Skovjord. Tidsskr. f. Skovbrug iii og vii. (Studien über die natürlichen Humusformen und deren Einwirkung auf Vegetation und Boden. German Ed., Berlin, 1887.)
—— 1886. Bemerkungen über die Mycorhiza der Buche. Bot. Centralbl., xxvi.
—— 1887. Om Bjergfyrren. Pinus montana Mill. Tidsskr. f. Skovbrug, viii, ix, xi.
—— 1894. Om Regnormenes Forhold til Rhizomplanterne. Oversigt Kongl. Danske Vid. Selsk.
—— 1899. Zur Theorie der Ortsteinbildung. Engler's Jahrb., xxvii, Beibl. 63.
—— 1902. Sur deux formes de Mycorhizes chez le pin de montagne. Oversigt Kongl. Danske Vid. Selsk. Forh.
—— 1903. Om Bjergfyrrens Forhold til Rödgranen i de jydske Hedekulturer. Tidsskr. f. Skovbrug, Suppl.-Hefte.
—— and F. WEIS. 1906. Studier over Skov- og Hedejord. I. Om Kalkens indvirkning paa Bögemor. Det forstlige Forsögsvæsen, i.
NADSON, G. 1900. Die perforierenden (kalkbohrenden) Algen und ihre Bedeutung in der Natur. Script. Bot. Hort. Univ. Petropolitanae, xviii.

NÄGELI, C. 1865. Bedingungen des Vorkommens von Arten und Varietäten innerhalb ihres Verbreitungsbezirkes. Sitzungsber. Münchener Akad.
—— 1872. Verdrängung der Pflanzenformen durch ihre Mitbewerber. Ibid.
NATHANSOHN, A. 1906. Über die Bedeutung vertikaler Wasserbewegungen für die Produktion des Planktons im Meere. Abhand. Math.-Phys. Kl. K. Sächs. Ges. Wiss., xxix, 5.
NATHORST, A. G. 1883 a. Nya bidrag till kännedomen om Spetsbergens kärlväxter. Sv. Vet. Akad. Handl., xx.
—— 1883 b. Studien über die Flora Spitzbergens. Engler's Jahrb., iv.
NAZAROW, P. 1886. Recherches zoologiques des steppes des Kirguiz. Bull. Soc. Nat. Imp. Moscou, lxii.
NEGER, F. W. 1897 a. Die Vegetationsverhältnisse im nördlichen Araucanien. Engler's Jahrb., xxiii.
—— 1897 b. Zur Biologie der Holzgewächse im südlichen Chile. Ibid.
—— 1901. Pflanzengeographisches aus den südlichen Anden und Patagonien. Ibid. xxviii.
NEGRI, G. 1905. La vegetazione della collina di Torina. Accad. Real. Sci. Torino, 1904–5.
NEHRING. 1890. Über Tundren und Steppen.
NILSSON, A. 1887. Studier öfver stammen saasom assimilationsorgan. Göteborg Vet. Sällsk. Handl., xxii.
—— 1896. Om örtrika barrskogar. Tidsskr. f. Skogshushaallning.
—— 1897 a. Om Norbottens växtlighet med särskild hänsyn till dess skogar. Ibid.
—— 1897 b. Om Norbottens myrar och försumpade skogar. Ibid.
—— 1899. Naagra drag ur de svenska växtsamhällenas utvecklingshistoria. 2 pts. Bot. Notiser.
—— 1902 a. Svenska växtsamhällen. Tidskr. f. Skogshushaallning.
—— 1902 b. Zur Ernährungsökonomie der Pflanzen. Naturforskaremötet, Helsingfors.
NILSSON, H. 1898. Einiges über die Biologie der schwedischen Sumpfpflanzen. Bot. Centralbl., lxxvi.
NOHLDE. 1895. Zum Klima von Inner-Arabien. Meteorologische Zeitschrift.
NOLL, F. 1893. Vorlesungsnotiz zur Biologie der Succulenten. Flora, lxxvii.
NORÉN, C. O. 1906. Om vegetationen paa Vänerns Sandstränder. Bot. Stud. tillägnade Kjellman. Upsala.
NORMAN, J. M. 1894–1901. Norges Arktiske Flora, I, II. Kristiania.
NORTON, J. B. S. 1897. A bibliography of literature relating to the effects of wind on plants. Trans. Kansas Acad. Sci., xvi.
ÖRSTED, A. S. 1844. De regionibus marinis. Elementa topographiae historico-naturalis freti Oeresund. Havniae.
ÖTTLI, M. 1903. Beiträge zur Ökologie der Felsflora. Dissert., Zürich. Schröter, Botanische Exkursionen, Heft iii. Zürich, 1905.
OLIVER, F. W. 1907 a. An experiment in co-operative field-work in botany. Trans. S.E. Union Sci. Soc.
—— 1907 b. The Bouche d'Erquy in 1907. New Phytologist, vi.
—— and A. G. TANSLEY. 1904. Methods of surveying vegetation on a large scale. Ibid. iii.
OLSSON-SEFFER, P. 1905. The principles of phytogeographic nomenclature. Botan. Gazette, xxxix.
—— 1909 a. Relation of soil and vegetation on sandy sea-shores. Ibid. xlvii.
—— 1909 b. Hydrodynamic factors influencing plant-life on sandy shores. New Phytologist, viii.
OLTMANNS, F. 1887. Über die Wasserbewegung in der Moospflanze und ihren Einfluss auf die Wasservertheilung im Boden. Cohn's Beiträge, iv.
—— 1892. Über die Kultur- und Lebensbedingungen der Meeresalgen. Pringsh. Jahrb., xxiii.
—— 1904, 1905. Morphologie und Biologie der Algen, i, ii.
OSTENFELD, C. H. 1898–1900. Plankton, in Iagttagelser over Overfladevandet, etc. Köbenhavn.
—— 1899, 1905. Skildringer af vegetationen i Island, I–IV. Bot. Tidsskr., xxii, xxvii.

OSTENFELD, C. H. 1905. Preliminary remarks on the distribution and the biology of the Zostera of the Danish Seas. Ibid. xxvii.

—— 1908 a. Aalegræssets (Zostera marina) Vækstforhold og Udbredelse i vore Farvande. Ber. Danske Biol. Stat., xvi. On the Ecology and Distribution of the Grass-Wrack (Zostera marina) in Danish Waters. Ibid.

—— 1908 b. The Land-Vegetation of the Faeröes. Botany of the Faeröes, Copenhagen, iii.

OSTWALD, W. 1903 a. Theoretische Planktonstudien. Zool. Jahrb., xviii.

—— 1903 b. Über eine neue theoretische Betrachtungsweise in der Planktologie. Forschungsber. Biol. Stat. Plön, x.

OVERTON, E. 1899. Beobachtungen und Versuche über das Auftreten von rothem Zellsaft bei Pflanzen. Pringsh. Jahrb., xxxiii.

PACZOSKI, J. 1904. Vegetationsverhältnisse im Dnjeperschen Kreise des Taurischen Gouvernements. Ber. Neuruss. Naturforscherges., xxvi.

PARISH, S. B. 1903. A sketch of the Flora of Southern California. 2 pts. Botan. Gazette, xxxvi.

PASSARGE, S. 1895. Adamaua. Bericht über d. Expedition des deutschen Kamerun-Komitees. Berlin.

PAX, F. 1896. Über die Gliederung der Karpathenflora. Jahresber. Schles. Ges. Vaterl. Kult.

—— 1898, 1908. Grundzüge der Pflanzenverbreitung in den Karpathen. I. Engler u. Drude, Die Vegetation der Erde, ii, x.

PEARSON, H. H. W. 1899. Botany of the Ceylon Patanas. Jour. Linn. Soc. Lond., xxxiv.

PECHUEL-LOESCHE. 1882. Die Loango-Expedition Leipzig.

PEIRCE, G. J. 1898. On the mode of dissemination and on the reticulations of Ramalina reticulata. Botan. Gazette, xxv.

PENZIG, O. 1902. Die Fortschritte der Flora des Krakatau. Ann. Jard. Bot. Buitenzorg, 2ᵉ sér., iii.

—— and CHIABRERA. 1903. Malpighia, xvii.

PETERSEN, H. E. 1908. Anatomy of Arctic Ericineae. Meddel. om Grönland, xxxvi.

PETERSON, O. G. 1893. Bidrag til Scitam. Anatomi. Danske Vid. Selsk. Skrift., 6 R., vii.

—— 1896. Stivelsen hos vore Troer under Vinterhvilen. Danske Vid. Selsk. Oversigt.

PETHYBRIDGE, C. G., and R. L. PRAEGER. 1905. The vegetation of the district lying south of Dublin. Proc. Roy. Irish Acad., xxv.

PETRY. 1889. Die Vegetationsverhältnisse des Kyffhäusergebirges. Halle.

PFEFFER. Physiology of Plants. Engl. Ed., Oxford, i, ii, iii., 1900–6.

PFEIL, Graf J. 1888. Beobachtungen während meiner letzten Reise in Ostafrika. Petermann's Mittheil., xxxiv.

PFITZER, E. 1870, 1872. Beiträge zur Kenntnis der Hautgewebe der Pflanzen. 3 pts. Pringsh. Jahrb., vii, viii.

PHILIPPI, R. A. 1858. Botanische Reise nach der Provinz Valdivia. Bot. Zeitg., xvi.

PICK, H. 1881. Beiträge zur Kenntnis des assimilierenden Gewebes armlaubiger Pflanzen. Diss., Bonn.

—— 1882. Über den Einfluss des Lichtes auf die Gestalt und Orientirung der Zellen des Assimilationsgewebes. Bot. Centralbl., xi.

PIETERS, A. J. 1894. The plants of Lake St. Clair. Bull. Michigan Fish. Commiss., ii, Lansing.

—— 1901. The plants of western Lake Erie. U.S. Fish. Commiss. Bull., 1901. Washington.

PILGER, R. 1902. Beitrag zur Flora von Mattogrosso. Engler's Jahrb., xxx.

PIPER, C. V. 1906. Flora of the State of Washington. Contrib. U.S. Nat. Herb., xi.

POHLE, R. 1903. Pflanzengeographische Studien über die Halbinsel Kanin und das angrenzende Waldgebiet. Act. Hort. Petropolitani, xxi.

—— 1907. Vegetationsbilder aus Nord-Russland. Karsten u. Schenck, Vegetationsbilder, v, 3, 4, 5.

POND, R. H. 1905. The biological relation of aquatic plants to the substratum. U.S. Fish. Commiss. Rep., 1903.

PORSILD, M. 1902. Bidrag til en skildring af vegetationen paa Öen Disko. Meddel. om Grönland, xxv.

POST, H. v. 1862. Studier öfver nutidens koprogena jordbildningar, gyttja, torf och mylla. Sv. Vet. Akad. Handl., iv. Translated by Ramann in Landwirtsch. Jahrb., xvii.

POTONIÉ, H. 1905. Formation de la houille et des roches analogues. Congrès internat. des mines.

POUND, R., and F. E. CLEMENTS. 1898–1900. The Phyto-geography of Nebraska. I. General Survey. Lincoln, Nebr.

—— 1898. The vegetation regions of the Prairie Province. Botan. Gazette, xxv.

PURPUS, A. and C. A. 1907. Arizona. Karsten und Schenck, Vegetationsbilder, iv. 7.

PURPUS, C. A. 1907. Mexikanische Hochgipfel. Ibid. v. 8.

RACIBORSKI, M. 1898. Biologische Mittheilungen aus Java. Flora, lxxxv.

RADDE. 1899. Grundzüge der Pflanzenverbreitung in den Kaukasusländern. Leipzig.

RADLKOFER, K. 1875. Monographie der Gattung Serjania.

RAMANN, E. 1886. Der Oststein und ähnliche Secundärbildungen in den Alluvial- und Diluvialsanden. Jahrb. Preuss. Geol. Landesanst., 1885.

—— 1890. Die Waldstreu und ihre Bedeutung für Boden und Wald. Berlin.

—— 1893. Forstliche Bodenkunde und Standortslehre. Berlin.

—— 1895. Organogene Ablagerungen der Jetztzeit. N. Jahrb. Miner., Beil. x.

—— 1905. Bodenkunde. Berlin.

—— 1906. Einteilung und Benennung der Schlammablagerungen. Monatsber. Deut. Geol. Ges.

RAUNKIÆR, C. 1889. Vesterhavets Öst- og Sydkysts Vegetation. Borchs Kollegiums Festskr., Kjöbenhavn.

—— 1895-9. De danske blomsterplanters Naturhistorie, I. Kjöbenhavn.

—— 1901. Om Papildannelsen hos Aira caespitosa. Bot. Tidsskr., xxiv.

—— 1903, in Bot. Tidsskr., xxvi.

—— 1905. Types biologiques pour la géographie botanique. Bull. Acad. des Sci. Danemark.

—— 1907. Planterigets Livsformer og deres Betydning for Geografien. Kjöbenhavn.

—— 1908. Livsformernes Statistik som Grundlag for Biologisk Plantegeografi. Bot. Tidsskr., xxix.

RAVN, F. K. 1894. Om flydeevnen hos fröene af vore vand- og sumpplanter. Bot. Tidsskr., xix.

RECHINGER, K. 1908 a. Samoa. Karsten u. Schenck, Vegetationsbilder, vi. 1.

—— 1908 b. Vegetationsbilder aus dem Neu-Guinea Archipel. Ibid. vi. 2.

REDWAY, J. W. 1894. The treeless plains of the United States. Geogr. Jour., iii.

REED, H. S. 1902. A survey of the Huron River Valley. Botan. Gazette, xxxiv.

—— 1905. A brief history of ecological work in Botany. Plant World, viii.

REICHE, K. 1893. Über polster- und deckenförmig wachsende Pflanzen. Santiago.

—— 1907. Grundzüge der Pflanzenverbreitung in Chile. Engler und Drude, Die Vegetation der Erde, viii.

REID, C. Origin of the British Flora. London.

REIN, J. 1879. Der Fugi-no-yama und seine Besteigung. Petermann's Mittheil.

—— 1881. Japan. Leipzig.

REINHARDT, J. 1856. Nogle Bemærkninger om den Indflydelse de idelige Markbrande have udövet. Vid. Meddel. Naturh. Foren. Kjöbenhavn.

REINKE, J. 1889. Algenflora der westlichen Ostsee. Ber. Kommiss. wiss. Unters. Deut. Meere, vi.

—— 1903 a. Die zur Ernährung der Meeresorganismen disponiblen Quellen an Stickstoff. Ber. Deut. Bot. Ges., xxi.

—— 1903 b. Die Entwickelungsgeschichte der Dünen der Westküste von Schleswig. Sitzungsber. Berliner Akad.

—— 1903 c. Botanisch-geologische Streifzüge an den Küsten des Herzogtums Schleswig. Wiss. Meeresunters., N. F., viii, Kiel u. Leipzig.

—— 1904. Zur Kenntnis der Lebensbedingungen von Azotobacter. Ber. Deut. Bot. Ges., xxii.

REITER, H. 1885. Die Consolidation der Physiognomik. Graz.
RESVOLL, T. 1903. Den nye vegetation paa lerfaldet i Værdalen. Nyt Mag. f. Naturvid., xli.
RICOME, H. 1903. Influence du chlorure de sodium sur la transpiration et l'absorption de l'eau chez les végétaux. Comptes Rendus, Paris.
RIKLI, M. 1899. Der Säckinger-See und seine Flora. Mitteil. Bot. Mus. Eidgenöss. Polytech. Zürich.
——— 1903. Botanische Reisestudien auf einer Frühlingsfahrt durch Korsika. Zürich.
——— 1907 a. Das Lägerngebiet. Mitteil. Bot. Mus. Eidgenöss. Polytech. Zürich, ix.
——— 1907 b. Botanische Reisestudien von der spanischen Mittelmeerküste. Vierteljahrschr. Naturforsch. Ges. Zürich, lii.
——— 1907 c. Kultur- und Naturbilder von der spanischen Riviera. Zürich.
——— 1907 d. Zur Kenntniss der Pflanzenwelt des Kantons Tessin. Ber. Zürcherischen Bot. Ges., x.
——— 1907 e. Spanien. Karsten und Schenck, Vegetationsbilder, v. 6.
ROBINSON, B. L. 1904. The problems of Ecology. Congress of Arts and Science, Universal Exposition, St. Louis, v.
ROSENBERG, O. 1897. Über die Transpiration der Halophyten. Öfvers. K. Sv. Vet. Akad. Förh.
ROSENVINGE, L. K. 1889–90. Vegetationen i en sydgrönlandsk Fjord. Geogr. Tidsskr., x.
———1898. Om Algevegetationen ved Grönlands Kyster. Meddel. om Grönland, xx.
——— 1903. Sur les organes piliformes des Rhodomélacées. Overs. K. Danske Vid. Selsk.
——— 1905. Sur les algues étrangères rejetées sur la côte occidentale du Jutland. Bot. Tidsskr., xxvii.
ROSS, H. 1887. Beiträge zur Kenntnis des Assimilationsgewebes und der Korkentwicklung armlaubiger Pflanzen. Diss., Freiburg i. B.
ROUX, C. 1900. Traité des rapports des plantes avec le sol. Montpellier.
ROUX, M. LE. 1907. Recherches biologiques sur le lac d'Annecy. Annales de Biologie lacustre, ii.
RÜBEL, E. 1908. Untersuchungen über das photochemische Klima des Berninahospizes. Vierteljahrschr. Naturforsch. Ges. Zürich, liii.
RYDBERG, P. A. 1895. Flora of the sand-hills of Nebraska. Contrib. U.S. Nat. Herb., iii.
SACHS, C. 1888. Aus den Llanos. Leipzig.
SACHS, J. VON. 1859. Über den Einfluss der chemischen und der physikalischen Beschaffenheit des Bodens auf die Transpiration. Landw. Vers.-Stat., i.
——— 1865. Handbuch der Experimental-Physiologie der Pflanzen.
SACHSSE, R. 1888. Lehrbuch der Agrikulturchemie.
SAINT-LAGER. 1895. L'appétence chimique des plantes et la concurrence vitale. Lyon.
SARAUW, G. F. L. 1893. Rodsymbiose og Mykorrhizer. Bot. Tidsskr., xviii. Contains an extensive bibliography.
——— 1898. Lyngheden i Oldtiden. Aarb. f. Nord. Oldkynd. og Hist., Kjöbenhavn.
——— 1903-4. Sur les mycorrhizes des arbres forestiers et sur le sens de la symbiose des racines. Rev. Mycol.
SAUVAGEAU. 1890. Observations sur la structure des feuilles des plantes aquatiques. Jour. de Bot., iv.
   See also many papers upon the Anatomy and Morphology of the Potamogetonaceae, Hydrocharideae and other families in the same Journal, 1888, 1890, 1891, 1894.
SCHACHT, H. 1859. Madeira und Tenerife. Berlin.
SCHENCK, A. 1903. Südwest-Afrika. Karsten und Schenck, Vegetationsbilder, i. 5.
SCHENCK, H. 1884. Über Strukturänderungen submers vegetierender Landpflanzen. Ber. Deutsch. Bot. Ges., ii.
——— 1886 a. Vergleichende Anatomie der submersen Gewächse. Biblioth. Bot., i.

SCHENCK, H. 1886 b. Die Biologie der Wassergewächse. Bonn.
—— 1889 a. Über die Luftwurzeln von Avicennia tomentosa und Laguncularia racemosa. Flora, lxxii.
—— 1889 b. Über das Aerenchym. Pringsh. Jahrb., xx.
—— 1892, 1893 a. Beiträge zur Biologie und Anatomie der Lianen. Bot. Mittheil. a. d. Tropen, iv, v.
—— 1893 b. Über die Bedeutung der Rheinvegetation für die Selbstreinigung des Rheines. Centralbl. f. allgem. Gesundheitspfl.
—— 1903 a. Vegetationsbilder aus Brasilien. Karsten und Schenck, Vegetationsbilder, i, 1.
—— 1903 b. Tropische Nutzpflanzen. Ibid. i. 3.
—— 1905 a. Mittelmeerbäume. Ibid. iii. 4.
—— 1905 b. Vergleichende Darstellung der Pflanzengeographie der subantarktischen Inseln. Wiss. Ergeb. Deut. Tiefsee-Exped., ii.
—— 1905 c. Über Flora und Vegetation von St. Paul und Neu-Amsterdam. Ibid.
—— and G. KARSTEN. 1903–8. Vegetationsbilder. Jena. (Südbrasilien, Tropische Nutzpflanzen, Strandvegetation Brasiliens, Mittelmeerbäume.)
SCHILLING, A. J. 1894. Anatomisch-biologische Untersuchungen über die Schleimbildung der Wasserpflanzen. Flora, lxxviii.
SCHIMPER, A. F. W. 1884. Über Bau und Lebensweise der Epiphyten Westindiens. Bot. Centralbl., xvii.
—— 1888 a. Die epiphytische Vegetation Amerikas. Bot. Mitth. a. d. Tropen, i.
—— 1888 b. Die Wechselbeziehungen zwischen Pflanzen und Ameisen. Bot. Mittheil. Tropen, i.
—— 1890. Über Schutzmittel des Laubes gegen Transpiration, vornehmlich in der Flora Java's. Monatsber. Berliner Akad., vii.
—— 1891. Die indo-malayische Strandflora. Bot. Mittheil. a. d. Tropen, iii.
—— 1893. Die Gebirgswälder Java's. Forstl.-Naturw. Zeitschr., ii.
—— 1898. Plant geography upon a physiological basis. Engl. Ed., Oxford, 1903.
SCHINZ, H. 1893. Die Vegetation des deutschen Schutzgebiets in Südwest-Afrika. Coloniales Jahrb., vi.
SCHIRMER, H. 1893. Le Sahara. Paris.
SCHLECHTER. 1904. Die Vegetationsformationen von Neu-Caledonien. Engler's Jahrb., xxxiii. Beibl. 73.
SCHMIDT, J. 1899. Om ydre faktorers Indflydelse paa Lövbladets anatomiske Bygning hos en af vore Strandplanter. Bot. Tidsskr., xxii.
—— 1903. Bidrag til Kundskab om Skuddene hos den gamle Verdens Mangrovetræer. Ibid. xxvi.
—— 1906. Vegetationstypen von der Insel Koh Chang im Meerbusen von Siam. Karsten und Schenck, Vegetationsbilder, iii. 7, 8.
SCHOMBURGK, R. 1841. Reisen in Guiana am Orinoco.
SCHOUW, J. F. 1822. Grundtræk til en almindelig Plantegeografie. (German Ed., Berlin, 1823.)
SCHRENK, H. v. 1898. On the mode of dissemination of Usnea barbata. Trans. Acad. Sci. St. Louis, iii.
SCHRÖDER, B. 1903. Über den Schleim und seine biologische Bedeutung. Biol. Centralbl., xxiii.
SCHRÖTER, C. 1895. Das St. Antönienthal im Prätigau in seinen landwirthschaftlichen und pflanzengeographischen Verhältnissen. Landw. Jahrb. d. Schweiz, ix.
—— 1897. Die Schwebeflora unserer Seen. Neujahrsbl. Naturf. Ges. Zürich, xcix.
—— 1902. See SCHRÖTER und KIRCHNER.
—— 1904. See also KIRCHNER, LOEW, und SCHRÖTER.
—— 1904–8. Das Pflanzenleben der Alpen. Zürich.
—— 1908. Eine Exkursion nach den Canarischen Inseln. Zürich.
—— und O. KIRCHNER, 1896–1902. Die Vegetation des Bodensees. Bodensee-Forschungen, 9. Abschn., Lindau, 6. Th., i, ii.
—— und J. FRÜH. 1904. Die Moore der Schweiz. Bern.
—— und M. RIKLI. 1904. Botanische Exkursionen im Bedretto-, Formazza- und Bosco-Tal. Botan. Exkurs. u. Pflanzengeogr. Stud. Zürich.

SCHRÖTER, C., et E. WILCZECK. 1904. Notice sur la flore littorale de Locarno. Boll. Soc. Ticinese Sci. Nat., i.

SCHUBE, T. 1885. Beiträge zur Kenntnis der Anatomie blattarmer Pflanzen. Breslau.

SCHÜBELER, F. C. 1886–8. Norges væxtrige. Christiania.

SCHUMANN, K. 1888. Einige neue Ameisenpflanzen. Pringsh. Jahrb., xix.

—— 1889. Einige weitere Ameisenpflanzen. Abhandl. Bot. Ver. Prov. Brandenburg, xxxi.

—— 1891 a. Rubiaceae. In Engler u. Prantl, Natürl. Pflanzenfamilien.

—— 1891 b. Über afrikanische Ameisenpflanzen. Ber. Deut. Bot. Ges., ix.

SCHÜTT, F. 1892. Analytische Planktonstudien. Kiel.

—— 1893. Das Pflanzenleben der Hochsee. Kiel.

SCHWAB, F. 1904. Über das photochemische Klima von Kremsmünster. Denkschr. Wiener Akad., lxxiv.

SCHWARZ, F. 1883. Die Wurzelhaare der Pflanzen. Inaug.-Diss., Breslau.

SCHWEINFURTH, G., und L. DIELS. Vegetationstypen aus der Kolonie Eritrea. Karsten und Schenck, Vegetationsbilder, ii. 8.

SCHWENDENER, S. 1874. Das mechanische Prinzip im anatomischen Bau der Monocotylen. Leipzig.

—— 1889. Die Spaltöffnungen der Gramineen und Cyperaceen. Sitzungsber. Berliner Akad.

SCOTT-ELLIOT, G. F. 1900. The formation of new land by various plants. Ann. Andersonian Nat. Soc., ii.

—— 1905. Acacias in various places. A study in associations. Trans. Bot. Soc. Edinb., 23.

—— 1906. The geographical functions of certain water-plants in Chile. Geogr. Journ.

SCRIBNER. See LAMSON-SCRIBNER.

SENDTNER, O. 1854. Die Vegetationsverhältnisse Südbayerns. München.

—— 1860. Die Vegetationsverhältnisse des Bayerischen Waldes. München.

SENFT. 1888. Der Erdboden.

SERNANDER, R. 1894. Studier öfver den gotländska vegetationens utvecklings historia. Diss., Upsala.

—— 1896. Naagra ord med anledning af Gunnar Andersson's Svenska växtvärldens historia. Bot. Notiser.

—— 1898. Studier öfver vegetationen i mellerste Skandinaviens fjälltrakter. I. Om tundra formationer i svenska fjälltrakter. Övers. K. Sv. Vet. Akad. Handl.

—— 1899. II. Fjällväxter i barrskogsregionen. Ibid., Bihang xxiv.

—— 1900. Studier öfver de sydsvenska Barrskogarnes Utvecklingshistoria. Ibid., Bihang xxv.

—— 1901. Den skandinaviska vegetationens spridningsbiologi. Zur Verbreitungsbiologie der skandinavischen Pflanzenwelt. Berlin u. Upsala.

—— 1906. Entwurf einer Monographie der europäischen Myrmekochoren. K. Sv. Vet. Akad. Handl., xli.

—— 1908. Stipa pennata : Västergötland. Svensk Botan. Tidskr., ii.

SHANTZ, H. LE R. 1905. A study of the vegetation of the Mesa Region east of Pike's Peak : The Bouteloua Formation. Botan. Gazette, xlii, 1906.

—— 1907. A biological study of the lakes of the Pike's Peak Region. Trans. Amer. Micros. Soc., xxvii.

SKOTTSBERG, C. 1904. On the zonal distribution of South Atlantic and Antarctic vegetation. Geogr. Jour.

—— 1905. Some remarks upon the geographical distribution of vegetation in the colder Southern Hemisphere. Ymer. Stockholm.

—— 1906. Vegetationsbilder aus Feuerland, von den Falklandsinseln und von Südgeorgien. Karsten und Schenck, Vegetationsbilder, iv. 2, 3.

SMITH, F. G. 1901. On the distribution of red color in vegetative parts in the New England Flora. Botan. Gazette, xxxii.

SMITH, R. 1898. Plant associations of the Tay basin. Proc. Perthshire Soc. Nat. Hist., ii.

—— 1899. On the study of plant associations. Nat. Sci., xiv. Edinburgh.

—— 1900 a. Botanical Survey of Scotland. I. Edinburgh District; II. North Perthshire District. Scott. Geogr. Mag., xvi.

SMITH, R. 1900 b. On the seed dispersal of Pinus sylvestris and Betula alba. Ann. Scott. Nat. Hist.

SMITH, W. G. 1902. The origin and development of heather moorland. Scott. Geogr. Mag., xviii.

—— 1903. Notes on the vegetation of ponds. The Naturalist.

—— and R. SMITH. 1904–5. Botanical Survey of Scotland. III, IV, Forfar and Fife. Scott. Geogr. Mag., xx, xxi.

——, C. E. Moss and W. M. RANKIN. 1903. Geographical distribution of vegetation in Yorkshire. I. Leeds and Halifax District; II. Harrogate and Skipton District. Geogr. Jour., xxi.

SNOW, L. M. 1902. Some notes on the ecology of the Delaware coast. Botan. Gazette, xxxiv.

SOKOLOW, N. A. 1894. Die Dünen. Bildung, Entwickelung und innerer Bau. Berlin.

SOLMS-LAUBACH, H. 1905. Die leitenden Gesichtspunkte einer allgemeinen Pflanzengeographie. Leipzig.

SORAUER, P. 1886. Handbuch der Pflanzenkrankheiten. I. Die nicht parasitären Krankheiten. 2. Aufl., Berlin.

SPALDING, V. M. 1904. Biological relations of certain desert shrubs. I. The creosote bush in its relation to water supply. Botan. Gazette, xxxviii.

STAHL, E. 1880 a. Über den Einfluss von Richtung und Stärke der Beleuchtung auf einige Bewegungserscheinungen im Pflanzenreiche. Bot. Zeitg., xxxviii.

—— 1880 b. Über den Einfluss der Lichtintensität auf Structur und Anordnung des Assimilationsparenchyms. Ibid.

—— 1881. Über sogenannte Kompasspflanzen. Jena. Zeitschr. f. Naturw., xv.

—— 1883. Über den Einfluss des sonnigen oder schattigen Standortes auf die Ausbildung der Laubblätter. Ibid. xvi.

—— 1893. Regenfall und Blattgestalt. Ann. Jard. Bot. Buitenzorg, xi.

—— 1894. Einige Versuche über Transpiration und Assimilation. Bot. Zeitg., lii.

—— 1896. Über bunte Laubblätter. Ein Beitrag zur Pflanzenbiologie. Ann. Jard. Bot. Buitenzorg, xiii.

—— 1900. Der Sinn der Mycorhizenbildung. Eine vergleichend-biologische Studie. Pringsh. Jahrb., xxxiv.

—— 1904 a. Die Schutzmittel der Flechten gegen Tierfrass. Häckel-Festschr., Jena.

—— 1904 b. Mexikanische Nadelhölzer. Mexikanische Xerophyten. Karsten und Schenck, Vegetationsbilder, ii. 3, 4. See also Karsten, G.

STANGE, B. 1892. Beziehungen zwischen Substratconcentration, Turgor und Wachsthum bei einigen phanerogamen Pflanzen. Bot. Zeitg., i.

STAPF, O. 1894. On the flora of Mount Kinabalu in North Borneo. Trans. Linn. Soc. Lond., ix.

STEBLER, F. G. 1897. Die Streuwiesen der Schweiz. Landwirthschaftliches Jahrbuch der Schweiz.

—— 1899. Die Unkräuter der Alpweiden und Alpmatten. Ibid.

STEBLER und ·SCHROETER. 1889, 1892. Beiträge zur Kenntnis der Matten und Weiden der Schweiz. X. Die Wiesentypen der Schweiz. Ibid. vi.

STEBLER, F. G., und A. VOLKART. 1904. Der Einfluss der Beschattung auf den Rasen. Beiträge zur Kenntnis der Matten und Weiden der Schweiz, xv. Landwirthsch. Jahrb. d. Schweiz.

STEENSTRUP, J. J. S. 1841. Geognostik-geologisk undersögelse af Skovmoserne Vidnesdam og Lillemose i det nordlige Sjælland, ledsaget af sammenlignende Bemærkninger, hentede fra Danmarks Skov-, Kjær- og Lyngmoser i Almindelighed. Danske Vid. Selsk. Afhandl., ix, 1842.

STEENSTRUP, K. J. V. 1877–8. Overfladevandets Varmegrad, Saltmængde og Farve i Atlanterhavet. Vid. Meddel. Naturh. For. Kjöbenhavn.

—— 1901. Om Bestemmelsen af Lysstyrken og Lysmongden. Meddelelser om Grönland, xxv.

STEFÁNSSON, S. 1894. Fra Islands Væxtrige, ii. Ibid.

STENSTRÖM, R. O. E. 1895. Über das Vorkommen derselben Arten in verschiedenen Klimaten an verschiedenen Standorten mit besonderer Berücksichtigung der xerophil ausgebildeten Pflanzen. Flora, lxxx.

STENSTRÖM, R. O. E. 1905. Studier öfver expositionens inflytande på vegetationen. Red. af H. Hesselman. Ark. f. Bot., iv.

STEPPUHN. 1895. Bot. Centralbl., lxii.

STÖHR. 1879. Über das Vorkommen von Chlorophyll in der Epidermis der Phanerogamen-Laubblätter. Sitzungsber. Wiener Akad., lxxix.

STOPES, M. C. 1907. The 'xerophytic' character of the Gymnosperms. Is it an 'ecological' adaptation? New Phytologist, vi.

STRODTMANN, S. 1895. Bemerkungen über die Lebensverhältnisse des Süsswasserplanktons. Forschungsber. Biol. Station Plön, iii.

SVEDELIUS, N. 1904. On the life-history of Enalus acoroides. Ann. Roy. Bot. Gard. Peradeniya, ii.

—— 1906. Ecological and systematic studies of the Ceylon species of Caulerpa. Ceylon Marine Biolog. Rep., No. 4.

SWELLENGREBEL, N. 1905. Über niederländische Dünenpflanzen. Beihefte z. Bot. Centralbl., xviii.

SYLVÉN, N. 1904. Studier öfver vegetationen i Torne Lappmarks björkregion. Ark. f. Bot., iii.

SZABÓ, Z. 1907. Eine pflanzengeographische Skizze der Sudeten. Bull. Soc. Hong. de Géographie, xxxvii.

TANFILJEW, G. 1894. Die Waldgrenzen in Südrussland. St. Petersburg.

—— 1898. Pflanzengeographische Studien im Steppengebiete. St. Petersburg.

—— 1902. Die Baraba und die Kulundinsche Steppe im Bereiche des Altai-Bezirkes. St. Petersburg.

—— 1905. Die südrussischen Steppen. Résult. Sci. Congr. Internat. Bot. Vienne, 1905.

TANNER-FULLEMANN. 1907. Contribution à l'étude des lacs alpins. Le Schœnenbodensee. Bull. Herb. Boissier, 2e sér., vii.

TANSLEY, A. G. 1904. The problems of ecology. New Phytologist, iii.

—— and F. E. FRITSCH. 1905. Sketches of vegetation at home and abroad. I. The Flora of the Ceylon Littoral. Ibid. iv.

THODE. 1890. Die Küstenvegetation von Britisch Kaffrarien. Engler's Jahrb., xii.

—— 1894. Die botanischen Höhenregionen Natals. Ibid. xviii.

THORNBER, J. 1901. Studies in the vegetation of the State. I. The prairie grass formation in Region I. Bot. Surv. of Nebraska, v, Lincoln, Nebr.

THURET, G. 1873. Expériences sur des graines de diverses espèces plongées dans de l'eau de mer. Arch. Sci. Phys. Nat., xlvii.

THURMANN, J. 1849. Essai de phytostatique, appliqué à la chaîne du Jura. Berne.

TIEGHEM, P. VAN. 1870–7. Recherches sur la symétrie de structure des plantes vasculaires. Ann. Sc. Nat., 5e sér., xiii.

TILDEN, J. E. 1898. Observations on some West American thermal algae. Botan. Gazette, xxv.

TISCHLER, G. 1905. Über die Beziehungen der Anthocyanbildung zur Winterhärte der Pflanzen. Beihefte z. Bot. Centralbl., xxviii.

TRABUT. 1888. Les zones botaniques de l'Algérie. Assoc. Française pour l'Avancem. d. Sci., Congrès d'Oran.

TRANSEAU, E. N. 1903. On the geographic distribution and ecological relations of the bog plant societies of Northern North America. Botan. Gazette, xxxvi.

—— 1905 a, 1906. The bogs and bog flora of the Huron River Valley. Ibid. xl, xli, 1905–6.

—— 1905 b. Forest centers of Eastern America. Amer. Nat., xxxix.

TREUB, M. 1888. Notice sur la nouvelle flore de Krakatau. Ann. Jard. Bot. Buitenzorg, vii.

TSCHAPLOWITZ. 1892. Humus und Humuserden. Oppeln.

TSCHIRCH, A. 1881. Über einige Beziehungen des anatomischen Baues der Assimilationsorgane von Klima und Standort, mit specieller Berücksichtigung des Spaltöffnungsapparates. Linnaea, xliii.

—— 1882. Beiträge zu der Anatomie und dem Einrollungsmechanismus einiger Grasblätter. Pringsh. Jahrb., xiii.

TSCHUDI, J. J. v. 1868. Reisen durch Süd-Amerika. Leipzig.

LITERATURE

?. 1901 a. Beobachtungen üb. d. Bodenstetigkeit der Arten im Gebiet Albulapasses. Ber. Schweiz. Bot. Ges., xi.

01 b. Über die Verbreitungsmittel der schweizerischen Alpenpflanzen. ra, lxxxix.

——— 1903. Hamburgische Elb-Untersuchung. Mitteil. Naturhist. Mus. 01–8). Hamburg.

, G. 1884. Zur Kenntnis der Beziehungen zwischen Standort und atomischem Bau der Vegetationsorgane. Jahrb. K. Bot. Gart. Berlin, iii.

887. Die Flora der ägyptisch-arabischen Wüste. Berlin.

890. Über Pflanzen mit lackirten Blättern. Ber. Deut. Bot. Ges., viii.

897. Der Kilimandscharo. Berlin.

903. Der Laubwechsel tropischer Bäume. Vortrag im Verein zur Beförrung des Gartenbaues zu Berlin.

I. DE. 1901, 1903. Die Mutationstheorie, i, ii.

, A. 1892. Zur Kenntniss des Blattbaues der Alpenpflanzen und dessen iologischer Bedeutung. Sitzungsber. Wiener Akad., ci.

, N. 1905. Pond vegetation. Naturalist.

E, A. R. 1880. Island Life. London.

1891. Natural Selection and Tropical Nature.

ZEK, H. 1893. Studien über die Membranschleime vegetativer Organe. Pringsh. Jahrb., xxv.

RG, O. 1893. Vegetationsschilderungen aus Südostasien. Engler's Jahrb., xvii.

1900. Einführung einer gleichmässigen Nomenclatur in die Pflanzengeographie. Ibid. xxix, Beibl. 66.

NG, E. 1869. En Udflugt til Brasiliens Bjärge. Tidsskr. f. Pop. Fremst. af Naturv. Translated by Zeise in Die Natur, 1881, and by Fonsny in La Belgique Horticole, 1883.

1875. Om nogle ved Danmarks Kyster levende Bakterier. Vid. Meddel. Naturh. For. Kjöbenhavn.

- 1881–1901. Familien Podostemaceae, I–VI. K. Danske Vid. Selsk. Skrift., 6. R., ii, iv, vii, ix, xi ; also in Engler and Prantl., Natürl. Pflanzenfam., iii. 2 a.

- 1883. Rhizophora Mangle L. Tropische Fragmente, ii. Engler's Jahrb., iv.

- 1884. Om Skudbygning, Overvintring og Foryngelse. Festskr. Naturh. Foren. Kjöbenhavn.

- 1887 a. Om Grönlands Vegetation. Meddel. om Grönland, xii.

- 1887 b. Tabellarisk Oversigt over Grönlands, Islands og Foröernes Flora. 1887. Vid. Meddel. Naturh. For. Kjöbenhavn.

- 1890. Fra Vesterhavskystens Marskegne. Vid. Meddel. Naturh. For. Kjöbenhavn.

- 1891. De psammofile Vegetationer i Danmark. Ibid. 1891.

- 1892. Lagoa Santa. Et Bidrag til den biologiske Plantegeografi. K. Danske Vid. Selsk. Skrift., 6. R., vi.

- 1893. Sur la biologie et l'anatomie de la feuille des Velloziacées. Bull. Acad. Sci. Copenhague.

- 1894. Exkursionen til Fanö og Blaavand i Juli 1893. Bot. Tidsskr., xix.

- 1895. Plantesamfund. Grundträk af den ökologiske Plantegeografi. Kjöbenhavn. German translation 1896 by Knoblauch ; new German edition by Graebner, 1902.

- 1896. P. E. Müller, nicht E. Ramann, hat die Entstehung des Ortsteins entdeckt. Engler's Jahrb., xxi, Beibl. 53.

- 1897 a. Botaniske Excursioner. 3. Skarridsö. Vid. Meddel. Naturh. For. Kjöbenhavn.

- 1897 b. Halofytstudier. K. Danske Vid. Selsk. Skrift., 6. R., viii.

- 1899 a. Planters og Plantesamfunds Kampe om Pladsen. Skandinav. Naturforskaremötets Forh., xv. Stockholm.

- 1899 b. On the vegetation of tropical America. Botan. Gazette, xxvii.

- 1901. Om Lövbladformer. I. Lianer. II. Skovbundsplanter. Overs. K. Danske Vid. Selsk. Forh.

- 1903. The History of the Flora of the Faröes. Botany of the Faröes. Copenhagen.

ULE, E. 1900. Die Verbreitung der Torfmoose
   Jahrb., xxvii.
——— 1901. Die Vegetation von Cabo Frio
   Ibid. xxviii.
——— 1903. Das Übergangsgebiet der Hylaea ￼
——— 1904. Epiphyten des Amazonengebietes.
   tionsbilder, ii. 1.
——— 1905. Blumengärten und Ameisen des An
——— 1906. Ameisenpflanzen des Amazonengebi
——— 1908. Das Innere von Nordost-Brasilien.
UNGER, F. 1836. Über den Einfluss des Bodens au
   nachgewiesen in der Vegetation des nordöstli
USTERI, A. 1905. Beiträge zur Kenntniss der Ph
   Arb. Bot. Mus. Polytech. Zürich, xiv.
VAHL, M. 1904 a. Notes on the summer-fall of tl
   Bot. Tidsskr., xxvi.
——— 1904 b. Madeira's Vegetation. Kjöbenhav
   Madeiras, Engler's Jahrb., xxxvi.
——— 1907. Om Vegetationen i Dobrogea. Geogr
VALLOT, J. 1883. Recherches physico-chimiques su
VANHÖFFEN. 1897. Die Fauna und Flora Grönlands
   Expedition der Gesellschaft der Erdkunde in B
VAUPELL, C. 1851. De nordsjällandske Skovmoser.
——— 1857. Bögens Indvandring i de danske Skov(
——— 1858. Nizzas Vinterflora. Vid. Meddel. Nat
——— 1863. De danske Skove. Kjöbenhavn.
VESQUE, J. 1878. De l'influence de la température
   l'eau par les racines. Ann. Sc. Nat., 6e sér., vi.
——— 1882 a. L'espèce végétale considérée au p(
   comparée. Ibid. xiii.
——— 1882 b. Essai d'une monographie anatomiqu(
   des Capparées (Capparidées ligneuses). Ibid.
——— 1883-4. Sur les causes et sur les limites d(
   des végétaux. Ann. Agron., ix, x.
——— 1886. Études microphysiologiques sur les rés(
   Ibid. xii.
——— 1889-92. Epharmosis sive materiae ad instruend
   naturalis.
   I. Folia Capparearum, Tab. i–lxxvii. Vincennes.
   II. Genitalia foliaque Garciniearum et Calophylle
     cennes.
   III. Genitalia foliaque Clusiearum et Moronobear
     cennes.
——— et C. VIET. 1881. De l'influence du milieu sur
   des végétaux. Ann. Sc. Nat., 6e sér., xii.
VESTERGREN, T. 1902. Om den olikförmiga snöbetäcl
   vegetationen i Sarjekfjällen. Bot. Notiser.
VIERHAPPER, F., und H. VON HANDEL-MAZZETTI. 1905.
   alpen. 2. Internat. Bot. Kongr. Wien.
VÖCHTING, H. 1874. Beiträge zur Morphologie und Anat
   Pringsh. Jahrb., ix.
——— 1878-84. Über Organbildung im Pflanzenreich. 2
——— 1888. Über die Lichtstellung der Laubblätter. Bo
——— 1891. Über die Abhängigkeit des Laubblattes von
   thätigkeit. Ibid. xlix.
——— 1893. Über den Einfluss des Lichtes auf die Gesta:
   Blüthen. Pringsh. Jahrb., xxv.
——— 1894. Über die Bedeutung des Lichtes für die Ges:
   Cakteen. Ibid. xxvi.
——— 1898. Über den Einfluss niedriger Temperatur auf
   Ber. Deut. Bot. Ges., xvi.

VOGLER,
  d￼
———￼
  F
VOLK, ￼
  (:
VOLKEN
  a

VRIES,
WAGNE

WALKE
WALLA
———
WALLI

WARB

WARN

WARMING, E. 1904. Den Danske Planteverdens Historie efter Istiden. Universitets-program, Kjöbenhavn.

———— 1906. Dansk Planteväkst. I. Strandvegetation. Kjöbenhavn.

———— 1907–9. Dansk Planteväkst. II. Klitterne. Kjöbenhavn.

———— 1908 a. The structure and biology of Arctic plants. I. Ericineae. Meddel. om Grönland, xxxvi.

———— 1908 b. Om Planterigets Livsformer. Festskr. udg. af Universitetet, Kjöbenhavn.

———— C. WESENBERG-LUND and others. 1904. Sur les ' vads ' et les sables maritimes de la mer du Nord. K. Danske Vid. Selsk. Skrift., 7. R., ii.

WEBBER, H. J. 1897. The water hyacinth and its relation to navigation in Florida. Bull. U.S. Dept. Agric., xviii.

WEBER, C. A. 1892. Über die Zusammensetzung des natürlichen Graslandes in Westholstein, Dithmarschen und Eiderstedt. Schrift. Naturw. Ver. Schleswig-Holstein, ix.

———— 1894 a. Vegetation des Moores von Augstumal. Mitteil. Ver. Förd. Moorkult. im Deut. Reiche, xii.

———— 1894 b. Veränderungen in der Vegetation der Hochmoore, etc. Ibid. xii.

———— 1900. Jahresber. der Männer vom Morgenstern. Heimatbund an Elb- und Wesermündung.

———— 1902. Über die Vegetation und Entstehung des Hochmoors von Augstumal im Memeldelta. Berlin.

———— 1903. Über Torf, Humus, und Moor. Abhand. Naturw. Ver. Bremen, xvii.

———— 1907. Aufbau und Vegetation der Moore Norddeutschlands. Engler's Jahrb., xl. Beibl. 90.

WEBERBAUER, A. 1905. Anatomische und biologische Studien über die Vegetation der Hochanden Perus. Engler's Jahrb., xxxvii.

WEED, W. H. 1887–8. Formation of Travertine and Siliceous Sinter by the Vegetation of Hot Springs. Ann. Rep. U.S. Geol. Surv., ix.

WEINROWSKY. 1898. Untersuchungen über die Scheitelöffnungen bei Wasserpflanzen. Fünfstück's Beitr. z. wiss. Bot.

WEISS, J. E., and R. H. YAPP. 1906. Sketches of vegetation at home and abroad. III. ' The Karroo ' in August. New Phytologist, v.

WERY, J. 1906. Sur le littoral belge. Excursions scientifiques, par Jean Massart. Rev. Univ. Bruxelles, 1905–6.

WESENBERG-LUND, C. 1900. Von dem Abhängigkeitsverhältnis zwischen dem Bau der Planktonorganismen und dem specifischen Gewicht des Süsswassers. Biol. Centralbl., xx.

———— 1901. Studier over Sökalk, Bönnemalm og Sögytje. Meddel. Dansk Geol. For.

———— 1905. A comparative study of the lakes of Scotland and Denmark, I. Proc. Roy. Soc. Edin., xxv.

———— 1906–8. Plankton Investigations of the Danish Lakes. I, II. General Part with 46 tables. Copenhagen.

WESSELY, J. 1873. Der europäische Flugsand und seine Kultur. Wien.

WEST, G. 1905. A comparative study of the dominant Phanerogamic and Higher Cryptogamic Flora of aquatic habit, in three Lake areas of Scotland. II. Proc. Roy. Soc. Edin., xxv.

———, W., and G. S. WEST. 1905. A further contribution to the freshwater plankton of the Scottish lochs. Trans. Roy. Soc. Edin., xli.

WESTERMAIER, M. 1884. Über Bau und Funktion des pflanzlichen Hautgewebesystems. Pringsh. Jahrb., xiv.

WETTSTEIN, R. v. 1893. Die geographische und systematische Anordnung der Pflanzenarten.

———— 1895. Der Saison-Dimorphismus. Ber. Deut. Bot. Ges., xiii.

———— 1897–8. Die Innovationsverhältnisse von Phaseolus coccineus L. (Ph. multiflorus Willd.). 2 pts., Österr. Bot. Zeitschr., xlvii, xlviii.

———— 1898. Grundzüge der geograph.-morpholog. Methode der Pflanzensystematik. Jena.

———— 1900. Untersuchungen über den Saisondimorphismus bei den Pflanzen. Denkschriften Wiener Akad., lxx. Wien.

404 LITERATURE

WETTSTEIN, R. V. 1902. Bemerkungen zur Abhandlung E. Heinrichers: Die grünen Halbschmarotzer. Pringsh. Jahrb., xxxvii.
—— 1905. Sokotra. Karsten und Schenck, Vegetationsbilder, iii. 5.
WHIPPLE, G. 1894. Some observations on the growth of diatoms in surface waters. Technology Quarterly, vii.
—— 1895. Some observations on the temperature of surface water. Journ. of the New Engl. Water Work Ass., ix.
—— 1899. Some observations on the relation of light to the growth of diatoms. Ibid., xiv.
WHITFORD, H. N. 1901. The genetic development of the forests of Northern Michigan. Botan. Gazette, xxxi.
—— 1905. The forest of the Flathead Valley, Montana. Ibid. xxxix.
—— 1906. The vegetation of the Lamao Forest Reserve. I, II. Philippine Jour. Sci., i, Nos. 4 and 6.
WHITNEY, M. 1897. The division of soils. U.S. Dept. Agric. Bur. Soils, Bull. x.
—— and F. K. CAMERON. 1904. Investigations in soil fertility. U.S. Dept. Agric. Bur. Soils, Bull. xxiii.
WIESNER, J. 1871. Untersuchungen über die herbstliche Entlaubung der Holzgewächse. Sitzungsber. Wiener Akad., lxiv.
—— 1876 a. Untersuchungen über den Einfluss des Lichtes und der strahlenden Wärme auf die Transpiration der Pflanze. Ibid. lxxiv.
—— 1876 b. Die natürlichen Einrichtungen zum Schutze des Chlorophylls. Festschr. Zool.-Bot. Ges. Wien.
—— 1887. Grundversuche üb. d. Einfluss der Luftbewegung auf die Transpiration der Pflanzen. Sitzungsber. Wiener Akad., xcvi.
—— 1893 a. Ombrophile und ombrophobe Pflanzen. Ibid. cii.
—— 1893 b. Photometrische Untersuchungen auf pflanzen-physiologischem Gebiete. Ibid. cii.
—— 1894 a. Pflanzenphysiologische Mittheilungen aus Buitenzorg. Ibid. ciii.
—— 1894 b. Über den vorherrschend ombrophilen Character des Laubes der Tropengewächse. Ibid. ciii.
—— 1894 c. Beobachtungen über die fixe Lichtlage der Blätter tropischer Gewächse. Ibid. ciii.
—— 1895 a. Beiträge zur Kenntniss des tropischen Regens. Ibid. civ.
—— 1895 b. Untersuchungen über den Lichtgenuss der Pflanzen, mit Rücksicht auf die Vegetation von Wien, Kairo, und Buitenzorg (Java). Ibid.
—— 1897. Untersuchungen über die mechanische Wirkung des Regens auf die Pflanze, &c. Ann. Jard. Bot. Buitenzorg, xiv.
—— 1898. Beiträge zur Kenntniss des photochemischen Klimas im arktischen Gebiete. Denkschr. Wiener Akad., lxvii.
—— 1899. Über die Formen der Anpassung der Blätter an die Lichtstärke. Biol. Centralbl., xix.
—— 1900. Untersuchungen über den Lichtgenuss der Pflanzen im arktischen Gebiete. Sitzungsber. Wiener Akad., cix.
—— 1903. Wiesner und seine Schule. Von Linsbauer u. a. Wien.
—— 1904. Über den Einfluss des Sonnen- u. des diffusen Tageslichtes auf die Laubentwicklung sommergrüner Holzgewächse. Sitzungsber. Wiener Akad., cxiii.
—— 1905. Untersuchungen über den Lichtgenuss der Pflanzen im Yellowstonegebiete und in anderen Gegenden Nordamerikas. Ibid. cxiv.
—— 1907. Der Lichtgenuss der Pflanzen. Leipzig.
——, FIGDOR, F. KRASSER, und L. LINSBAUER. 1896. Untersuchungen über d. photochemische Klima von Wien, Kairo, und Buitenzorg, Java. Denkschr. Wiener Akad., lxiv.
WILHELM, K. 1883. Über eine Eigenthümlichkeit der Spaltöffnungen bei Coniferen. Ber. Deut. Bot. Ges., i.
WILL, H. 1890. Vegetationsverhältnisse in Süd-Georgien. Die internationale Polarforschung, 1882–3. Die deutschen Expeditionen und ihre Ergebnisse. II. Hamburg.
WILLE, N. 1885. Bidrag til Algernes physiologiske Anatomi. Sv. Vet. Akad. Handl., xxi.

LITERATURE

405

WILLE, N. 1887. Kritische Studien über die Anpassungen der Pflanzen an Regen und Thau. Cohn's Beiträge zur Biologie der Pflanzen, iv.

—— 1897. Om Færöernes Ferskvandsalger og om Ferskvandsalgernes Spredningsmaader. Bot. Notiser.

—— 1904 a. Schizophyceen. Nordisches Plankton, Heft xx.

—— 1904 b. Die Schizophyceen der Plankton-Expedition. Ergebnisse der in dem Atlantischen Ocean . . . 1899 ausgeführten Plankton-Expedition der Humboldt-Stiftung, hrsg. von V. Hensen.

WILLIS, J. C., and T. H. BURKILL. 1895. Observations on the flora of the pollard willows near Cambridge. Proc. Cambridge Philos. Soc., viii.

WILLKOMM, M. 1852. Vegetation der Strand- und Steppengebiete der iberischen Halbinsel.

—— 1896. Grundzüge der Pflanzenverbreitung auf der iberischen Halbinsel. Engler u. Drude, Vegetation der Erde. Leipzig.

WILSON, W. P. 1889. The production of aerating organs on the roots of swamp and other plants. Proc. Acad. Nat. Sc. Philadelphia.

WINKLER, H. 1901. Pflanzengeographische Studien über die Formation des Buchenwaldes. Diss., Breslau.

WITTE, H. 1906. Till de Svenska Alfvaväxternas ekologi. Uppsala.

WITTROCK, V. B. 1883. Om snöns och isens flora. Nordenskiöld, Studier och Forskningar. Stockholm.

—— 1891. Biologiska Ormbunkstudier. Acta Hort. Bergiani, I.

—— 1894. Om den högre Epifyt-vegetationen i Sverige. Ibid. ii.

WOEIKOF, A. 1887. Die Klimate der Erde. 2 Thle. Jena.

—— 1889. Der Einfluss einer Schneedecke auf Boden, Klima und Wetter. Wien.

WOLLNY. Forschungen auf dem Gebiete der Agriculturphysik. München.

WOODHEAD, T. W. 1906. Ecology of woodland plants in the neighbourhood of Huddersfield. Jour. Linn. Soc. Lond., xxxvii.

WOODS, J. E. T. 1878. The forests of Tasmania. Journ. Roy. Soc. N. S. Wales, xii.

WULFF, T. 1902. Botanische Beobachtungen aus Spitsbergen. Lund.

YAPP, R. H. 1908. Sketches of vegetation at home and abroad. IV. Wicken Fen. The New Phytologist, vii.

YOKOYAMA, J. TANAKA. 1887. Untersuchungen über die Pflanzenzonen Japans. Petermann's Mittheil.

ZACHARIAS, O. 1891. Die Tier- und Pflanzenwelt des Süsswassers, i, ii. Leipzig.

ZEDERBAUER, E. 1904. Ceratium hirundinella in den österreichischen Alpenseen. Oesterr. Botan. Zeitschr., liv.

—— 1906. Vegetationsbilder aus Kleinasien. Karsten und Schenck, Vegetationsbilder, iii. 6.

ZINGELER, C. 1873. Die Spaltöffnungen der Carices. Pringsh. Jahrb., ix.

ZUKAL, H. 1895. Morphologische und biologische Untersuchungen über Flechten. Sitzungsber. Wiener Akad., civ.

INDEX

OXFORD
PRINTED AT THE CLARENDON PRESS
BY HORACE HART, M.A.
PRINTER TO THE UNIVERSITY

HISTORY OF ECOLOGY
An Arno Press Collection

Abbe, Cleveland. **A First Report on the Relations Between Climates and Crops.** 1905

Adams, Charles C. **Guide to the Study of Animal Ecology.** 1913

American Plant Ecology, 1897-1917. 1977

Browne, Charles A[lbert]. **A Source Book of Agricultural Chemistry.** 1944

Buffon, [Georges-Louis Leclerc]. **Selections from Natural History, General and Particular, 1780-1785.** Two volumes. 1977

Chapman, Royal N. **Animal Ecology.** 1931

Clements, Frederic E[dward], John E. Weaver and Herbert C. Hanson. **Plant Competition.** 1929

Clements, Frederic Edward. **Research Methods in Ecology.** 1905

Conard, Henry S. **The Background of Plant Ecology.** 1951

Derham, W[illiam]. **Physico-Theology.** 1716

Drude, Oscar. **Handbuch der Pflanzengeographie.** 1890

Early Marine Ecology. 1977

Ecological Investigations of Stephen Alfred Forbes. 1977

Ecological Phytogeography in the Nineteenth Century. 1977

Ecological Studies on Insect Parasitism. 1977

Espinas, Alfred [Victor]. **Des Sociétés Animales.** 1878

Fernow, B[ernhard] E., M. W. Harrington, Cleveland Abbe and George E. Curtis. **Forest Influences.** 1893

Forbes, Edw[ard] and Robert Godwin-Austen. **The Natural History of the European Seas.** 1859

Forbush, Edward H[owe] and Charles H. Fernald. **The Gypsy Moth.** 1896

Forel, F[rançois] A[lphonse]. **La Faune Profonde Des Lacs Suisses.** 1884

Forel, F[rançois] A[lphonse]. **Handbuch der Seenkunde.** 1901

Henfrey, Arthur. **The Vegetation of Europe, Its Conditions and Causes.** 1852

Herrick, Francis Hobart. **Natural History of the American Lobster.** 1911

History of American Ecology. 1977

Howard, L[eland] O[ssian] and W[illiam] F. Fiske. **The Importation into the United States of the Parasites of the Gipsy Moth and the Brown-Tail Moth.** 1911

Humboldt, Al[exander von] and A[imé] Bonpland. **Essai sur la Géographie des Plantes.** 1807

Johnstone, James. **Conditions of Life in the Sea.** 1908

Judd, Sylvester D. **Birds of a Maryland Farm.** 1902

Kofoid, C[harles] A. **The Plankton of the Illinois River, 1894-1899.** 1903

Leeuwenhoek, Antony van. **The Select Works of Antony van Leeuwenhoek.** 1798-99/1807

Limnology in Wisconsin. 1977

Linnaeus, Carl. **Miscellaneous Tracts Relating to Natural History, Husbandry and Physick.** 1762

Linnaeus, Carl. **Select Dissertations from the Amoenitates Academicae.** 1781

Meyen, F[ranz] J[ulius] F. **Outlines of the Geography of Plants.** 1846

Mills, Harlow B. **A Century of Biological Research.** 1958

Müller, Hermann. **The Fertilisation of Flowers.** 1883

Murray, John. **Selections from *Report on the Scientific Results of the Voyage of H.M.S. Challenger During the Years 1872-76.*** 1895

Murray, John and Laurence Pullar. **Bathymetrical Survey of the Scottish Fresh-Water Lochs.** Volume one. 1910

Packard, A[lpheus] S. **The Cave Fauna of North America.** 1888

Pearl, Raymond. **The Biology of Population Growth.** 1925

Phytopathological Classics of the Eighteenth Century. 1977

Phytopathological Classics of the Nineteenth Century. 1977

Pound, Roscoe and Frederic E. Clements. **The Phytogeography of Nebraska.** 1900

Raunkiaer, Christen. **The Life Forms of Plants and Statistical Plant Geography.** 1934

Ray, John. **The Wisdom of God Manifested in the Works of the Creation.** 1717

Réaumur, René Antoine Ferchault de. **The Natural History of Ants.** 1926

Semper, Karl. **Animal Life As Affected by the Natural Conditions of Existence.** 1881

Shelford, Victor E. **Animal Communities in Temperate America.** 1937

Warming Eug[enius]. **Oecology of Plants.** 1909

Watson, Hewett Cottrell. **Selections from *Cybele Britannica.*** 1847/1859

Whetzel, Herbert Hice. **An Outline of the History of Phytopathology.** 1918

Whittaker, Robert H. **Classification of Natural Communities.** 1962

VOGLER, P. 1901 a. Beobachtungen üb. d. Bodenstetigkeit der Arten im Gebiet des Albulapasses. Ber. Schweiz. Bot. Ges., xi.
—— 1901 b. Über die Verbreitungsmittel der schweizerischen Alpenpflanzen. Flora, lxxxix.
VOLK, R. 1903. Hamburgische Elb-Untersuchung. Mitteil. Naturhist. Mus. (1901–8). Hamburg.
VOLKENS, G. 1884. Zur Kenntnis der Beziehungen zwischen Standort und anatomischem Bau der Vegetationsorgane. Jahrb. K. Bot. Gart. Berlin, iii.
—— 1887. Die Flora der ägyptisch-arabischen Wüste. Berlin.
—— 1890. Über Pflanzen mit lackirten Blättern. Ber. Deut. Bot. Ges., viii.
—— 1897. Der Kilimandscharo. Berlin.
—— 1903. Der Laubwechsel tropischer Bäume. Vortrag im Verein zur Beförderung des Gartenbaues zu Berlin.
VRIES, H. DE. 1901, 1903. Die Mutationstheorie, i, ii.
WAGNER, A. 1892. Zur Kenntniss des Blattbaues der Alpenpflanzen und dessen biologischer Bedeutung. Sitzungsber. Wiener Akad., ci.
WALKER, N. 1905. Pond vegetation. Naturalist.
WALLACE, A. R. 1880. Island Life. London.
—— 1891. Natural Selection and Tropical Nature.
WALLICZEK, H. 1893. Studien über die Membranschleime vegetativer Organe. Pringsh. Jahrb., xxv.
WARBURG, O. 1893. Vegetationsschilderungen aus Südostasien. Engler's Jahrb., xvii.
—— 1900. Einführung einer gleichmässigen Nomenclatur in die Pflanzengeographie. Ibid. xxix, Beibl. 66.
WARMING, E. 1869. En Udflugt til Brasiliens Bjärge. Tidsskr. f. Pop. Fremst. af Naturv. Translated by Zeise in Die Natur, 1881, and by Fonsny in La Belgique Horticole, 1883.
—— 1875. Om nogle ved Danmarks Kyster levende Bakterier. Vid. Meddel. Naturh. For. Kjöbenhavn.
—— 1881–1901. Familien Podostemaceae, I–VI. K. Danske Vid. Selsk. Skrift., 6. R., ii, iv, vii, ix, xi ; also in Engler and Prantl., Natürl. Pflanzenfam., iii. 2 a.
—— 1883. Rhizophora Mangle L. Tropische Fragmente, ii. Engler's Jahrb., iv.
—— 1884. Om Skudbygning, Overvintring og Foryngelse. Festskr. Naturh. Foren. Kjöbenhavn.
—— 1887 a. Om Grönlands Vegetation. Meddel. om Grönland, xii.
—— 1887 b. Tabellarisk Oversigt over Grönlands, Islands og Foröernes Flora, 1887. Vid. Meddel. Naturh. For. Kjöbenhavn.
—— 1890. Fra Vesterhavskystens Marskegne. Vid. Meddel. Naturh. For. Kjöbenhavn.
—— 1891. De psammofile Vegetationer i Danmark. Ibid. 1891.
—— 1892. Lagoa Santa. Et Bidrag til den biologiske Plantegeografi. K. Danske Vid. Selsk. Skrift., 6. R., vi.
—— 1893. Sur la biologie et l'anatomie de la feuille des Velloziacées. Bull. Acad. Sci. Copenhague.
—— 1894. Exkursionen til Fanö og Blaavand i Juli 1893. Bot. Tidsskr., xix.
—— 1895. Plantesamfund. Grundträk af den ökologiske Plantegeografi. Kjöbenhavn. German translation 1896 by Knoblauch ; new German edition by Graebner, 1902.
—— 1896. P. E. Müller, nicht E. Ramann, hat die Entstehung des Ortsteins entdeckt. Engler's Jahrb., xxi, Beibl. 53.
—— 1897 a. Botaniske Excursioner. 3. Skarridsö. Vid. Meddel. Naturh. For. Kjöbenhavn.
—— 1897 b. Halofytstudier. K. Danske Vid. Selsk. Skrift., 6. R., viii.
—— 1899 a. Planters og Plantesamfunds Kampe om Pladsen. Skandinav. Naturforskaremötets Forh., xv. Stockholm.
—— 1899 b. On the vegetation of tropical America. Botan. Gazette, xxvii.
—— 1901. Om Lövbladformer. I. Lianer. II. Skovbundsplanter. Overs. K. Danske Vid. Selsk. Forh.
—— 1903. The History of the Flora of the Faröes. Botany of the Faröes. Copenhagen.

LITERATURE

ULE, E. 1900. Die Verbreitung der Torfmoose und Moore in Brasilien. Engler's Jahrb., xxvii.

———— 1901. Die Vegetation von Cabo Frio an der Küste von Brasilien. Ibid. xxviii.

———— 1903. Das Übergangsgebiet der Hylaea zu den Anden. Ibid. xxxiii.

———— 1904. Epiphyten des Amazonengebietes. Karsten und Schenck, Vegetationsbilder, ii. 1.

———— 1905. Blumengärten und Ameisen des Amazonengebietes. Ibid. iii.1.

———— 1906. Ameisenpflanzen des Amazonengebietes. Ibid. iv. 1.

———— 1908. Das Innere von Nordost-Brasilien. Ibid. vi. 3.

UNGER, F. 1836. Über den Einfluss des Bodens auf die Vertheilung der Gewächse, nachgewiesen in der Vegetation des nordöstlichen Tirols. Wien.

USTERI, A. 1905. Beiträge zur Kenntniss der Philippinen und ihrer Vegetation. Arb. Bot. Mus. Polytech. Zürich, xiv.

VAHL, M. 1904 a. Notes on the summer-fall of the leaf on the Canary Islands. Bot. Tidsskr., xxvi.

———— 1904 b. Madeira's Vegetation. Kjöbenhavn. See Über die Vegetation Madeiras, Engler's Jahrb., xxxvi.

———— 1907. Om Vegetationen i Dobrogea. Geogr. Tidskr., xix.

VALLOT, J. 1883. Recherches physico-chimiques sur la terre végétale. Paris.

VANHÖFFEN. 1897. Die Fauna und Flora Grönlands, in E. v. Drygalski, Grönland-Expedition der Gesellschaft der Erdkunde in Berlin, ii.

VAUPELL, C. 1851. De nordsjällandske Skovmoser. Kjöbenhavn.

———— 1857. Bögens Indvandring i de danske Skove. Kjöbenhavn.

———— 1858. Nizzas Vinterflora. Vid. Meddel. Naturh. For. Kjöbenhavn.

———— 1863. De danske Skove. Kjöbenhavn.

VESQUE, J. 1878. De l'influence de la température du sol sur l'absorption de l'eau par les racines. Ann. Sc. Nat., 6e sér., vi.

———— 1882 a. L'espèce végétale considérée au point de vue de l'anatomie comparée. Ibid. xiii.

———— 1882 b. Essai d'une monographie anatomique et descriptive de la tribu des Capparées (Capparidées ligneuses). Ibid.

———— 1883–4. Sur les causes et sur les limites des variations de structure des végétaux. Ann. Agron., ix, x.

———— 1886. Études microphysiologiques sur les réservoirs d'eau des plantes. Ibid. xii.

———— 1889–92. Epharmosis sive materiae ad instruendam anatomiam systematis naturalis.

I. Folia Capparearum, Tab. i–lxxvii. Vincennes.

II. Genitalia foliaque Garciniearum et Calophyllearum. Tab. clxii. Vincennes.

III. Genitalia foliaque Clusiearum et Moronobearum. Tab. cxiii. Vincennes.

———— et C. VIET. 1881. De l'influence du milieu sur la structure anatomique des végétaux. Ann. Sc. Nat., 6e sér., xii.

VESTERGREN, T. 1902. Om den olikförmiga snöbetäckningens inflytande paa vegetationen i Sarjekfjällen. Bot. Notiser.

VIERHAPPER, F., und H. VON HANDEL-MAZZETTI. 1905. Exkursion in die Ostalpen. 2. Internat. Bot. Kongr. Wien.

VÖCHTING, H. 1874. Beiträge zur Morphologie und Anatomie der Rhipsalideen. Pringsh. Jahrb., ix.

———— 1878–84. Über Organbildung im Pflanzenreich. 2 Bde., Bonn.

———— 1888. Über die Lichtstellung der Laubblätter. Bot. Zeitg., xlvi.

———— 1891. Über die Abhängigkeit des Laubblattes von seiner Assimilationsthätigkeit. Ibid. xlix.

———— 1893. Über den Einfluss des Lichtes auf die Gestaltung und Anlage der Blüthen. Pringsh. Jahrb., xxv.

———— 1894. Über die Bedeutung des Lichtes für die Gestaltung blattförmiger Cakteen. Ibid. xxvi.

———— 1898. Über den Einfluss niedriger Temperatur auf die Sprossrichtung. Ber. Deut. Bot. Ges., xvi.

D d